Energy, Environment, and Sustainability

This books series publishes cutting edge monographs and professional books focused on all aspects of energy and environmental sustainability, especially as it relates to energy concerns. The Series is published in partnership with the International Society for Energy, Environment, and Sustainability. The books in these series are editor or authored by top researchers and professional across the globe. The series aims at publishing state-of-the-art research and development in areas including, but not limited to:

- Renewable Energy
- Alternative Fuels
- Engines and Locomotives
- Combustion and Propulsion
- Fossil Fuels
- Carbon Capture
- Control and Automation for Energy
- Environmental Pollution
- Waste Management
- Transportation Sustainability

More information about this series at http://www.springer.com/series/15901

Tarun Gupta · Avinash Kumar Agarwal
Rashmi Avinash Agarwal
Nitin K. Labhsetwar
Editors

Environmental Contaminants

Measurement, Modelling and Control

Springer

Editors
Tarun Gupta
Department of Civil Engineering
Indian Institute of Technology Kanpur
Kanpur, Uttar Pradesh
India

Rashmi Avinash Agarwal
Department of Civil Engineering
Indian Institute of Technology Kanpur
Kanpur, Uttar Pradesh
India

Avinash Kumar Agarwal
Department of Mechanical Engineering
Indian Institute of Technology Kanpur
Kanpur, Uttar Pradesh
India

Nitin K. Labhsetwar
Energy and Resource Management Division
CSIR-National Environmental Engineering Research Institute (NEERI)
Nagpur, Maharashtra
India

ISSN 2522-8366 ISSN 2522-8374 (electronic)
Energy, Environment, and Sustainability
ISBN 978-981-10-7331-1 ISBN 978-981-10-7332-8 (eBook)
https://doi.org/10.1007/978-981-10-7332-8

Library of Congress Control Number: 2017958621

Printed on acid-free paper

This Springer imprint is published by Springer Nature
The registered company is Springer Nature Singapore Pte Ltd.
The registered company address is: 152 Beach Road, #21-01/04 Gateway East, Singapore 189721, Singapore

Preface

Energy demand has been rising remarkably due to increasing population and urbanization. Global economy and society are significantly dependent on the energy availability because it touches every facet of human life and its activities. Transportation and power generation are major examples of energy. Without the transportation by millions of personalized and mass transport vehicles and availability of 24×7 power, human civilization would not have reached contemporary living standards.

First international conference on 'Sustainable Energy and Environmental Challenges' (SEEC-2017) was organized under the auspices of 'International Society for Energy and Environmental Sustainability' (ISEES) by the 'Center of Innovative and Applied Bioprocessing' (CIAB), Mohali, held from 26 to 28 February 2017. ISEES was founded at IIT Kanpur in January 2014 with the aim of spreading knowledge in the fields of energy, environment, sustainability and combustion. The society's goal is to contribute to the development of clean, affordable and secure energy resources and a sustainable environment for the society and to spread knowledge in the above-mentioned areas and spread awareness about the environmental challenges, which the world is facing today. ISEES is involved in various activities such as conducting workshops, seminars, conferences in the domains of its interest. The society also recognizes the outstanding works done by the young scientists and engineers for their contributions in these fields by conferring them with awards under various categories.

This conference provided a platform for discussions between eminent scientists and engineers from various countries including India, USA, South Korea, Norway, Malaysia and Australia. In this conference, eminent speakers from all over the world presented their views related to different aspects of energy, combustion, emissions and alternative energy resource for sustainable development and a cleaner environment. The conference started with four mini-symposiums on very topical themes, which included (i) New Fuels and Advanced Engine Combustion, (ii) Sustainable Energy, (iii) Experimental and Numerical Combustion and (iv) Environmental Remediation and Rail Road Transport. The conference had 14 technical sessions on topics related to energy and environmental sustainability and a panel discussion on 'Challenges, Opportunities and Directions of Technical Education & Research in the Area of Energy, Environment and Sustainability' to wrap up the three-day technical extravaganza. The conference included 2 plenary talks, 12 keynote talks, 42 invited talks from prominent scientists, 49 contributed talks and 120 posters. A total of 234 participants and speakers attended this three-day conference, which hosted Dr. V. K. Saraswat, Member NITI Ayog, India, as a chief guest for the award ceremony of ISEES. This conference laid out the road map for technology development, opportunities and challenges in this technology domain. The technical sessions in the conference included Advances in IC Engines and Fuels; Conversion of Biomass to Biofuels; Combustion Processes; Renewable Energy: Prospects and Technologies; Waste to Wealth—Chemicals and Fuels; Energy Conversion Systems; Numerical Simulation of Combustion Processes; Alternate Fuels for IC Engines; Sprays and Heterogeneous Combustion of Coal/Biomass; Biomass Conversion to Fuels and Chemicals—Thermochemical Processes; Utilization of Biofuels; and Environmental Protection and Health. All these topics are very relevant for the country and the world in the present context. The society is grateful to Prof. Ashok Pandey for organizing and hosting this conference, which led to germination of this series of monographs, which included 16 books related to different aspects of energy, environment and sustainability. This is the first time that such significant and high-quality outcome has been achieved by any society in India from one conference.

The editors would like to express their sincere gratitude to the authors for submitting their work in a timely manner and revising it appropriately at short notice. We would like to express our special thanks to Prof. Rajesh Sathiyamoorthy, Prof. V. Ganesan, Dr. P. A. Laxminarayanan, Prof. Amritanshu Shriwastav, Dr. Nivedita Kaul, Dr. Rohit Goyal, Dr. Shalini Gupta, Dr. Heidi Salonen, Dr. Jennifer Lynch, Dr. Swatantra Pratap Singh, Dr. Kavita Gandhi, Dr. Pravin Mankar, Dr. Rita Dhodapkar, Dr. Prashant Rajput, Mr. Ankit Gupta, Mr. Gyanesh Singh, Ms. Pradhi Rajeev, Mr. Harshit Mishra and Mr. Saifi Izhar, who reviewed various chapters of this monograph and provided their valuable suggestions to improve the manuscripts. We acknowledge the support received from various funding agencies and organizations for the successful conduct of the first ISEES conference SEEC-2017, where these monographs germinated. These include Department of Science and Technology, Government of India (special thanks to Dr. Sanjay Bajpai); TSI, India (special thanks to Dr. Deepak Sharma); Tesscorn, India (special thanks to Sh. Satyanarayana); AVL India; Horiba, India; Springer (special thanks to Swati Mehershi); CIAB (special thanks to Dr. Sangwan).

The book covers different aspects of environmental contaminants in terms of their measurement in three different media, namely water, air and soil. In addition, it has two separate parts on modelling and control of different existing and emerging pollutants. Major topics include pharmaceutical wastes, paper and pulp waste, poly-aromatic hydrocarbons, mining dust, bioaerosols, endosulphan, biomass combustion and landfill design aspects. The book also contains chapters on environmental and human health modelling exposure to various contaminants; one chapter discusses the effect of indoor air pollutants on health and another one models the expected deposition of outdoor aerosols in different regions of our lungs. Various modifications and improvement in existing control technologies for remediation of environmental contaminants via better design of wastewater system and/or innovation in designing newer membranes for water treatment are also presented here through a series of chapters.

Kanpur, India Tarun Gupta
Kanpur, India Avinash Kumar Agarwal
Kanpur, India Rashmi Avinash Agarwal
Nagpur, India Nitin K. Labhsetwar

Contents

About the Editors

Dr. Tarun Gupta is Professor and P. K. Kelkar Research Fellow at the Department of Civil Engineering, IIT Kanpur, Kanpur, India. He holds Doctor of Science (2004) in Environmental Health, Harvard University, USA, and Master of Technology (2000) in Environmental Science and Engineering, Indian Institute of Technology Bombay (9-month research at TU Dresden, Germany). He is teaching and carrying out research at IIT Kanpur since June 2006. He has graduated 6 Ph.D. and 31 M.Tech. students. He has published more than 90 articles in ISI indexed journals, 4 chapters, numerous conference papers and filed 4 applications for Indian patent and 3 design applications. A submicron aerosol sampler designed, developed and evaluated at IIT Kanpur has been commercialized by Envirotech, Delhi. He has developed several low-flow-rate as well as high-flow-rate impaction-based samplers and a non-selective membrane-based diffusion denuder. He is a member of INYAS (2016), PK Kelkar Research Fellow (2015), NASI Scopus Young Scientist (2015), INSA Young Scientist (2011), INAE Young Engineer (2009), IEI Young Engineer (2008).

Professor Avinash Kumar Agarwal joined IIT Kanpur in 2001 and is currently a Poonam and Prabhu Goyal Endowed Chair Professor. He was at ERC, University of Wisconsin, Madison, USA, as a Postdoctoral Fellow (1999–2001). His areas of interest are IC engines, combustion, alternative fuels, hydrogen, conventional fuels, lubricating oil tribology, optical diagnostics, laser ignition, HCCI, emission and particulate control, and large bore engines. He has published more than 160 peer-reviewed international journals and conference papers. He is Associate Editor of ASME Journal of Energy Resources Technology and International Journal of Vehicle Systems Modelling and Testing. He has edited 'Handbook of Combustion' (5 volumes; 3168 pages), published by Wiley VCH, Germany. He was a Fellow of SAE (2012), Fellow of ASME (2013) and a Fellow of INAE (2015). He is the recipient of several prestigious awards such as NASI-Reliance Industries Platinum Jubilee Award-2012; INAE Silver Jubilee Young Engineer Award-2012; Dr. C. V. Raman Young Teachers Award-2011; SAE International's Ralph R. Teetor Educational Award-2008; INSA Young Scientist Award-2007; UICT Young Scientist Award-2007; INAE Young Engineer Award-2005. He is the recipient of Prestigious Shanti Swarup Bhatnagar Award-2016 in Engineering Sciences. He is the first combustion/IC engine researcher to get this honour.

Dr. Rashmi Avinash Agarwal is a Young Scientist Fellow at Indian Institute of Technology Kanpur, Kanpur, 208016, India. She completed her doctoral degree in Inorganic Chemistry from July 2008 to May 2014 at Indian Institute of Technology Kanpur under the supervision of Prof. P. K. Bhardwaj. The Ph.D. thesis title was 'Coordination Polymers of Transition Metal Ions with Benzimidazole Based Ligands: Single-Crystal to Single-Crystal (SC-SC) Transformations, Gas Storage and Magnetic Studies'. Prior to her doctoral degree, she completed her M.Sc. in Organic Chemistry from Rajasthan University, Jaipur, in 2002 and B.Sc. in Chemistry from Kanoria College, Rajasthan University, Jaipur, in 2000. She has expertise in supramolecular chemistry, coordination chemistry, coordination polymers, organic synthesis, inorganic synthesis, hydrothermal synthesis of coordination polymers, crystallization, crystal structure determination, topology, hydrogen bonding, structural analysis, fluorescence, SC (single crystal)-to-SC transformation, FTIR, thermo-gravimetric, NMR, mass and PXRD analysis, column chromatography, synthesis of nano-particles utilizing porous materials, TEM analysis. She has published over 20 research papers in leading international journals.

Dr. Nitin K. Labhsetwar is a Professor at AcSIR and Senior Principal Scientist and Head of the Energy and Resource Management Division at CSIR-National Environmental Engineering Research Institute (NEERI-CSIR). He completed his Ph.D. in Chemistry and has 31 years of research experience in environmental and energy-related research. He has worked as STA/JSPS Fellow and Visiting Overseas Researcher at NIMS, Tsukuba, Japan, and as a Visiting Professor at Kyushu University, Japan, under the GCOE programme. He has also worked at other international laboratories on the development of materials and processes including low cost and nano-materials for their applications in energy and environment, vehicular emissions, photocatalysis, GHG emission control, cleaner energy generation, heterogeneous catalyst, etc. He has over 155 research publications with over 3800 citations and 22 international patents in addition to a few contributions in books. He has received nine awards for excellence in research and also received various fellowships in India and abroad. He is a reviewer for more than ten SCI journals and supervised Ph.D. of 14 students. He is currently involved in more than 15 R&D projects.

Introduction

Abstract Accurate measurement of any contaminant in a given medium can be the right step taken in the direction of its eventual control or remediation. Contaminants may manifest phase change in terms of gaseous to liquid form as a result of condensation under the effect of dilution and cooling, for instance, particle formation as a result of tail-pipe emissions emanating from exhaust of a diesel car. This calls for adequate tools to measure such species in both gaseous and liquid phases. In practice, environmental modelling aids limited set of measurements of contaminants made in any given medium (air, water and soil) and is essential before appropriate control methods can be implemented for either complete or partial elimination of such contaminants. This book provides a holistic approach in terms of measurement, modelling and control of environmental contaminants.

Keywords POPs · Bioaerosol · Endosulphan · Biomass combustion Pharmaceuticals

Currently, there is an increasing awareness regarding the effect of antibiotics on various organisms in the ecosystem even though they may occur at a relatively low concentration. Thus, their concentration needs to be quantified precisely in various matrices including groundwater, surface water and municipal wastewater. The use of liquid chromatography together with high-resolution mass spectrometry (HR-LC-MS/MS) has enhanced the reliability, accessibility and accuracy with which antibiotics can be detected. In the first chapter, an HR-LC-MS/MS-based method was developed for four antibiotics (ciprofloxacin, norfloxacin, azithromycin and sulphamethoxazole). Initially, various mobile phases and LC columns were evaluated. The effect of solution pH, drying time of cartridge and solvent used in the concentration step on the recovery of antibiotics during solid-phase extraction (SPE) was also evaluated. The scientific community is facing a challenge in terms of the accurate measurement of pharmaceutical waste which is clearly highlighted in this chapter.

Arsenic (As) being a class A carcinogen is of interest to both environmental scientists and analytical chemists. Thus, sensitive and adept determination of As and speciation of different forms of As in various environmental matrices are indispensable. The second chapter in this book provides a comprehensive review of As concentrations and an overview of latest tools available for its accurate measurement. The oxidation state of As determines its toxicity and mobility in the environment. Thus, quantification and speciation of As are critical in assessing the overall risk. Most of the countries around the world have relevant official regulations on permissible levels of As in drinking water. In India, the permissible limit of As in drinking water is set at 10 μg/L by Bureau of Indian Standards (BIS) and WHO guideline value is also 10 μg/L. In this review, authors focus on extraction of As from various environmental samples, As speciation, sample treatment and determination of As in various matrices. Analytical methods for the determination of various forms of As, such as atomic absorption spectrophotometer (AAS), hydride generation AAS (HG_AAS), atomic fluorescence spectrometry (AFS), inductively coupled plasma mass spectrometry (ICPMS), electrochemical methods, capillary electrophoresis (CE), high-performance liquid chromatography (HPLC), HPLC coupled with mass spectrometer (HPLC-MS), neutron activation analysis and biosensors, are also summarized.

While moving on from water to air media, we recognize that organics remain the most complex and important contaminants to be tackled in the current times. Polycyclic aromatic hydrocarbons (PAHs) are organic compounds with two to seven fused benzene rings in a linear or angular arrangement. PAHs having ubiquitous presence are extensively reported, have carcinogenic potential, low aqueous solubility and semi-volatile nature and have been recognized as persistent toxic substances (PTS). Therefore, they are of considerable environmental concern, and we have a dedicated chapter to discuss them in detail. Combustion of fuels of all types including wood, coke, gas is the major anthropogenic activity that produces PAHs, while forest fires and volcanic eruptions are the natural sources. PAHs with low molecular weight dominate in the gaseous phase and are considered less toxic to humans, whereas PAHs with high molecular weight due to their low vapour pressures remain in particulate phase and are carcinogenic and/or mutagenic. PAHs with low molecular weight are much more abundant and can react with other pollutants such as O_3 and NO_x to form highly toxic nitrated and oxy-PAH compounds. The various factors that govern the atmospheric partitioning of PAHs are ambient temperature, relative humidity, aerosol nature and its properties, and interactions between the compound and the aerosol. The sources of PAHs and their quantitative contributions to a particular region are a matter of concern. In the environment, they occur as complex mixtures and not as single compounds. The development of appropriate strategies for adequate and effective control measures requires identification of sources and their quantitative contributions to total PAHs in any given region.

In continuation, we have one chapter that describes the measurement of fractional deposition of ambient particulate matter (PM) in the human respiratory system. The adverse impact of particulate matter (PM) on human health can be best explained in terms of PM deposition inside respiratory organs rather than the typically reported ambient exposure concentration. This study measured the PM mass concentration (MC) at various microenvironments inside a residential campus and estimated the corresponding PM mass deposition in human respiratory regions such as head (H), tracheobronchial (TB) and pulmonary (P) zones. This was done using the multiple path particle dosimetry lung model. One more chapter discusses the sources, distribution and prevalence of bioaerosols within an academic campus. Bioaerosols (particles of biological origin) can be produced from living or dead plants and animals. They can potentially serve as the cloud condensation and ice nuclei (CCN and IN). Their role in global carbon cycle further highlights the importance of studying their variability to link up with climate relevant parameters. From the Indian region, production estimates of bioaerosols from human population (current: $\sim$1.25 billion; of which over 45% resides in Indo-Gangetic Plain—IGP) and wildlife sanctuaries and national parks (100 in numbers, situated from north to south and east to west) are not accurately known. Hence, studying bioaerosols and establishing their linkage to health and climate appear to be of utmost importance.

Exposure of mine workers to particulate matter (PM) in opencast mines is of major concern because of associated adverse health impacts. Dispersion studies of PM generated during opencast mining until now have mostly been confined to the estimation of emissions from individual mining operations as well as total emission from an opencast mine. This chapter focuses on PM dispersion inside the mine since its generation until it escapes the mine as no field data is currently available on this topic. However, this is important particularly for deep mines where mine works are confined to bottom benches and emission from it passes across all benches before it reaches the surface, thus affecting the exposure level of workers at higher benches.

Towards the end of the part on measurement of contaminants, one chapter focuses on measurement of endosulphan in soil medium. Endosulphan is a highly toxic-chlorinated organic compound that has widely been used throughout the world as a pesticide. Commercial endosulphan is a mixture of two stereoisomers, α-endosulphan and β-endosulphan, in a ratio of 7:3. Due to its recalcitrant nature, it is highly persistent in the environment and has been included in the list of persistent organic pollutants (POPs) by Stockholm Convention in 2011. Further, high residual levels of endosulphan and their metabolites have gradually built up in contaminated environmental matrices at the point of application in all three phases—soil, water and air. In addition, numerous studies have reported that due to its high chemical stability, semi-volatile and hydrophobic nature, endosulphan contamination is frequently found in the environment at considerable distances from the point of its original application. Due to its high toxicity to humans, its application has been discontinued in most countries including India. However, the issues of legacy contamination still persist. The partitioning of endosulphan and its metabolites complicates their accurate quantification in the environment and requires complex analytical procedure as well as sophisticated instrumentation. This chapter provides a detailed review of available detection protocols for endosulphan and their advantages and limitations. Further, the chapter also reviews the existing efforts for the removal of this persistent compound from a variety of environmental matrices.

The second part provides glimpses of various rigorous modelling techniques that are currently employed to know the spatial as well as temporal extent of contamination of air, water and soil media. In the last few decades, due to the overuse of groundwater resources and human mismanagement, groundwater quality has been severely affected. If the water quality is not safe, even if it is abundant, it is not useful. This chapter essentially deals with the nature and movement of contaminants in groundwater due to the transport mechanisms like advection, dispersion, retardation and other chemical reactions and numerical modelling. The numerical modelling refers to the process of simulation where approximate solutions to the governing equation of contaminant transport are obtained. This chapter gives an insight into the groundwater contamination process, from discussing contaminant transport mechanisms to challenges faced in its numerical modelling. The details of the governing partial differential equations and the evolution of the methodology in obtaining these approximate solutions using different numerical techniques are presented. Finally, a field-based case study is given in which the numerical simulation provides an integrated framework wherein the conceptual principles are applied to the observed data and investigative and predictive calculations are done, which helps in developing management and protection policies.

In order to protect the health of population, health risk analysis is carried out. It involves various uncertainties and highly variable parameters like multiple routes (ingestion, dermal and inhalation), complex environmental contaminants, various pathways and different exposure to population, which make the risk estimation procedure extremely challenging and rigorous. Each step of the risk assessment process involves various assumptions, both quantitative and qualitative in nature, which must be evaluated through uncertainty analysis. However, it is necessary that risk process of evaluation must treat uncertainty and variability scientifically and in a robust manner. Moreover, addressing uncertainties in health risk assessment is a critical issue while evaluating the effects of environmental contaminants on public health. The uncertainty propagation in health risk can be assessed and quantified using probability theory, possibility theory or a combination of both. This chapter reports the development of various methodologies and frameworks to address the uncertainties that are intrinsic to health risk estimation.

Environmental models are often used to identify key sources polluting our environment. We have included one chapter that employs advanced isotopic tracer technique to identify dominant air pollution sources. This chapter assesses stable carbon isotopic composition (δ13C) of total carbon (TC) in ambient aerosols ($PM_{2.5}$) collected during wintertime from Kanpur in northern India. Chemical constituents, viz. organic carbon (OC), elemental carbon (EC) and water-soluble ions, in $PM_{2.5}$ have also been measured. Back trajectories of air masses arriving at the sampling site have been utilized to infer the air mass transport. Most of the trajectories showed their origin from north-western region during the study period. The average δ13C values of TC in the integrated 24-hour samples were centred around −25‰. Integrated data analyses of chemical constituents and stable C isotope suggest the influence of mixed emission sources.

One chapter deals with the emerging topic of indoor air pollution and covers the basics to model human exposure of the residents spending time in various indoor microenvironments. Human exposure to indoor air contaminants may be several times higher than outdoor levels. People belonging to rural areas in developing countries are usually exposed to high levels of household pollution daily owing to use of biomass for cooking. The situation becomes worse in winter as the consumption of household biomass fuel for cooking goes up. The unprocessed biofuel is related to asthma, respiratory tract infections, chronic obstructive pulmonary disease (COPD), cataract and in severe cases lung cancer. The aim of the study was to assess indoor air quality through measurement of CO, CO_2, NO_2, SO_2 and particulate matter (PM_{10} and $PM_{2.5}$) and health symptoms of the exposed subjects. A questionnaire survey was conducted in 125 households in 15 villages in the vicinity of Lucknow City. After analysing the survey results, five villages were shortlisted from where 15 houses were selected for monitoring from November 2013 to February 2014. The survey presented a comprehensive picture of a rural household along with socio-economic structure. Many other confounding factors including smoking, poor ventilation practices were also identified in the course of the study revealing poor household conditions. High particulate levels and health symptoms were likely to be associated mainly to use of crude fuel.

The final chapter in the second part deals with regional air quality modelling. It reports a comprehensive evaluation of the impacts of biomass burning on regional ozone and fine particulate matter ($PM_{2.5}$) over the continental USA, southern Canada and northern Mexico during 2012–2014 using the Community Multiscale Air Quality (CMAQ) chemical transport model. Inputs included the Fire Inventory from National Center for Atmospheric Research (FINN) for fire emissions, Biogenic Emission Inventory System (BEIS) for biogenics, the United States Environmental Protection Agency (USEPA)'s National Emissions Inventory of 2011 (NEI2011) for anthropogenic sources and Weather Research and Forecasting (WRF) model fields for meteorology. In situ data were taken from the Texas Commission on Environmental Quality (TCEQ)'s Continuous Ambient Monitoring Stations (CAMS) and the USEPA's Air Quality System (AQS) networks.

The final part is dedicated to various pollution control strategies that are currently used or are evolving and are tested at bench-scale laboratory setup. Effective treatment of wastewater of different origin has increased worldwide, whereas membrane bioreactor (MBR) technology appears as the most appropriate treatment system. Membrane provides solid–liquid separation in MBRs and releases high-quality effluents which could be further reused for different purposes. The conventional biological treatment system often faces the challenge of sludge disposal and requires larger footprint for their operation. These issues can be combated by the use of MBR systems. However, MBR system is associated with the membrane fouling problems, which limits its widespread application in wastewater treatment systems. Membrane fouling requires repeated cleaning that may reduce the membrane lifespan and frequent membrane replacement which incurs high cost. Membrane fouling leads to increase in trans-membrane pressure (TMP), rapid declination of permeate flux and increase in filtration resistance. The main factors that influence the membrane fouling are membrane characteristics, biomass characteristics, operating conditions and membrane configuration. To address the recent advances made in MBR fouling, this chapter reviews the mechanism, influencing factors and methods for controlling and preventing membrane fouling. In addition, this chapter also deliberates upon the use of various adsorbents for controlling membrane fouling in MBRs.

Water pollution is a major environmental issue faced by the modern world, which leads to ecological disequilibrium which can cause harmful effect on flora and fauna of the ecosystem. Heavy metals and synthetic dyes are the major pollutants to be prioritized in wastewater treatment because of their lethal toxicity. Most synthetic dyes are toxic and highly resistant to degradation due to their complex chemical structures. Adsorption is a promising technology for treating wastewater containing heavy metal contaminants. Recovery and reusability of the adsorbents make adsorption an eco-friendly and cost-effective technology. Photocatalysis is a highly proven treatment technique for dye-intoxicated wastewater. Conversion of non-biodegradable and complex organic dyes into simple biodegradable molecules by photocatalysis is a valuable addition in wastewater treatment. Concentration of heavy metal ions could be measured throughout the treatment process using atomic absorption spectrophotometer (AAS), and for dyes, UV–visible spectrophotometer could be employed. Kinetic modelling and adsorption isotherms would pave way for the better understanding of the rate and nature of the adsorption of heavy metals.

Designing of a sewage treatment plant (STP) and related pipework to collect sewage especially at remote sites and places having extreme environmental conditions is a challenge even for technically trained armed forces. Planning and design of sewer networks and STPs are outsourced by public works department-Military Engineering Services (MES) due to shortage of designers in-house. The peculiarities of defence establishments, which are small townships with sparse population density, i.e. low initial discharges, large distances, wide fluctuations in discharge, high ingress of rainwater and extreme climatic conditions, are generally ignored by consultants, and DPRs are prepared based on the average parameters normally prevalent with civil population. Further, consultants do not carry out cost–benefit analysis in totality including cost of conveyance, treatment and reuse. All these issues lead to delays in clearance of DPRs/estimates when they are sent for scrutiny. An attempt has been made to list out such mistakes, analyse and suggest methods to mitigate these. Suggestions like higher factor—4 to 6 for peak flows, restrictive use of flushing tanks, reuse of treated wastewater for flushing, use of energy-efficient DWCPE sewers and restrictions on depth of sewers for all stations are suggested. For extreme cold conditions, insulation of sewers and manholes has been reiterated, and for hot climates, higher self-cleansing velocities, >0.8 m/s, have been suggested. Cost analysis of STPs in defence establishments has also been done to facilitate designers for selecting appropriate capacity of STP.

The next chapter talks about solid waste management. The biodegradation of municipal solid wastes (MSWs) within waste containment system (engineered landfill) generates various gases, primarily methane and carbon dioxide, commonly known as landfill gases or biogas. Methane generation in landfills and the resulting emissions to the atmosphere have become the most important anthropogenic methane source. The generation of landfill gas could create pressure gradient within the waste containment system that forces the gas to migrate. As reported by a few case studies, the migration of landfill gas through landfill covers is unavoidable even with the provision of gas drainage layer in the landfill capping system. Hence, the performance of waste containment system, in specific cap barriers under gas flow condition, needs to be evaluated, monitored and understood. This chapter essentially deals with the nature and movement of gases in various cap barriers due to transport mechanisms like advection and diffusion. The mechanism involved in the generation of landfill gas due to biodegradation of MSW, gas transport through cap barrier and development of test apparatus for the determination of transport parameters has been explained. This chapter highlights the need for the evaluation of gas permeability and the diffusion coefficient of the cap barriers experimentally. The factors governing the performance of various cap barriers under gas flow conditions have been explained.

The paper industry releases huge quantity of toxic effluent into the surrounding environment which inhibits the photosynthetic activity of aquatic biota, crop growth, etc., and thus have adverse effects on health, soil properties and relative mobility in the soil profile. Hence, several aspects for the management of pulp and paper effluent, such as chemical, biological and mechanical treatments, recycling, monitoring, toxicology, environmental impact and sustainable practice, have been carried out. Further, various strategies need to be developed to determine the nature and extent of contaminants spread in the natural resources. Presently, bioremediation methods using terrestrial and aquatic plants as well as microorganisms are considered more favourable for treating various industrial effluents. Although all bioremediation processes using microorganisms are appropriate for wastewater treatment, the integrated approach of yeast treatment process showed higher contaminant removal efficiencies with reduced energy costs and sustainable production of biodiesel. Thus, the utmost need of the hour is to synchronize environmental friendliness by using these methods over conventional methods owing to their reduced chemical and biological sludge production, higher efficiency and low cost.

Towards the end of the control part, one chapter has been kept to sensitize the readers to another emerging important pollution category. Waste heat dispersion from power plants and heavy industries is always a major concern. Every industry needs a cheap source for cooling its necessary components, and water serves this purpose. This is due to ease of availability and high specific heat capacity of water. But after industrial use, the heated effluent is again discharged in the same waterbody from where it was taken. This not only disturbs the aquatic life but also affects the balance of the ecosystem. This chapter presents the background of the thermal pollution, modelling approach and analysis methods. For primary analysis of thermal pollution, an analytical solution of two-dimensional thermal dispersion is discussed. Dispersion is considered over a surface with velocity in only one direction, i.e. in the direction of the wind. A parabolic partial differential equation is solved analytically to predict temperature contours over a surface. Due to lack of adequate boundary condition, this solution is only capable of predicting far-field temperatures. For prediction of near-field temperatures, the same parabolic equation or a full three-dimensional energy and momentum equations can be solved numerically. A numerical problem formulation methodology is discussed for accurate prediction of thermal pollution. Finally, a scaling analysis is shown to develop an experimental model for proper validation of the numerical code and laboratory-scale experimental study.

The topics are organized into three different parts: (i) Measurement, (ii) Modelling, (iii) Control. Specific topics covered for the three media, namely water, air and soil, in this book include:

- Challenges in detection of antibiotics in wastewater matrix
- A comprehensive review on various analytical methods for the determination of inorganic and organic arsenic in environmental samples
- PAHs in gas and particulate phases: measurement and control
- Study of environmental particle levels, its effects on lung deposition, and relationship with human behaviour
- Bioaerosols over the Indo-Gangetic plain: Influence of biomass burning emission and ambient meteorology
- Particulate matter dispersion in Indian non-coal opencast mines
- Measurement and removal of endosulphan from contaminated environmental matrices
- Groundwater contamination problems and numerical simulation
- A systematic development of uncertainty modelling in the assessment of health risk to environmental contaminants
- Stable carbon isotope and bulk composition of wintertime aerosols from Kanpur
- Indoor air quality assessment as related to household conditions in rural houses during winter season
- Effects of biomass burning emissions on air quality over the continental USA: a 3-year comprehensive evaluation accounting for sensitivities due to boundary conditions and plume rise height
- A review on membrane fouling in membrane bioreactors: control and mitigation
- Eco-friendly treatment strategies for wastewater containing dyes and heavy metals
- Common mistakes in planning of sewer networks and STPs
- Waste containment system to limit landfill gas emission—mechanism, measurement and performance assessment
- Effect of emerging contaminants from paper mill industry into the environment and their control
- Thermal pollution: mathematical modelling and analysis

To summarize, this book contains information about existing and emerging pollutants from combustion, pesticide residue and pharmaceutical industries. Various novel analytical and diagnostic techniques like GC-MS, IRMS, real-time aerosol monitors have been discussed to measure and model the concentration of various contaminants. Finally, novel control and remediation techniques for the clean-up of various media are described in this book. We sincerely hope that you will enjoy its content!

Part I
Measurement

Chapter 1
Challenges in Detection of Antibiotics in Wastewater Matrix

Sanjeeb Mohapatra, Lokesh P. Padhye and Suparna Mukherji

Abstract Currently, there is increasing awareness regarding the effect of antibiotics on various organisms in the ecosystem even though they may occur at relatively low concentration. Thus, their concentration needs to be quantified precisely in various matrices including ground water, surface water, and municipal wastewater. The use of liquid chromatography together with high-resolution mass spectrometry (HR-LC-MS/MS) has enhanced the reliability, accessibility, and accuracy with which antibiotics can be detected. In the present study, a HR-LC-MS/MS-based method was developed for four antibiotics (ciprofloxacin, norfloxacin, azithromycin, and sulfamethoxazole). Initially, various mobile phases and LC columns were evaluated. The effect of solution pH, drying time of cartridge, and solvent used in the concentration step on recovery of antibiotics during solid-phase extraction (SPE) was also evaluated. Subsequently, the recovery of antibiotics from influent and effluent wastewater samples, obtained from a wastewater treatment plant (WWTP) in India, was determined based on the matrix spike method. The results revealed that for these wastewater samples even the use of structurally similar, isotopically labelled (deuterated) standards could not adequately improve the recovery of target antibiotics.

Keywords Pharmaceuticals · Antibiotics · Solid-phase extraction (SPE) Matrix effect · Ion suppression

S. Mohapatra · S. Mukherji (✉)
Centre for Environmental Science and Engineering, IIT Bombay, Mumbai, Maharashtra, India
e-mail: mitras@iitb.ac.in

L. P. Padhye
Department of Civil and Environmental Engineering, The University of Auckland, Auckland, New Zealand

T. Gupta et al. (eds.), *Environmental Contaminants*, Energy, Environment, and Sustainability, https://doi.org/10.1007/978-981-10-7332-8_1

1.1 Introduction

Antibiotics are a significant group of "contaminants of emerging concern" that have been detected in various environmental compartments throughout the world (Gros et al. 2006; Kasprzyk-Hordern et al. 2007; Mohapatra et al. 2016; Padhye et al. 2014; Sun et al. 2016; Ternes 1998; Watkinson et al. 2007; Zorita et al. 2009). Even though they are found in very low concentration, receptors in non-target organisms may be sensitive to individual antibiotics or their residues or to the synergistic action of a combination of drug residues (Diwan et al. 2010; Huber et al. 2005; Oaks et al. 2004). Some of these antibiotics are carcinogenic, mutagenic, and are also reported to cause endocrine disruption (Bevan et al. 2012). The production and consumption of antibiotics in developing countries are relatively high owing to their state of economic development and high population (Van Boeckel et al. 2014). Pharmaceuticals and their metabolites may enter wastewater treatment plants (WWTPs) as a result of direct excretion, discharges from pharmaceutical facilities and dumping of unused or expired medicines. It has been reported that pharmaceuticals are often poorly removed in the conventional WWTPs and ultimately they get released into the environment (Balakrishna et al. 2016; Mohapatra et al. 2016; Ternes 1998).

In view of the large number of listed pharmaceuticals (>3000), there is increasing interest in development of multigroup analytical methods. The majority of multigroup methods generally include more than one extraction step, using two or more different sorbents and elution solvents (EPA 2007). Thus, the target compounds are fractionated into groups according to their physicochemical properties and pharmaceutical classes. However, the sample preparation methods are laborious and time-consuming. Derivatization becomes an additional limiting step for analysis of acidic pharmaceuticals using gas chromatography–mass spectrometry (GC–MS) (Vanderford et al. 2003). Thus, liquid chromatography together with high-resolution mass spectrometry (HR-LC-MS/MS) is the preferred technique for ensuring accuracy and precision in detection of pharmaceuticals in the environment (Gros et al. 2006; Zhang and Zhou 2007). In a real wastewater matrix, various classes of antibiotics are present in varying proportion. Thus, a common analytical methodology for measurement of multiple classes of antibiotics is important for obtaining reliable estimates regarding their occurrence and fate in WWTPs. Concurrent detection of several antibiotics with diverse physicochemical characteristics involves extensive method development, and the method developed may differ significantly from the optimal detection method for any specific compound of interest. In addition to hydrophobic interaction, the intermolecular association between polar pharmaceuticals and dissolved organic matter (DOM) present in the wastewater may give rise to matrix effect during electrospray ionization (ESI) (Pan et al. 2009; Tolls 2001). Hence, in the presence of DOM, the concentration of pharmaceuticals may be over- or underestimated.

The objective of this work was to develop a multiresidue analytical method for a set of antibiotics commonly found in wastewater by solid-phase extraction (SPE) and HR-LC-MS/MS. The antibiotics selected include: azithromycin (AZIT), ciprofloxacin (CIPRO), norfloxacin (NORF), and sulfamethoxazole (SULF). The structure and properties of the selected antibiotics are shown in the Table 1.1. These antibiotics were chosen based on their rate of consumption and frequency of occurrence in WWTPs. Selection of suitable column, mobile phase, and parameters that may potentially affect the extraction efficiency of target antibiotics, such as, the pH and drying time during SPE and solvent used in the concentration step were optimized based on the response achieved during analysis in HR-LC-MS/MS. At the prevalent DOM concentration, antibiotics may remain bound to DOM present in wastewater which can alter their extraction efficiency during SPE and contribute to ion suppression during HR-LC-MS/MS analysis (Hernandez-Ruiz et al. 2012). Hence, isotopically labelled standards (deuterated) were used during analysis of real wastewater.

Table 1.1 Properties and structure of the antibiotics selected for the study

Pharmaceutical	pK_a	Charge at pH 7	Molecular structure
Azithromycin	pK_1 = 8.7, pK_2 = 9.5	Positive	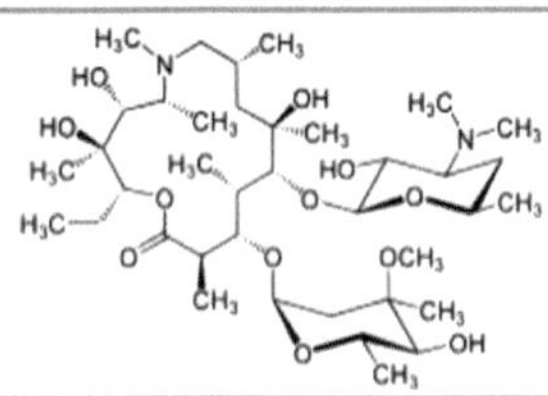
Ciprofloxacin	6.38	Positive/ neutral	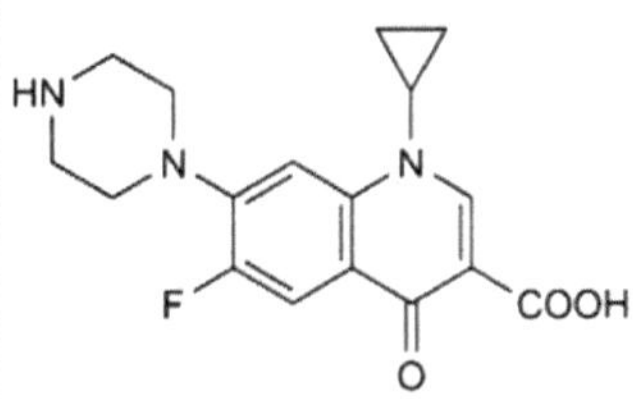
Norfloxacin	pK_1 = 6.3, pK_2 = 8.4	Positive	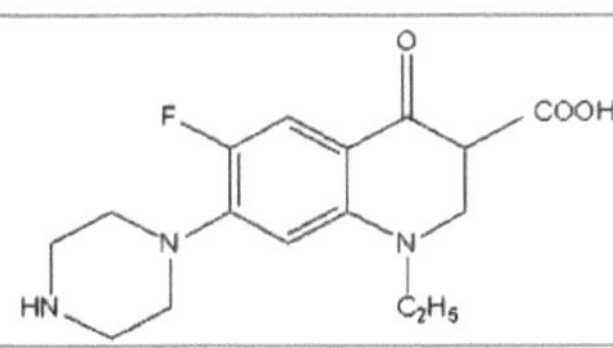
Sulfamethoxazole	5.7	Neutral/ negative	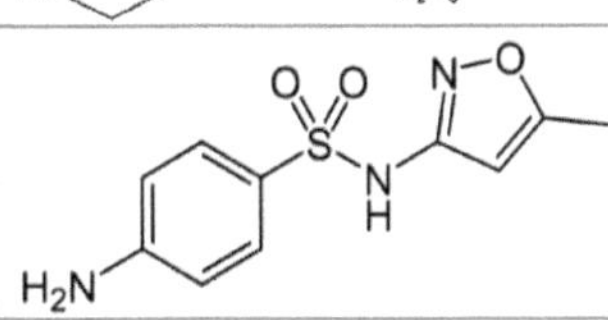

1.2 Materials and Methodology

1.2.1 Reagents Used

The antibiotic standards used were purchased from Sigma-Aldrich, India, and were of the highest available purity (>98%). Nano pure water (0.1 μS/cm at 25 °C) from nano pure water purification system (Thermo Fischer, Pune, India) was used for quality control purposes. Mixed standards of varying concentration were prepared from the stock solutions using HPLC-grade methanol obtained from Merck, India. The SPE cartridges used for analysis of synthetic wastewater were 60-mg Oasis HLB cartridges (Waters Corporation, USA). Appropriate isotope-labelled standards were chosen for each of the antibiotics (Table 1.2), and these standards were purchased from Toronto Research Chemicals (Toronto, Canada).

1.2.2 HR-LC-MS/MS Analysis

Initially, antibiotic standards of 1000 μg/L were prepared for each of the four antibiotics for the method optimization studies. Samples were analysed using high-resolution liquid chromatography coupled with quadrupole time-of-flight (Q-TOF) mass spectrometer (1290 Infinity UHPLC System, 1260 infinity Nano HPLC with Chipcube, 6550 iFunnel, Agilent Technologies, USA) in positive ESI mode. No SPE was performed for the samples analysed in HR-LC-MS during method development. Preliminary LC analysis was performed using four different columns (Table 1.3) with various combinations of mobile phases (Table 1.4).

ZORBAX SB-C18 column is explicitly designed for use with ultra-high performance liquid chromatographs for a variety of samples in basic, neutral, and acidic ranges. In contrast, the stationary phase of pinnacle PAH column has C-18 bonding which is recommended for detection of polynuclear aromatic hydrocarbons and other neutral compounds. Pinnacle C-18 column with hydrophobic C-18 phase is usually recommended for analysis of acidic compounds. Phenomenex MAX-RP column is characterized by densely bonded C-12 which is expected to improve hydrophobic retention and also the peak sharpness for basic compounds. The

Table 1.2 Isotope-labelled compounds used as internal standard for selected antibiotics

Pharmaceuticals	Isotope-labelled standard	References
Azithromycin[a]	Erythromycin ^{13}C-d_3	
Sulfamethoxazole	Sulfamethoxazole-d_4	Gros et al. (2013), Padhye et al. (2014)
Ciprofloxacin	Ciprofloxacin-d_8	Gros et al. (2013), Padhye et al. (2014)
Norfloxacin[a]	Ciprofloxacin-d_8	

[a]Erythromycin ^{13}C-d_3 and Ciprofloxacin-d_8 were chosen as suitable internal standard for azithromycin and norfloxacin, respectively, based on their structural similarity

Table 1.3 Columns used for analysis of the antibiotics

Column features	Zorbax SB-C18 (Zorb)	Pinnacle PAH (Pin_PAH)	Pinnacle C-18 (Pin_C18)	Phenomenex MAX-RP (Phen)
Dimension (mm)	50 × 2.1	250 × 4.6	250 × 4.6	250 × 4.6
Particle size (μm)	1.8	4	5	5

Table 1.4 Various combination of mobile phases used for the study

Mobile phase	Composition of A	Composition of B	References
MP_1	0.1% acetic acid + 0.1% ammonium formate aqueous solution	Acetonitrile: MeOH::1:1	Comero et al. (2013)
MP_2	0.1% ammonium acetate aqueous solution	100% acetonitrile	Lacey et al. (2008)
MP_3	0.1% formic acid aqueous solution	100% acetonitrile	Lin and Tsai (2009)
MP_4	0.1% formic acid aqueous solution	MeOH with 0.1% formic acid	Nelson et al. (2011)

mobile-phase gradient consisted of a combination of solvents A and B with a common ramp rate, i.e. increase in solvent B from 10 to 90% over 60 min. After screening, LC analysis was performed using the Phenomenex MAX-RP column and mobile phase consisting of water supplemented with 0.1% formic acid (FA), and acetonitrile (ACN) with gradient elution at a flow rate of 0.3 ml/min over a 60 min run-time. The final mobile phase program included five stages with variable ramp rate: increase in ACN from 10 to 30% over 10 min and held constant at 30% up to 30 min, increase in ACN up to 90% at 50 min, reduction in ACN to 10% at 52 min and held constant at this level up to 60 min.

Nitrogen and argon were used for nebulization and collision, respectively. Capillary voltage was set at 3.2 kV and the source temperature was set at 120 °C. The mass spectrometer was operated under optimized ionization potential and collision energies (CE) for each of the target pharmaceuticals as shown in Table 1.5.

Both retention time (±2% shifts with respect to the standard) and abundance of the precursor ion and product ions were used as the positive identification criteria for the target analytes (Gómez et al. 2006; Gros et al. 2006; Zhang and Zhou 2007). A separate study was conducted to obtain the limit of detection (LOD) and limit of quantification (LOQ). In this study, the concentration range used was 50–5000 μg/L, 1–100 μg/L, 1–100 μg/L, and 5–100 μg/L for azithromycin, ciprofloxacin, norfloxacin, and sulfamethoxazole, respectively. Linear regression was performed, and standard error in intercept (σ) and the magnitude of slope (S) were used to compute LOD and LOQ as $3.3(\sigma/S)$ and $10(\sigma/S)$, respectively, as reported in the proceedings

Table 1.5 Optimized parameter values for detection of antibiotics based on the screening studies

Pharmaceuticals	Azithromycin	Ciprofloxacin	Norfloxacin	Sulfamethoxazole
m/z	749.5	332.1	320.1	254.1
Daughter ion	591.4	231.1	233.1	156
Retention time (Rt)	16.4	15.7	15.4	39.7
Capillary voltage (V)	3500	2500	4500	2500
Fragmentor voltage (V)	150	250	200	300
Collision energy (V)	20	15	10	10
R^2	0.98	0.94	0.92	0.98
Inter-day variation RSD (%)	13.8	16.1	18.5	15.3
Intra-day variation RSD (%)	6.2	8.9	14.2	5
LOD (μg/L)	139	10	18	14
LOQ (μg/L)	422	29	54	42

of International Conference on Harmonisation (ICH 1994). The LOD and LOQ values are listed in Table 1.5. For analysis of antibiotics in wastewater samples concentrated using SPE, five-point calibration plots over the concentration range 100–1000 μg/L were prepared for each of the antibiotics. The LOD and LOQ for these samples would depend on the concentration factor. For 1 L of water concentrated to 1 mL prior to injection, the detection limits would be 1000 times lower than those listed in Table 1.5. Inter- and intra-day precisions were expressed as relative standard deviation (% RSD) based on of duplicate measurements. For each sample, duplicates were set-up and quality assurance/quality control (QA/QC) samples were also analysed using HR-LC-MS.

1.2.3 Optimization of Solid-Phase Extraction

The method developed by EPA (2007) was adopted with some modifications as suggested by Padhye et al. (2014). Initially, recovery of each pharmaceutical was determined by loading the Oasis HLB cartridge (60 mg) with synthetic wastewater containing 10 μg/L of each of the antibiotics. Since pH of wastewater can influence recovery, three pH values, i.e. 2.5, 4.0, and 7.0 were used during optimization of this step. Preconditioning of the cartridges was done with 5-mL HPLC-grade methanol followed by nano pure water (5 ml) at a flow rate of 5 mL/min. After the preconditioning step, 100 mL of synthetic wastewater (adjusted to the desired pH value with H_2SO_4) was loaded onto the cartridge at a flow rate of 5 ml/min. Rinsing of the loaded cartridge was done with 5 mL of nano pure water. Subsequently, the cartridges were dried and eluted with solvent. The impact of cartridge drying time under vacuum prior to elution was tested, using 5- and 30-min drying time. Finally, the antibiotics retained in the cartridge were eluted with 8 mL of methanol at a flowrate of 5 mL/min under vacuum followed by gravity flow. The methanol extracts were concentrated by gentle purging with nitrogen gas. The concentration step was further optimized (after adjusting the pH and drying time to optimal values) using three different conditions: (i) concentration of 8 mL eluate to 1 mL; (ii) complete removal of solvent and dissolution in 1 mL of methanol supplemented with 0.1% FA; and (iii) complete removal of solvent and dissolution in 1 mL methanol supplemented with 0.1% FA and 5% ACN.

1.2.4 Wastewater Sampling and SPE

The influent and effluent wastewater samples were collected from a WWTP located in western India. Prior to sampling, all the amber borosilicate bottles were cleaned with appropriate detergent, rinsed with water followed by acetone and deionised water and subsequently baked in an oven (>300 °C). Upon arrival in the laboratory, the ice-cooled wastewater samples were immediately filtered using 0.45-μm glass fibre filters (Pall, India) to remove suspended particles. The filtered wastewater samples were adjusted to pH 2–2.5 using H_2SO_4 and was stored in pre-cleaned amber borosilicate glass bottles with teflon-lined screw caps at 4 °C. Within two days of sample collection, SPE of duplicate samples was conducted using 500-mg HLB cartridges (Waters Corp., Milford, MA) eluted using 6 mL methanol at a flow rate of 5–10 mL/min. Subsequently, the methanol eluted from the cartridge was blown down to dryness by a gentle stream of pure nitrogen gas and reconstituted to 1.0 mL. The concentration factor used was 500 and 1000, for the influent and effluent samples, respectively. Isotope dilution method (IDM) employing isotope-labelled (deuterated) standards was employed for wastewater analysis. For IDM, calibration plots were plotted as peak area of an antibiotic normalized to that of the corresponding labelled isotope versus concentration of the antibiotic. Before analysis, each wastewater sample was divided into two sets. In one set, each of the isotope-labelled antibiotics were spiked (1000 and 500 ng/L were spiked in the influent and effluent, respectively, such that after 500- and 1000-fold concentrations, the expected concentration of these labelled standards would be similar, i.e. 500 μg/L in both influent and effluent, if no losses occurred). In the other set, along with isotope-labelled antibiotics, 2000 and 1000 ng/L of each of the target antibiotics were added in the influent and effluent wastewater samples, respectively. The difference in concentration of antibiotics between these two sets (C_d, μg/L) as determined using the calibration plot was corrected with the concentration factor and compared with the concentration spiked so as to obtain a measure of percentage recovery. To illustrate the need for IDM, another set of wastewater analysis was conducted using a similar approach, but without the addition of isotope-labelled standards (WIDM). In this study, standards were prepared similarly and a plot of peak area versus antibiotic concentration was used as the calibration plot.

In addition to the above studies, another study was conducted by spiking 50 and 100 µg/L of the labelled and native antibiotics, respectively, to 10 ml of influent and effluent wastewater samples where the dissolved organic carbon (DOC) level was kept constant at 10 mg/L by diluting appropriately. Another set was similarly prepared without the labelled antibiotics. Both the sets were kept for 48 h in an orbital shaker set at 100 rpm. Finally, SPE and LC-MS/MS analysis were conducted using a concentration factor of 10. The recovery was determined based on both IDM and WIDM calibration plots.

1.3 Results and Discussion

1.3.1 Selection of Suitable Column

The selection of the most suitable column was based on peak symmetry, peak resolution, and better peak separation. The symmetry of the peak was computed as the ratio between the front half-width and the back half-width of the analyte peak. A symmetrical peak will have a value of 1.0. The variation in peak symmetry value for each compound is shown in Fig. 1.1a where the mobile phase consisted of 0.1% FA aqueous solution and 100% ACN. For elution through Phenomenex MAX-RP column, all antibiotics except ciprofloxacin were found to have peak symmetry close to 1. The area under each analyte peak determined for the Zorbax SB-C18 column showed better resolution (Fig. 1.1b), possibly due to the smaller particle size (1.8 µm) of the packing material (Berridge 1985). Although the Phenomenex MAX-RP and Zorbax SB-C18 columns were both found to be equally suitable for mobile phase having high ionic strength and low pH (<2), overlapping of analyte chromatograms was observed in case of the Zorbax SB-C18 column, possibly due to its smaller column length. Additional packing of C-12 bonded phase in the

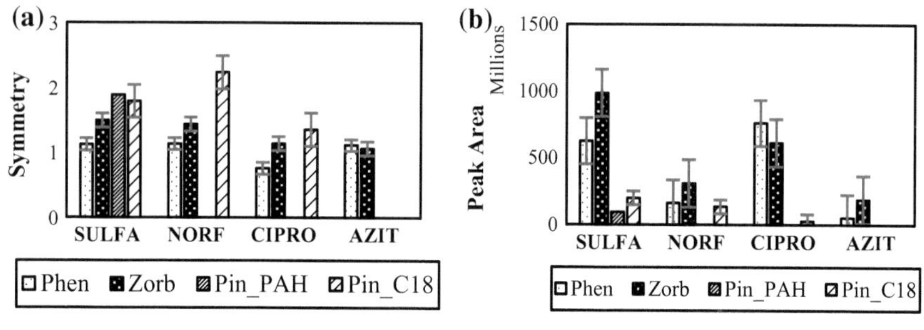

Fig. 1.1 Variation in **a** peak symmetry and **b** peak area of the antibiotics analysed using various columns. The mobile phase (MP_3) consisted of a combination of 0.1% formic acid aqueous solution and acetonitrile

Phenomenex Max-RP column reduced the steric hindrance caused by the C-18 bonded phase. Hence, hydrophobic retention, steric selectivity, and additional hydrogen bonding resulted in better resolution for the antibiotics analysed through the Phenomenex MAX-RP column (Fig. 1.1).

1.3.2 Selection of Suitable Mobile Phase

In reversed phase chromatography, more polar analytes are eluted faster compared to less polar analytes. A combination of solvents (A and B) was used for elution. The initial study was conducted with gradient elution using four different mobile phases (Table 1.4) and Phenomenex MAX-RP column. A mobile phase giving good separation for polar compounds will take longer time to elute the non-polar ones and vice versa. Hence, gradient optimization was further improved for achieving good separation for the mixture of antibiotics of varying polarity. The choice of solvents and buffers is critical for successful ionization with ESI. Hence, the mobile phase pH was adjusted according to the polarity of ions and pH of the sample (Kasprzyk-Hordern et al. 2008). Finally, the selection of the best suitable mobile phase was based on peak symmetry, peak resolution, and better peak separation. The variation in the peak symmetry value for each compound is shown in the Fig. 1.2a. For mobile phase 3, all the antibiotics except azithromycin demonstrated peak symmetry close to 1. Similarly for each combination of mobile phases, area under the curve for each of the antibiotics was determined after good resolution was achieved as shown in the Fig. 1.2b. Mobile phase 3 yielded higher peak area for most of the antibiotics except ciprofloxacin and azithromycin. In contrast to basic additives (ammonium acetate and ammonium formate), acidic additives (acetic acid and formic acid) are known to promote better protonation of analytes, thus increasing their signal intensity in positive ESI source (Kasprzyk-Hordern et al. 2007). Hence, mobile phase 3 was selected as the optimized mobile phase for analysis of the selected antibiotics. The deconvoluted chromatograms and MS/MS

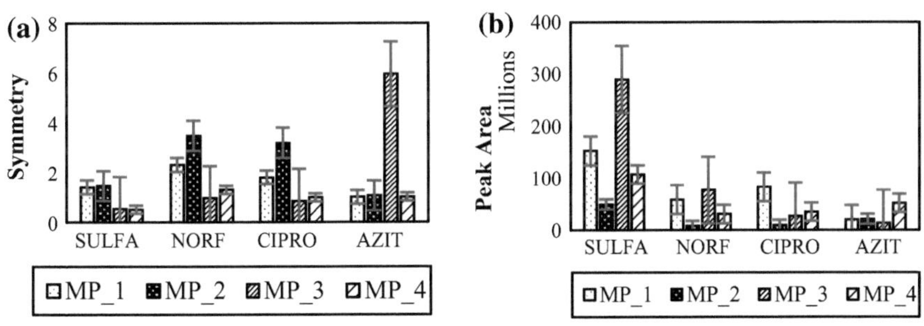

Fig. 1.2 Variation in **a** peak symmetry and **b** peak area of the antibiotics analysed using various mobile-phase composition. Phenomenex MAX-RP column was used

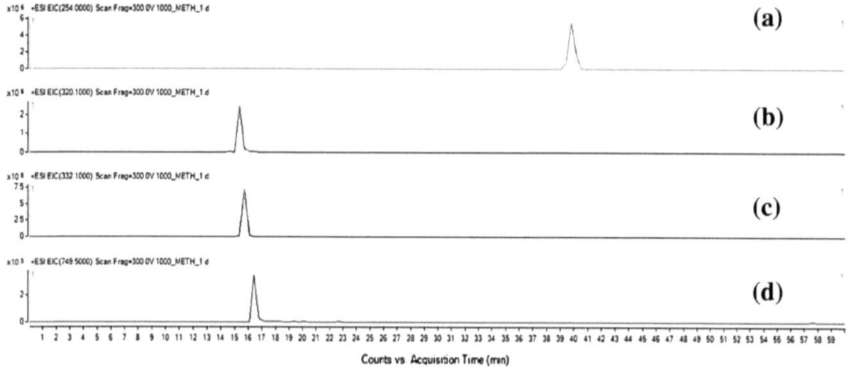

Fig. 1.3 Chromatograms showing **a** sulfamethoxazole, **b** norfloxacin, **c** ciprofloxacin, and **d** azithromycin for Phenomenex MAX-RP column and mobile phase consisting of a combination of 0.1% formic acid aqueous solution and acetonitrile (MP_3)

spectra for each of the antibiotics for the optimal column and mobile phase are shown in the Figs. 1.3 and 1.4, respectively.

1.3.3 Effect of pH on Recovery of Antibiotics

Beside the column and mobile phase that affects HR-LC-MS/MS analysis, the SPE process also needs to be optimized. The pH of the wastewater matrix can affect the recovery of target antibiotics using SPE since it can affect the adsorption–desorption interactions of the target analytes and the other matrix components. The other matrix components if retained should not elute out from the column during SPE (Mutavdžić Pavlovic et al. 2012). The matrix pH determines the degree of speciation of both weakly acidic/basic compounds. The recovery was relatively high for azithromycin. Decrease in pH increased the recovery for azithromycin, ciprofloxacin, and norfloxacin unlike sulfamethoxazole (Fig. 1.5).

At pH 2.5, adsorption of the ionized form of sulfamethoxazole on the packing material of the SPE cartridge may have been much lower than that of the other antibiotics as also suggested based on studies reported by other researchers (Zhang and Zhou 2007). Acidification of the aqueous solution increased the extraction and recovery of the other target antibiotics due to the strong association of the non-dissociated form of the analyte with the functional groups on the sorbent packed in the SPE cartridge. Difference in ionization and the presence of diverse structural groups is reported to affect the recovery of antibiotics during SPE (Liu et al. 2004; Weigel et al. 2004).

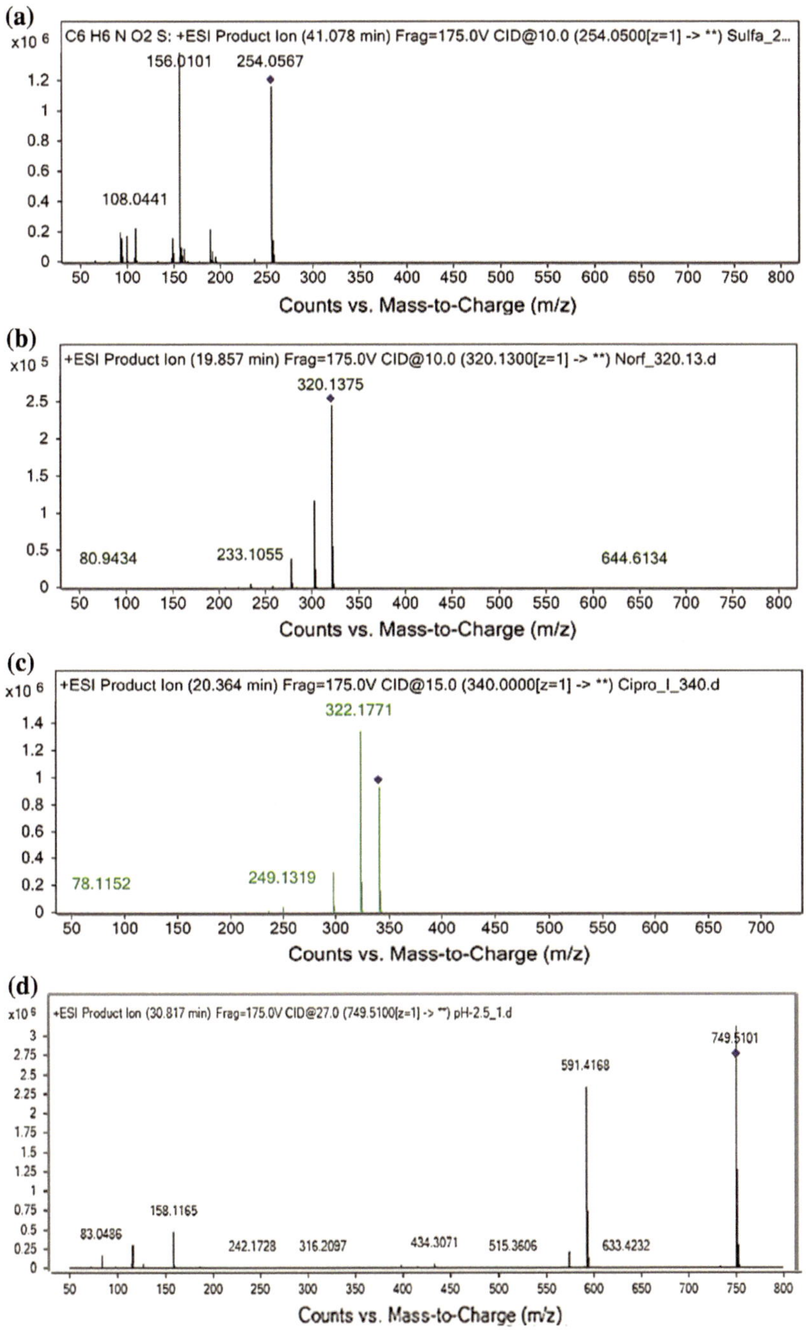

Fig. 1.4 Mass spectra of **a** sulfamethoxazole, **b** norfloxacin, **c** ciprofloxacin, and **d** azithromycin obtained for the standards

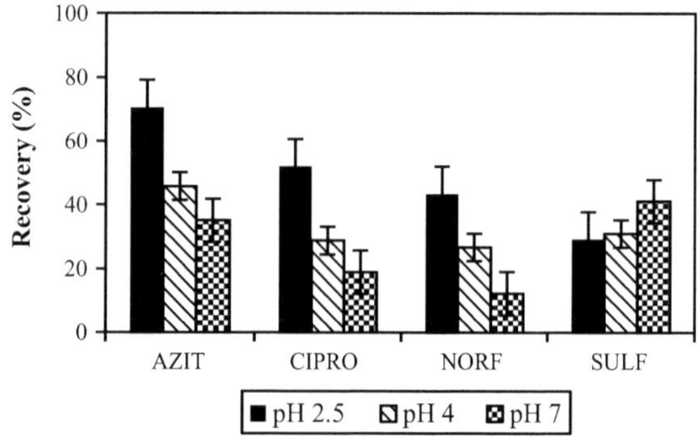

Fig. 1.5 Effect of pH on recovery of antibiotics Phenomenex MAX-RP column was used. The mobile phase (MP_3) consisted of a combination of 0.1% formic acid aqueous solution and acetonitrile (MP_3)

1.3.4 Effect of Drying Time and Concentration Step on Recovery of Antibiotics

In the literature, a drying step to reduce residual water from the SPE cartridge after loading the aqueous sample is widely reported (Gros et al. 2006; Kasprzyk-Hordern et al. 2007). Residual water may hinder the elution of sorbed analytes by a non-polar solvent and reduce its recovery. Residual water may also influence the elution strength of the solvent by mixing and may cause interference when solvent exchange with a water-immiscible solvent is performed. Increase in drying time from 5 to 30 min had insignificant effect on the recovery of antibiotics (Fig. 1.6a). In some studies, addition of FA to the elution solvent used for recovery of antibiotics from the SPE cartridge has been shown to improve the shape of chromatographic peaks, reduce peak tailing, and enhance sensitivity (Kasprzyk-Hordern et al. 2007). The effect is primarily due to enhanced ionization. Incorporation of ACN in the sample is also reported to increase the recovery for some pharmaceuticals. In contrast, in this study the recovery of ciprofloxacin, norfloxacin, azithromycin, and sulfamethoxazole were found to decrease as a result of ACN incorporation (Fig. 1.6b). The direct concentration approach (Dir) yielded better or comparable recovery for most of the antibiotics compared to when 0.1% FA and 0.1% FA with 5% ACN (FA and FA_ACN, respectively) were used.

1.3.5 Application of the Developed Method to Wastewater Samples and Interpretation of Matrix Effect

The influent and effluent samples obtained from a WWTP were analysed using IDM and WIDM, and the percentage recovery values obtained were compared as shown in Fig. 1.7. High R^2 values in the range of ~ 0.99 were observed for all the standard plots prepared with and without adding isotope-labelled standards. The percentage recovery of all the antibiotics was significantly reduced when isotope-labelled

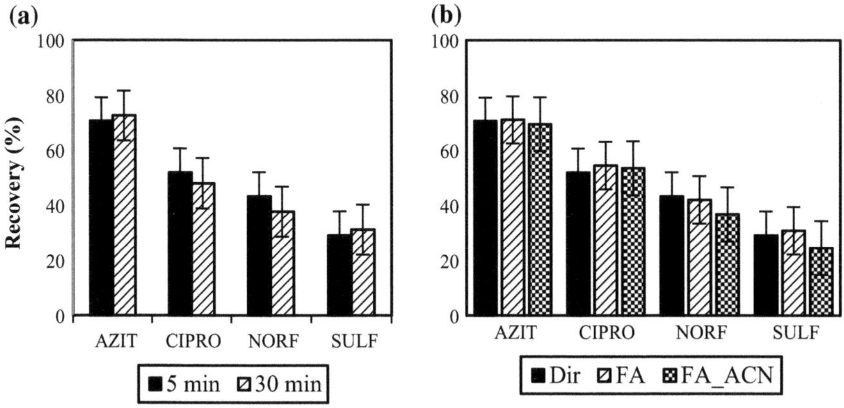

Fig. 1.6 Effect of **a** drying time and **b** concentration step on recovery of antibiotics. During optimization of the drying time, the pH was kept constant at 2.5 and direct concentration approach was adopted. During optimization of the concentration step, the pH was held constant at 2.5 and 5-min drying time was provided

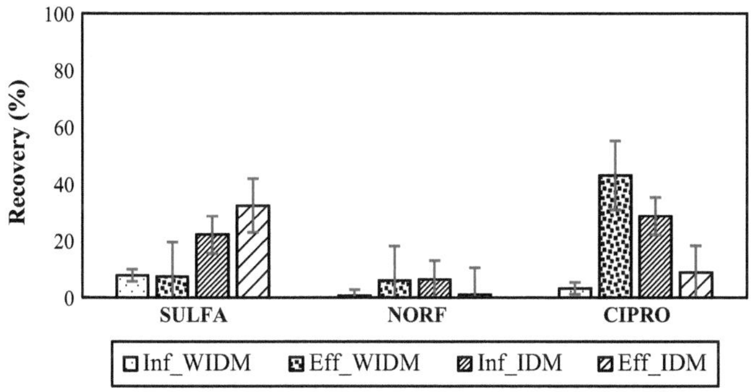

Fig. 1.7 Recovery of selected antibiotics in influent and effluent samples from a WWTP with (IDM) and without (WIDM) isotope dilution method. The concentration factor for influent and effluent were 500 and 1000, respectively

standards were not used for analysis. The WIDM method is thus inadequate for wastewater samples.

Although IDM was applied to improve the recovery and compensate for the matrix effect, the matrix effect could not be completely eliminated for most of the antibiotics, as shown by the recovery values in Fig. 1.7. When analysed using IDM, the extraction efficiency for sulfamethoxazole in the influent and effluent was markedly improved compared to WIDM. However, the recovery values for norfloxacin remained less than 10% even when IDM was used. Ciprofloxacin in influent wastewater sample showed improved recovery with IDM, however, in effluent wastewater ciprofloxacin showed higher recovery with WIDM.

Components in wastewater, such as, DOM may have been responsible for the low recovery values observed. The diverse physicochemical properties of both antibiotics and DOM in the wastewater matrix may have been responsible for the observed differences in recovery.

Variation in the recovery values for the antibiotics in the study with fixed DOC level in influent and effluent is shown in Fig. 1.8. Here, the recovery values for each of the antibiotics were significantly improved for IDM and the recoveries were comparable in both influent and effluent samples. While the recovery values in IDM were in the range of 50–90%, the recoveries based on WIDM were still less than 25%.

The DOC values for undiluted influent and effluent samples from the WWTP were 23.6 and 17.3 mg/L, respectively. After 500- and 1000-fold concentration in the influent and effluent, respectively, the extracts may have had correspondingly higher DOC. For the later study with DOC adjusted to 10 mg/L and concentration factor of 10, the DOC values in the extract would be 100 mg/L or less. Under these conditions, the recoveries using IDM improved significantly. The nature and composition of DOC in the wastewater used may have resulted in the severely low recovery values when concentration factors of 500 and above were used. The recoveries were very low (<28%) even when IDM was employed for analysis. Researchers from other countries have reported recoveries in the range of 81–90, 77.5, and 78–87% for sulfamethoxazole (Park et al. 2017; Xu et al. 2007), norfloxacin (Park et al. 2017), and ciprofloxacin (Karthikeyan and Meyer 2006; Park et al. 2017), respectively, with IDM employing labelled standards.

DOC in wastewater may influence both the loading of antibiotics on the SPE cartridge and their subsequent elution, although the contribution of antibiotics to DOC in wastewater may be negligible. Moreover, DOC may interfere with HR-LC-MS-MS analysis. Thus, these effects may have hindered the recovery of antibiotics. Researchers have reported that interaction of DOM with polar analytes

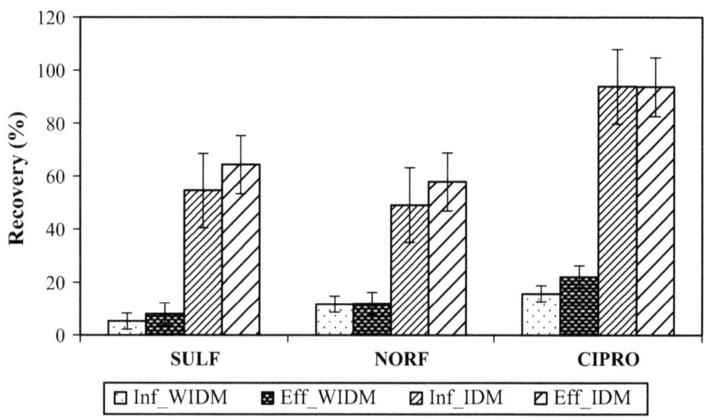

Fig. 1.8 Recovery of selected antibiotics in influent and effluent samples from a WWTP with (IDM) and without (WIDM) when DOC of wastewater influent and effluent was held constant at 10 mg/L. A concentration factor of 10 was used for both influent and effluent

of interest may interfere with their quantification (Bonfiglio et al. 1999). Competition for charge between the analyte of interest and co-eluting substances during ESI (King et al. 2000; Matuszewski et al. 2003), the presence of less volatile solutes (Annesley 2003), and higher concentration of interfering compounds can decrease the efficiency of droplet formation during mass spectrometry-based analysis (Mallet et al. 2004). Molecules with higher mass, such as, DOM, are the major factors responsible for matrix effect (Sterner et al. 2000).

1.4 Conclusions

A multiresidue analytical method based on SPE and LC-MS/MS for the detection of antibiotics was developed for the analysis of antibiotics in wastewater. The usage of the HR-LC coupled with Q-TOF MS system and Phenomenex Max-RP column (5-μm particle size) with mobile phase consisting of 0.1% FA aqueous solution and 100% ACN allowed effective separation of antibiotics in the wastewater. For the pH range tested, matrix pH of 2.5 showed relatively higher recovery for most of the antibiotics except sulfamethoxazole. The optimal drying time was 5 min. In addition, direct concentration of the eluent to 1 ml under nitrogen purging yielded better recoveries. This method was also applied to evaluate the percentage recovery of antibiotics in wastewater samples obtained from a WWTP in India. Although IDM improved the recovery of sulfamethoxazole from the wastewater matrix, antibiotics, such as, ciprofloxacin and norfloxacin showed poor recoveries even with IDM. Lowering the concentration factor and thereby the DOC in the wastewater extract improved the recoveries based on IDM. More research is required to further improve the recovery of pharmaceuticals during SPE.

Acknowledgements We acknowledge IRCC, IIT Bombay, for providing partial financial support. The LC-MS/MS facility in Sophisticated Analytical Instrument Facility (SAIF), IIT Bombay, was used for analysis of the antibiotics. Department of Science and Technology (DST), New Delhi, is acknowledged for providing INSIPRE fellowship to Sanjeeb Mohapatra.

References

Annesley TM (2003) Ion suppression in mass spectrometry. Clin Chem 49:1041–1044. https://doi.org/10.1373/49.7.1041

Balakrishna K, Rath A, Praveenkumarreddy Y, Guruge KS, Subedi B (2016) A review of the occurrence of pharmaceuticals and personal care products in Indian water bodies. Ecotoxicol Environ Saf 137:113–120. https://doi.org/10.1016/j.ecoenv.2016.11.014

Berridge JC (1985) Techniques for the automated optimization of HPLC separation. Wiley

Bevan R, Harrison P, Youngs L, Whelan M, Goslan E, Macadam J, Holmes P, Persich T (2012) A review of latest endocrine disrupting chemicals research implications for drinking water. Institute of Environment and Health (IEH), UK

Bonfiglio R, Bonfiglio R, King R, King R, Olah T, Olah T, Merkle K, Merkle K (1999) The effects of sample preparation methods on the variability of the electrospray ionization response for model drug compounds. Rapid Commun Mass Spectrom 13:1175–1185. https://doi.org/10.1002/(SICI)1097-0231(19990630)13:12<1175:AID-RCM639>3.0.CO;2-0

Comero S, Loos R, Carvalho R, Anto DC, Locoro G, Tavazzi S, Paracchini B, Ghiani M, Lettieri T, Blaha L, Jarosova B, Voorspoels S, Servaes K, Haglund P, Fick J, Lindberg RH, Schwesig D, Gawlik BM, António DC, Comero S, Locoro G, Tavazzi S, Paracchini B, Ghiani M, Lettieri T, Blaha L, Jarosova B, Voorspoels S, Servaes K, Haglund P, Fick J, Lindberg RH, Schwesig D, Gawlik BM (2013) EU-wide monitoring survey on emerging polar organic contaminants in wastewater treatment plant effluents. Water Res 47:6475–6487. https://doi.org/10.1016/j.watres.2013.08.024

Diwan V, Tamhankar AJ, Khandal RK, Sen S, Aggarwal M, Marothi Y, Iyer RV, Sundblad-Tonderski K, Stålsby-Lundborg C (2010) Antibiotics and antibiotic-resistant bacteria in waters associated with a hospital in Ujjain, India. BMC Public Health 10:1–8. https://doi.org/10.1186/1471-2458-10-414

EPA (2007) Method 1694: pharmaceuticals and personal care products in water soil, sediment, and biosolids by HPLC/MS/MS. EPA Method. https://doi.org/10.1002/etc.3451

Gómez MJ, Petrović M, Fernández-Alba AR, Barceló D (2006) Determination of pharmaceuticals of various therapeutic classes by solid-phase extraction and liquid chromatography-tandem mass spectrometry analysis in hospital effluent wastewaters. J Chromatogr A 1114:224–233. https://doi.org/10.1016/j.chroma.2006.02.038

Gros M, Petrović M, Barceló D (2006) Development of a multi-residue analytical methodology based on liquid chromatography-tandem mass spectrometry (LC-MS/MS) for screening and trace level determination of pharmaceuticals in surface and wastewaters. Talanta 70:678–690. https://doi.org/10.1016/j.talanta.2006.05.024

Gros M, Rodríguez-Mozaz S, Barceló D (2013) Rapid analysis of multiclass antibiotic residues and some of their metabolites in hospital, urban wastewater and river water by ultra-high-performance liquid chromatography coupled to quadrupole-linear ion trap tandem mass spectrometry. J Chromatogr A 1292:173–188. https://doi:10.1016/j.chroma.2012.12.072

Hernandez-Ruiz S, Abrell L, Wickramasekara S, Chefetz B, Chorover J (2012) Quantifying PPCP interaction with dissolved organic matter in aqueous solution: combined use of fluorescence quenching and tandem mass spectrometry. Water Res 46:943–954. https://doi.org/10.1016/j.watres.2011.11.061

Huber MM, Göbel A, Joss A, Hermann N, Löffler D, McArdell CS, Ried A, Siegrist H, Ternes TA, Von Gunten U (2005) Oxidation of pharmaceuticals during ozonation of municipal wastewater effluents: a pilot study. Environ Sci Technol 39:4290–4299. https://doi.org/10.1021/es048396s

ICH (1994) Validation of analytical procedures : text and methodology Q2(R1) [WWW Document]. Int Conf Harmon. https://www.ich.org/fileadmin/Public_Web_Site/ICH_Products/Guidelines/Quality/Q2_R1/Step4/Q2_R1__Guideline.pdf. Accessed on 9th Aug 2017

Karthikeyan KG, Meyer MT (2006) Occurrence of antibiotics in wastewater treatment facilities in Wisconsin, USA. Sci Total Environ 361:196–207. https://doi.org/10.1016/j.scitotenv.2005.06.030

Kasprzyk-Hordern B, Dinsdale RM, Guwy AJ (2008) The effect of signal suppression and mobile phase composition on the simultaneous analysis of multiple classes of acidic/neutral pharmaceuticals and personal care products in surface water by solid-phase extraction and ultra performance liquid chromatogra. Talanta 74:1299–1312. https://doi.org/10.1016/j.talanta.2007.08.037

Kasprzyk-Hordern B, Dinsdale RM, Guwy AJ (2007) Multi-residue method for the determination of basic/neutral pharmaceuticals and illicit drugs in surface water by solid-phase extraction and ultra performance liquid chromatography-positive electrospray ionisation tandem mass spectrometry. J Chromatogr A 1161:132–145. doi:10.1016/j.chroma.2007.05.074

King R, Bonfiglio R, Fernandez-Metzler C, Miller-Stein C, Olah T (2000) Mechanistic investigation of ionization supression in electrospray ionization. J Am Soc Mass Spectrom 11:942–950. https://doi.org/10.1016/S1044-0305(00)00163-X

Lacey C, McMahon G, Bones J, Barron L, Morrissey A, Tobin, JM (2008) An LC-MS method for the determination of pharmaceutical compounds in wastewater treatment plant influent and effluent samples. Talanta 75:1089–1097. doi:10.1016/j.talanta.2008.01.011

Lin AY-CC, Tsai Y-TT (2009) Occurrence of pharmaceuticals in Taiwan's surface waters: impact of waste streams from hospitals and pharmaceutical production facilities. Sci Total Environ 407:3793–3802. https://doi.org/10.1016/j.scitotenv.2009.03.009

Liu R, Zhou J, Wilding A (2004) Simultaneous determination of endocrine disrupting phenolic compounds and steroids in water by solid-phase extraction–gas chromatography–mass spectrometry. J Chromatogr A 1022:179–189. doi:10.1016/j.chroma.2003.09.035

Mallet CR, Lu Z, Mazzeo JR (2004) A study of ion suppression effects in electrospray ionization from mobile phase additives and solid-phase extracts. Rapid Commun Mass Spectrom 18:49–58. https://doi.org/10.1002/rcm.1276

Matuszewski BK, Constanzer ML, Chavez-Eng CM (2003) Strategies for the assessment of matrix effect in quantitative bioanalytical methods based on HPLC-MS/MS. Anal Chem 75:3019–3030. https://doi.org/10.1021/ac020361s

Mohapatra S, Huang C-H, Mukherji S, Padhye LP (2016) Occurrence and fate of pharmaceuticals in WWTPs in India and comparison with a similar study in the United States. Chemosphere 159:526–535. https://doi.org/10.1016/j.chemosphere.2016.06.047

Mutavdžić Pavlovic D, Pinušić T, Periša M, Babić S (2012) Optimization of matrix solid-phase dispersion for liquid chromatography tandem mass spectrometry analysis of 12 pharmaceuticals in sediments. J Chromatogr A 1258:1–15. https://doi.org/10.1016/j.chroma.2012.08.025

Nelson ED, Do H, Lewis RS, Carr SA (2011) Diurnal variability of pharmaceutical, personal care product, estrogen and alkylphenol concentrations in effluent from a tertiary wastewater treatment facility. Environ Sci Technol 45:1228–1234. doi:10.1021/es102452f

Oaks JL, Gilbert M, Virani MZ, Watson RT, Meteyer CU, Rideout BA, Shivaprasad HL, Ahmed S, Chaudhry MJI, Arshad M, Mahmood S, Ali A, Khan AA (2004) Diclofenac residues as the cause of vulture population decline in Pakistan. Nature 427:630–633. https://doi.org/10.1038/nature02317

Padhye LP, Yao H, Kung'u FT, Huang C-H (2014) Year-long evaluation on the occurrence and fate of pharmaceuticals, personal care products, and endocrine disrupting chemicals in an urban drinking water treatment plant. Water Res 51:266–276. doi:10.1016/j.watres.2013.10.070

Pan B, Ning P, Xing B (2009) Part V-Sorption of pharmaceuticals and personal care products. Environ Sci Pollut Res Int 16:106–116. https://doi.org/10.1007/s11356-008-0052-x

Park J, Yamashita N, Park C, Shimono T, Takeuchi DM, Tanaka H (2017) Removal characteristics of pharmaceuticals and personal care products: comparison between membrane bioreactor and various biological treatment processes. Chemosphere 179:347–358. https://doi.org/10.1016/j.chemosphere.2017.03.135

Sterner J, Johnston M, Nicol G, Ridge D (2000) Signal suppression in electrospray ionization Fourier transform mass spectrometry of multi-component samples. J Mass Spectrom 35:385–391. https://doi.org/10.1002/(SICI)1096-9888(200003)35:3<385:AID-JMS947>3.0.CO;2-O

Sun Q, Li M, Ma C, Chen X, Xie X, Yu CP (2016) Seasonal and spatial variations of PPCP occurrence, removal and mass loading in three wastewater treatment plants located in different urbanization areas in Xiamen, China. Environ Pollut 208:371–381. https://doi.org/10.1016/j.envpol.2015.10.003

Ternes TA (1998) Occurrence of drugs in german sewage treatment plants and rivers. Water Res 32:3245–3260. https://doi.org/10.1016/S0043-1354(98)00099-2

Tolls J (2001) Sorption of veterinary pharmaceuticals in soils: a review. Environ Sci Technol 35:3397–3406. https://doi.org/10.1021/es0003021

Van Boeckel TP, Gandra S, Ashok A, Caudron Q, Grenfell BT, Levin SA, Laxminarayan R (2014) Global antibiotic consumption 2000 to 2010: an analysis of national pharmaceutical sales data. Lancet Infect Dis 14:742–750. doi:10.1016/S1473-3099(14)70780-7

Vanderford BJ, Pearson RA, Rexing DJ, Snyder SA (2003) Analysis of endocrine disruptors, pharmaceuticals, and personal care products in water using liquid chromatography/tandem mass spectrometry. Anal Chem 75:6265–6274. doi:10.1021/ac034210g

Watkinson AJ, Murby EJ, Costanzo SD (2007) Removal of antibiotics in conventional and advanced wastewater treatment: implications for environmental discharge and wastewater recycling. Water Res 41:4164–4176. doi:10.1016/j.watres.2007.04.005

Weigel S, Kallenborn R, Hühnerfuss H (2004) Simultaneous solid-phase extraction of acidic, neutral and basic pharmaceuticals from aqueous samples at ambient (neutral) pH and their determination by gas chromatography-mass spectrometry. J Chromatogr A 1023:183–195. https://doi.org/10.1016/j.chroma.2003.10.036

Xu W, Zhang G, Li X, Zou S, Li P, Hu Z, Li J (2007) Occurrence and elimination of antibiotics at four sewage treatment plants in the Pearl River Delta (PRD), South China. Water Res 41:4526–4534. https://doi.org/10.1016/j.watres.2007.06.023

Zhang ZL, Zhou JL (2007) Simultaneous determination of various pharmaceutical compounds in water by solid-phase extraction-liquid chromatography-tandem mass spectrometry. J Chromatogr A 1154:205–213. https://doi.org/10.1016/j.chroma.2007.03.105

Zorita S, Mårtensson L, Mathiasson L (2009) Occurrence and removal of pharmaceuticals in a municipal sewage treatment system in the south of Sweden. Sci Total Environ 407:2760–2770. https://doi.org/10.1016/j.scitotenv.2008.12.030

Chapter 2
A Comprehensive Review on Various Analytical Methods for the Determination of Inorganic and Organic Arsenic in Environmental Samples

Nalini Sankararamakrishnan and Shruti Mishra

Abstract In natural environment, Arsenic (As) occurs in four oxidation states, namely, As(V), As(III), As(0), and As(−III). The oxidation state of As determines its toxicity and mobility in the environment. Thus, quantification and speciation of As are critical in assessing the overall risk. Further, As being a class A carcinogen, it is of interest to both environmental scientists and analytical chemists. Thus, sensitive and adept determination of As and speciation of different forms of As in various environmental matrices are indispensable. Most of the countries around the world have relevant official regulations on permissible levels of As in drinking water. In India, the permissible limit of As in drinking water is set at 10 μg/L by Bureau of Indian Standards (BIS) and WHO guideline value is also 10 μg/L. In this review, we focus on extraction of As from various environmental samples, As speciation, sample treatment, and determination of As in various matrices. Analytical methods for the determination of various forms of As like atomic absorption spectrophotometer (AAS), hydride generation AAS (HG_AAS), atomic fluorescence spectrometry (AFS), inductively coupled mass spectrometry (ICPMS), electrochemical methods, capillary electrophoresis (CE), high-performance liquid chromatography (HPLC), HPLC coupled with mass spectrometer (HPLC-MS), neutron activation analysis, and biosensors are also summarized. Determination of As in the field using various field test kits available in the market is also detailed.

Keywords Arsenic · Speciation · Extraction · Determination · Sensitivity

N. Sankararamakrishnan (✉) · S. Mishra
Centre for Environmental Science and Engineering, Indian Institute of Technology Kanpur, Kanpur 208016, UP, India
e-mail: nalini@iitk.ac.in

T. Gupta et al. (eds.), *Environmental Contaminants*, Energy, Environment, and Sustainability, https://doi.org/10.1007/978-981-10-7332-8_2

2.1 Introduction

Arsenic (As) is one of the most abundant elements in earth's crust (Ahmed 2009). As contamination in groundwater has been reported in more than 20 countries, out of which Bangladesh and India, inner Mongolia and Taiwan are severely affected (Chakraborti et al. 2002; Rahman et al. 2001). Around nine districts in West Bengal, India, and fifty districts of Bangladesh have reported to contain As levels above WHO permissible limit of 10 μg/L in groundwater (World Health Organisation 1993). Long-term exposure to As leads to various kinds of cancer including skin, lung, liver, bladder, and kidney (World Health Organization 2001). The four predominant oxidation states of As found in natural environment are: As(III), As(V), As(0), and As(−3). The mobility of As in water and its toxicity toward living being is dependent upon the oxidation state of As (Meng et al. 2003). As predominantly exists as As(III) and As(V) with a minor amount of monomethyl arsenic (MMA) and dimethyl Arsenic (DMA) in groundwater (Fig. 2.1). Generally, inorganic As compounds are considered to be more acutely toxic compared to organic arsenicals and the toxicity of organic arsenicals decreases with methylation. Thus, the order of arsenic toxicity is As(III) > As(V) ⋙ MMA > DMA (Shiomi 1994).

Hence, it becomes extremely important to speciate arsenic and determine with sensitivity below 10 μg/L. Speciation of arsenic is reported using various techniques including hydride generation (HG) (Holak 1969), chromatographic techniques like ion chromatography (IC) (Ammann 2011), high-performance liquid chromatography (HPLC) (Jia et al. 2016), capillary electrophoresis (CE) (Qu et al. 2015) etc. The most commonly used detection method for As are inductively coupled mass spectrometry (ICP MS), graphite furnace atomic absorption spectrophotometer (GFAAS), hydride generation atomic absorption spectrophotometer

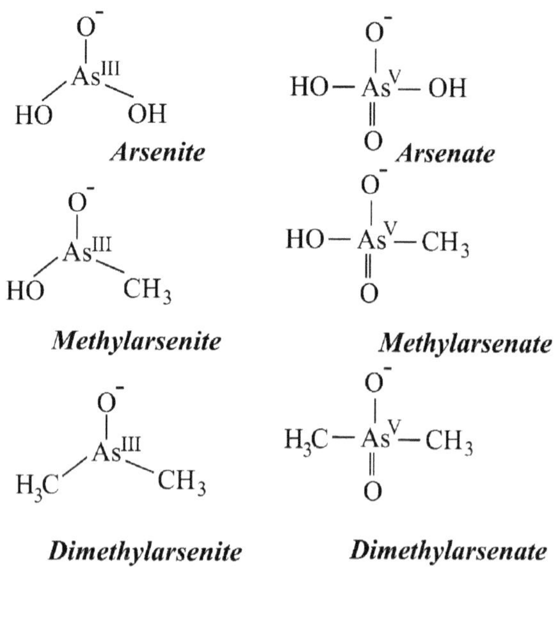

Fig. 2.1 Arsenic species found in water (Leermakers et al. 2006)

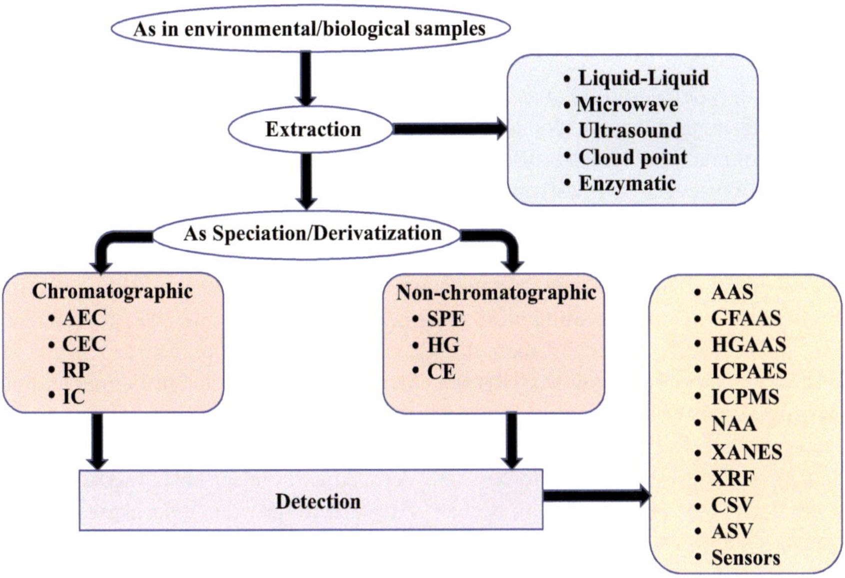

Fig. 2.2 Schematic representation of various stages for determination of different forms of As. *AEC* anion exchange chromatography, *CEC* cation exchange chromatography, *IC* ion exchange chromatography, *RP* reversed-phase chromatography, *HG* hydride generation, *CE* capillary electrophoresis, *SPE* solid-phase extraction, *AAS* atomic absorption spectroscopy, *GFAAS* graphite furnace atomic absorption spectroscopy, *HGAAS* hydride generation atomic absorption spectrophotometer, *ICPAES* inductively coupled plasma atomic emission spectrometry, *NAA* neutron activation analysis, *XANES* X-ray absorption near edge spectroscopy, *XRF* X-ray fluorescence, *CSV* cathodic stripping voltammetry, *ASV* anodic stripping voltammetry

(HGAAS), which have detection limits ranging from 0.5 to 50 μg/L (United States Environmental Protection Agency 1999a). Thus, this chapter focuses on the comprehensive review of extraction of arsenic from various matrices, speciation, and detection using various methods. Determination of As in the field using various field test kits available in the market is also detailed. The schematic representation of various stages of As determination is detailed in Fig. 2.2.

2.2 Extraction of Arsenic

With the advancement of various technologies, assessment of As in various environmental samples could be carried out directly without resorting to a preconcentration or extraction step. In general, HG techniques are quite selective as the interference from matrix elements are removed as the analyte is volatilized as gaseous arsine. However, the applicability of this method to seawater is a challenge due to the presence of high amounts of chloride ions. Reviews have been reported on As determination in environmental samples (Akter et al. 2005), water (Llorente-Mirandes et al. 2017), and food (Guell et al. 2010) samples.

2.2.1 *Water*

Guell et al. (2010) used Aliquat 336 in 4% dodecane to extract ppb levels of As(V) and As(III) at pH 13. It was found that As(V) extraction by aliquat-336 is kinetically more favorable, and the adsorbed As(V) was desorbed using 0.1 M HCl. Using this method, ppb levels of As(V) could be separated from As(III). Experiments were performed with As-spiked groundwater, and the results obtained were satisfactory. Preconcentration of trace amounts of As using ultrasonic bath and ionic liquid-based microextraction methodology has been reported (Amjadi et al. 2011). Initially, As(III) was extracted as ammonium pyrrolidine dithiocarbamate (APDC) complex into ionic liquid, 1-hexyl-3-methylimidazolium hexafluorophosphate, using an ultrasonic bath. Extracted As(III) was determined using electrothermal atomic absorption spectrophotometer (ETAAS). A detection limit (3s) of 0.01 μg/L has been reported. For six replicate determinations of 0.2 μg/L of As(III), the relative standard deviation was found to be 4.6%. As speciation in seawater was studied by collecting both As species in cobalt (III) oxide powder (Narukawa 2000). Quantitative extraction of both inorganic forms of As was achieved using cobalt (III) oxide powder as collector in the pH range of 1.0–11.0. Dimethyl arsine (DMA) was extracted from seawater into benzene after addition of potassium iodide. After extraction with benzene, the solution was acidified with 7.2 M HCl, and DMA was collected on cobalt (III) oxide powder.

2.2.2 *Soils and Sediments*

The main challenge is to extract both forms of inorganic arsenic quantitatively and minimize the conversion from one form to another. Georgiadis et al. (2006) reported that that phosphate solutions with 0.5% sodium diethyldithiocarbamate trihydrate (NaDDC) could be used to extract arsenic species and restrict the transformation between As(III) and As(V) in the samples analyzed. Recoveries of As(III) spiked samples ranged from 80 to 120% which As(V) concentration remained constant. A similar procedure is reported by Garcia-Manyes et al. (2002) using the combination of ascorbic acid and phosphoric acid. Other extraction chemicals like 1,3-propanedithiol or 1,2-ethanedithiol (Szostek and Aldstadt 1998) are methanol/hydrochloric acid/water (Yehl and Tyson 1997), acetone and hydrochloric acid (Yehl et al. 2001) have also been reported. Extraction of phenyl arsenic compounds have been carried out using supercritical fluid extraction with carbon dioxide and methanol containing 15% dichloromethane as modifier. The efficiency of the method is reported to be 40% (Thurow et al. 2002).

2.2.3 *Plants and Marine Organisms*

Kroukamp et al. (2016) recently reviewed speciation of As along with other metalloids in plants. Extraction of arsenic from biological samples is always carried out by shaking, heating, microwave digestion, or sonication. Generally, binary mixtures containing ethanol/water (Zhao et al. 2015), methanol/water, or methanol/chloroform are used (Alberti et al. 1995) for extraction purposes. Zhao et al. (2015) reported various extractants for the quantiative extraction of arsenic species from three different plants and subsequent determination by HPLC-ICP-MS. Satisfactory extraction of >78% was obtained using a mixture of 25% ethanol in water with sonication time of 0.5 h for fronds and 2.0 h for roots. Further, the method preserved the arsenic species during extraction. The three solvent systems, namely water, methanol, and 1:1 methanol-water were compared for the extraction of As from Molluscs (McSheehy et al. 2001). It was reported that the extraction of arsenic was found to be same in all the three systems. In marine organism, the kind of arsenic species is arsenobetaine (AB) and is completely soluble in water and >90% efficiency is found in all the three systems.

2.2.4 *Biological Samples of Humans*

To ascertain the seriousness of As contamination in human beings, As is analyzed in human nails, urine and blood samples. Ultratrace As determination was carried out in urine and blood samples by FI-HG-AAS after preconcentration and speciation using dispersive liquid–liquid microextraction (Shirkhanloo et al. 2011). Initially, As(III) was complexed with ammonium pyrrolidone dithiocarbamate at pH 4 and extracted with ionic liquid (IL). Then the extracted As(III) was back extracted from IL using HCl, and stripped As(III) was determined using FI-HG-AAS. Total As was determined by initially reducing As(V) to As(III) using KI and ascorbic acid in HCl solution. As(V) was determined by the difference between total As and As(III) concentration. The method reported a LOD of 5 ng/L. A good precision with RSD <5% was reported by using this method on biological samples of multiple sclerosis patients.

In another method, a water-miscible ionic liquid (IL) ([Hmim][BF_4]) was added to the sample followed by the addition of an ion exchange reagent ($NaPF_6$) to obtain the hydrophobic IL ([Hmim][PF_6]) (Howard 1997). The formed product acted as extractant to extract trace amounts of As(III) and As(V) which was then determined by ETAAS. In situ solvent formation microextraction exhibited a limit of detection (3s), and the enrichment factor were 6 ng/L and 198, respectively. The method was applied for the determination of total As in biological samples. Further, the application of this method was found successful in food salts and water samples as well.

2.3 Arsenic Speciation and Determination

2.3.1 Hydride Generation

Speciation of both inorganic forms of As was achieved by hydride generation technique. It is the most popular sample derivatization method used for determination of As(III) and As(V) (Howard 1997). Both sodium and potassium tetrahydroborate (III) are reliable reducing reagents for the conversion of As to volatile arsine (Holak 1969). The ability of As(III) to react with tetrahydroborate at a higher pH than As(V) was used for the differentiation of As(III) and As(V) species. Thus, total arsenic could be determined by the reduction to arsine at acidic conditions and at pH 4. As(III) was converted to arsine using tetrahydroborate. Concentration of As(V) was acquired from the difference between the total As and As(III) concentration. Inclusion of the flow injection technique along with HG decreased the interference from transition elements (Yamamoto et al. 1985). Gonzalvez et al. (2009) reported a method for the determination of As(III) and As(V) in food samples using FI-HG-AFS. As(III) was measured directly by feeding the sample, and total As was determined by reducing it with KI and ascorbic acid for 30 min. The method reported a detection limit of 5.0 ng/g. Musil et al. (2014) reported a method or the speciation of inorganic As from organic As using HG-ICPMS. Initially, the samples were treated with HCl and $NaBH_4$ and the generated arsine was sent to ICPMS detector using Ar gas. Under these conditions, only inorganic As underwent reduction to volatile arsine gas. The method was applied to the determination of inorganic As from rice and marine organisms. HG is also used for pre-column or post-column derivatization. In the pre-column derivatization, initially, volatile arsine is formed which is cryogenically trapped and sequentially desorbed and carried to the detector (Gomez-Ariza et al. 2000). In the post-column method, separation of As species is carried out by HPLC followed by HG to improve the sensitivity of detection (Sloth et al. 2003; Wangkarn and Pergantis 2000).

2.3.2 Chromatographic Method

Chromatography is a very powerful tool to separate various forms of As. Further, coupling the chromatography with various element-specific detectors improves the sensitivity and selectivity of As determination. As speciation by various chromatographic methods are adequately reviewed (Larsen 1998; Ammann 2011; Feldmann et al. 2004). Among the various liquid chromatographic (LC) techniques, ion exchange chromatography (IEC) and ion interaction chromatography (IIC) have been widely used. In IEC, the analyte is transported into the column via a mobile phase where the various species compete for stationary phase containing oppositely charged functional groups and the separation of As species occurs by displacement of mobile phase ions. The hyphenated IEC-ICPMS offers detection levels in the range of sub-nanogram levels, with great linear range.

Sheppard et al. (1992) reported a method for the speciation of inorganic and organic As using ion chromatography(IC) coupled with ICP. A Wescan Anion/R-IC, 250 ~ 4.1 mm i.d. was used as the ion exchange column with 2% propan-1-ol and 50 mmol dm^{-3} carbonate buffer, pH 7.5 as eluants 1 and 2, respectively. A gradient elution program was used to resolve the various peaks. The order of the elution of various As species was As(III), DMA, MMA and As(V). Total run time of the program was 12 min. The method was used for the determination of As in urine, and limits of detection were found to be 4.9, 6.0, 1.2, 3.6 μg/L for As(III), As(V), DMA, MMA, respectively.

Reversed-phase ion pair HPLC with mobile phases such as tetramethylammonium cation or heptanesulfonate anion has also been reported. Seven As species have been separated using hexanesulfonate as mobile phase using C18 column as stationary phase (Le et al. 1996). Kohlmeyer et al. (2002) separated 17 organo-arsenical in single chromatographic run using ion exchange chromatography. Recently, Rasheed et al. (2017) assessed 228 samples of groundwater from Pakistan for inorganic and organic As using IC–ICPMS. It was found that As was present mainly as As(V) and the levels of MMA, DMA, and AB were well below the permissible limits.

Both inorganic and organic As species in various water samples were preconcentrated using on-line solid-phase extraction (SPE) (Jia et al. 2016). In this technique, phosphine functionalized polymer microspheres have been used as solid phase extractant. The inorganic and organic species, namely As(V), As(III), DMA, and MMA species, have been quantitatively retained on the solid phase column and were eluted with a mixed solution of ammonium nitrate and ammonium dihydrogen phosphate. After preconcentration and separation, speciation and determination of all the four As species were achieved using HPLC–ICP-MS technique. The enrichment factor of 28 was obtained for DMA and As(III) in 25 mL sample solution while a factor of 30 was obtained As(V) and MMA. The low detection limits of 0.91, 0.82, 0.96, 1.2 ng/L were obtained for As(V), MMA, DMA, and As(III), respectively.

Gas chromatography (GC) in conjunction with mass spectrometry (MS) (GC-MS) and GC-MS/MS has been used for the speciation of As. As speciation was achieved using derivatization agents like thiols, thioglycol methylate (TGM), thioglycol ethylate (TGE) and British anti-Lewisite (BAL; 1,3-dimercapto-1-propanol) (Namera et al. 2012; Campillo et al. 2008; Takeuchi et al. 2012; Kang et al. 2016). After derivatization, the samples were extracted in dichloromethane and injected in GC/MS in SIM mode to quantify the As species. Using GC-MS/MS technique, a very low detection limit of 0.08 pg was achieved with very high precision and accuracy.

2.3.3 Capillary Electrophoresis (CE)

CE is a robust separation technique possessing a high resolving power and used for the separation of various As species based on its charge (Qu et al. 2015). In CE, various separation processes are possible including capillary electrochromatography (CEC), micellar electrokinetic capillary chromatography (MECC), isotachophoresis (ITP), isoelectric focussing (IEF), and capillary zone electrophoresis (CZE) using the same instrument (Michalke 2003). The interaction between the analyte and the stationary phase is eliminated as the process is carried out without stationary phase. Coupling this technique to MS and ICPMS has been reported. Qu et al. (2015) reported a method for the quantification and speciation of As using CE-ICPMS technique. In this technique, α-amylase enzyme facilitated the water-phase microwave extraction of As species, including DMA, MMA, As(V), and As(III) from rice matrices. Usually, capillary columns are as the nebulizer to couple CE to ICP-MS. Some nebulizers may introduce backpressure which might affect the electrophoretic separation due to production of laminar flow in the capillary, hence the choice of the nebulizer is extremely important (Dressler et al. 2011). Yang et al. (2009) introduced an improved sheath flow interface to facilitate the coupling of CE to ICPMS. This sheath flow technique avoided laminar flow completely in the capillary and enabled stable electric supply in CE and efficient transport of the sample from CE to ICPMS. Sheath flow technique reduced the dead volume of interface to approximately zero which led to lower detection limit and better electrophoretic resolution. Thus, well-resolved peaks with lower detection limits (0.030–0.042 μg/L) were obtained. Using a 100 cm length_50 μm ID fused-silica capillary as the nebulizer, separation of various As species including As(III),As(V),DMA, MMA, AsB, AsC, 3-NHPAA,4-NPAA,o-ASA(o-arsanilic acid), and p-UPAA was achieved using CE-ICPMS technique (Liu et al. 2013). To ensure quality control in CE, methods like internal standard, standard addition are often employed (Michalke 2003). Due to application of high voltage during electrophoresis, alterations of the chemical species may occur and the detection limits, reproducibility, and peak resolution are found to be inferior to LC (Liu et al. 2013).

2.3.4 Detection Methods

The USEPA has recently reviewed the methods available for monitoring As (United States Environmental Protection Agency 2004) Inductively coupled mass spectrometry (ICPMS), inductively coupled atomic emission spectrometry (ICP-AES), hydride generation atomic absorption spectrophotometer (HG-AAS), graphite furnace AAS (GFAAS) are the methods approved by USEPA. These methods report detection limit in the range of 0.5–50 μg/L (United States Environmental Protection Agency 1999b). In the last six years, lots of comprehensive reviews have been reported regarding arsenic speciation and analysis in various matrices (Radke et al. 2012; Ammann 2011; Leermakers et al. 2006; Akter et al. 2005; Llorente-Mirandes et al. 2017; Tyson 2013; Rajakovic et al. 2013; Chen et al. 2014; Sadee et al. 2015; Nearing et al. 2014; Quinaia and Rollemberg 2001).

The method chosen for the determination of arsenic depends on several factors including detection limit, precision, selectivity, cost-effectiveness. Usually, the sensitivity, selectivity, and speciation of arsenic are achieved by hyphenating the various detection techniques with solid phase extraction (SPE), flow injection, hydride generation, or liquid chromatography. Table 2.1 lists the various methods available for arsenic and their corresponding detection limits.

Table 2.1 List of various analytical methods and their detection limits for As

Analytical method	Detection limit (μg/L)	Strength/weakness of the method	References
HG-AAS	As(III) 0.6 As(V) 0.5	Strength: low detection limits compared to AAS and GFAAS and speciation Weakness: single element analysis, strictly controlled reaction conditions, time consuming	Quinaia and Rollemberg (2001)
FI-SE-HG-AAS	As(III) 0.05 As(V) 0.20	S: very low detection limit. Applicable to seawater	Karthikeyan et al. (1999)
FI-HG-AAS	As(III) 0.037	S: very low detection limit. Applicable to groundwater samples	Naykki et al. (2001)
FI-HG-AFS	As(III) 0.023	S: very low detection limit. Speciation	Yan et al. (2002)
SPE-GF-AAS	As(III) 0.11 As(V) 0.15	W: high detection limits compared to other techniques	Hsieh et al. (1999)
HG-ICP-AES	As(III) 0.7	S: multielement analysis W: high detection limits compared to other techniques	Hueber and Winefordner (1995)
TRXRF	As(III) 0.65	W: high detection limit, preconcentration and separation procedure is necessary	Cataldo (2012)
FI-ICPMS	As(III) 0.021 As(V) 0.029	S: low detection limit, speciation, multielement analysis W: high operational cost	Yan et al. (1998)

(continued)

Table 2.1 (continued)

Analytical method	Detection limit (μg/L)	Strength/weakness of the method	References
SPE-ICPMS	As(III) 5.0×10^{-6}	S: low detection limit, speciation, multielement analysis W: high operational cost	Chen et al. (2009)
	As(V) 2.4×10^{-4}		
HG-ICPMS	AsB 0.0301	S: low detection limit, speciation of InAs and organic As, multielement analysis W: high operational cost	Sengupta and Dasgupta (2009)
	DMA 0.0022		
	As(III) 0.0021		
	MMA 0.0021		
	As(V) 0.00208		
FI-HG-ICPMS	AsB 0.0192	S: low detection limit, speciation of InAs and organic As, multielement analysis W: high operational cost	Sengupta and Dasgupta (2009)
	DMA 0.0145		
	As(III) 0.0177		
	MMA 0.0192		
	As(V) 0.0321		
Capillary microextraction-ICPMS	As(III) 0.0109	S: low detection limit, multielement analysis W: in environmental samples only As(V) could be detected	Zheng and Hu (2009)
	As(V) 0.0062		
PDC-NAA	As(III) 0.001	S: sample not destroyed W: preconcentration with PDC is necessary to achieve low detection limit	Sun and Yang (1999)
LC-NAA	As(III) 0.9	S: sample not destroyed W: preconcentration using liquid chromatography is necessary to achieve speciation and low detection limit	Sanchez et al. (2009)
	As(V) 1.7		
	MMA 1.6		
	DMA 3.8		
	Total As 16		

(continued)

Table 2.1 (continued)

Analytical method	Detection limit (μg/L)	Strength/weakness of the method	References
HG-Gas diffusion amperometry	As(Total) 5.0	W: high detection limit, matrix effect	Lolic et al. (2008)
HG-pervaporation-amperometry	As(Total) 1.0	W: high detection limit	Rupasinghe et al. (2009)
CSV	As(III) 0.035	S: low operational cost W: high detection limit, interference from matrix elements	Gibbon-Walsh et al. (2010)
DP-ASV	As(V) 0.07	S: low operational cost W: high detection limit, interference from matrix elements	AlvesGMS et al. (2011)
Modified Au electrode, CV	As(III) 0.047	S: low operational cost W: high detection limit, interference from matrix elements	Giacomino et al. (2011)

S strength; *W* weakness

2.3.5 *AAS, AFS, ETAAS/GFAAS*

In general, for the determination of arsenic in various environmental samples, both AAS and AFS are coupled with hydride generation technique. Hydride generation technique improves the selectivity as the analyte is separated from the matrix to volatile hydride, and further sensitivity of the analysis is also increased. By controlling the acidity of the reaction mixture during hydride generation, speciation is also achieved. A LOD of 12 ng/L was achieved for the determination of As(V) by HG-AAS (Tuzen et al. 2009) while HG-AFS reported LOD as 14 ng/L (Chen et al. 2013). The main interference in HGAAS technique is the interference from trace elements like iron which could conveniently overcome by the addition of L-cysteine (Howard and Salou 1996). Further, inclusion of flow injection technique eliminates the interference from transition metals (Yamamoto et al. 1985). Electrothermal or graphite furnace AAS (ETAAS/GFAAS) is based on the atomization of arsenic at very high temperature using electric furnace. In general, the LOD of As by GFAAS is 1 μg/L.

2.3.6 Total Reflection X-ray Fluorescence Spectroscopy (TRXRF)

The major advantages of TRXRF are high sensitivity, hence with very low detection limit, highly selective, and very low sample requirement. Sitko et al. (2015) recently reported a method for the preconcentration of Arsenic using mercapto-modified graphene oxide nanosheets and determination using TRXRF. A LOD of 0.064 μg/L is reported for As(III). Speciation of As in cucumber (*Cucumis sativus* L.) xylem sap was achieved using synchrotron radiation-induced TRXRF. Arsenic speciation in xylem sap down to 30 μg/L (30 ppb) was achieved using the above technique.

2.3.7 Electrochemical Methods

Several electrochemical methods are reported for the determination of arsenic. The main disadvantage of these methods is the interference from matrix elements. A detection limit of 20 μg/L is achieved for arsenite determination using differential pulse polarography (Greschonig and Irgolic 1992). Higher sensitivity and selectivity is achieved by stripping voltammetric procedures. In this process, arsenic is electrochemically deposited on a suitable electrode followed by the oxidation to metal back to the solution by a reverse potential scan. The oxidation or stripping current is recorded as a function of the analyte concentration. Mays and Hussam (2009) reviewed the various voltammetric methods for the determination of arsenic. The most popular method for the determination of arsenic is anodic stripping voltammetry (ASV). It is reported that Au electrodes are more sensitive than Pt electrode, and a LOD of 0.2 μg/L was achieved (Forsberg et al. 1975). The most commonly used method for the determination of As in ground waters of Bangladesh involves the ASV using thin gold film-deposited glassy carbon electrode. More than 950 samples of groundwater were analyzed using this method (Aggarwal et al. 2001). Similar to ASV, cathode stripping voltammetry (CSV) are also used for the determination of anions. Hanging drop mercury electrode (HDME) is the most commonly used electrode; however, Cu and Bi electrodes are also being used. A very low LOD of 0.01 and 0.02 μg/L for As(III) and As(V), respectively, is reported using HBr as electrolyte by CSV (Profumo et al. 2005). Most of the electrochemical methods reported are applicable only for As(III).

2.3.8 Neutron Activation Analysis (NAA)

In this method, the analyte is bombarded with neutrons to form radioactive nuclides which further decay via β and/or γ emission. The γ rays emitted during the decay is detected by a multichannel γ-ray spectrometer (Tulasi et al. 2013). Detection of As in seawater was found to be difficult using NAA owing to the spectral interference caused by the salt content of seawater (Shi and Chatt 2014). This interference was overcome by adding lead nitrate and titanium chloride as carrier and reducing agent, respectively, (Rottschafer et al. 1972). Analysis of total As and As speciation was monitored in water and sediments from the Kwabrafo stream, in southwestern Ghana. Total As content was determined by NAA, and ion pair reverse phase high-performance liquid chromatography-neutron activation analysis (HPLC-NAA) was used for speciation of As species. A solvent extraction preconcentration method involving ammonium pyrrolidine dithiocarbamate (APDC) and 4-methyl-2-pentanone (MIBK) in conjunction with NAA was developed for the simultaneous measurement of low levels of inorganic arsenic along with antimony and selenium species in natural waters (Sun and Yang 1999). The LOD of As was in the range 0.026 µg/L. Sun and Yang (1999) used lead nitrate and titanium chloride as carrier and reducing agent in analysis of As in seawater by NAA. Sanchez et al. (2009) combined column chromatography with NAA to separate and analyze MMA, DMA, As(III), and As(V). The LOD for As(III), As(V), MMA, DMA, and total As was found to be 0.9, 1.7, 1.6, 3.8, and 16 µg/L.

2.3.9 Inductively Coupled Plasma (ICP) Techniques

This technology was employed since the beginning of 1960s (Dickinson 1969). Plasma is used in this technique to atomize and ionize all forms of arsenic in the acidified sample. Generally, ICP is used in conjunction with atomic emission (AES) (Sansoni et al. 1988) or mass spectrometric (MS) detectors (Stetzenbach et al. 1994). ICP-AES is less commonly used technique compared to ICPMS. Low sensitivity for arsenic using ICP could be attributed to poor ionization efficiency. High precision, robust analysis, wide linear range, isotope analysis are the advantages of ICPMS over ICPAES. In ICPMS, during direct nebulization, possible interference from Ar arises due to the formation of dimer molecule ($^{40}Ar^{35}Cl$) in the plasma which coincides with the mass of As (^{75}As) (Colon et al. 2011). In recent years, these interferences are overcome by the use of collision cell or diffusion cell technology (Colon et al. 2011). In this technology, after the initial ion selection, the dimers are allowed to collide with small molecular weight gases such as He, H_2, CH_4, and O_2 to break the polyatomic species and a second quadrupole is employed to collect As species (Zhu et al. 2009).

Generally, ICPMS techniques are hyphenated with solid phase extraction (SPE) (Zhu et al. 2009; Chandrasekaran et al. 2010; Popp et al. 2010; Kempahanumakkagari et al. 2017; Profumo et al. 2006) capillary microextraction (Sun and Yang 1999), hydride generation (HG) (Zheng and Hu 2009; Popp et al. 2010), HPLC-IC (Popp et al. 2010).

2.3.10 Sensors

Generally, arsenic sensors are based on either stripping voltammetry or fluorescence or electrochemical detection with enzymatic inhibition. Recently, Kempahanumakkagari et al. (2017) have adequately reviewed the nanomaterial-based sensors for arsenic determination. Different electrode systems with surface modification have been developed to improve the detection limit and selectivity. As(V) is initially reduced to As(III) which is then electrochemically reduced to As(0). After deposition, the electrode potential is increased to oxidize and strip the deposited As(0) from the electrode. The oxidation current is used to quantify the arsenic oxyanions. Various electrode modifications are carried to facilitate the metal binding to the electrode and subsequent electron transfer. Nanomaterials like carbon nanotubes (Profumo et al. 2006), graphene oxide (Dreyer et al. 2010), metal nanoparticles (Aragay et al. 2011; Yavuz et al. 2016), and enzymes (Male et al. 2007) have been found useful as electrode material. In a study by Ramsesha and Sampath (2011), reduced graphene oxide–lead oxide composite has been used as the electrode to detect As(III) up to 10 nM. A selective and sensitive fluorescent sensor has been reported (Ezeh and Harrop 2012) for the determination of As(III) using ArsenoFluor1 (AF1) as fluorescent chemical probe. The nonconjugated AF1 is non-fluorescent. However, on reaction with As(III) salts, leads to 25-fold increased fluorescent intensity owing to the formation of Coumarin C6-CF_3 (Scheme 2). The method reports a sub-ppb detection limit and selective for As(III) over other ions such as Fe^{2+}, Zn^{2+}, Na^{+}, and Mg^{2+}. Though progress has been made in arsenic sensors with very low LOD of 0.1 ppb (Xiao et al. 2008)

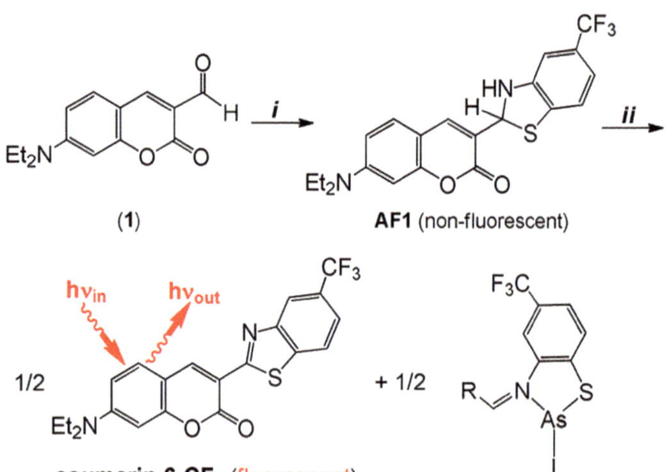

Fig. 2.3 Synthesis of AF1 and proposed As^{3+} response (Ezeh and Harrop 2012). (i) 4-(Trifluoromethyl)-2-aminothiophenol•HCl, EtOH, Et3N, 298 K, 6 h. (ii) AsI3, THF, Et3N, 298 K. R in the As^{3+} complex represents the (diethylamino) coumarin moiety

issues like cost-effectiveness, portability, reproducibility, and interference effect need to be addressed (Fig. 2.3).

2.3.11 *Field Kits*

Field kits developed for As are generally based on Gutzeit reaction. Initially, the sample is acidified and treated with powerful reducing agent like $NaBH_4$, and As(V) is reduced to As(III) and the reduced As(III) further undergoes reduction to release arsine gas. The liberated gas is passed through a lead acetate-soaked cotton filter to remove H_2S followed by $HgBr_2$ impregnated test strip. The test relies on the color stain reaction of arsine generated with $HgBr_2$ impregnated test strip to give a yellow color that then becomes progressively browner as arsenic level increases. Though the method is inexpensive and portable, toxic Hg- and Pb-containing waste is generated and care must be taken to prevent arsine from leaking. Several brands are available in the market which uses this technology including Wagtech Arsenator,[1] Arsenic Quick,[2] EZ Arsenic,[3] PurTest Arsenic,[4] Merckoquant arsenic.[5] Determination of arsenic by these kits is either based on color chart comparison or LED-based photometers calibrated to read the strip and provide a digital output. Recently, Bralatei et al. (2017) reported a method for rapid screening of inorganic As in seaweed samples. The method involved extraction using diluted HNO_3 to quantitatively extract inorganic As without decomposing the organoarsenicals to inorganic As followed by the selective volatilization of inorganic As to arsine and subsequent chemotrapping on a filter paper soaked in mercury bromide ($HgBr_2$) solution. The method reported a reproducibility with an average error of $\pm 19\%$, and further, the method was also validated with HPLC-ICPMS.

2.4 Conclusions

This review summarizes various available methods for extraction, speciation, and analysis of arsenic at ultra-trace levels. Various methods have been explored for speciation of various As compounds. With suitable extraction procedures, As determination with high sensitivity and precision is plausible in both environmental and biological samples. ICPMS coupled with HPLC or HG technique demonstrated

[1]www.wagtech.co.uk. Accessed on 09/08/17.

[2]www.sensafe.com. Accessed on 09/08/17.

[3]http://www.hach.com/arsenic-low-range-test-kit/product?id=7640217303. Accessed on 09/08/17.

[4]http://www.vitasalus.net/purtest-arsenic-water-test-kit-1-test-kit. Accessed on 09/08/17.

[5]http://www.merckmillipore.com/food-analytics/rapid-arsenic-tests/c_Hzib.s1OprIAAAEbFfcXP9oy. Accessed on 09/08/17.

separation, speciation, and robust analysis of As with high sensitivity. However, sample storage, preservation of various arsenic species during sample pretreatment is still a challenge. Though field kits offer the advantage of portability and cost-effectiveness, accuracy of determination is questionable.

References

Aggarwal PK, Dargie M, Groening M, Kulkarni KM, Gibson JJ (2001) Draft report: an inter-laboratory comparison of arsenic analysis in Bangladesh, isotope hydrology section. International Atomic Energy Agency, Vienna, Austria. http://www.iaea.org/programmes/aqcs/icpt/arsenicpt.pdf

Ahmed MF (2009) Periodic table. Chemical & Engineering News. http://pubs.acs.org/cen/80th/arsenic.html

Akter KF, Chen Z, Smith L, Davey D, Naidu R (2005) Speciation of arsenic in groundwater samples: a comparative study of CE-UV, HG-AAS and LC-ICP-MS. Talanta 68:406–415

Alberti J, Rubio R, Rauret G (1995) Extraction method for arsenic speciation in marine organisms. Fresenius J Anal Chem 351:420–425

Alves GMS, Magalhães JMCS, Salaün P, Berg, CMG, Soares HMVM (2011) Simultaneous electrochemical determination of arsenic, copper, lead and mercury in unpolluted fresh waters using a vibrating gold microwire electrode. Anal Chim Acta 703(1):1–7

Amjadi M, Manzoori JL, Abulhassani J (2011) Ultrasound-assisted ionic liquid-based microextraction for preconcentration of arsenic prior to its determination by electrothermal atomic absorption spectrometry. Curr Anal Chem 7:262–268

Ammann AA (2011) Arsenic speciation analysis by ion chromatography—a critical review of principles and applications. Am J Analyt Chem 2:27–45

Aragay G, Pons J, Merkoci A (2011) Recent trends in macro, micro, and nanomaterial based tools and strategies for heavy-metal detection. Chem Rev 111:3433–3458

Bralatei E, Nekrosiute K, Ronan J, Rob A, Stengel DM, McGovern E, Krupp EM, Feldmann J (2017) A field deployable method for a rapid screening analysis of inorganic arsenic in seaweed. Microchim Acta 184:1701–1709

Campillo N, Peñalver R, Viñas P, López-García I, Hernandez-Córdoba M (2008) Speciation of arsenic using capillary gas chromatography with atomic emission detection. Talanta 77:793–799

Cataldo F (2012) Multielemental analysis of municipal landfill leachate with total reflection xray fluorescence (TXRF). A comparison with ICP-OES analytical results. J Radioanal Nuc Chem 293:119–126

Chakraborti D, Rahman MM, Raul K, Chowdhyry UK, Sengupta MK, Lodh D, Chanda CR, Saha KC, Mukherjee SC (2002) Arsenic calamity in the Indian subcontinent. What lessons have been learned? Talanta 58:3–22

Chandrasekaran K, BalaramaKrishna MV, Karunasagar D (2010) On-line speciation of inorganic arsenic in natural waters using polyaniline (PANI) with determination by flow injection-hydride generation-inductively coupled plasma mass spectrometry at ultra-trace levels. J Anal At Spectrom 25:1348–1353

Chen S, Zhan X, Lu D, Liu C, Zhu L (2009) Speciation analysis of inorganic arsenic in natural water by carbon nanofibers separation and inductively coupled plasma mass spectrometry determination. Anal Chim Acta 634:192–196

Chen ML, Lin YM, Gu CB, Wang JH (2013) A green sorbent of esterified egg-shell membrane for highly selective uptake of arsenate and speciation of inorganic arsenic. Talanta 104:53–57

Chen ML, Ma LY, Chen XW (2014) New procedures for arsenic speciation: a review. Talanta 125:78–86

Colon M, Hidalgo M, Iglesias M (2011) Arsenic determination by ICP-QMS with octopole collision/reaction cell. Overcome of matrix effects under vented and pressurized cell conditions. Talanta 85:1941–1947
Cosnier S, Mousty C, Cui X, Yang X, Dong S (2006) Specific determination of As(V) by an acid phosphatase-polyphenol oxidase biosensor. Anal Chem 78(14):4985–4989
Dickinson GW (1969) Applications of the induction coupled plasma to analytical spectroscopy. Ames Lab, Ames, IA, USA, p 118
Dressler VL, Antes FG, Moreira CM, Pozebon D, Duarte FA (2011) As, Hg, I, Sb, Se and Sn speciation in body fluids and biological tissues using hyphenated-ICP-MS techniques: a review. Int J Mass Spectrom 307:149–162
Dreyer DR, Park S, Bielawski CW, Ruoff RS (2010) The chemistry of graphene oxide. Chem Soc Rev 39(1):228–240
Ezeh VC, Harrop CT (2012) A sensitive and selective fluorescence sensor for the detection of arsenic(III) in organic media. Inorg Chem 51:1213–1215
Feldmann J, Devalla S, Raab A, Hansen HR (2004) Analytical strategies for arsenic speciation in environmental and biological samples. In: Hirner AV, Emons H (eds) Organic metal and metalloid species in the environment. Springer, Heidelberg, p 41
Forsberg G, O'Laughlin JW, Megargle RG (1975) Determination of arsenic by anodic stripping voltammetry and differential pulse anodic stripping voltammetry. Anal Chem 47(9):1586–1592
Garcia-Manyes S, Jiminez G, Padro A, Rubio R, Rauret G (2002) Arsenic speciation in contaminated soils. Talanta 58(1):97–109
Georgiadis M, Cai Y, Solo-Gabriele HM (2006) Extraction of arsenate and arsenite species from soils and sediments. Environ Pollut 141(1):22–29
Giacomino A, Abollino O, Lazzara M, Malandrino M, Mentasti E (2011) Determination of As(III) by anodic stripping voltammetry using a lateral gold electrode: experimental conditions, electron transfer and monitoring of electrode surface. Talanta 83:1428–1435
Gibbon-Walsh K, Salaün P, Berg CMG (2010) Arsenic speciation in natural waters by cathodic stripping voltammetry. Anal Chim Acta 662(1):1–8
Gomez-Ariza JL, Sanchez-Rodas D, Giráldez I, Morales E (2000) Comparison of biota sample pretreatments for arsenic speciation with coupled HPLC-HG-ICP-MS. Analyst 125:401–407
Gonzalvez A, Llorens A, Cervera ML, Armenta S, Guardia M (2009) Non-chromatographic speciation of inorganic arsenic in mushrooms by hydride generation atomic fluorescence spectrometry. Food Chem 115:360–364
Greschonig H, Irgolic KJ (1992) Electrochemical methods for the determination total arsenic and arsenic compounds. Appl Organomet Chem 6:565–577
Guell R, Fontas C, Salvado V, Antico E (2010) Modeling of liquid-liquid extraction and liquid membrane separation of arsenic species in environmental matrices. Sep Purif Technol 72:319–325
Holak W (1969) Gas-sampling technique for arsenic determination by atomic absorption spectrophotometry. Anal Chem 41(12):1712–1713
Howard AG (1997) (Boro)Hydride techniques in trace element speciation. J Anal At Spectrom 12:267–272
Howard AG, Salou C (1996) Cysteine enhancement of the cryogenic trap hydride AAS determination of dissolved arsenic species. Anal Chim Acta 333:89–96
Hsieh CJ, Yen CH, Kuo MS (1999) Determination of trace amounts of arsenic(III) and arsenic(V) in drinking water and arsenic(III) vapor in air by graphite-furnace atomic absorption spectrophotometry using 2,3-dimercaptopropane-1-sulfonate as a complexing agent. Anal Sci 15(7):669–673
Hueber DM, Winefordner JD (1995) A flowing electrolytic hydride generator for continuous sample introduction in atomic spectrometry. Anal Chim Acta 316:129–144
Jia X, Gong D, Wang J, Huang F, Duan T, Zhang X (2016) Arsenic speciation in environmental waters by a new specific phosphine modified polymer microsphere preconcentration and HPLC–ICP-MS determination. Talanta 160:437–443

Kang JH, Jung HJ, Jung MY (2016) One step derivatization with British Anti-Lewsite in combination with gas chromatography coupled to triple-quadrupole tandem mass spectrometry for the fast and selective analysis of inorganic arsenic in rice. Anal Chim Acta 934:231–238

Karthikeyan S, Rao TP, Iyer CSV (1999) Determination of arsenic in sea water by sorbent extraction with hydride generation atomic absorption spectrometry. Talanta 49(3):523–530

Kempahanumakkagari S, Deep A, Kim KH, Kumar Kailasa S, Yoon HO (2017) Nanomaterial-based electrochemical sensors for arsenic—a review. Biosens Bioelectron 95:106–116

Kohlmeyer U, Kuballa J, Jantzen E (2002) Simultaneous separation of 17 inorganic and organic arsenic compounds in marine biota by means of high-performance liquid chromatography/ inductively coupled plasma mass spectrometry. Rapid Commun Mass Spectrom 16:965–974

Kroukamp EM, Wondimu T, Forbes PBC (2016) Metal and metalloid speciation in plants: overview, instrumentation, approaches and commonly assessed elements. Trends Anal Chem 77:87–99

Larsen EH (1998) Method optimization and quality assurance in speciation analysis using HPLC with detection ICPMS. Spectrochim Acta Part B At Spectrosc 53(2):253–265

Le XC, Mingsheng MA, Wong NA (1996) Speciation of arsenic compounds using high performance liquid chromatography at elevated temperature and selective hydride generation atomic fluorescence detection. Anal Chem 68:4501–4506

Leermakers M, Baeyens W, De Gieter M, Smedts B, Meert C, De Bisschop HC, Morabito R, Quevauviller Ph (2006) Toxic arsenic compounds in environmental samples: speciation and validation. Trends Anal Chem 25:1–10

Liu LH, Hen B, Yun Z, Sun J, Jiang GB (2013) Speciation analysis of arsenic compounds by capillary electrophoresis on-line coupled with inductively coupled plasma mass spectrometry using a novel interface. J Chromatogr 1304:227–233

Llorente-Mirandes T, Rubio R, López-Sánchez JF (2017) Inorganic arsenic determination in food: a review of analytical proposals and quality assessment over the last six years. Appl Spectrosc 71:25–69

Lolic A, Nikolic S, Mutic J (2008) Optimization of a flow injection system with amperometric detection for arsenic determination. Anal Sci 24:877–880

Majidi B, Shemirani F (2011) In situ solvent formation microextraction in the presence of ionic liquid for preconcentration and speciation of arsenic in saline samples and total arsenic in biological samples by electrothermal atomic absorption spectrometry. Biol Trace Elem Res 143:579–590

Male KB, Hrapovic S, Santini JM, Luong JH (2007) Biosensor for arsenite using arsenite oxidase and multiwalled carbon nanotube modified electrodes. Anal Chem 79(20):7831–7837

Mays DE, Hussam A (2009) Voltammetric methods for determination and speciation of inorganic arsenic in the environment—a review. Anal Chim Acta 646:6–16

McSheehy S, Pohl P, Lobinski R, Szpunar J (2001) Analyst 126:1055

Meng X, Jing C, Korfiatis GP (2003) A review of redox transformation of arsenic in aquatic environments. In: ACS symposium series 835, Washington, pp 70–83

Michalke B (2003) Element speciation definitions, analytical methodology, and some examples. Ecotoxicol Environ Saf 56:122–139

Mondal P, Balomajumder C, Mohanty B (2007) Quantitative separation of As(III) and As(V) from a synthetic water solution using ion exchange columns in the presence of Fe and Mn ions. Clean-Soil Air Water 35:255–260

Musil S, Petursdottir AH, Raab A, Gunnlaugsdottir H, Krupp E, Feldmann J (2014) Speciation without chromatography using selective hydride generation: inorganic arsenic in rice and samples of marine Origin. Anal Chem 86(2):993–999

Namera A, Takeuchi A, Saito T, Miyazaki S, Oikawa H, Saruwatari T, Nagao M (2012) Sequential extraction of inorganic arsenic compounds and methyl arsenate in human urine using mixed-mode monolithic silica spin column coupled with gas chromatography-mass spectrometry. J Sep Sci 35:2506–2513

Narukawa T (2000) Abstract of the 7th frontiers in education: computer science and computer engineering conference. Environ Sci Pollut Resvol 40

Naykki T, Perämäki P, Kujala J, Mikkonen A (2001) Optimization of a flow injection hydride generation atomic absorption spectrometric method for the determination of arsenic, antimony and selenium in iron chloride/sulfate-based water treatment chemical. Anal Chim Acta 439:229–238

Nearing MM, Koch I, Reimer KJ (2014) Complementary arsenic speciation methods: a review. Spectrochim Acta B 99:150–162

Popp M, Hann S, Koellensperger G (2010) Environmental application of elemental speciation analysis based on liquid or gas chromatography hyphenated to inductively coupled plasma mass spectrometry—a review. Anal Chim Acta 668(2):114–129

Profumo A, Merli D, Pesavento M (2005) Voltammetric determination of inorganic As(III) and total inorganic As in natural waters. Anal Chim Acta 539(1–2):245–250

Profumo A, Fagnoni M, Merli D, Quartarone E, Protti S, Dondi D, Albini A (2006) Multiwalled carbon nanotubes chemically modified gold electrode for inorganic as speciation and Bi(III) determination. Anal Chem 78:4194–4199

Qu H, Mudalige TK, Linder SW (2015) Arsenic speciation in rice by capillary electrophoresis/ inductively coupled plasma mass spectrometry: enzyme-assisted water-phase microwave digestion. J Agric Food Chem 63:3153–3160

Quinaia SP, Rollemberg ME (2001) Selective reduction of arsenic species by hydride generation-atomic absorption spectrometry. Part 2-sample storage and arsenic determination in natural waters. J Braz Chem Soc 12:37–41

Radke B, Jewell L, Namiesnik J (2012) Analysis of arsenic species in environmental samples. Crit Rev Anal Chem 42:162–183

Rahman MM, Chowdhury UK, Mukherjee SC, Mandal BK, Paul K, Lodh D, Biswas BK, Chanda CR, Basu GK, Saha KS, Roy S, Das R, Palit SK, Quamruzzaman Q, Chakraborti D (2001) Chronic arsenic toxicity in Bangladesh and West Bengal, India—a review and commentary. J Toxicol Clin Toxicol 39:683–700

Rajakovic L, Todorovic Z, Rajakovic-Ognjanovic V, Onjia A (2013) Analytical methods for arsenic speciation analysis. J Serbian Chem Soc 78(10):1461–1479

Ramesha G, Sampath S (2011) In-situ formation of graphene–lead oxide composite and its use in trace arsenic detection. Sens Actuators B 160:306–311

Rasheed H, Kay P, Slack R, Gong YY, Carter A (2017) Human exposure assessment of different arsenic species in household water sources in a high risk arsenic area. Sci Tot Environ 584–584:631–641

Rottschafer JM, Boczkowski RJ, Mark HB Jr (1972) Preconcentration techniques for trace analysis *via* neutron activation. Talanta 19:163–172

Rupasinghe TWT, Cardwell TC, Cattral RW, Kolev SD (2009) Determination of arsenic in industrial samples by pervaporation flow injection with amperometric detection. Anal Chim Acta 652:266–271

Sadee B, Foulkes ME, Hill SJ (2015) Coupled techniques for arsenic speciation in food and drinking water: a review. J Anal At Spectrom 30:102–118

Sanchez WM, Zwicker B, Chatt A (2009) Determination of As(III), As(V), MMA and. DMA in drinking water by solid phase extraction and neutron activation. J Radioanal Nucl Chem 282:133–138

Sansoni B, Brunner W, Wolff G, Ruppert H, Dittrich R (1988) Comparative instrumental multielement determination. I: comparison of ICP source mass spectrometry with ICP atomic emission spectrometry, ICP atomic fluorescence spectrometry for the analysis of natural waters from a granite region. Fresenius Z Anal Chem 331(2):154–169

Sengupta MK, Dasgupta PK (2009) An automated hydride generation interface to ICPMS for measuring total arsenic in environmental samples. Anal Chem 81:9737–9743

Sheppard BS, Carusot JA, Heitkernper DT, Wolnik KA (1992) Arsenic speciation by ion chromatography with inductively coupled plasma mass spectrometric detection. Analyst 117:971–976

Shi Y, Chatt A (2014) Simultaneous determination of inorganic As(III), As(V), Sb(III), Sb(V), and Se(IV) species in natural waters by APDC/MIBK-NAA. J Radioanal Nucl Chem 299:867–877
Shiomi K (1994) Arsenic in the environment. Part II: human health and ecosystem effects. In: Nriagu JO (ed) vol 27. Wiley, New York, pp 261–293
Shirkhanloo H, Rouhollahi J, Mousavi HZ (2011) Ultra-trace arsenic determination in urine and whole blood samples by flow injection-hydride generation atomic absorption spectrometry after preconcentration and speciation based on dispersive liquid-liquid microextraction. Korean Chem Soc 32:3923–3927
Sitko R, Janik P, Zawisza B, Talik E, MarguiE Queralt I (2015) Green approach for ultratrace determination of divalent metal ions and arsenic species using total-reflection X-ray fluorescence spectrometry and mercapto-modified graphene oxide nanosheets as a novel adsorbent. Anal Chem 87(6):3535–3542
Sloth JJ, Larsen EH, Julshamn K (2003) Determination of organoarsenic species in marine samples using gradient elution cation exchange HPLC-ICP-MS. J Anal At Spectrom 18(5):452–459
Stetzenbach KJ, Amano M, Kreamer DK, Hodge VF (1994) Testing the limits of ICP-MS: determination of trace elements in groundwater at the part-per trillion level. Ground Water 32 (6):976–985
Sun YC, Yang JY (1999) Simultaneous determination of arsenic(III,V), selenium(IV,VI), and antimony(III,V) in natural water by coprecipitation and neutron activation analysis. Anal Chim Acta 395:293–300
Szostek B, Aldstadt JH (1998) Determination of organoarsenicals in the environment by solid-phase microextraction-gas chromatography-mass spectrometry. J Chromatogr A 807 (2):253–263
Takeuchi A, Namera A, Kawasumi Y, Imanaka T, Sakui N, Ota H, Endo Y, Sumino K, Endo G (2012) Development of an analytical method for the determination of arsenic in urine by gas chromatography-mass spectrometry for biological monitoring of exposure to inorganic arsenic. J Occup Health 54(6):434–440
Thurow K, Koch A, Stoll N, Haney CA (2002) Environmental aspects of converting CW facilities to peaceful purposes. In: McGuire RR, Compton JC (Eds) Kluwer Academic Publishers, Dordrecht, pp 123–138
Tulasi D, Adotey D, Affum A, Carboo D, Serfor-Armah Y (2013) Speciation of As(III) and As(V) in water and sediment using reverse-phase ion-pair high-performance liquid chromatography-neutron activation analysis (HPLC-NAA). Env Monitor Assess 185:7979–7991
Tuzen M, Citak D, Mendil D, Soylak M (2009) Arsenic speciation in natural water samples by coprecipitation-hydride generation atomic absorption spectrometry combination. Talanta 78:52–56
Tyson J (2013) The determination of arsenic compounds: a critical review. ISRN Anal Chem 2013:1–24
United States Environmental Protection Agency (1999a) Analytical methods support document for arsenic in drinking water. http://www.epa.gov/safewater/arsenic/pdfs/methods.pdf. Accessed on 09 Aug 17
United States Environmental Protection Agency (1999b) Analytical methods support document for arsenic in drinking water. Standard and risk management division, Washington. http://www.epa.gov/safewater/arsenic/pdfs/methods.pdf. Accessed on 09 Aug 17
United States Environmental Protection Agency (2004) Monitoring arsenic in the environment: a review of science and technologies for field measurements and sensors. http://www.epa.gov/tio/download/char/arsenic_paper.pdf
Wangkarn S, Pergantis SA (2000) High-speed separation of arsenic compounds using narrow-bore high-performance liquid chromatography on-line with inductively coupled plasma mass spectrometry. J Anal At Spectrom 15:627–633
World Health Organisation (1993) Guidelines for drinking-water quality, Geneva, Switzerland. http://www.who.int/mediacentre/factsheets/fs372/en/. Accessed on 09 Aug 17

World Health Organization (2001) United Nations synthesis report on arsenic in drinking water. http://www.bvsde.paho.org/bvsacd/who/arsin.pdf. Accessed on 09 Aug 17

Xiao L, Wildgoose GG, Compton RG (2008) Sensitive electrochemical detection of arsenic (III) using gold nanoparticle modified carbon nanotubes via anodic stripping voltammetry. Anal Chim Acta 620:44–49

Yamamoto M, Yasuda M, Yamamoto Y (1985) Hydride-generation atomic absorption spectrometry coupled with flow injection analysis. Anal Chem 57:1382–1385

Yan XP, Kerrich R, Hendry MJ (1998) Determination of (ultra)trace amounts of arsenic(III) and arsenic(V) in water by inductively coupled plasma mass spectrometry coupled with flow injection on-line sorption preconcentration and separation in a knotted reactor. Anal Chem 70 (22):4736–4742

Yan XP, Yin XB, He XW, Jiang Y (2002) Flow injection on-line sorption preconcentration coupled with hydride generation atomic fluorescence spectrometry for determination of (ultra)trace amounts of arsenic(III) and arsenic(V) in natural water samples. Anal Chem 74 (9):2162–2166

Yang GD, Xu JH, Zheng JP, Xu XQ, Wang W, Xu LJ, Chen GN, Fu FF (2009) Speciation analysis of arsenic in Mya arenaria Linnaeus and Shrimp with capillary electrophoresis-inductively coupled plasma mass spectrometry. Talanta 78:471–476

Yavuz CTM, Yu JT, Prakash WW, Falkner A, Yean JC, Cong S, Shipley LL, Kan HJ, Tomson M, Natelson, Zhang X, Zeng T, Hu C, Hu S, Tian Q (2016) Anal Methods 8:1162–1169

Yehl PM, Tyson JF (1997) Towards speciation of arsenic in a standard reference river sediment by high-performance ion chromatography coupled with plasma source mass spectrometry. Anal Commun 34:49–51

Yehl PM, Gurleyuk H, Tyson JF, Uden PC (2001) Microwave-assisted extraction of monomethyl arsenic acid from soil and sediment standard reference materials. Analyst 126:1511–1518

Zhao D, Li HB, Xu JY, Luo J, Ma LQ (2015) Arsenic extraction and speciation in plants: method comparison and development Science. Sci Total Environ 523:138–145-425

Zheng F, Hu B (2009) Dual silica monolithic capillary microextraction (CME) on-line coupled with ICP-MS for sequential determination of inorganic arsenic and selenium species in natural waters. J Anal At Spectrom 24:1051–1061

Zhu L, Chen S, Lu D, Cheng X (2009) Single-wall carbon nanotubes for speciation of arsenic in environmental samples by inductively coupled plasma mass spectrometry. At Spectrom 30 (6):218–222

Chapter 3
PAHs in Gas and Particulate Phases: Measurement and Control

K. Maharaj Kumari and Anita Lakhani

Abstract Polycyclic aromatic hydrocarbons (PAHs) are organic compounds with two to seven fused benzene rings in a linear or angular arrangement. PAHs, having ubiquitous presence, are extensively reported and have carcinogenic potential, low aqueous solubility and semi-volatile nature and have been recognized as persistent toxic substances (PTS). Therefore, they are of considerable environmental concern. Combustion of fuels of all types including wood, coke and gas is the major anthropogenic activities that produce PAHs, while forest fires and volcanic eruptions are the natural sources. PAHs with low molecular weight dominate in the gaseous phase and are considered less toxic to humans, whereas PAHs with high molecular weight due to their low vapour pressures remain in particulate phase and are carcinogenic and/or mutagenic. PAHs with low molecular weight are much more abundant and can react with other pollutants such as O_3 and NO_x to form highly toxic nitrated and oxy-PAH compounds. The various factors that govern the atmospheric partitioning of PAHs are ambient temperature, relative humidity, aerosol nature and its properties, interactions between the compound and the aerosol. The sources of PAHs and their quantitative contributions to a particular region are a matter of concern. In the environment, they occur as complex mixtures and not as single compounds. The two important parameters that provide a basis for identifying PAH sources are the differences in the pattern of PAH mixtures and their concentration ratios. The development of appropriate strategies for adequate and effective control measures requires identification of sources and their quantitative contributions in a region.

3.1 Introduction

In recent years, research has focused on possible adverse effects of ambient chemical carcinogens on human health. Polycyclic aromatic hydrocarbons (PAHs) were the first class of atmospheric organic pollutants to be recognized as carcino-

K. Maharaj Kumari (✉) · A. Lakhani
Department of Chemistry, Dayalbagh Educational Institute, Agra, India
e-mail: maharajkumari.k@rediffmail.com

T. Gupta et al. (eds.), *Environmental Contaminants*, Energy, Environment, and Sustainability, https://doi.org/10.1007/978-981-10-7332-8_3

gens (Singla et al. 2012; Keyte et al. 2013). Several studies have been conducted on PAH due to their presence in both urban atmosphere and rural atmosphere as well as their adverse role in health.

PAHs include organic compounds with two or more fused benzene rings. The US Environmental Protection Agency (EPA) recognizes 16 of them as priority pollutants because of their mutagenic and carcinogenic potential. World Health Organization (WHO) has added 17 more PAHs to the priority PAHs totalling to 33 PAHs (Pandey et al. 2011). Although the carcinogenic potential of PAHs released into the atmosphere varies greatly, they are considered as toxic compounds. The guidelines recommended by World Health Organization are in terms of carcinogenic slope factor. The European Union has set a limit value of benzo[α]pyrene at 1 ngm^{-3} (WHO 2006 and European Union 2005), while the air quality standard set by UK is 0.25 ngm^{-3} benzo[α]pyrene. The standards and guidelines for PAH mixtures are set with respect to the levels of benzo[α]pyrene as this compound typically represents a substantial proportion of the total carcinogenic potential. Earlier studies have reported 16 PAHs selected by USEPA and are presented in Table 3.1, and they are acenaphthene (Ace), acenaphthylene (Acy), fluorene (Fln), naphthalene (Nap), anthracene (Ant), fluoranthene (Flt), phenanthrene (Phe), benzo [a]anthracene (BaA), benzo[b]fluoranthene (BbF), benzo[k]fluoranthene (BkF), chrysene (Chr), pyrene (Pyr), benzo[ghi]perylene (BgP), benzo[a]-pyrene (BaP), dibenzo[a,h]anthracene (DBahA) and indeno[1,2,3-cd]-pyrene (IPy). In 2004, PAH emissions were highest in China (114 kt $year^{-1}$) followed by India (90 kt $year^{-1}$) and the USA (32 kt $year^{-1}$) (Zhang and Tao 2009).

PAHs are coloured crystalline solids (Masih et al. 2012) with 2–6 rings and molecular weight ranging between 128 and 252 (Table 3.2). The PAHs having less than four rings are classified as low molecular weight PAHs, whereas those with four or more than four rings are called high molecular weight PAHs. PAHs are present both in atmosphere and hydrosphere as well as in biosphere. Their distribution is dependent upon their solubility and volatility. The MW and structure of PAHs significantly affect physical properties of PAHs; for example, HMW PAHs have low vapour pressure (Akyuz and Cabuk 2010). PAHs are soluble in organic solvents because of their lipophilic nature, and the solubility of PAHs in aqueous solutions decreases with increase in ring size. Size distribution of PAHs shows that the PAHs have a bimodal distribution with one peak in the fine particle size range (0.4–2.1 μm) and another in the coarse particle size range (3.3–9.0 μm). It has been suggested that both adsorption and absorption control the particle size distribution of PAHs; the intensity of the fine mode peak generally increases with increase in ring number of PAHs, while the coarse mode peak decreases with increase in ring numbers (Lv et al. 2016).

The aqueous solubility of PAHs shows a wide range and varies from 0.003 to 31 mg/L for BgP and Nap, respectively. Among PAHs, Nap has high volatility, while DBahA is almost non-volatile. PAHs also have other properties like corrosion resistance, heat resistance and sensitivity towards light (Masih et al. 2012). PAHs

Table 3.1 Physical properties of 16 PAHs

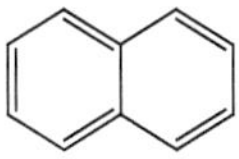 Naphthalene (Nap) MW-128 MP-80.5 °C VP-11.9 BP-218 °C	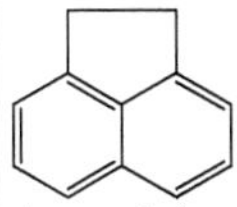 Acenaphthene (Ace) MW-154 MP-95 °C VP-0.50 BP-279 °C	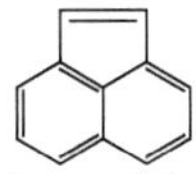 Acenaphthylene (Acy) MW-1528 MP-92–93 °C VP-3.86 BP-280 °C	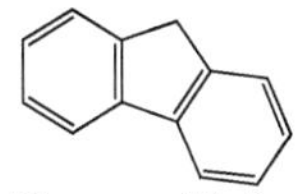 Fluorene (Fln) MW-166 MP-116–117 °C VP-0.432 BP-295 °C
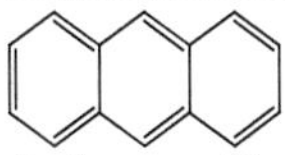 Anthracene (Ant) MW-178 MP-218 °C VP-3.4 $\times$ 10^{-3} BP-340 °C	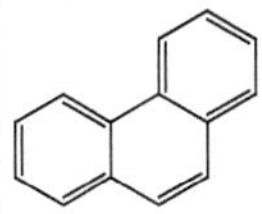 Phenanthrene (Phe) MW-178 MP-100 °C VP-9.07 $\times$ 10^{-2} BP-340 °C	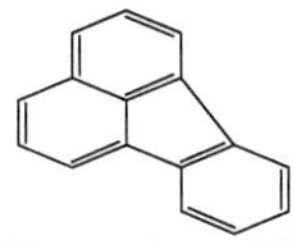 Fluoranthene (Flt) MW-202 MP-110.5 °C VP-1.08 $\times$ 10^{-3} BP-375 °C	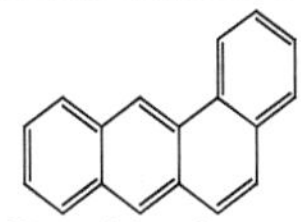 Benz(a)anthracene (BaA) MW-228 MP-158 °C VP-6.52 $\times$ 10^{-7} BP-438 °C
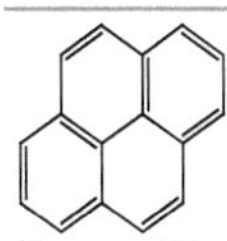 Pyrene (Pyr) MW-202 MP-156 °C VP-5.67 $\times$ 10^{-4} BP-393-404 °C	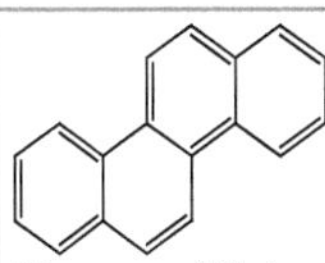 Chrysene (Chr) MW-228 MP-254 °C VP-1.04 $\times$ 10^{-6} BP-448 °C	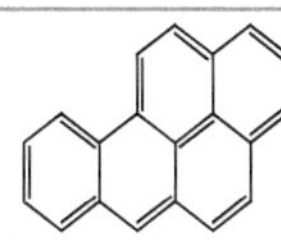 Benzo[a]-pyrene (BaP) MW-252 MP-179°C VP-6.52 $\times$ 10^{-7} BP-495 °C	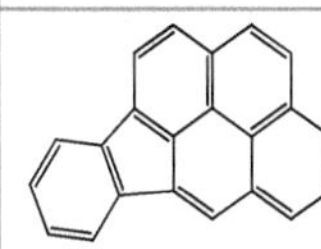 Indeno(1,2,3,cd)-pyrene (IPy) MW-128 MP-163.6 °C VP - < 1 $\times$ 10^{-3} BP-530 °C
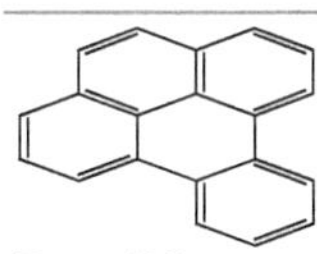 Benzo[b] fluoranthene (BbF MW-252 MP-168 °C VP-1.07 $\times$ 10^{-5} BP-168 °C	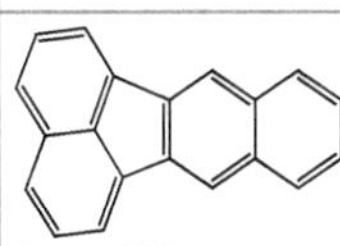 Benzo[k] fluoranthene (BkF) MW-252 MP-215 °C VP-1.28 $\times$ 10^{-8} BP-480 °C	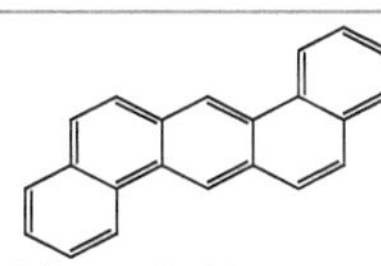 Dibenzo[a,h] anthracene (DBahA) MW-278 MP-262 °C VP-2.80 $\times$ 10^{-9} BP-262 °C	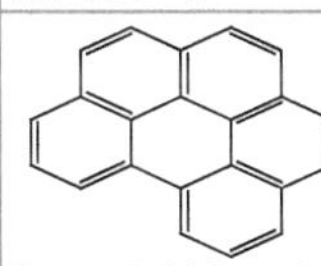 Benzo[ghi]perylene (BgP) MW-276 MP-273 °C VP-1.33 $\times$ 10^{-8} BP-550 °C

MW—molecular weight, *MP*—melting point, *VP*—vapour pressure at 25 °C (Pa), *BP*—boiling point

show characteristic UV spectra, and each ring structure has its unique spectrum. In addition, isomers also exhibit different spectrum; therefore, it is a useful technique to identify their structures. On excitation, PAHs emit characteristic wavelengths of light and exhibit fluorescence (Masih et al. 2010).

Table 3.2 Classification of PAHs on the basis of molecular weight and aromatic rings

Molecular weight	No. of rings	PAHs
Low molecular weight (LMW, molecular weight between 128–178)	2	Nap
	3	Acy, Ace, Flu, Phen, Anth
High molecular weight (HMW, molecular weight between 202–252)	4	Pyr, Fla, BaA, Chy
	5	BbF, BkF and Bap

On the basis of their volatility and MW, PAHs can exist in both particulate and gaseous phases: LMW PAHs are more volatile, occur mainly in the gas phase and are less toxic (Kameda 2011). HMW PAHs show insignificant vapourization under all environmental conditions and are predominantly emitted in the particulate phase. LMW PAHs react with other pollutants in the atmosphere (such as O_3, NO_x and SO_2) to form diones, nitro- and dinitro-PAHs, and sulphuric acid, respectively (Park et al. 2001). The gas-to-particle phase partitioning of PAHs in air is a function of both ambient temperature and also volatility of the PAH in question. Therefore, as temperature increases, the equilibrium shifts towards PAHs in the vapour phase. The gas-to-particle partitioning ratio is also dependent on the emission source, atmospheric chemistry and meteorological conditions (Kim et al. 2013). In the atmosphere, PAHs have widespread sources and persistent characteristics. PAHs present in the atmosphere in both gaseous and particulate phases are inhaled during respiration by human beings. This causes adverse health effects, and long-term exposure may lead to lung cancer. In recent years, this instigated research to evaluate PAHs in particulate matter (PM) both in PM_{10} and in $PM_{2.5}$. Table 3.3 presents average concentration of ΣPAH and BaP (in ngm^{-3}) in $PM_{2.5}$ at various sites. PAHs tend to be found in greater concentrations in urban environments than in rural environments because most PAH sources are located in or near urban centres. In urban and rural areas, PAHs show seasonal variation having the highest levels in winter season (Ravindra et al. 2006a). Residential and commercial heating emits PAHs with distinct seasonal variation with relatively enhanced concentrations in winters. This is because the fuels used for heating have high PAH emission factors. However, in industrial areas, seasonal variation is not distinct because emissions from industries are almost constant throughout the year. Brown et al. (2013) reported average BaP concentration during December–January was 3.0 ngm^{-3} compared to 0.19 ngm^{-3} during June–August at Northern Ireland. At remote sites, i.e. far from industrial and transport sources, PAH levels may be attributed to long-range transport. Some PAHs, BaP, BbF, BkF and IPy are bioaccumulative and hence due to resistance to degradation are considered as persistent organic pollutants (POPs).

Table 3.3 Average concentration of ΣPAH and BaP (in ng m^{-3}) in $PM_{2.5}$ at various sites

Site	Avg ΣPAH ng m^{-3}	Avg BaP ng m^{-3}	References
Delhi	110[a]		Kannan and Kapoor (2004)
Flanders, Belgium	114		Ravindra et al. (2006a)
Araraquara, South America	1.2	0.019	Silva et al. (2010)
Chennai, India	517.1	6.8–16.4	Mohanraj et al. (2011a)
Tiruchirappalli, India	202.6–333.7[b]	8.5–25	Mohanraj et al. (2011b)
Patiala, India			Rajput et al. (2011)
Coimbatore, India	4.1–1632.3[b]	0.1–40.7[b]	Mohanraj et al. (2012)
Kanpur, India	14.1		Rajput et al. (2013)
Urumqi, China	54.1 (winter)	2.2	Limu et al. (2013)
	11.9 (autumn)	0.5	
Guanzhou	18.7	0.54	Wang et al. (2013)
Lulang, Tibet	0.59[a]	0.03[a]	Chen et al. (2014)
Nanjing, China	30.7–102.2[b]	3.73	He et al. (2014)
Kaunas, Lithuania	20.3–131.1[b]		Krugly et al. (2014)
Oporto, Portugal	16.6		Slezakova et al. (2014)
Gipuzkoa, Spain	0.3–8.2[b]	0.15	Villar-Vidal et al. (2014)
Beijing, China	42.3 ± 52.4	0.9–4.8	Wu et al. (2014)
Taiyuan, China	119.8	13.8	Li et al. (2014)
Agra, India	880 ± 2.7	32.1	Dubey et al. (2015)

[a]TSP, [b]Range

3.2 Sources and Emission of PAHs

PAHs in urban atmosphere are products of incomplete combustion of coal, oil and biomass (Bandowe et al. 2014). Further, coke and aluminium production, vehicular transport and waste incineration also emit PAH (Zhang and Tao 2009; Okuda et al. 2010). Besides parent-PAHs, significant concentrations of oxy-PAHs and nitro-PAHs are also released during combustion processes. (Karavalakis et al. 2010; Shen et al. 2012). Oxy-PAHs and nitro-PAHs are known to have toxic, carcinogenic and mutagenic potential which are even more than parent-PAHs. Furthermore, oxy-PAHs and nitro-PAHs are direct acting mutagens and also produce reactive oxygen species (ROS) both of which are toxic (Chung et al. 2006; Wenger et al. 2009; Benbrahim-Talla et al. 2012). PAHs in atmosphere are present in PM $_{2.5}$ particles (Albinet et al. 2008; Ringuet et al. 2012). However, very few studies report O-PAHs and nitro-PAHs in the atmosphere. Although industrial scale synthesis of PAHs has not been reported, however, few PAHs are synthesized chemically to be used as intermediates in industries like photography, lubricants, pharmaceuticals and other chemical industries; for example, Pyr and Phen are used in the manufacturing of pigments and pesticides, respectively.

The major PAH sources are pyrogenic, petrogenic and biological. At high temperatures and under low or no oxygen conditions, pyrogenic PAHs are formed, e.g. distillation of coal into coke and coal tar, or generation of lighter hydrocarbons from petroleum residuals through thermal cracking process. PAHs formed during crude oil refining and similar processes are called petrogenic which are common due to the widespread transportation, storage and use of crude oil and crude oil products. On the other hand, it is not well known that PAHs can be produced biologically. Natural sources of PAHs include forest and brush fires, volcanoes, bacterial and algal synthesis, erosion of sedimentary rocks (Abdel-Shafy and Mansour 2016).

The distribution and toxic nature of PAHs are significantly influenced by emission sources of PAHs. The major sources of PAHs can be divided into three categories: stationary sources, mobile sources and natural sources (Fig. 3.1).

3.2.1 Stationary Sources

3.2.1.1 Domestic Sources

Combustion of any kind of fuel leads to PAH emissions. Heating and cooking are the main activities releasing PAH in domestic sector. The amounts of PAH emitted

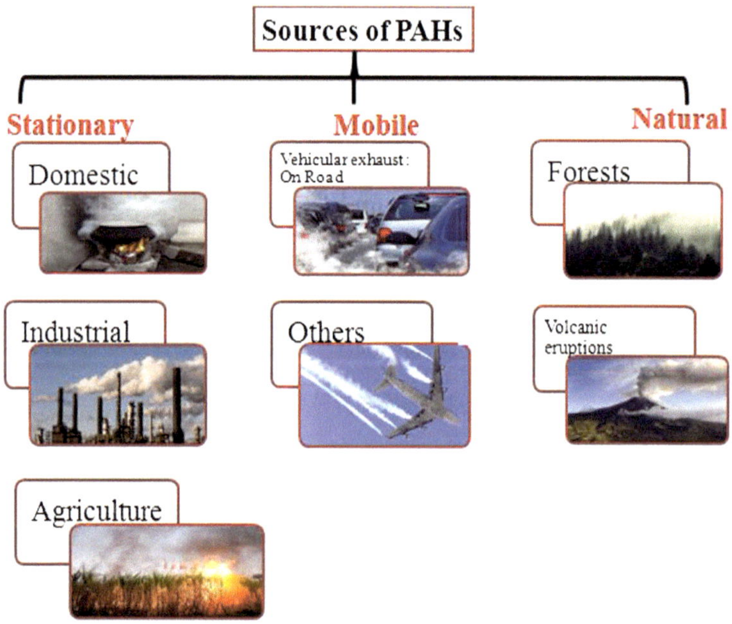

Fig. 3.1 Sources of polyaromatic hydrocarbons

Table 3.4 Amount of PAHs emitted by combustion of different domestic fuels

Fuel type	Sampling period (min)	PAH type	PAH emission	References
Wood and dung cake	15–30	BaP	1.3–9.3	Aggarwal et al. (1982)
Wood	45–60	Total 16 PAHs	2.0	Raiyani et al. (1993)
Wood/dung			3.5	
Dung cake			3.6	

from combustion of different fuels during cooking as reported from earlier studies can be summarized as (Table 3.4):

There are variations in the domestic emissions of PAHs due to difference in climate patterns and fuels used in domestic heating. As the domestic emissions predominantly affect indoor environment, therefore, these sources are a major health issue (Ravindra et al. 2006a). Among domestic emissions, cooking accounts for 32.8% of total indoor emission of PAHs. Cigarette smoke, incense burning, candles and mosquito coils are also a dominant source of PAHs in indoor environments (Dubey et al. 2014a). In rural areas of North China, wood and coal stoves are widely used for heating and cooking purpose (Liu et al. 2008). Although on a fuel-weight basis, emission factors (EF) for wood and coal are same (110 mg/kg), but wood burning emits almost double the amount of genotoxic PAHs.

3.2.1.2 Industrial Sources

Emission of PAHs from industries is mainly from:

(a) Combustion of various fuels (gas, coal and oil) (Rajput et al. 2010).
(b) Industrial Activities

   (i) Aluminium processing
   (ii) Coke production
   (iii) Petrochemical industries
   (iv) Rubber tyre manufacturing
   (v) Cement industry
   (vi) Power generation

The pyrolysis of scrap tires results in emission of 42.3 g day^{-1} of total PAHs with an EF of 4 mg kg^{-1} (Chen et al. 2007). Fabbri and Vassura (2006) reported PAH emissions from pyrolysis of organic material and found EFs of PAHs ranged from 0.4 ± 0.13 mg g^{-1} for cellulose to 9.0 ± 0.5 mg g^{-1} for tire (Table 3.5).

Table 3.5 Emission factors of PAHs for different fuels reported by the other researchers

Fuel type	PAHs ($mgkg^{-1}$)			References
	Gas phase	Particulate phase	Total	
Fuel wood		2–3.2		Venkataraman et al. (2002)
	104.6	5.2	109.7	Oanh et al. (2005)
		0.65–7.14		Hays et al. (2003)
Fireplace/softwood		39.8		McDonald et al. (2000)
Pyrolysis of cellulose			0.4 ± 0.13	Fabbri and Vassura (2006)
Pyrolysis of tyres			9.0 ± 0.5	
Pyrolysis of scrap tyres			4.0	Chen et al. (2007)
Firewood		40.9		Gadi et al. (2012)
Crop residue	198.4	144.8	343.3	Zhang et al. (2008a)
	2.6–6.4	30.5–39.6	36.3–45.1	Singh et al. (2013)
	0.65–3.62	0.72–1.64	0.12–4	Zhang et al. (2011)
	13–40	12–58	25.8–100.5	Shen et al. (2011)
Dung cake	2.2–3.7	47.5–66.6	49.7–69.9	Singh et al. (2013)
		45.7		Gadi et al. (2012)

3.2.1.3 Agricultural Sources

Biomass burning of agricultural products after harvest (straw and stubble) produces a significant amount of PAHs. The concentrations of released PAHs depend on the type of wood and combustion temperature. The biomass burning results in emission of large fraction (80–90%) of LMW PAHs including Nap, Acy, Flt and Pyr. Open burning of rice and bean straw emits 9.29–23.6 and 3.13–49.9 μgg^{-1} of PAHs, respectively (Lu et al. 2009).

3.2.2 Mobile Sources

Most studies report emissions from vehicular exhaust as largest contributors of PAHs in urban areas (Miguel et al. 1998; Marchand et al. 2004; Marr et al. 2006; Ravindra et al. 2006b). PAHs can be formed in vehicular exhaust in three ways:

(a) Synthesis from compounds in fuel
(b) Storage in engine deposits
(c) Pyrolysis (breakdown of lubricants)

The emission of PAH is dependent upon fuel/air ratio; Ravindra et al. 2006b report lesser PAH emission from engine exhaust with low fuel ratio. Earlier studies report diesel vehicles emit more PAH than petrol vehicles (Rajput et al. 2009; Jain et al. 2012). Non-road vehicles—ships, trains and aircrafts—also emit PAH. In engine exhaust of a helicopter, Chen et al. (2006) reported PAH concentration to be 843 mgm^{-3}.

3.2.3 *Natural Sources*

Natural sources of PAHs include burning of forests, woodlands and highlands; in addition, decaying of organic matter and volcanic eruptions is other possible natural sources. Several meteorological factors like wind speed, temperature and humidity affect emission rate of PAHs. Other factors which influence PAH levels are fuel characteristics like moisture content, green wood and seasonal wood. Table 3.5 shows emission factors of PAHs for different biomass fuels reported by the other researchers.

3.3 Source Identification of PAHs

Molecular diagnostic ratios, i.e. the ratio of individual PAH congeners, are the most widely employed diagnostic tool to identify sources of PAHs in the ambient air. Some specific PAHs have been suggested as indicators for certain processes that release PAHs into the environment, and these PAHs are termed as source markers, tracers or signatures (Lakhani 2012; Rajput and Lakhani 2012; Dubey et al. 2015). Sources for PAH can be derived by comparing ratios of selected pairs of particulate phase PAHs with the same ratios reported in the literature to be characteristic of different sources. The contribution of petroleum as a combustion source of PAHs is assessed through the ratio of Phen/Anth and Fla/Pyr. A Phen/Anth ratio <10 and Fla/Pyr > 1 indicate pyrogenic origins of PAHs (Baumard et al. 1998). The Anth/(Anth + Phen) is another important indicator of combustion sources. It indicates a combustion source if the ratio is >0.1, whereas it indicates a petroleum source if the ratio is <0.1 (Yunker et al. 2002). Similarly, the ratio of Fla/(Fla + Pyr) < 0.4 is characteristic of petroleum source, and a ratio between 0.4 and 0.5 is characteristic of liquid fossil fuel combustion, whereas a ratio >0.5 is a tracer of PAHs derived from the combustion of grass, wood or coal. COMPAHs are the combustion-derived PAHs including Flu, Pyr, Chy, BbF, BkF, BaA, BaP, IP and B(ghi)P. To discriminate the petrogenic contribution from combustion sources, the ratio of combustion PAH (COMPAH) to total PAH (ΣPAH) can be used (Rogge et al. 1993). Rogge et al. (1993) reported COMPAH/ΣPAH ratios for non-catalyst

(0.41) and catalyst-equipped (0.51) automobiles as well as of heavy-duty diesel trucks (0.30). Cecinato et al. (1998) reported that high values (0.8–0.9) of COMPAH/ΣPAH ratio are a characteristic of coal burning. A value of 0.35–0.7 of Ipy/(Ipy + BgP) ratio has been used for diesel emissions (Kavouras et al. 2001). Khalili et al. (1995) and Guo et al. (2003) reported that the ratio of BaP/(BaP + Chr) was 0.49 and 0.73 for diesel and gasoline engines, respectively. The BaP/BgP ratio higher than 0.6 refers to the presence of traffic emission and contribution from other PAH sources (Pandey et al. 1999; Park et al. 2002). Ipy/BgP ratio for gasoline engines is about 0.4, while the ratio for diesel engines is nearly 1.0.

3.4 Gas Phase Reactions of PAHs

Lower molecular weight (LMW) PAHs have high vapour pressure which results in their high concentration in gas phase in troposphere. The key sink for LMW PAHs in the atmosphere is chemical reactions with atmospheric oxidants, i.e. OH, NO_3 and O_3.

3.4.1 Gas Phase Reactions of PAHs with OH Radicals

The OH radical is one of the most reactive radicals in the atmosphere with a concentration of 1.0×10^6 radicals cm^{-3} during the daytime. The OH radicals can react with PAHs by two ways: (1) abstraction of a proton forming water molecule and (2) addition of OH radical to the double bond in the aromatic ring. For example, Ace reacts with OH radical in both ways as shown below, whereas Acy reacts with OH only by proton abstraction.

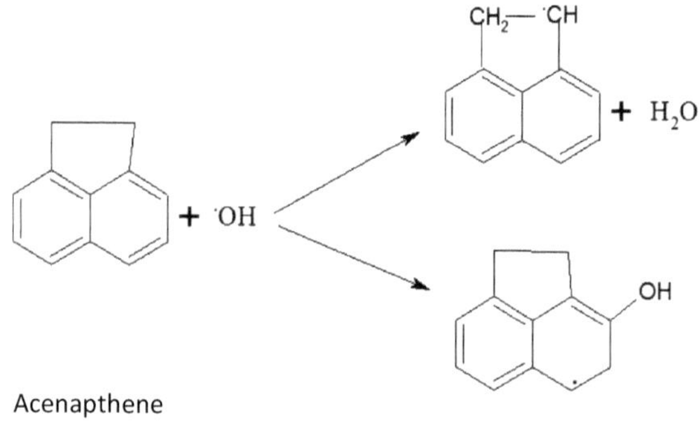

3.4.2 Gas Phase Reactions of PAHs with NO_3 Radicals

PAHs undergo nitration to form nitro-derivatives; direct nitration as in combustion process or OH radical initiated reactions. Different isomers are formed under both conditions. PAHs like Pyr and Fln undergo these reactions forming: 1-nitropyrene and 2-nitrofluorene as products of direct reaction, while 2-nitropyrene and 3-nitrofluorene are formed through OH radical initiated reactions.

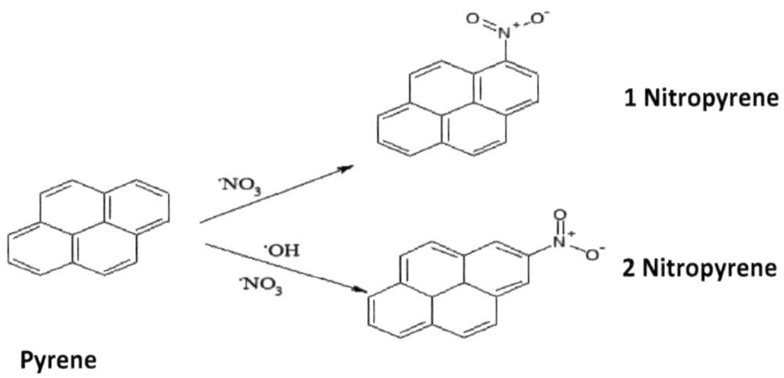

3.4.3 Gas Phase Reactions of PAHs with O_3

Reaction between ozone and PAHs yields oxygenate intermediates. These compounds are less volatile than parent-PAHs and tend to be more associated with particles. Oxy-PAHs relevantly contribute to the adverse health effects associated with atmospheric particulate exposure.

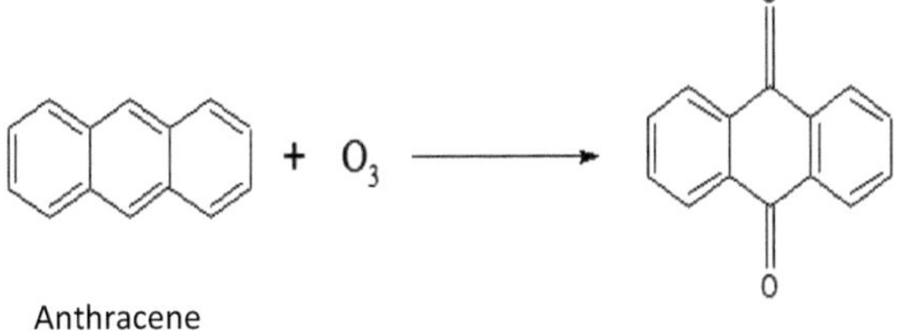

3.5 Sampling of PAHs

As PAHs released into the atmosphere may be associated with the particle and/or gas phase, the selection of the sampling method and the collection efficiency depend on the physical state of PAHs. Also, large volumes of air must be sampled for the collection of both vapour and particulate phase PAHs as the concentration of PAHs is relatively low (order of ngm^{-3}). Both gas and particulate phases are collected by active and passive sampling. Active samplers include high volume and low volume, and cascade samplers employ a suitable sorbent, filter or plug to collect PAHs through deposition or adsorption of PAH compounds using a pump. The limitations of active sampling are high costs of samplers and loss of PAHs during sampling due to volatilization of PAHs. These limitations can be overcome by using passive sampling methods. In passive sampling, the differences in chemical potentials of the PAH molecules in the sampled air and the collecting substrate allow free flow of PAH molecules from the sampled air onto the collecting medium. Passive sampling is usually carried out using semi-permeable membrane devices (SPMDs) and diffusion samplers, in which vapour is allowed to diffuse into the adsorbent. However, the use of passive samplers is limited as they cannot be calibrated for quantitative purposes. Most of the studies aiming to monitor PAHs in air have focused only on either the gas phase or the particulate phase (Rajput and Lakhani 2009a, b, 2010; Dubey et al. 2014b, c; Dubey et al. 2015) showing a preference for the particle phase, while several studies have attempted to measure both the gas and particle phase PAHs simultaneously (Verma et al. 2017) (Fig. 3.2).

3.5.1 Sampling of PAHs in the Gas Phase

The gas phase PAHs can be collected by passing a known volume of air through adsorbent cartridges filled with XAD-2 (styrene–divinylbenzene polymer) or polyurethane foam (PUF) plugs with a filter located upstream for the retention of particulate phase PAHs. The sampling efficiency of XAD-2 resin is >90% for two-

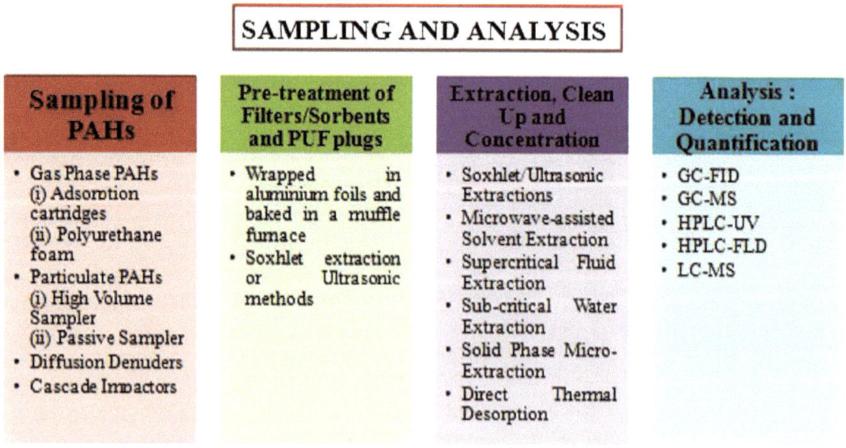

Fig. 3.2 Sampling and analysis of PAHs

and three-ring compounds (Possanzini et al. 2004). PUF plugs have been widely employed for a wide range of PAHs and their nitro derivatives as they give low blank values and exert low pressure drop, besides being cheap and easy to use (Gustafson and Dickhut 1997; Dimashki et al. 2000; Mandalakis et al. 2002; Tsapakis and Stephanou 2005). The retention efficiency of PUF plugs towards volatile PAHs decreases at higher temperatures. Some studies have also used Tenax–GC as an adsorbent support for sampling volatile PAHs and analyses by subsequent thermal desorption (Pankow et al. 1988).

An ideal adsorbent should quantitatively remove the analyte from the gas phase and should also not release the analyte during sampling. Practically, no adsorbent system behaves in an ideal way and suffers from breakthrough problems (unadsorbed fraction of the analyte). Breakthrough of an analyte on an adsorbent depends on the air sampling volume, the mass or volume of adsorbent, the affinity of the analyte for the adsorbent at the sampling temperature and the chromatographic efficiency of the adsorbing system (measured by the number of theoretical plates). Breakthrough is estimated by using a sampling train with two or more adsorbent beds in series. In a two-bed system, breakthrough (B%) can be evaluated as B (%) = $[S/(P + S)] \times 100$, where S is the concentration found on the secondary bed and P is the concentration on the primary bed. PAHs less volatile than Phen have been observed to be well trapped with measured breakthrough being <10% (Hart and Pankow 1994). XAD-2 adsorption cartridges and PUF plugs may contain significant levels of residues which should be removed prior to use to minimize sample contamination. Rigorous clean-up procedures are often employed; however, they can leave residual organic constituents that make the identification and quantification of compounds difficult. Modified cartridges comprising of mixed sorbents like XAD-2 resin placed between layers of PUF (Lee et al. 1995) or polydimethylsiloxane (PDMS) foam, PDMS particles and Tenax TA (Wauters et al. 2008) have also been employed to sample gas phase PAHs.

3.5.2 *Sampling of Particulate Phase PAHs*

Particulate phase PAHs as total suspended particulate matter (TSPM), $PM_{2.5}$, PM_{10}, PM_1 are collected using high-volume samplers or in different size fraction by cascade impactors employing filters as the collection substrate. Glass fibre filter (GFF) and quartz filters have largely been employed to sample the 16 priority PAHs and the oxy- and nitro-derivatives of PAHs. Teflon-coated GFFs (also called as Emfab) have also been used in one study (Dimashki et al. 2000). Particle-bound PAHs are also collected on quartz fibre filters by low-volume samplers, and these filters are also a suitable collection media for PAH analysis by solvent-free methods like thermal desorption. Teflon filters of 37 mm diameter have been frequently used as sampling substrate in personal environmental monitors for the collection of PAHs. Aluminium foils have also been tested as sampling medium for particle phase PAHs due to its high thermal conductivity and very high efficiency for thermal desorption in comparison with quartz filters. Long-chain compounds on quartz filters may undergo pyrolysis during the thermal desorption process. Some studies have also used PTFE-coated polystyrene and polycarbonate membrane as the sampling substrate (Soo et al. 2016).

In order to assess personal exposure at traffic intersections (Liu et al. 2006), kitchen (Chen et al. 2016) and industrial workplaces (Chaspoul et al. 2005), personal samplers have been frequently employed to collect particle-bound PAHs and their derivatives. These samplers draw air at a flow rate between 1 and 3 L/min.

Particulate-bound PAHs become immobilized on filters following sampling and can undergo degradation by light and by reacting with oxidizing compounds, particularly ozone and NO_2 and sulphur-containing compounds present in the sampled air. The levels of PAHs in air may be underestimated if the PAHs collected on the filter react with ozone and NO_2 to produce oxy- and nitro-PAHs (Schauer et al. 2003). Thus, it is also essential to analyse the oxy- and nitro-derivatives of PAHs to estimate the extent of degradation. However, the derivative compounds may already be present in ambient air formed in the atmosphere through secondary reactions. PAHs collected on the filters and adsorbents are prone to sampling artefacts. Negative particulate artefacts can occur due to volatilization of PAHs from the particles collected on filters. Loss by volatilization (also called as blow off) is caused due to pressure gradient existing through the filter. Due to this pressure gradient, particles deep within the filters are exposed to gas phase concentrations of PAHs that are lower than at the front of the filter. The compounds are removed from the filtered particles and collected on the gas sampling adsorbent placed downstream of the filter. For semi-volatile PAHs, a shift in the partitioning of PAHs towards the gas phase may occur with increase in temperature, resulting in decrease in the contaminant levels. To minimize this artefact, it is preferred to replace the filter frequently during the sampling event as a result of which the collected particles are exposed to the initial conditions of temperature and concentration. Losses by degradation and volatilization can be prevented and minimized by reduced sampling times (<24 h) and flow rates. Adsorption of vapour phase compounds on the filters and the collected particles may lead to positive artefacts. Gas sorption to filters depends on the surface area of the filters; this can be corrected by using a second backup filter.

3.5.3 Diffusion Denuders

PAHs in the atmosphere may undergo oxidation by ozone, nitrate and hydroxyl radicals which may lead to incorrect measurements of gas and particle phase concentrations. Hence, to minimize these artefacts, diffusion denuders were developed which can selectively sample the gas phase compounds on the basis of the difference between the diffusion coefficients of particles and vapour phase molecules which vary by a factor of 10^3–10^6 (Liu et al. 2006; Delgado-Sabori et al. 2010). A diffusion denuder consists of either a single tube or set of tubes through which air is drawn. The inner surface of the tube is coated with a material to collect vapour phase compounds and also remove oxidizing gases from the airstream. Large particles are removed from the airstream by a size-selective inlet, while the finer particles are collected on a filter placed downstream. A sorbent or another denuder placed after the filter can collect any material desorbed from the filter. Artefacts may originate due to <100% gas sorption or particle transmission by the denuder. However, contamination of the denuder coating material and difficulty in removing the adsorbed species from the coating material may lead to problems in direct determination of gas phase concentrations by denuders, and artefacts may arise due to inefficient sorption of gas and transference of particles. Thus, in some studies, sampling trains with denuders have been used to remove the gas phase compounds and only particulate phase concentrations have been measured with denuders (Schauer et al. 2003). An annular denuder coated with the adsorbent resin XAD-4 was used by Gundel et al. (1995) to determine the gas phase PAHs. He found that gas phase PAH concentrations determined with the coated denuder agreed well with the conventional sampler results with lower limits of detection. In comparison with adsorption cartridges containing resin beds, the coated denuders have reproducible blanks and lower limits of detection as a large surface area is exposed to clean-up solvents.

3.5.4 *Impactors*

PAHs are generally associated with fine particles which can reach the lungs more easily. Therefore, the amount of particle-associated compounds that reaches the lungs is determined by the distribution of the compounds within the particle size range. Particle size-dependent concentrations of PAHs can be obtained through single- or multi-staged high-volume, low-volume and low-pressure impactors (Lung and Liu 2015; Alves et al. 2016; Zhang et al. 2016). Comparative studies of high-volume and low-pressure impactors for the collection of particle-associated PAHs show a good agreement of the results (differences < 20%) for all PAHs. Low-pressure impactors allow a size segregation of the aerosols <0.5 μm because at low pressure the main free path in the air is comparable to the diameter of the aerosol, reducing the drag on the particles and thus enabling their collection.

3.6 Pretreatment of Filters/Sorbents and PUF Plugs

To avoid and minimize contamination of the samples, it is recommended to pretreat the filters, adsorption cartridges and the PUF plugs. The filters must be wrapped in aluminium foils and baked in a muffle furnace at 450 °C for 6 h, before sampling. Similarly, after sampling the filters, adsorption cartridges and PUFs should be refrigerated at low temperature to protect them from thermal degradation and loss by volatilization. XAD-2 resins and PUF plugs should be cleaned by either Soxhlet extraction or ultrasonic methods before use.

3.7 Extraction, Clean-up and Concentration

For quantitative and qualitative determination of PAHs, multi-step procedures including extraction of the compounds from particulate matter trapped in filters and adsorbents and clean-up of the extracted sample are performed to reduce the interferences and make the identification easier after chromatographic separation. For extracting PAHs from air particulates or adsorbents, solvent-based methods involving Soxhlet (Larson and Baker 2003) or ultrasonic extractions (Tang and Isacsson 2008) using various pure organic solvents like dichloromethane (Miguel and De Andrade 1989; Castells et al. 2003; Rajput and Lakhani 2010) or solvent mixtures such as benzene–methanol–acetonitrile (Gundel et al. 1995) in dichloromethane (Miguel and De Andrade 1989) and dichloromethane–methanol mixture (Miguel and De Andrade 1989) are the most popular. The use of Soxhlet extraction is limited as it requires large amounts of toxic and expensive solvents and is also time-consuming, while sonication can be performed with a minimum amount of extracting solvent and is less time-consuming.

Several new extraction techniques have been developed to reduce the extraction time and volume of organic solvents required for the extraction. Among these, accelerated solvent extraction (ASE, also known as pressurized liquid extraction, PSE) using either pure solvents like DCM (Alves et al. 2016) or solvent mixtures like toluene and methanol (Ahmed et al. 2015) and microwave-assisted solvent extraction (MSE) procedures involving different solvent mixtures like acetone–hexane mixture (Alexandrou et al. 2001; Tutino et al. 2016; Chen et al. 2016) have been employed recently for PAH determination in airborne particulate matter, soils and sediments. MSE procedures are less time-consuming, require less solvent and give good and reproducible recoveries in comparison with soxhlet extraction.

Supercritical fluid extraction (SFE) is another useful and alternative technique to conventional liquid solvent extractions that has received significant attraction in recent years. CO_2 is the most common fluid for SFE applications due to its low toxicity, reduced cost, chemical inertia and critical properties. Modified fluids like DCM (Shimmo et al. 2004) and toluene (Castells et al. 2003) have been used to increase the extraction efficiencies and recoveries of environmental pollutants obtained with pure CO_2. Another approach has been used to increase the extraction temperature. A further advantage of SFE techniques is that they can be coupled with chromatographic techniques. Shimmo et al. (2004) developed an online supercritical fluid extraction–liquid chromatography–gas chromatography–mass spectrometry (SFE-LC-GC-MS) method for the analysis of polycyclic aromatic hydrocarbons on particulates. Subcritical water extraction (SWE) is another promising approach for low ppb measurements of organic pollutants from environmental solids as it requires less volume of solvent for sample preparation and also reduces the analysis time (Hawthorne et al. 2000).

Extraction techniques like solid-phase microextraction (SPME) methods which are solvent-free methods have also been employed to extract PAHs from particulate matter (Tang and Isacsson 2008). Another method involves thermal desorption of PAHs at 550 °C from filters or PM in a thermogravimetric analyser and its recovery in a sampling bag where the vapourized phase is in equilibrium with a SPME fibre. PAHs are then desorbed from the SPME fibre into a gas chromatograph (Ballesteros et al. 2009). Direct thermal desorption (TD) of PAHs from PM at 300–340 °C can also be achieved without pyrolytic degradation. The desorbed PAHs are cryogenically trapped and then released into a gas chromatograph for subsequent analysis (Van Drooge et al. 2009; Gil-Molto et al. 2009). Laser desorption–ionization hyphenated to mass spectrometer has also been employed to characterize PAHs immobilized on solid particles (Bente et al. 2008). This technique is more sensitive for the analysis of PAHs and more rapid than solvent extraction.

3.7.1 Online Mass Spectrometry Analysis

Time-of-flight (TOF) mass spectrometers have been used to directly analyse PAHs desorbed from particles by IR or UV lasers. This online analysis has the advantage that very small amounts of PM (as small as 10^{-12}–10^{-11} g) are required and avoid the time-consuming sample preparation procedures (Kalberer et al. 2002).

3.7.2 Clean-up Procedures

During the extraction of filters and/or adsorption cartridges by Soxhlet, ultrasonication or ASE procedures by organic solvents, several organic compounds may be coextracted along with PAHs. Hence, to determine trace levels of PAHs requires clean-up procedures to be adopted for the removal of such matrix organics as they may exert interfering effects, contamination risks of the injector port liner or column and result in loss of separation reproducibility. The clean-up procedures to be adopted mainly depend on the selectivity of the extraction technique employed. Conventional thin layer chromatography (TLC) has been used as clean-up procedures after extraction by sonication, wherein PAHs are isolated on silica gel and/or silica gel and alumina. After elution, the section corresponding to the PAH fraction is scraped off the plate and dissolved in dichloromethane or toluene. TLC reduces contamination risks and eliminates the time-consuming column preparation step. Liquid–liquid separation has also been adopted to isolate the PAH fraction after Soxhlet extraction by using cyclohexane–dimethyl sulfoxide and cyclohexane–dichloromethane liquid–liquid extraction methods (You et al. 2012).

3.8 Analysis: Detection and Quantification

The detection and quantification of PAHs and their derivatives in the purified extracts of filters and adsorbents (particle and gas phase) have usually been attempted through chromatographic techniques using both gas and liquid mobile phases coupled with different detection devices. PAHs and nitro-PAHs in environmental matrix are mostly analysed by gas chromatography with flame ionization detection (GC-FID) and mass spectrometry (MS) and high-performance liquid chromatography (HPLC) with UV and fluorescence detectors (FLD). Separation of PAHs and nitro-PAHs using gas chromatography is achieved on columns with a length of 30 m and an internal diameter of 0.25 mm; the open tube is generally coated with a stationary phase composed of cross-linked phenyl (5%) methyl (95%) siloxane (film thickness of 0.25 μm). Kawanaka et al. used a less polar stationary phase for the separation of PAHs (100% methyl) and a more polar one for nitro-PAHs (50% phenyl) (Kawanaka et al. 2007). Longer columns (50–60 m long) or smectic-liquid crystalline polysiloxane phases, however, are reported to give a better isomer resolution (Schubert et al. 2003). Mass spectrometer (MS) is a widely used detector for detecting PAHs or nitro-PAHs as compared to the flame ionization detector as it is more sensitive and provides information for identification. Among the mass spectrometers, quadrupole mass analysers are the most used for PAH detection, with electron ionization energies of 70 eV. For more accurate structural information for isomer identification, ion traps are used in the MS-MS mode. The electron capture detector (ECD) is sensitive and selective detector employed for identification and quantification of nitro-PAHs. The detection of nitro-PAHs can also be done by using MS detectors in the negative chemical ionization (NICI) mode with methane as a reagent gas. Electron monochromator–mass spectrometry (EM-MS) is another more selective and specific detector employed for detection of nitro-PAHs. However, the analysis of compounds with molecular weight greater than 300 amu is difficult by GC due to their low volatilities. Fourier transform infrared spectrometer (FT-IR), laser-induced molecular fluorescence detector (LIMF), diode array detector (DAD) and gas phase fluorescence detector (GPFDA) are some of the other detection devices used for quantification by GC.

Another widely used analytical method for determining PAHs is high-performance liquid chromatography (HPLC) accompanied with ultraviolet (HPLC-UV) and fluorescence detector (HPLC-FLD). PAH being lipophilic compounds, their separation is achieved using reversed phase liquid chromatography on a polar octadecyl (C18) column and a mobile phase composed of acetonitrile and water. Letzel et al. used phenyl-modified stationary phase and methanol and water as the mobile phase for the separation of PAHs and their oxidized derivatives (Letzel et al. 2001). Nitro-PAHs are usually detected by fluorescence detectors; they are first derivatized by reducing them to amines so that they become fluorescent. Chemiluminescence detectors, mass spectrometers and coulometric detectors can also be employed for the detection of nitro-PAHs (Yang et al. 2010). PAHs have also been analysed using an atmospheric pressure chemical ionization interface

(APCI) between the separation column and a single quadrupole detection system (Letzel et al. 2001). Atmospheric pressure photoionization interface (APPI) can also be coupled to triple quadrupole mass spectrometers which allow the analysis of PAHs in the positive ion mode and nitro-PAHs in the negative ion mode and also provide structural information about the metabolites (Hutzler et al. 2011). Two-dimensional HPLC with online derivatization and separation, electrospray ionization source (ESI) and a triple-quadrupole mass spectrometer have been developed to improve the sensitivity of nitro-PAHs analysis over chemiluminescence detection (Miller-Schulze et al. 2007). HPLC-UV and HPLC-FLD suffer from uncertainty of identification due to possible interference from other compounds. This can be overcome by the technology of LC/MS using either atmospheric pressure photoionization interface (APPI), electrospray ionization (ESI) or atmospheric pressure chemical ionization (APCI), which effectively ionize the non-polar PAH compounds; APPI is superior to ESI and APCI.

Synchronous luminescence spectroscopy (SLS), resonant (R)/non-resonant (NR) synchronous scan luminescence (SSL) spectrometry, room temperature phosphorescence (RTP), ultraviolet resonance Raman spectroscopy (UV-RRS), X-ray excited optical luminescence spectroscopy (XEOL), laser-induced molecular fluorescence (LIMF), supersonic jet/laser-induced fluorescence (SSJ/LIF), low-temperature fluorescence spectroscopy (LTFS), high-resolution low-temperature spectrofluorometry, low-temperature molecular luminescence spectrometry (LT-MLS) and supersonic jet spectroscopy/capillary supercritical fluid chromatography (SJS/SFC) are some of the less commonly used analytical techniques for determining PAHs (Soper et al. 1994; Dijkstra et al. 2001).

3.9 Strategies for Remediation of PAHs

Due to the genotoxicity, carcinogenicity and mutagenicity of PAHs, extensive attention has been focused on methods to control them by developing remediation techniques. The low molecular weight (LMW) PAHs containing two or three benzene rings are relatively water soluble, but the high molecular weight (HMW) PAHs containing four or more rings are hydrophobic, generally insoluble and can adsorb on particulate matter and soil organic matter. This tendency of being adsorbed on particulate matter makes them less available and susceptible to degradation techniques. Moreover, due to the presence of pi-electrons surrounding the aromatic rings, the HMW PAHs have high-resonance energies. Therefore, they are less prone to degradation and persistent in the environment. Substantial research efforts have been made to adopt sustainable technologies for the remediation of PAHs particularly from soil. These techniques include chemical degradation, biological degradation, phytodegradation as well as combined degradation methods. Most recently, novel approaches like use of solar ultraviolet radiation, sonochemical degradation using high-frequency ultrasound, photocatalytic degradation and

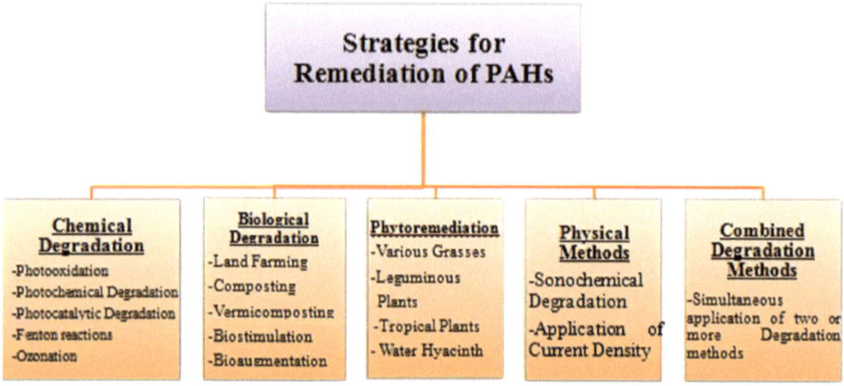

Fig. 3.3 Techniques for remediation of PAHs

current density degradation have been developed and successfully applied for degradation of these compounds (Zheng et al. 2007; Koua et al. 2010) (Fig. 3.3).

3.9.1 Chemical Degradation

Chemical strategies for remediation of PAHs include photo-oxidation and photochemical reactions, photocatalytic degradation, Fenton reactions, treatment with ozone. Photo-oxidation in the presence of sunlight is an important process for the degradation of surface-adsorbed and atmospheric PAHs. PAHs absorb radiation above 290–335 nm from sunlight and undergo direct photo-oxidation. Rate of photo-oxidation depends on the intensity of sunlight. Indirect photo-oxidation also occurs if the sunlight energy is absorbed by substances like clay, organic and inorganic matter and through electron orbital interactions is transmitted to the PAHs (Pierzynski et al. 2000). PAHs in the environment can undergo photo-oxidations initiated by oxidants like ozone, peroxides (H_2O_2) and OH radicals produced by photochemical processes involving complex pathways. The reactions proceed through the formation of various intermediates finally producing mixtures of ketones, quinines, aldehydes, phenols and carboxylic acids (Lee et al. 2001; Rivas et al. 2000; Reisen and Arey 2002). Ozone may also attack double bonds directly, or it can react by forming OH radicals by decomposing water (Legrini et al. 1993; Gurol and Singer 1982).

Photocatalytic degradation of PAHs on soil surfaces has been successfully attempted. Zhang et al. (2008b) investigated the photocatalytic degradation of Phe, Pyr and B[α]P on soil surfaces in the presence of TiO_2 and in the presence of ultraviolet light at 30 °C. They observed that the photodegradation was accelerated with the catalyst TiO_2 and the rates were greater in acidic or alkaline conditions compared to neutral conditions; degradation rates were also enhanced in the presence of humic acid which acted as a sensitizer. The photocatalytic degradation of Phe and Pyr on soil surfaces has also been studied in the presence of TiO_2 (rutile form) (Dong et al. 2010). Solid solution GaN:ZnO showed excellent activity for degradation of four PAHs, Phe, Ant, Ace and B(a)A (Koua et al. 2010).

Treatment with Fenton's reagent (hydrogen peroxide at concentrations between 3 and 35% along with ferrous iron (Fe II)) has also been used for degradation of PAHs on soil (Gan et al. 2013; Choi et al. 2014). Use of gaseous ozone for remediation of PAHs is an attractive alternative as PAHs react rapidly with ozone with the added ability of use of higher concentrations of the gas and its higher diffusivity facilitates the delivery of ozone to contaminated sites (Masten and Davies 1997; Haapea and Tuhkanen 2006; Rivas 2006; Rivas et al. 2009; Luster-Teasleya et al. 2009; Gomez-Alvarez et al. 2012).

The efficiency of degradation of PAHs by chemical means is limited by their low aqueous solubility and vapour pressure. To overcome these limitations, use of surfactants has been encouraged which increase the solubility of hydrophobic PAHs by lowering the interfacial tension between the PAHs and soil/water interphase (Li and Chen 2009; Zhou and Zhu 2007).

3.9.2 Biological Degradation

Biological degradation or biodegradation is a bioremediation technique applied to treat sediments, water and soils contaminated with PAHs using micro-organisms like bacteria, fungi, earthworms. The micro-organisms degrade the PAHs by multiplying in population, and their population subsequently declines on the degradation of the PAH. The contaminants are degraded to CO_2, H_2O and cell biomass. There are different approaches of application of these methods like land farming, composting, vermicomposting, biostimulation and bioaugmentation (Das and Das 2015). Various micro-organisms of the *Pseudomonas, Mycobacterium, Sphingomonas* genera and bacterial isolates of bacteria and fungi have been investigated for bioremediation of PAH-contaminated soils and municipal wastes (Kuyukina et al. 2005).

Land farming involves the use of indigenous micro-organisms to degrade PAHs by addition of a carbon source and nutrients, making available oxygen at greater depths of soil and increasing probability of microbial contacts with PAHs. It is an inexpensive method preferably used for removal of PAHs from contaminated soils (Gray et al. 2000; Straube et al. 2003). Composting is a similar technique involving addition of nutrients, moisture and controlled oxygen for the remediation of PAHs in municipal solid wastes (Potter et al. 1999; Canet et al. 2001; Stegmann et al. 1991; Dooley et al. 1995; Antizar-Ladislo et al. 2006a, b). An environmentally friendly approach has been the application of earthworms for the removal of PAHs from soil (Contreras-Ramos et al. 2008). The biodegradation activity of the indigenous population of microbes already present at a contaminated site can be increased by addition of nutrients and/or a terminal electron acceptor. This method is often referred as biostimulation (Das and Das 2015). Bioaugmentation is another similar method which involves the introduction of a specific microbe or group of microbes to improve the metabolic activity of microbial population.

3.9.3 Phytoremediation

Phytoremediation is the use of plants in situ and their associated micro-organisms for the remediation of contaminated systems. The internal mechanisms and physiological processes of plants either extract sequester or detoxify the contaminants from the substrates. It is believed probably that the plants increase the microbial activity in the rhizosphere by breaking down the organic compounds. Various grasses, leguminous plants, tropical plants, water hyacinth have been reported to potentially remove specific PAHs from contaminated substrates. Table 3.6 lists some of the plant species and grasses used for phytoremediation.

Table 3.6 Plant species and grasses used for phytoremediation

Plant/grass species	Conditions and PAHs removed	Reference
Fescue grass (*Festuca arundinaceae*) and Switch grass (*Pannicum virgatum*)	38% Pyr in 190 days	Chen et al. (2003)
Corn (Zea mays) Alfalfa (Medicago sativa) Rapeseed (Brassica napus)	Phe and Pyr in acidic soil, spiked soil by single and combined plant cultivation	Chouychai et al. (2009), Cheema et al. (2009)
Fescue grass (*Festuca arundinaceae*)		Cheema et al. (2008)
Rice (*Oryza sativa*)		Du et al. (2011)
Switch grass (*Pannicum virgatum*) Bluestem grass (*Schizachyrium scoparium*)		Pradham et al. (1998)
Industrial hemp (*Cannabis sativa*)	Reduction in concentration of B[a]P and Chy in PAH-contaminated soil	Campbell et al. (2002)
Rye grass (*Lolium multiflorum*) Bermuda grass (*Cynodon dactylon*)	Alkylated naphthalenes	White et al. (2006)
Rye grass (*Lolium multiflorum*)	More adsorption of Pyr in the roots than other PAHs	Kang et al. (2010)
Water hyacinth (*Eichhornia crassipes*)	More adsorption of PAHs >5 rings as opposed to two- and three-ring PAHsl	Moustafa and Shara (2009)
Water hyacinth (*Eichhornia crassipes*)	Plant devoid of rhizospheric bacteria reduced about 45% of naphthalene in 7 days	Nesterenko et al. (2012)

3.9.4 Physical Methods of Degradation

Sonochemical methods employing ultrasound frequencies have been successfully employed for degradation of PAHs (Wheat and Turneo 1997; Wang et al. 2003; Psillakis et al. 2004; Manariotis et al. 2011). More recently, current density has also been successfully applied for the degradation of PAHs (Alshawabkeh and Sarahney 2005).

3.9.5 Combined Degradation

Two or more degradation techniques can be simultaneously employed to remove PAHs from contaminated systems. These combined degradation techniques are more efficient, cost-effective and leave no dead-end products. Table 3.7 lists some of the combined degradation techniques employed for degradation of PAHs.

3.10 Conclusion

PAHs are organic compounds formed from the incomplete combustion and/or pyrolysis of organic materials like oil, gas, coal and wood used in energy production. Other contributors are indoor smoking and heating. They are a major class of chemical carcinogens present in the environment and play a significant role in human health. Many PAH compounds are identified as probable human

Table 3.7 Combined degradation techniques employed for degradation of PAHs

Combined methods	References
Chemical preoxidation and bioremediation	Kulik et al. (2005)
Anaerobic digestion and ozonation	Bernal-Martinez et al. (2005)
Biodegradation and Fenton reagent	Nam et al. (2001)
Biological, chemical and electrochemical treatment	Zheng et al. (2007)
Fenton reagent and ozonation	Goi and Trapido (2004)
Phytoremediation using *L. arundinacum* and *L. multiflorum* mixture with fertilizer	White et al. (2006)
Pressure-assisted ozonation and soil washing	Hong et al. (2008)
Ozonation and biological treatment	Haapea and Tuhkanen (2006)
Phytodegradation with E.crassipes and chemical degradation with inorganic nutrients such as $NaNO_3$, Na_2SO_4, Na_3PO_4	Ukiwe et al. (2013)

carcinogens. They exist in the atmosphere in gas and/or particle phases and are removed by wet and dry deposition.

Considering the potential health impact of airborne PAHs, accurate measurements of their concentration levels are crucial for their management, assessment of their health risks and development of abatement techniques. As PAHs exhibit significant differences in concentration levels and existence between gas and particle phases in line with their molecular weights, methods have been developed to sample them in separate phases simultaneously and also in different size fractions. These include employment of high- and low-volume samplers, impactors using adsorbent cartridges filled with XAD-2 or PUF plugs for gas phase sampling and glass or quartz filters for collection of particle phase PAHs. The adsorbent and filters are extracted in suitable solvents by various methods like solvent extraction, ultrasonication, microwave-assisted extraction or supercritical fluid and subcritical water extraction. Before analysis, the extracts are cleaned on silica gel and/or silica gel and alumina and concentrated. The determination of PAHs in both the gas and particle phases is performed by GC with FID, HPLC with UV detection or HPLC with fluorescence detection or more commonly by GC-MS. Diffusion denuders have also been used to sample gas phase PAHs which can be directly analysed by thermal desorption and mass spectrometry. Due to the genotoxicity, carcinogenicity and mutagenicity of these compounds, attention has been focused on methods to control them. Substantial efforts have been made to develop methods for remediation of PAHs. These include chemical and biological degradation involving micro-organisms, bioremediation, biostimulation, phytodegradation as well as combined degradation methods involving the use of two or more degradation methods in a collective manner. Novel approaches like use of solar ultraviolet radiation, sonochemical degradation, photocatalytic degradation and current density degradation have also been developed and successfully applied.

Acknowledgements This work was supported by DST Project Grant (No. SB/S4/AS-150/2014). The authors also thank the Head, Department of Chemistry, and Director of the institute for support and encouragement.

References

Abdel-Shafy HI, Mansour MSM (2016) A review on polycyclic aromatic hydrocarbons: source, environmental impact, effect on human health and remediation. Egypt J Petrol 25:107–123

Aggarwal AL, Raiyani CV, Patel PD, Shah PG, Chatterjee SK (1982) Assessment of exposure to benzo(a)-pyrene in air for various population groups in Ahmedabad. Atmos Environ 16:867–870

Ahmed TM, Bergvall C, Aberg M, Westerholm R (2015) Determination of oxygenated and native polycyclic aromatic hydrocarbons in urban dust and diesel particulate matter standard reference materials using pressurized liquid extraction and LC-GC/MS. Anal Bioanal Chem 407:427–438

Akyuz M, Cabuk H (2010) Gas-particle partitioning and seasonal variation of polycyclic aromatic hydrocarbons in the atmosphere of Zonguldak, Turkey. Sci Total Environ 408:5550–5558

Albinet A, Leoz -GE, Budzinski H, Villenave E, Jaffrezo JL (2008) Nitrated and oxygenated derivatives of polycyclic aromatic hydrocarbons in the ambient air of two French alpine valleys Part 2: particle-size distribution. Atmos Environ 42:55–64

Alexandrou N, Smith N, Park R, Lumb K, Brice K (2001) The extraction of polycyclic aromatic hydrocarbons from atmospheric particulate matter samples by accelerated solvent extraction (ASE). Int J Environ Anal Chem 81:257–280

Alshawabkeh AN, Sarahney H (2005) Effect of current density on enhanced transformation of naphthalene. Environ Sci Technol 39:5837–5843

Alves CA, Vicente AMP, Gomes J, Nunes T, Duarte M, Bandowe BAM (2016) Polycyclic aromatic hydrocarbons (PAHs) and their derivatives (oxygenated-PAHs, nitrated-PAHs and azaarenes) in size-fractionated particles emitted in an urban road tunnel. Atmos Res 180:128

Antizar-Ladislao B, Lopez -RJ, Beck AJ (2006a) Bioremediation of polycyclic hydrocarbons (PAH) in an aged coal tar contaminated soil using in-vessel composting approaches. J Hazard Mater 137:1583–1588

Antizar-Ladislao B, Lopez -RJ, Beck AJ (2006b) Degradation of polycyclic aromatic hydrocarbons (PAHs) in an aged coal-tar contaminated soil under in-vessel composting conditions. Environ Pollut 141:459–468

Ballesteros R, Hernandez J, Lyons L (2009) Determination of PAHs in diesel particulate matter using thermal extraction and solid phase micro-extraction. Atmos Environ 43:655–662

Bandowe BAM, Meusel H, Huang R, Ho K, Cao J, Hoffmann T, Wilcke W (2014) $PM_{2.5}$-bound oxygenated PAHs, nitro-PAHs and parent-PAHs from the atmosphere of a Chinese megacity: seasonal variation, sources and cancer risk assessment. Sci Total Environ 473–474:77–87

Baumard P, Budzinski H, Garrigues P (1998) Polycyclic aromatic hydrocarbons in sediments and mussels of the western Mediterranean Sea. Environ Toxicol Chem 17(5):765–776

Benbrahim-Talla L, Baan RA, Grosse Y, Lauby-Secretan B, Ghissassi El, Bouvard F, Guha N, Loomis D, Straif K (2012) Carcinogenicity of diesel-engine and gasoline-engine exhausts and some nitroarenes. Lancet Oncol 13:663–664

Bente M, Sklorz M, Streibel T, Zimmermann R (2008) Online laser desorption- multiphoton postionization mass spectrometry of individual aerosol particles: molecular source indicators for particles emitted from different traffic-related and wood combustion sources. Anal Chem 80:8991–9004

Bernal-Martinez A, Carriere H, Patureau D, Delgene JP (2005) Combining anaerobic digestion and ozonation to remove PAH from urban sludge. Process Biochem 40:3244–3250

Brown AS, Brown RJC, Coleman PJ, Conolly C, Sweetman AJ, Jones KC et al (2013) Twenty years of measurement of polycyclic aromatic hydrocarbons (PAHs) in UK ambient air by nationwide air quality networks. Environ Sci Proces Impacts 15(1):199–215

Campbell S, Paquin D, Awaya JD, Li QX (2002) Remediation of benzo (a) pyrene and chrysene-contaminated soil with industrial hemp (*Cannabis sativa*). Int J Phytorem 4:157–168

Canet R, Birnsting JG, Malcolm DG, Lopez -RJM, Beck AJ (2001) Biodegradation of polycyclic aromatic hydrocarbons (PAHs) by native microflora and combinations of white-rot fungi in a coal-tar contaminated soil. Biores Technol 76:113–117

Castells P, Santos FJ, Gaiceran MT (2003) Development of a sequential supercritical fluid extraction method for the analysis of nitrated and oxygenated derivatives of polycyclic aromatic hydrocarbons in urban aerosols. J Chromatogr A 1010:141–151

Cecinato A, Repetto M, Guerriero E, Allegrini I (1998) Levels and sources of polynuclear aromatic hydrocarbons in the Genoa–Cornigliano area. In Brebbia CA, Ratto CF, Power H (eds) Proceedings of "Air PollutionVI", WITPress, Southampton, Genoa, pp 587–596

Chaspoul F, Barban G, Gallice P (2005) Simultaneous GC/MSs analysis of polycyclic aromatic hydrocarbons and their nitrated derivatives in atmospheric particulate matter from workplaces. Polycyclic Aromat Compd 25:157–167

Cheema SA, Khan MI, Shen C, Tang X, Farooq M, Chen L, Chen Y (2009) Degradation ofphenanthrene and pyrene in spiked soils by single and combined plants cultivation. J Hazard Mater 16:207–211

Cheema SA, Khan MT, Tang X, Zhang C, Shen C, Malik Z, Chen Y (2008) Enhancement of phenanthrene and pyrene degradation in rhizosphere of tall fescue (*Festica arundinacea*). J Hazard Mater 9:191–195

Chen SJ, Su HB, Chang JE, Lee WJ, Huang KL, Hsieh LT, Huang YC, Lin WY, Lin CC (2007) Emissions of polycyclic aromatic hydrocarbons (PAHs) from the pyrolysis of scrap tires. Atmos Environ 41:1209–1220

Chen Y, Cao J, Zhao J, Xu H, Arimoto R, Wang G, Han Y, Shen Z, Li G (2014) n-Alkanes and polycyclic aromatic hydrocarbons in total suspended particulates from the southeastern Tibetan Plateau: concentrations, seasonal variations, and sources. Sci Total Environ 470:9–18

Chen Y, Du W, Shen G, Zhuo S, Zhu X, Shen H, Huang Y, Su S, Lin N, Pei L, Zheng X, Wu J, Duan Y, Wang X, Liu W, Wong M, Tao S (2016) Household air pollution and personal exposure to nitrated and oxygenated polycyclic aromatics (PAHs) in rural households: influence of household cooking energies. Indoor Air 27(1):169–178

Chen YC, Banks MK, Schwab AP (2003) Pyrene degradation in the rhizosphere of tall fescue (*Festuca arundinacea*) and switchgrass (*Panicum Virgatum*). Environ Sci Technol 37:5778–5782

Chen Y-C, Lee W-J, Uang S-N, Lee S-H, Tsai P-J (2006) Characteristics of polycyclic aromatic hydrocarbon (PAH) emissions from a UH-1H helicopter engine and its impact on the ambient environment. Atmos Environ 40:7589–7597

Choi K, Bae S, Lee W (2014) Degradation of pyrene in cetylpyridinium chloride-aided soil washing wastewater by pyrite fenton reaction. Chem Eng J 249:34–41

Chouychai W, Thongkukiatkul A, Upatham S, Lee H, Pokethitiyook P, Kruatrachue M (2009) Plant-enhanced phenanthrene and pyrene biodegradation in acidic soil. J Environ Biol 30:139–144

Chung MY, Lazaro RA, Lim D, Jackson J, Lyon J, Rendulic D, Hasson AS (2006) Aerosol-borne quinines and reactive oxygen species generation by particulate matter extracts. Environ Sci Technol 40:4880–4886

Contreras-Ramos SM, lvarez BDA, Dendooven L (2008) Removal of polycyclic aromatic hydrocarbons from soil amended with biosolid or vermicompost in the presence of earthworms (*Eisenia fetida*). Soil Biol Biochem 40:1954–1959

Das N, Das D (2015) Strategies for remediation of polycyclic aromatic hydrocarbons from contaminated soil-an overview. J Crit Rev 2(1):20–25

Delgado-Sabori JM, Aquilina N, Baker S, Harrad S, Meddings C, Harrison RM (2010) Determination of atmospheric particulate-phase polycyclic aromatic hydrocarbons from low volume air samples. Anal Methods 2:231–242

Dijkstra RJ, Martha CT, Ariese F, Brinkman UATh, Gooijer C (2001) On-line identification method in column liquid chromatography: UV resonance raman spectroscopy. Anal Chem 73:4977–4982

Dimashki M, Harrad S, Harrison RM (2000) Measurements of nitro-PAH in the atmospheres of two cities. Atmos Environ 34(15):2459–2469

Dong D, Li P, Li X, Xu C, Gong D, Zhang Y, Zhao Q, Li P (2010) Photocatalytic degradation of phenanthrene and pyrene on soil surfaces in the presence of nanometer rutile TiO_2 under UV-irradiation. Chem Eng J 158:378–383

Dooley MA, Taylor K, Allen B (1995) Composting of herbicide-contaminated soil in bioremediation of recalcitrant organics. Battelle Press, Columbus, Ohio

Du W, Sun Y, Cao L, Huang J, Ji R, Wang X, Wu J, Zhu J, Guo H (2011) Environmental fate of phenanthrene in lysimeter planted with wheat and rice in rotation. J Hazard Mater 188:408–413

Dubey J, Kumari KM, Lakhani A (2015) Chemical characteristics and mutagenic activity of PM 2.5 at a site in the Indo-Gangetic plain, India. Ecotoxicol Environ Saf 114:75–83

Dubey J, Banerjee A, Meena RK, Kumari KM, Lakhani A (2014a). Characterization of polycyclic aromatic hydrocarbons in emissions of different mosquito coils. Bull Environ Contam Toxicol 92:650–654. doi:10.1007/s00128-014-1278-6

Dubey J, Kumari KM, Lakhani A (2014c) Identification of polycyclic aromatic hydrocarbons in atmospheric particles of PM_{10} at Agra, India. Int J Eng Tech Res 2, ISSN: 2321-0869

Dubey, J., Singla, V., Kumari, K.M., Lakhani, A. (2014b). Polycyclic aromatic hydrocarbons in atmospheric particles of PM_{10} at Yamuna Nagar, Haryana, India. Int J Eng Tech Res ISSN: 2321-0869, Special Issue, 333–336

European Union (2005) Directive 2004/107/EC of the European Parliament and of the Council of 15 December 2004 relating to arsenic, cadmium, mercury, nickel and polycyclic aromatic hydrocarbons in ambient air. Official J 24 Eur Union L23:3–16

Fabbri D, Vassura I (2006) Evaluating emission levels of polycyclic aromatic hydrocarbons from organic materials by analytical pyrolysis. J Anal Appl Pyrol 75(2):150–158

Gadi R, Singh DP, Mandal TK, Saud T, Saxena M (2012) Emission estimates of particulate PAH from biomass fuels used in Delhi, India. Int J Hum Ecol Risk Assess 18(4):871–887

Gan S, Yap CL, Ng HK (2013) Investigation of the impacts of ethyl lactate based Fenton treatment on soil quality for polycyclic aromatic hydrocarbons (PAHs)-contaminated soils. J Hazard Mater 262:691–700

Gil-Molto J, Varea M, Galindo N, Crespo J (2009) Application of an automatic thermal desorption-gas chromatography-mass spectrometry system for the analysis of polycyclic aromatic hydrocarbons in airborne particulate matter. J Chromatogr A 1216:1285–1289

Goi A, Trapido M (2004) Degradation of polycyclic aromatic hydrocarbons in soil: the Fenton reagent versus ozonation. Environ Technol 25:155–164

Gomez-Alvarez M, Poznyak T, Rios-Leal E, Silva-Sanchez C (2012) Anthracene decomposition in soils by conventional ozonation. J Environ Manage 113:545–551

Gray MR, Banerjee DK, Dudas MJ, Pickard MA (2000) Protocols to enhance biodegradation of hydrocarbon contaminants in soil. Bioremediat J 4(4):249–257

Gundel LA, Lee VC, Kariawasam RRM, Stevens RK, Daisey JM (1995) Direct determination of the phase distributions of semi-volatile polycyclic aromatic hydrocarbons using annular denuders. Atmos Environ 29(14):1719–1733

Guo H, Lee SC, Ho KF, Wang XM, Zou SC (2003) Particle-associated polycyclic aromatic hydrocarbons in urban air of Hong Kong. Atmos Environ 37:5307–5317

Gurol MD, Singer PC (1982) Kinetics of ozone decomposition: a dynamic approach. Environ Sci Technol 16:377–383

Gustafson KE, Dickhut RM (1997) Gaseous exchange of polycyclic aromatic hydrocarbons across the air—water interface of Southern Chesapeake Bay. Environ Sci Technol 31(6):1623–1629

Haapea P, Tuhkanen T (2006) Integrated treatment of PAH contaminated soil by soil washing, ozonation and biological treatment. J Hazard Mater B136:244–250

Hart KM, Pankow JF (1994) High-volume air sampler for particle and gas sampling. 2.Use of backup filters to correct for the adsorption of gas-phase polycyclic aromatic hydrocarbons to the front filter. Environ Sci Technol 28(4):655–661

Hawthorne SB, Grabanski CB, Martin E, Miller DJ (2000) Comparisons of Soxhlet extraction, pressurized liquid extraction, supercritical fluid extraction and subcritical water extraction for environmental solids: recovery, selectivity and effects on sample matrix. J Chromatogr A 892 (1–2):421–433

Hays MD, Smith ND, Kinsey J, Dong Y, Kariher P (2003) Polycyclic aromatic hydrocarbon size distributions in aerosols from appliances of residential wood combustion as determined by direct thermal desorption e GC/MS. J Aerosol Sci 34:1061–1084

He J, Fan S, Meng Q, Sun Y, Zhang J, Zu F (2014) Polycyclic aromatic hydrocarbons (PAHs) associated with fine particulate matters in Nanjing, China: distributions, sources and meteorological influences. Atmos Environ 89:207–215

Hong PK, Nakra S, Jimmy KCM, Hayes DF (2008) Pressure-assisted ozonation of PCB and PAH contaminated sediments. Chemosphere 72:1757–1764

Hutzler C, Luch A, Filser JG (2011) Analysis of carcinogenic polycyclic aromatic hydrocarbons in complex environmental mixtures by LC-APPI-MS/MS. Anal Chim Acta 702:218–224

Jain N, Singla V, Satsangi A, Pachauri T, Kumari KM, Lakhani A (2012) Polycyclic aromatic hydrocarbon emissions and mutagenicity assessment of exhaust from a diesel generator. J Hazard Toxic Radioactive Waste Manag 15(1):18–25

Kalberer M, Morrical BD, Sax M, Zenobi R (2002) Picogram quantitation of polycyclic aromatic hydrocarbons adsorbed on aerosol particles by two-step laser mass spectrometry. Anal Chem 74:3492–3497

Kameda T (2011) Atmospheric chemistry of polycyclic aromatic hydrocarbons and related compounds. J Health Sci 57(6):504–511

Kang F, Chen D, Gao Y, Zhang Y (2010) Distribution of polycyclic aromatic hydrocarbons in subcelluar root tissue of ryegrass (*Lolium multiflorum*). BMC Plant Biol 10:1471–1477

Kannan GK, Kapoor SC (2004) Analysis of particles size fraction (PM_{10} and $PM_{2.5}$) and PAH of urban ambient air. DRDO, Ministry of Defence, Delhi, India

Karavalakis G, Deves G, Fontaras G, Stournas S, Samaras Z, Bakeas E (2010) The impact of soy-based biodiesel on PAH, nitro-PAH and oxy-PAH emissions from a passenger car operated over regulated and nonregulated driving cycles. Fuel 89:3876–3883

Kavouras IG, Koutrakis P, Tsapakis M, Lagoudaki E, Stephanou EG, VonBaer D, Oyola P (2001) Source apportionment of urban particulate aliphatic and polynuclear aromatic hydrocarbons (PAHs) using multivariate methods. Environ Sci Technol 35: 2288–2294

Kawanaka Y, Sakamoto K, Wang N, Yun SJ (2007) Simple and sensitive method for determination of nitrated polycyclic aromatic hydrocarbons in diesel exhaust particles by gas chromatography-negative ion chemical ionisation tandem mass spectrometry. J Chromatogr A 1163:312–317

Keyte IJ, Harrison RM, Lammel G (2013) Chemical reactivity and long-range transport potential of polycyclic aromatic hydrocarbons. Chem Soc Rev 42:9333

Khalili NR, Scheff PA, Holsen TM (1995) PAH source fingerprints for coke ovens, diesel and gasoline engines, highway tunnels, and wood combustion emissions. Atmos Environ 29 (4):533–542

Kim KH, Jahan SA, Kabir E, Brown RJC (2013) A review of airborne polycyclic aromatic hydrocarbons (PAHs) and their human health effects. Environ Int 60:71–80

Koua J, Li Z, Guo Y, Gao J, Yang M, Zoua Z (2010) Photocatalytic degradation of polycyclic aromatic hydrocarbons in GaN: ZnO solid solution-assisted process: Direct hole oxidation mechanism. J Mol Catal A: Chem 325:48–54

Krugly E, Martuzevicius D, Sidaraviciute R, Ciuzas D, Prasauskas T, Kauneliene V, Stasiulaitiene I, Kliucininkas L (2014) Characterization of particulate and vapor phase polycyclic aromatic hydrocarbons in indoor and outdoor air of primary schools. Atmos Environ 82:298–306

Kulik N, Goi A, Trapido M, Tuhkanen T (2005) Degradation of polycyclic aromatic hydrocarbons by combined chemical pre-oxidation and bioremediation in creosote contaminated soil. J Environ Manage 78:382–391

Kuyukina MS, Ivshina IB, Makarov SO, Litvinenko LV, Cunningham CJ, Philip CJ (2005) Effect of biosurfactants on crude oil desorption and mobilization in a soil system. Recent Adv Bioremediat 31:155–161

Lakhani A (2012) Source apportionment of particle bound polycyclic aromatic hydrocarbons at an industrial location in Agra, India. Sci World J 781291. doi:10.1100/2012/781291

Larsen RK, Baker JE (2003) Source apportionment of polycyclic aromatic 624 hydrocarbons in the urban atmosphere: a comparison of three methods. Environ Sci Technol 37:1873–1881

Lee BD, Iso M, Hosomi M (2001) Prediction of Fenton oxidation positions in polycyclic aromatic hydrocarbons by Frontier electron density. Chemosphere 42:431–435

Lee WJ, Wang YF, Lin TC, Chen YY, Lin WC, Ku CC, Cheng JT (1995) PAH characteristics in the ambient air of traffic-source. Sci Total Environ 159(2–3):185–200

Legrini O, Oliveros E, Braun AM (1993) Photochemical processes for water treatment. Chem Rev 93:671–698

Letzel T, Poschl U, Wissiack R, Rosenberg E, Grasserbauer M, Niessner R (2001) Phenyl-modified reversed-phase liquid chromatography coupled to atmospheric pressure chemical ionization mass spectrometry: a universal method for the analysis of partially oxidized aromatic hydrocarbons. Anal Chem 73:1634–1645

Li JL, Chen BH (2009) Effects of non-ionic surfactants on biodegradation of phenanthrene by marine bacteria of Neptunomnas naphthovorans. J Hazard Mater 162:66–73

Li R-J, Kou X-J, Geng H, Dong C, Cai Z-W (2014) Pollution characteristics of ambient $PM_{2.5}$-bound PAHs and NPAHs in a typical winter time period in Taiyuan. Chin Chem Lett 25:663–666

Limu YLMABD, LiFu DLNT, Miti ABLY, Wang X, Ding X (2013) Autumn and wintertime polycyclic aromatic hydrocarbons in $PM_{2.5}$ and $PM_{2.5-10}$ from Urumqi, China. Aerosol Air Qual Res 13:407–414

Liu WX, Dou H, Wei ZW, Chang B, Qui WX, Liu Y, Shu T (2008) Emission characteristics of polycyclic aromatic hydrocarbons from combustion of different residential coals in North China. Sci Total Environ 407:1436–1446

Liu Y, Sklorz M, Schnelle -KJ, Orasche J, Ferge T, Kettrup A, Zimmerman R (2006) Oxidant denuder sampling for analysis of polycyclic aromatic hydrocarbons and their oxygenated derivates in ambient aerosol: evaluation of sampling artifact. Chemosphere 62(11):1889–1898

Lu H, Zhu L, Zhu N (2009) Polycyclic aromatic hydrocarbon emission from straw burning and the influence of combustion parameters. Atmos Environ 43(4):978–983

Lung SCC, Liu CH (2015) Fast analysis of 29 polycyclic aromatic hydrocarbons (PAHs) and nitro-PAHs with ultra-high performance liquid chromatography-atmospheric pressure photoionization-tandem mass spectrometry. Sci Rep 5. doi:10.1038/srep12992

Luster-Teasleya S, Ubaka-Blackmoorea N, Masten SJ (2009) Evaluation of soil pH and moisture content on in-situ ozonation of pyrene in soils. J Hazard Mater 167:701–706

Lv Y, Li X, Xu TT, Cheng TT, Yang X, Chen JM, Iinuma Y, Herrmann H (2016) Size distributions of polycyclic aromatic hydrocarbons in urban atmosphere: sorption mechanism and source contributions to respiratory deposition. Atmos Chem Phys 16(5):2971–2983

Manariotis ID, Karapanagioti KH, Chrysikopoulo CY (2011) Degradation of PAHs by high frequency ultrasound. Water Res 45:2587–2594

Mandalakis M, Tsapakis M, Tsoga A, Stephanou EG (2002) Gas-particle concentrations and distribution of aliphatic hydrocarbons, PAHs, PCBs and PCDD/Fs in the atmosphere of Athens (Greece). Atmos Environ 36:4023–4035

Marchand N, Besombes J-L, Chevron N, Masclet P, Aymoz G, Jaffrezo J-L (2004) Polycyclic aromatic hydrocarbons (PAHs) in the atmospheres of two French alpine valleys: sources and temporal patterns. Atmos Chem Phys 4:1167–1181

Marr LC, Dzepina K, Jimenez JL, Riesen F, Bethel HL, Arey J, Gaffney JS, Marley NA, Molina LT, Molina MJ (2006) Sources and transformations of particle-bound polycyclic aromatic hydrocarbons in Mexico City. Atmos Chem Phys 6:1733–1745

Masih J, Masih A, Kulshrestha A, Singhvi R, Taneja A (2010) Characteristics of polycyclic aromatic hydrocarbons in indoor and outdoor atmosphere in the north central part of India. J Hazard Mater 177:190–198

Masih J, Singhvi R, Kumar K, Jain VK, Taneja A (2012) Seasonal variation and sources of polycyclic aromatic hydrocarbons (PAHs) in indoor and outdoor air in a semi arid tract of Northern India. Aerosol Air Qual Res 12:515–525

Masten SJ, Davies SHR (1997) Efficacy of in-situ ozonation for the remediation of PAH contaminated soils. J Contam Hydrol 28:327–335

McDonald JD, Zielinska B, Fujita EM, Sagebiel JC, Chow JC, Watson JG (2000) Fine particle and gaseous emission rates from residential wood combustion. Environ Sci Technol 34:2080–2091

Miguel AH, De Andrade JB (1989) Rapid quantitation of ten polycyclic aromatic hydrocarbons in atmospheric aerosols by direct HPLC separation after ultrasonic acetonitrile extraction. Int J Environ Anal Chem 35:35–41

Miguel AH, Kirchstetter TW, Harley RB, Hering RA (1998) On-road emissions of particulate polycyclic aromatic hydrocarbons and black carbon from gasoline and diesel vehicles. Environ Sci Technol 32:450–455

Miller-Schulze JP, Paulsen M, Toriba A, Hayakawa K, Simpson CD (2007) Analysis of 1-nitropyrene in air particulate matter standard reference materials by using two dimensional high

performance liquid chromatography with online reduction and tandem mass spectrometry detection. J Chromatogr A 1167:154–160
Mohanraj R, Dhanakumar S, Solaraj G (2012) Polycyclic aromatic hydrocarbons bound to PM 2.5 in urban Coimbatore, India with emphasis on source apportionment. Sci World J 2012. doi:10.1100/2012980843
Mohanraj R, Solaraj G, Dhanakumar S (2011a) Fine particulate phase PAHs in ambient atmosphere of Chennai metropolitan city, India. Environ Sci Pollut Res 18:764–771
Mohanraj R, Solaraj G, Dhanakumar S (2011b) $PM_{2.5}$ and PAH concentrations in urban atmosphere of Tiruchirappalli, India. Bull Environ Contam Toxicol 87:330–335
Moustafa YM, Shara SI (2009) Studies of seasonal variations on polycuclear aromatic hydrocarbons along the Nile River, Egypt. J Appl Sci Res 5:2349–2356
Nam K, Rodriguez W, Kukor JJ (2001) Enhanced degradation of polycyclic aromatic hydrocarbons by biodegradation combined with a modified Fenton reaction. Chemosphere 45:11–20
Nesterenko MA, Kirzhner F, Zimmels Y, Armon R (2012) Eichhornia crassipes capability to remove naphthalene from waste water in the absence of bacteria. Chemosphere 87:1186–1191
Oanh NTK, Albina DO, Ping L, Wang XK (2005) Emission of particulate matter and polycyclic aromatic hydrocarbons from select cookstove-fuel systems in Asia. Biomass Bioenerg 28:579–590
Okuda T, Okamoto K, Tanaka S, Shen Z, Han Y, Huo Z (2010) Measurement and source identification of polycyclic aromatic hydrocarbons (PAHs) in the aerosol in Xi'an, China, by using automated column chromatography and applying positive matrix factorization (PMF). Sci Total Environ 408:1909–1914
Pandey PK, Patel KS, Lenicek J (1999) Polycyclic aromatic hydrocarbons: need for assessment of health risks in India? Study of an urban-industrial location in India. Environ Monit Assess 59:287–319
Pandey SK, Kim KH, Brown RJC (2011) A review of techniques for the determination of polycyclic aromatic hydrocarbons in air. Trends Anal Chem 30(11):1716–1739
Pankow JF, Ligocki MP, Rosen ME, Isabelle LM, Hart KM (1988) Adsorption/thermal desorption with small cartridges for the determination of trace aqueous semivolatile organic compounds. Anal Chem 60(1):40–47
Park JS, Wade TL, Sweet S (2001) Atmospheric distribution of polycyclic aromatic hydrocarbons and deposition to Galveston Bay, Texas, USA. Atmos Environ 35:3241–3249
Park SS, Kim YJ, Kang CH (2002) Atmospheric polycyclic aromatic hydrocarbons in Seoul, Korea. Atmos Environ 36:2917–2924
Pierzynski GM, Sims JT, Vance GF (2000) Soils and environmental quality. CRC Press, Boca Raton, FL
Possanzini M, Palo VD, Gigliucci P, Sciano MCT, Cecinato A (2004) Determination of phase-distributed PAH in Rome ambient air by denuder/GC-MS method. Atmos Environ 38:1727–1734
Potter CL, Glaser JA, Chang LW, Meier JR, Dosani MA, Herrmann RF (1999) Degradation of polynuclear aromatic hydrocarbons under bench-scale compost conditions. Environ Sci Technol 33:1717–1725
Pradham SP, Conrad JR, Paterek JR, Srivastava VJ (1998) Potential of phytoremediation for treatment of PAHs in soil at MGP sites. J Soil Contam 7:467–480
Psillakis E, Goula G, Kalogerakis N, Mantzavinos D (2004) Degradation of polycyclic aromatic hydrocarbons in aqueous solutions by ultrasonic irradiation. J Hazard Mater B108:95–102
Raiyani CV, Shah SH, Desai NM, Venkaiah K, Patel JS, Parikh DJ, Kashyap SK (1993) Characterization and problems of indoor air pollution due to cooking stove smoke. Atmos Environ 27A:1643–1655
Rajput N, Lakhani A (2009a) Particle associated polycyclic aromatic hydrocarbons in urban air of Agra. Indian J Radio Space Phys 38:98–104
Rajput N, Lakhani A (2009b) Measurement of polycyclic aromatic hydrocarbons at an industrial site in India. Environ Monit Assess 150:273–284

Rajput N, Lakhani A (2010) Measurements of polycyclic aromatic hydrocarbons in an urban atmosphere of Agra, India. Atmosfera 23(2):165–183

Rajput N, Lakhani A (2012) Particle associated polycyclic aromatic hydrocarbons (PAHS) in urban air of Agra. Polycyclic Aromat Compd 32(1):48–60

Rajput N, Khemani LD, Lakhani A (2009) PAHs and their carcinogenic potencies in diesel fuel and diesel generator exhaust. Hum Ecol Risk Assess 15:201–213

Rajput N, Pyari AA, Saini MK, Kumari KM, Lakhani A (2010) Assessment of PAH toxicity and mutagenicity in emissions from coal and biofuel combustion. J Environ Sci Eng 52(3):185–192

Rajput P, Sarin MM, Rengarajan R, Singh D (2011) Atmospheric polycyclic aromatic hydrocarbons (PAHs) from post-harvest biomass burning emissions in the Indo-Gangetic Plain: isomer ratios and temporal trends. Atmos Environ 45(37):6732–6740

Rajput P, Sarin M, Kundu SS (2013) Atmospheric particulate matter (PM 2.5), EC, OC, WSOC and PAHs from NE–Himalaya: abundances and chemical characteristics. Atmos Pollut Res 4 (2):214–221

Ravindra K, Bencs L, Wauters E, de Hoog J, Deutsch F, Roekens E, Bleux N, Bergmans P, Van Grieken R (2006a) Seasonal and site specific variation in vapor and aerosol phase PAHs over Flanders (Belgium) and their relation with anthropogenic activities. Atmos Environ 40:771–785

Ravindra K, Wauters E, Taygi SK, Mor S, Van Grieken R (2006b) Assessment of air quality after the implementation of CNG as fuel in public transport in Delhi, India. Environ Monit Assess 115:405–417

Reisen F, Arey J (2002) Reactions of hydroxyl radicals and ozone with acenaphthrene and acenaphthylene. Environ Sci Technol 36:4302–4311

Ringuet J, Leoz -GE, Budzinski H, Villenave E, Albinet A (2012) Particle size distribution of nitrated and oxygenated polycyclic aromatic hydrocarbons (NPAHs and OPAHs) on traffic and suburban sites of a European megacity: Paris (France). Atmos Chem Phys 12:8877–8887

Rivas FJ (2006) Polycyclic aromatic hydrocarbons sorbed on soils: a short review of chemical oxidation based treatments. J Hazard Mater 138(2):234–251

Rivas FJ, Beltran FJ, Acedo B (2000) Chemical and photochemical degradation of acenaphthylene; intermediate identification. J Hazard Mater 75:89–98

Rivas J, Gimeno O, de la Calle RG, Beltran FJ (2009) Ozone treatment of PAH contaminated soils: operating variables effect. J Hazard Mater 169:509–515

Rogge WF, Hildemann LM, Mazurek MA, Cass GR, Simoneit BRT (1993) Sources of fine organic aerosol. Natural gas home applications. Environ Sci Technol 27:2736–2744

Schauer C, Niessner R, Poschl U (2003) Polycyclic aromatic hydrocarbons in urban air particulate matter: decadal and seasonal trends, chemical degradation, and sampling artifacts. Environ Sci Technol 37(13):2861–2868

Schubert P, Schantz MM, Sander LC, Wise SA (2003) Determination of polycyclic aromatic hydrocarbons with molecular weight 300 and 302 in environmental-matrix standard reference materials by gas chromatography/mass spectrometry. Anal Chem 75:234–246

Shen G, Tao S, Wei S, Zhang Y, Wang R, Wang B et al (2012) Emissions of parent, nitro, and oxygenated polycyclic aromatic hydrocarbons from residential wood combustion in rural China. Environ Sci Technol 46:8123–8130

Shen G, Wang W, Yang Y, Ding J, Xue M, Min Y, Zhu C, Shen H, Li W, Wang B, Wang R, Wang X, Tao S, Russell AG (2011) Emissions of PAHs from indoor crop residue burning in a typical rural stove: emission factors, size distributions, and gas-particle partitioning. Environ Sci Technol 45:1206–1212

Shimmo M, Anttila P, Hartonen K, Hyotylainen T, Paatero J, Kulmala M, Riekkola ML (2004) Identification of organic compounds in atmospheric aerosol particles by on-line supercritical fluid extraction–liquid chromatography–gas chromatography–mass spectrometry. J Chromatogr A 1022:151–159

Silva FS, Cristale J, André PA, Saldiva PH, Marchi MR (2010) PM 2.5 and PM 10: the influence of sugarcane burning on potential cancer risk. Atmos Environ 44(39):5133–5138

Singh DP, Gadi R, Mandal TK, Saud T, Saxena M, Sharma SK (2013) Emissions estimates of PAH from biomass fuels used in rural sector of Indo-Gangetic Plains of India. Atmos Environ 68:120–126

Singla, V., Pachauri, T., Satsangi, A., Kumari, K.M., Lakhani, A. (2012). Characterization and mutagenicity assessment of $PM_{2.5}$ and PM_{10} PAH at Agra, India. Polycyclic Aromat Compd 32(2). doi:10.1080/10406638.2012.657740

Slezakova K, Castro D, Delerue-Matos C, Morais S, do Carmo Pereira M (2014) Levels and risks of particulate-bound PAHs in indoor air influenced by tobacco smoke: a field measurement. Environ Sci Pollut Res 21(6):4492–4501

Soo J-C, Monaghan K, Lee T, Kashon M, Harper M (2016) Air sampling filtration media: Collection efficiency for respirable size-selective sampling. Aerosol Sci Technol J Am Assoc Aerosol Res 50(1):76–87

Soper SA, McGown LB, Warner IM (1994) Molecular fluorescence, phosphorescence, and chemiluminescence spectrometry. Anal Chem 66:428R–444R

Stegmann R, Lotter S, Jeerenklage J (1991) Biological treatment of oil-contaminated soils in bioreactors. In: Hinchee RE, Olfenbuttel RF (eds) On-site bioreclamation processes for xenobiotic and hydrocarbon treatment. Butterworth-Heinemann, MA

Straube WL, Nestler CC, Hansen LD, Ringleberg D, Pritchard PH, Jones -MJ (2003) Remediation of polyaromatic hydrocarbons (PAHs) through landfarming with biostimulation and bioaugmentation. Acta Biotechnol 23(2–3):179–196

Tang B, Isacsson U (2008) Analysis of Mono- and polycyclic aromatic hydrocarbons using solid-phase microextraction: state-of-the-art. Energy Fuels 22:1425–1438

Tsapakis M, Stephanou EG (2005) Occurrence of gaseous and particulate polycyclic aromatic hydrocarbons in the urban atmosphere: study of sources and ambient temperature effect on the gas/particle concentration and distribution. Environ Pollut 133(1):147–156

Tsapakis M, Tsapakis M (2005) Occurrence of gaseous and particulate polycyclic aromatic hydrocarbons in the urban atmosphere: study of sources and ambient temperature effect on the gas/particle concentration and distribution. Environ Pollut 133:147–156

Tutino M, Di Gilio A, Laricchiuta A, Assennato G, de Gennaro G (2016) An improved method to determine PM-bound nitro-PAHs in ambient air. Chemosphere 161:463

Ukiwe LN, Egereonu UU, Njoku PC, Nwoko CIA (2013) Combined chemical and water hyacinth (*Eichhornia crassipes*) treatment of PAHs contaminated soil. Int J Sci Eng Res 4:1–12

van Drooge BL, Nikolova I, Ballesta PP (2009) Thermal desorption gas chromatography-mass spectrometry as an enhanced method for the quantification of polycyclic aromatic hydrocarbons from ambient air particulate matter. J Chromatogr A 1216:4030–4039

Venkataraman C, Negi G, Sardar SB, Rastogi R (2002) Size distributions of polycyclic aromatic hydrocarbons in aerosol emissions from biofuel combustion. Aerosol Sci 33:503–518

Verma PK, Sah D, Kumari KM, Lakhani A (2017) Atmospheric concentrations and gas-particle partitioning of Polycyclic Aromatic Hydrocarbons (PAHs) and Nitro-PAHs at Indo-Gangetic sites. Environ Sci Proces Impacts 19(8):1051–1060

Villar-Vidal M, Lertxundi A, de Dicastillo MDML, Alvarez JI, Marina LS, Ayerdi M, Basterrechea M, Ibarluzea J (2014) Air Polycyclic Aromatic Hydrocarbons (PAHs) associated with $PM_{2.5}$ in a North Cantabric coast urban environment. Chemosphere 99:233–238

Wang W, Huang M, Chan C-Y, Cheung KC, Wong MH (2013) Risk assessment of non-dietary exposure to polycyclic aromatic hydrocarbons (PAHs) via house $PM_{2.5}$, TSP and dust and the Implications from human hair. Atmos Environ 73:204–213

Wang XK, Chen GH, Yao ZY (2003) Sonochemical degradation of polychlorinated biphenyls in aqueous solution. Chin Chem Lett 14:205–208

Wauters E, Caeter PV, Desmet G, David F, Devos C, Sandra P (2008) Improved accuracy in the determination of polycyclic aromatic hydrocarbons in air using 24 h sampling on a mixed bed followed by thermal desorption capillary gas chromatography–mass spectrometry. J Chromatogr A 1190(1–2):286–293

Wenger D, Gerecke AC, Heeb NV, Schmid P, Hueglin C, Naegeli H, Zanobi R (2009) In vitro estrogenicity of ambient particulate matter: contribution of hydroxylated polycyclic aromatic hydrocarbons. J Appl Toxicol 29:223–232

Wheat PH, Tumeo MA (1997) Solid-phase micro-extraction to monitor the sonochemical degradation of polycyclic aromatic hydrocarbons in water. Ultrason Sonochem 4:55–59

White PM, Wolf DC, Thoma GJ, Reynolds CM (2006) Phytoremediaton of alkylated polycyclic aromatic hydrocarbons in a crude oil-contaminated soil. Water Air Soil Pollut 169:207–220

WHO (2006) Air quality guidelines, Global update 2005, World Health Organisation, Bonn

Wu D, Wang Z, Chen J, Kong S, Fu X, Deng H, Shao G, Wu G (2014) Polycyclic aromatic hydrocarbons (PAHs) in atmospheric PM 2.5 and PM 10 at a coal-based industrial city: implication for PAH control at industrial agglomeration regions, China. Atmos Res 149:217–229

Yang X, Igarashi K, Tang N, Lin JM, Wang W, Kameda T, Toriba A, Hayakawa K (2010) Indirect- and direct-acting mutagenicity of diesel, coal and wood burning-derived particulates and contribution of polycyclic aromatic hydrocarbons and nitro-polycyclic aromatic hydrocarbons. Mutat Res 695:29–34

You HT, Shen T, Yin CZ (2012) Determination of polycyclic aromatic hydrocarbons in non-carcinogenic rubber oil by gas chromatography-mass spectrometry. Shandong Chem Ind 1:010

Yunker MB, Macdonald RW, Vingarzan R, Mitchell RH, Goyette D, Sylvestre S (2002) PAHs in the Fraser River basin: a critical appraisal of PAH ratios as indicators of PAH source and composition. Org Geochem 33:489–515

Zhang H, Hu D, Chen J, Ye X, Wang SX, Hao JM, Wang L, Zhang N, An Z (2011) Particle size distribution and polycyclic aromatic hydrocarbons emissions from agricultural crop residue burning. Environ Sci Technol 45:5477–5482

Zhang H, Ye X, Cheng T, Chen JM, Yang X, Wang L, Zhang RY (2008a) A laboratory study of agricultural crop residue combustion in China: emission factors and emission inventory. Atmos Environ 42:8432–8441

Zhang L, Li P, Gong Z, Li X (2008b) Photocatalytic degradation of polycyclic aromatic hydrocarbons on soil surfaces using TiO_2 under UV light. J Hazard Mater 158:478–484

Zhang M, Xie JF, Wang ZT, Zhao LJ, Zhang H, Li M (2016) Determination and source identification of priority polycyclic aromatic hydrocarbons in $PM_{2.5}$ in Taiyuan, China. Atmos Res 178–179

Zhang Y, Tao S (2009) Global atmospheric emission inventory of polycyclic aromatic hydrocarbons (PAHs) for 2004. Atmos Environ 2009(43):812–819

Zheng XJ, Blais JF, Mercier G, Bergeron M, Drogui P (2007) PAH removal from spiked municipal wastewater sewage sludge using biological, chemical and electrochemical treatments. Chemosphere 68:1143–1152

Zhou W, Zhu L (2007) Efficiency of surfactant-enhanced desorption for contaminated soils depending on the component characteristics of soil-surfactant-PAHs system. Environ Pollut 147:66–73

Chapter 4
Study of Environmental Particle Levels, Its Effects on Lung Deposition and Relationship With Human Behaviour

Anubha Goel, Saifi Izhar and Tarun Gupta

Abstract The adverse impact of particulate matter (PM) on human health can be best explained in terms of PM deposition inside respiratory organs rather than the typically reported ambient exposure concentration. This study measured the PM mass concentration (MC) at various microenvironments inside a residential campus and estimated the corresponding PM mass deposition in human respiratory regions such as Head (H), tracheobronchial (TB) and pulmonary (P) organs. This was done using the Multiple Path Particle Dosimetry lung model. Particle size fractions in the range of 0.25–10 μm were measured by an optical particle counter during winter when the PM MC is significantly high. The results showed that MC varied widely at different microenvironments ranging from 169.7 to 604.9 $\mu g/m^3$ for PM_1 and 31.2 to 649 $\mu g/m^3$ for PM_{10-1}. The highest and lowest MC for both fine and coarse sizes were found in the canteen and library, respectively. Our model revealed that the total deposited fraction increased with higher physical activity (e.g. higher breathing rate). The deposition fraction increased consistently in the Head with increase in particle size for all physical activity levels, whereas it varied in the P region. Results for 24 h lung dosimetry computed for students displayed that the dose deposition for coarse size particles for all type of activities followed the same trend: H > P > TB. A similar trend was found for fine particles except during sleep and sit activity levels with dissimilar trend: P > H > TB. The total deposited dose was found highest at play court while lowest at canteen. This type of focused preliminary study is much needed as physical activity levels and time spent at particular microenvironment besides the PM exposure concentration and helps to determine the critical microenvironments.

Keywords PM · Physical activity · Multiple Path Particle Dosimetry
Deposition fraction · Microenvironments

A. Goel (✉) · S. Izhar · T. Gupta
Department of Civil Engineering, IIT Kanpur, Kanpur 208016, India
e-mail: anubha@iitk.ac.in

A. Goel · T. Gupta
APTL, Centre for Environmental Science and Engineering,
Indian Institute of Technology Kanpur, Kanpur 208016, India

T. Gupta et al. (eds.), *Environmental Contaminants*, Energy, Environment, and Sustainability, https://doi.org/10.1007/978-981-10-7332-8_4

4.1 Introduction

PM in the atmosphere consists of a concoction of particles ranging from a few micrometres to nanometres. This includes suspended water droplets that originate from multitude of sources like industry, vehicles, natural emissions, combustion and incinerators. The PM size and composition including organics, inorganic ions, trace metals, carbonaceous matter vary significantly both temporally and spatially depending on the source and meteorological parameters like temperature, wind speed, relative humidity. In addition, how this PM affects an individual further depends on the person's lifestyle, which in turn depends on the microenvironments most frequented by him/her (home, workplace, schools, outdoor, etc.) (Buonanno et al. 2011). Many studies linked to epidemiology and toxicology have found strong associations between adverse health effects and ambient PM exposure (Dominici et al. 2006; Pope and Dockery 2006; Schikowski et al. 2007). PM alone has been linked to around 2.1 million premature deaths per year globally (Kelly and Fussell 2012; Kim et al. 2015).

The PM size is one of the most important factors governing its deposition inside different regions of the human respiratory tract. Many studies have concluded that PM toxicity is dependent on its size distribution and associated chemical compositions. The evidences are concrete that inhalation of fine particles (PM_1: particle with aerodynamic diameter less than 1 μm) constitutes an excess adverse health impact compared to coarse particles (PM_{1-10}: particles with aerodynamic diameter between 1 and 10 μm) of similar chemical composition (Kreyling et al. 2006). This happens because finer size particles consist of higher insoluble fraction and surface area compared to coarse particles (Carvalho et al. 2011) which lead to lung overburden through accumulation.

The adverse health effects of PM can be best explained by their deposition inside the human respiratory regions rather than the ambient exposure concentration (Salma et al. 2002; Asgharian and Price 2007; Löndahl et al. 2009). The extent of PM deposition inside the human respiratory tract (HRT) can be characterised in terms of mass, surface area or number concentration of the particles. The values of these depend on the ambient PM concentration, exposure period, human physical characteristics like age, gender, health conditions and physical exertion levels including lung morphology and breathing rates (Heyder 2004; Glytsos et al. 2010; Hofmann 2011). Due to experimental constraints, most of the earlier studies estimate the dose deposition in the respiratory system through various mathematical dosimetry models (ICRP 1994; Asgharian et al. 2004; Klepeis 2006; Aleksandropoulou et al. 2008). Some of these modelled results are also well correlated with experimental data (Asgharian and Price 2007; Löndahl et al. 2008).

This study has been conducted at an Indian academic institute with a large residential campus. The aim was to assess students' exposure to different particle sizes in different microenvironments like their hostel room, canteen, library. We determined the representative fraction and mass of particle deposition indifferent parts of the HRT, and the results were interpreted in terms of the critical microenvironment based on daily deposited dose. To the best of our knowledge, this type of study has not been given due consideration in the Indian context until now.

4.2 Methodology

Area selection: The study was conducted within the premises of IIT Kanpur (latitude 26.5 °N and longitude 80.3 °E at 142 m above m.s.l (mean sea level) which is situated on the north westside in the industrial city of Kanpur. It is approximately 17 km away from the railway station on the Grand Trunk Road. IIT Kanpur is an educational institute with a large residential campus ($\sim$1000 acres) and student strength of around 6000. There are no major sources of PM present inside the campus except frequent building construction activities and vehicular movement. The vehicles are mainly bicycles, scooters, motorcycles and cars (Izhar et al. 2016).

4.2.1 Monitoring Exposure in Microenvironments

An optical particle counter (OPC; model 1.108, Grimm) with a flow rate of 1.2 LPM and 15 channels was used to measure the PM exposure in the size range of 0.23 to 10 μm in real time. OPC directly provides the mass-fractioned concentration for respective size bins. Sample collection was performed in the winter (from 10 to 24 Jan 2016) at each of the nine microenvironments including a hostel room, dining room, classroom, play court, gymnasium, canteen, environment engineering lab (EE Lab), library and outdoors with varying monitoring durations (Table 4.1). The number of readings taken in each microenvironment ranged from 3 to 5, and the furthermore calculations are based on the average readings for each microenvironments. In view of the ongoing discussion about the choice between PM_1 and $PM_{2.5}$ as the cut-off diameter between fine and coarse fraction particles (Hieu and Lee 2010; Izhar et al. 2016), we considered PM_1 (0.23 μm < diameter < 1 μm) as fine fraction and PM_{10-1} (1 μm < diameter < 10 μm) as coarse. These were computed by adding the concentration in their respective bins of varying sizes. The instrument was kept at a height near the breathing zone of the exposed students. It was ensured that the instrument was well away from disturbances inside the microenvironment to ensure that the values obtained were unbiased.

Quality Control: In the beginning of this study, the OPC instrument was factory calibrated and was repeatedly tested for zero count by means of a HEPA filter at its inlet. The sensitivity of the measuring instrument was about 1 μg/m^3, and its reproducibility was ±2%. The ambient temperature recorded during the period of study was 13.6 ± 4.8 °C and the relative humidity was below 75%, which is in accordance with the requirements needed for proper functioning of the OPC (Jai Devi et al. 2013).

Table 4.1 Diurnal activity pattern for students at different microenvironments within IIT Kanpur

Time (hh:mm)	Microenvironment	Activity
00:00–07:00	Hostel room	Sleep
07:00–08:00	Hostel room	Light exercise
08:00–08:40	Dining room	Light exercise
08:40–09:00	Outdoor	Heavy exercise
09:00–13:00	Classroom/library	Sit
13:00–13:20	Outdoor	Heavy exercise
13:20–14:00	Dining room	Light exercise
14:00–14:20	Outdoor	Heavy exercise
14:20–17:30	EE Lab	Light exercise
17:30–18:00	Canteen	Sit
18:00–20:00	Play court/gymnasium	Heavy exercise
20:00–20:40	Dining room	Light exercise
20:40–21:00	Outdoor	Light exercise
21:00–24:00	Hostel room	Sit

Selection of Microenvironments: This was based on the daily activity profile of the students (see Table 4.1). The microenvironment selections were based on three factors including frequency of visit, differences in related activities (based on physical exertion) and variability in PM concentration. Different microenvironments also differed in terms of occupancy and ventilation. For instance, microenvironments like the hostel room, classroom, library and gymnasium had constant occupancy and the window and doors were closed, whereas the canteen and dining room involved continuous in and out movement of students and the doors were kept open at all times. Mostly, the hostel rooms are single and double bed. The library, classrooms, dining room and gymnasium have large number of occupants, whereas in canteen, play court and EE lab have occupants very limited. These microenvironments vary with their respective type of PM sources like combustion and burning dominant at canteen and dining, whereas the play court and outdoor are affected by soil resuspension. The hostel room is affected by incense and mosquito coil burning. These microenvironments have been discussed in detail in earlier studies (Devi et al. 2013; Ashok et al. 2014). The outdoor microenvironment is linked to increased physical exertion as it refers to the cycling activity during commute between hostel and class and vice versa. The indoor places except the gymnasium were linked to decreased physical exertion levels since they were associated with activities like sitting, sleeping, walking or light exercising.

4.2.1.1 Daily Integrated Exposure

The daily integrated exposure was measured using the weighted average concentration concept as conveyed in earlier studies (Ott 1982; Devi et al. 2013).

$$E = \frac{\sum_{j=1}^{m} Cj * Tj}{\sum_{j=1}^{m} Tj} \tag{4.1}$$

Here, E is the time-weighted average exposure of the students, C_j is the average concentration of the pollutant measured in the jth microenvironment, T_j is the time spent by the students in the jth microenvironment such that the summation of total time equals 24 h.

4.2.2 Modelling Deposition in HRT

PM deposition inside human respiratory system [Head (extra thoracic), tracheobronchial (TB) and pulmonary (P) regions] was estimated by using the Multiple Path Particle Dosimetry model (MPPD 2.11) patented by the Chemical Industry Institute of Toxicology (CIIT, USA). Different mechanisms including diffusion, sedimentation and impaction within the respiratory tracts are taken into account in this model for the estimation of deposition results (Yeh and Schum 1980). This mathematical model has been used extensively in computing particle dosimetry applications (Oberdörster et al. 2004; Teeguarden et al. 2007; Nong et al. 2009; Rostami 2009). MPPD software requires pre-information about the particle morphology including size, density, mass mean aerodynamic diameter (MMAD), geometric standard deviation (GSD) and morphological characteristics of the exposed subjects—upper respiratory tract volume in mL (URT), functional residual capacity in mL (FRC), tidal volume in mL (TV) and BF (minute ventilation in $\min^{-1}$). The morphological parameters are different for different physical exertion levels like sleep, sit, light exercise and heavy exercise. The parameter values are only available for the adult male students which are taken from different sources (Vijayan et al. 1990; ICRP 1994; USEPA 2011; Sanchez-Soberon et al. 2015) are reported in Table 4.2 and thus furthermore results are derived for male adult only. Also, it was assumed that the inhalation of particles was completely through the nose for all types of activities. Particle size distribution data from each microenvironment were utilised to compute MMAD and GSD. Also, particles were assumed to be spherical (shape factor of 1) and of constant 1 gm/cm^3 density in spite of their varying chemical composition (Hussein et al. 2015). Finally, the size-wise deposition fractions were estimated for different HRT regions. These values were further used to evaluate the PM MC data particular microenvironment. Respective size bin dose values were then added to form PM_1 and PM_{10-1}. The dose can be expressed as (Yeh and Schum 1980):

$$\text{Dose} = \text{DF} * C * T * V \tag{4.2}$$

where DF is the deposition fraction of particles in different parts of respiratory system (dimensionless), C is the PM concentration (μg/m^3), T is the time spent

Table 4.2 Morphological parameters for adults under different activities

Subject	Activity	BF (min^{-1})	TV (cc)	VR (m^3/h)	URT (ml)	FRC (ml)
Adult	Sleeping	12	625	0.45	50	2418
	Sitting	12	750	0.54		
	Light exercise	20	1250	1.50		
	Heavy exercise	26	1920	3.00		

ICRP (1994), USEPA (2011), Vijayan et al. (1990), Sanchez-Soberon et al. (2015)

during exposure (h) and *V* is minute ventilation (m^3/h). This study does not consider health status or the gender of the subject which may influence the simulated deposition fractions results.

4.3 Results and Discussion

4.3.1 Trends Observed in MC at Different Microenvironments

The mean size segregated MC of particles measured at various microenvironments frequented by the students are presented in Fig. 4.1. These size segregated concentrations were further used as input towards deposition modelling outcomes. Total particle concentrations were found to be the lowest in the library, indoors and highest in the canteen. Notable difference was observed in the contribution of different size bins to the total particle concentrations recorded at different microenvironments. The notable difference in levels was directly linked to the probable sources of airborne particles from events such as burning (cooking, use of mosquito coils or incense), particle resuspension (walking, brushing and vacuum cleaning) and sampling duration time (day/night).

It was observed that the indoor locations like hostel room, canteen, dining room, classroom and EE lab had concentrations greater than outdoor locations. The hostel room interior showed high particle concentrations, which was probably due to the use of room heater, mosquito coils and dust from quilts which get accumulated due to improper maintenance. Samples were collected during night hours when the room was closed. The higher particle concentration at indoor locations maybe ascribed to leakages and ventilation via natural and mechanical causes (Hussein et al. 2015).

Distribution of fine and coarse particles. The mean mass PM_1 and PM_{10-1} concentrations for different microenvironments are presented in Fig. 4.2. The fine mass concentrations exceeded coarse ones at all microenvironments except the canteen. This is because fine particles have larger penetration and lesser deposition

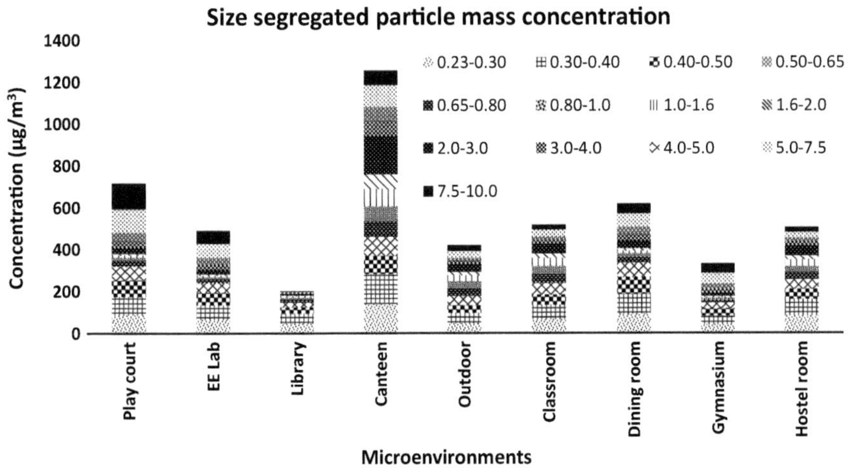

Fig. 4.1 Size-segregated mass concentration of particles at different microenvironments

rates compared to coarse fraction particles which may result into prolonged suspension of fine particle inside the indoor microenvironments (Monn 2001). Concentration varied widely at different microenvironments and ranged from 169.7 to 604.9 μg/m^3 (PM_1) and 31.2 to 649 μg/m^3(PM_{10-1}), respectively. Although the instrument reports mass concentration for particle size up to 0.23 μm, there are chances that particulate matter below 0.23 μm persists in the microenvironments. An earlier study reported that the maximum mass concentration loss below 0.23 μm is around 10% of the total particle mass concentration (Gupta et al. 2004; Singh and Gupta 2014). Hence, it can be said that use of this instrument will not make much difference to the PM_1 mass concentration. Levels of fines recorded outdoors were comparable to those reported in an earlier study at IIT Kanpur campus (Chakraborty and Gupta 2010). This study reported ambient PM_1 concentration during the winter season in 2009–10 as 199 ± 66 μg/m^3.

Both ventilation and size of the microenvironment were found to influence particle levels. Canteen showed the highest concentrations for both fine and coarse fractions while the library had the least. The observations may be attributed to factors such as continuous cooking, incense burning, poor exhaust facilities, limited space and constant opening and closing of windows and doors in the kitchen. On the contrary, Library was found to be the lowest in fine and coarse mass concentration because of its large size and centralised and air-sealed cooling system. The integrated daily mean personal exposure concentrations for PM_1 and PM_{10-1} using weighted exposure concept were found to be 306.9 and 200.9 μg/m^3, respectively, which were significantly higher compared to an earlier study done on IIT Kanpur campus (Devi et al. 2013). This study was conducted during summer (April-May 2010) and reported $PM_{2.5}$ and $PM_{10-2.5}$ of about 36.7 and 56.5 μg/m^3, respectively. We believe that these significant differences arise primarily due to the differences in

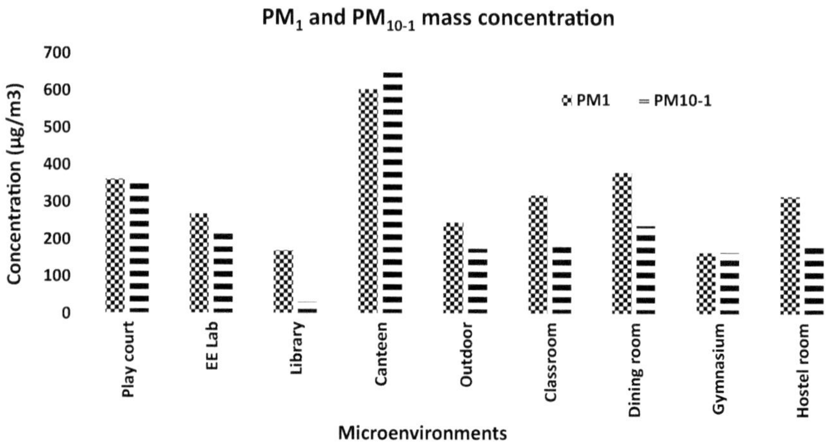

Fig. 4.2 Mean MC of PM_1 and PM_{10-1} in different microenvironments

the sampling months (summer vs winter) and hence, meteorological conditions and source emissions activities. Increased construction activities in the campus during our sampling period are also likely contributors to increased PM MC in comparison to the data collected six years earlier.

4.3.2 *Particle Fractions Deposited in HRT*

The average deposition fractions (DF) corresponding to each particle size bins' mid-point diameter were computed by using the MPPD software in the extra thoracic/Head(H), tracheobronchial (TB) and pulmonary/alveolar (P) regions of the human respiratory system (Fig. 4.3). The total DF variation was seen to be between 33.5 and 92.5% during sleep, 34.6–93.7% during sitting, 36.6–95.2% during light exercise and 42.2–96% during heavy exercise. This variation in DF was attributed to particle size, activity level-related breathing rate and pattern. For all particle size bins, the order in DF under different activity levels for the three regions of lungs was as follows:

$$\text{H : Heavy exercise} > \text{Light exercise} > \text{Sit} > \text{Sleep}$$
$$\text{TB : Sleep} > \text{Sit} > \text{Light exercise} > \text{Heavy exercise}$$
$$\text{P : Sit} > \text{Sleep} > \text{Light exercise} > \text{Heavy exercise.}$$

These results show that an increase in physical activity increased DF in the Head and subsequently decreased DFs in the TB and P regions.

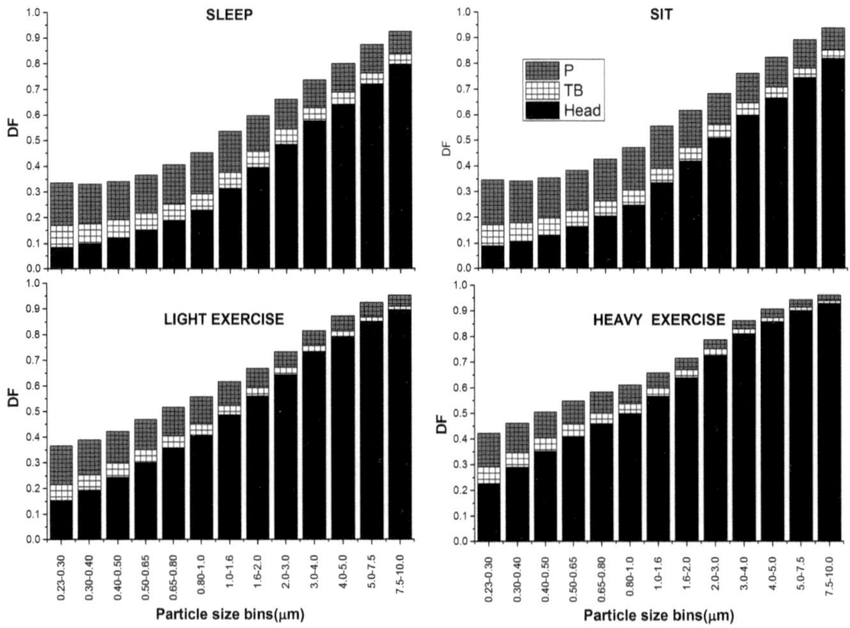

Fig. 4.3 Calculated average particle deposition fractions in lungs for students under various activity levels. (*Note* DF for particular activity level is computed by taking average value at different microenvironment)

4.3.2.1 Variability in DF

Influence of particle size and activity level. The DF at Head region and total DF increased with particle size, whereas an opposite trend was seen for DF at TB. The DF at pulmonary region initially decreased from 0.23 to 0.5 μm, then increased from 0.65 to 1.6 μm, and finally decreased for sitting and sleeping activities. However, activity level including light exercise and heavy exercise showed monotonous decrease in DF at pulmonary region. The percentage deposition for all activity levels was maximum in the Head region. Earlier studies have shown similar findings where DF at the extrathoracic region was found highest for coarse size particles (Behera et al. 2015; Sarigiannis et al. 2015). To this observation, mechanisms including sedimentation and the impaction of particles have been linked for deposition of coarse particles onto the upper respiratory regions (Behera et al. 2015). The results from Saber and Heydari (2012) also confirmed that increase in physical exertion level caused higher DF in Head region regardless of the particle size.

Influence of Ventilation Parameters and Mode of Inhalation. A study done in Budapest, Hungary for the PM_1, reported that mean DF in the H region decreased from 26% for sleeping to 9.4% for heavy exercise and increased in the P region from 14.7% for sleep to 34% for heavy exercise (Salma et al. 2015). In contrast, our study finds that light and heavy exercise experiences the highest DF in the H region, whereas maximum DF in the sitting and sleep activities occurs in the P region. One of the possible reasons for this difference could be due to the inhalation pathway considered. Salma et al. (2015) showed comparatively lesser deposition in the Head region by considering oral breathing during light and heavy exercise, as in our study nasal breathing has been consistent for all activity levels. Sanchez-Soberon et al. (2015) showed ventilation parameters like FRC and URT could also affect the deposition fraction values inside HRT. They reported that 65% of FRC increase might decrease DF by 25% in P while 40% URT increase might result in decrease of DF from 3 to 12%. Also, in most studies, DFs have been calculated by considering one constant value for tidal volume and minute ventilation for distinct physical activity levels (Ham et al. 2010; Sarigiannis et al. 2015) which are not the correct approach, as shown by our study, and are likely to give biased results.

4.3.3 Variation in Diurnal Deposited Dose

The estimated mass of PM_1 and PM_{10-1} deposited in different parts of the respiratory system for students under various activities at different microenvironments over 24 h is presented in Fig. 4.4. The dose deposition for coarse fraction (PM_{10-1}) particles in different parts of HRT for all activities followed the same trend: Head > P > TB. Trend for fine fraction (PM_1) particles was also similar except during lighter activity levels sleep and sit, for which the trend was: P > Head > TB. For PM_1, the total dose deposited was approximately at 54.8% at H, 14.2% at TB and 31% at P, whereas 88.5, 3.4 and 8.1%, respectively, for PM_{10-1}.

Influence of microenvironment conditions and duration of exposure: Table 4.3 shows the total time spent in each microenvironment, concentrations and deposited dose of PM_1 and PM_{10-1} daily deposited dose. The time spent at hostel room was maximum. Time spent was minimum at canteen where ambient particle concentrations were highest. Maximum deposition was recorded at playground. General trends for concentrations in ambient air and dose for fine and coarse particle fractions are noted below. The trends observed for PM_1:

- Concentration: Canteen > Dining room > Hostel room > EE Lab > Play court/ Gymnasium > Outdoor > Classroom/Library.
- Dose: Play court/Gymnasium > Hostel room > EE Lab > Outdoor > Classroom/ Library > Dining room > Canteen.

The trends observed for PM_{10-1}:

- Concentration: Canteen > Play court/Gymnasium > Dining room > EE Lab > Hostel room > Outdoor > Classroom/Library.
- Dose: Play court/Gymnasium > EE Lab > Hostel room > Outdoor > Dining room > Classroom/Library > Canteen.

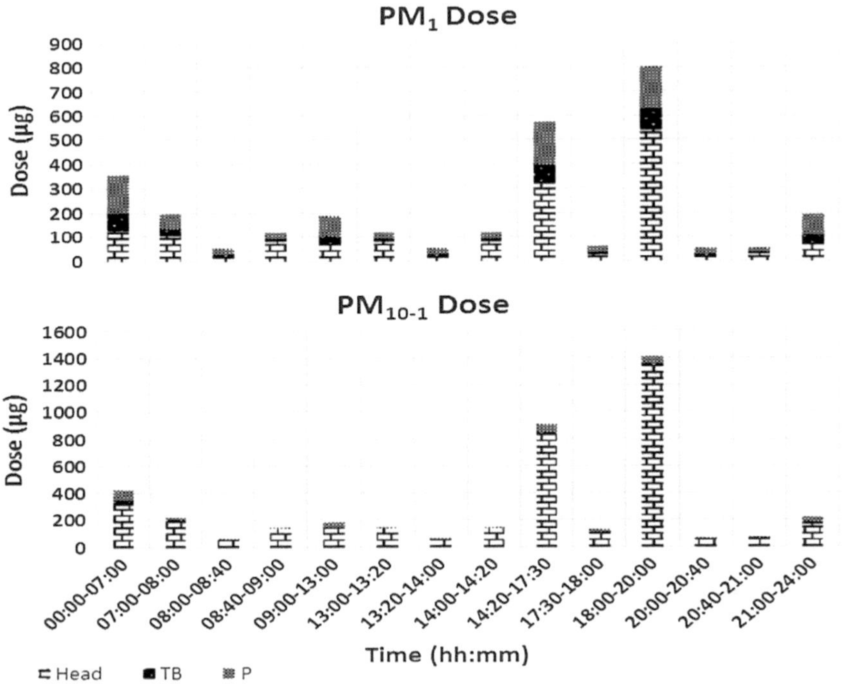

Fig. 4.4 Daily PM_1 and PM_{10-1} dose deposited in different region of HRT for students at IIT Kanpur

Table 4.3 Summary of time spent, $PM_1$and PM_{10-1} MC ($\mu g/m^3$) recorded and total dose (μg) calculated for each microenvironment

Microenvironment	Time exposure (h)	Daily time fraction (%)	Concentration ($\mu g/m^3$)		Total dose (μg)	
			PM_1	PM_{10-1}	PM_1	PM_{10-1}
Hostel room	11.0	45.8	317.2	187.4	744.9	868.1
Dining room	2.0	8.3	379.7	238.6	164.1	203.5
Classroom/library	4.0	16.7	244.3	114.7	188	181.4
EE lab	3.2	13.2	267.9	222.6	572.9	907.4
Canteen	0.5	2.1	604.9	649.0	63.2	129.2
Play court/gymnasium	2.0	8.3	262.5	270.7	803.4	1407.2
Outdoor	1.3	5.4	245.2	174.9	290.8	366.8

The above trend for PM_1 and PM_{10-1} showed that microenvironments have distinct order in terms of PM concentration and their deposited dose. This highlights the fact that the ambient air concentration cannot be used as a proxy for 'critical' microenvironments. Duration of exposure and activity level plays a critical role.

Critical microenvironment: From Fig. 4.4, it is seen that maximum total deposition for both fine and coarse particles occurs at the same time—6–8 p.m. (playground/dinner time) >2:20–5:30 p.m. (laboratory/class time) >midnight–7 a. m. (inside hostel room). Deposition amount for PM_1 and PM_{10-1} are maximum during evening hours while playing (heavy exercise), whereas minimum while eating (sit) in dining room. This is primarily because of high minute ventilation, DF, particles MC and time spent while playing, whereas vice versa during eating at dining room. Hostel room interior showed high-dose deposition despite lowest deposition fraction. This is primarily due to maximum time spent inside the room and also particle mass concentration was higher in wintertime due to use of heater, mosquito coils and lack of proper maintenance of quilt wherein dust gets accumulated. Since in our modelling most of the particle mass is deposited in the nasal Head region, diseases like rhinitis, pharyngitis or sinusitis are most probable to happen (Vincent 2005). To limit occurrence of health problems related to inhalation of particles, hostel rooms, where maximum time of students is spent, should be cleaned and maintained properly and use of mosquito coils and room heater should be controlled.

Sensitivity of results to parameter values considered. The deposition fraction and deposition mass have been calculated only for the winter period which can be higher than the average results due to higher PMC in winter season as compared to others (Chakraborty and Gupta 2010). Also, the activity patterns and time spent in different microenvironments can be different depending upon the prevailing season. For instance, time spent at room is likely to be more during winters than during early spring. Further ranges for FRC and URT for the exposed populations have not been taken into account. Hence, different outcomes are expected due to variability in the parameters values which can be subjected to future analyses.

4.4 Conclusions

We have simulated the deposition of size-segregated particles in HRT based on the exposure of the subjects to different microenvironments daily. Nine out of such microenvironments were selected taking into account time spent at and different physical activity levels students are engaged in. The new empirical evidence is based on monitoring at a university campus, IIT Kanpur, during a few weeks period in January 2016. The particulate matter size segregated mass concentrations varied significantly across the microenvironments considered. Likely factors for these observations were particles sources including combustion events, resuspension activities and ventilation effects. The MCs of PM_1 and PM_{10-1} were found to be highest and lowest at canteen and library, respectively. The integrated mean personal exposure concentrations over 24 h for PM_1 and PM_{10-1} were found to be 306.9 and 200.9 $\mu g/m^3$, respectively. Lung deposition calculation performed by

Multiple Path Particle Dosimetry model implied that particle size, mode of inhalation and activity level affect regional deposition fraction significantly. With increase in particle size, deposition fraction at Head airway region increased and decreased at TB region for all activity levels. P region, with increase in size of particles, deposition fraction showed first decreasing order (0.23–0.5 μm) followed by increasing order (0.65–1.6 μm) and then again decreasing order (particle size > 1.6 μm) during light activity levels, sleep and sit, whereas it was monotonous decreasing during light and heavy exercise activities. Increase in physical exertion resulted in higher lung deposition fraction at Head region. The dose deposition of fine and coarse fraction in different parts of respiratory tract for all activities followed the same trend: Head > P > TB. Influence of particle levels in ambient air and duration of exposure are factors that need further examination. The total deposited dose was found highest at play court/gymnasium while lowest at canteen. Also, hostel room, where students spend most of their time, showed high concentration of deposited dose of particles which is sufficient to affect health adversely. This preliminary research provides an overview of health impact, in terms of probable dose deposited of particles, due to exposure during winter at a university campus in northern India. Its scope can be extended to different seasons in order to investigate variation in dose and deposition rates and other similar locations can be considered for influence of geographical and meteorological conditions.

Acknowledgements The authors would like to acknowledge ARA organisation (www.ara.com) for providing the free link to Multiple Path Particle Dosimetry model software (MPPD 2.11); the tutorials were easy to comprehend. This study was performed using institute resources. No conflict of funding is claimed.

References

Aleksandropoulou V, Mitsakou C, Housiadas C, Lazaridis M (2008) Particulate matter exposure and dose relationships derived from realistic exposure scenarios. Indoor Built Environ 17 (3):237–246

Asgharian B, Menache MG, Miller FJ (2004) Modeling age-related particle deposition in humans. J Aerosol Med 17(3):213–224

Asgharian B, Price OT (2007) Deposition of ultrafine (nano) particles in the human lung. Inhalation Toxicol 19(13):1045–1054

Ashok V, Gupta T, Dubey S, Jat R (2014) Personal exposure measurement of students to various microenvironments inside and outside the college campus. Environ Monit Assess 186(2):735–750

Behera SN, Betha R, Huang X, Balasubramanian R (2015) Characterization and estimation of human airway deposition of size-resolved particulate-bound trace elements during a recent haze episode in Southeast Asia. Environ Sci Pollut Res 22(6):4265–4280

Buonanno G, Fuoco FC, Stabile L (2011) Influential parameters on particle exposure of pedestrians in urban microenvironments. Atmos Environ 45(7):1434–1443

Carvalho TC, Peters JI, Williams RO (2011) Influence of particle size on regional lung deposition–what evidence is there? Int J Pharm 406(1):1–10

Chakraborty A, Gupta T (2010) Chemical characterization and source apportionment of submicron (PM1) aerosol in Kanpur region India. Aerosol Air Qual. Res 10(5):433–445

Devi JJ, Gupta T, Jat R, Tripathi SN (2013) Measurement of personal and integrated exposure to particulate matter and co-pollutant gases. Environ Sci Pollut Res 20(3):1632–1648

Dominici F, Peng RD, Bell ML, Pham L, McDermott A, Zeger SL, Samet JM (2006) Fine particulate air pollution and hospital admission for cardiovascular and respiratory diseases. JAMA 295(10):1127–1134

Glytsos T, Ondráček J, Džumbová L, Kopanakis I, Lazaridis M (2010) Characterization of particulate matter concentrations during controlled indoor activities. Atmos Environ 44 (12):1539–1549

Gupta T, Demokritou P, Koutrakis P (2004) Effects of physicochemical properties of ultrafine particles on the performance of an ultrafine particle concentrator. Aerosol Sci Technol 38 (S2):37–45

Ham WA, Herner JD, Green PG, Kleeman MJ (2010) Size distribution of health-relevant trace elements in airborne particulate matter during a severe winter stagnation event: implications for epidemiology and inhalation exposure studies. Aerosol Sci Technol 44(9):753–765

Heyder J (2004) Deposition of inhaled particles in the human respiratory tract and consequences for regional targeting in respiratory drug delivery. Proc Am Thorac Soc 1(4):315–320

Hieu NT, Lee BK (2010) Characteristics of particulate matter and metals in the ambient air from a residential area in the largest industrial city in Korea. Atmos Res 98(2):526–537

Hofmann W (2011) Modelling inhaled particle deposition in the human lung—a review. J Aerosol Sci 42(10):693–724

Hussein T, Wierzbicka A, Löndahl J, Lazaridis M, Hänninen O (2015) Indoor aerosol modeling for assessment of exposure and respiratory tract deposited dose. Atmos Environ 106:402–411

ICRP (1994) Human respiratory tract model for radiological protection. ICRP Publication 66. Ann ICRP 24(1–3)

Izhar S, Goel A, Chakraborty A, Gupta T (2016) Annual trends in occurrence of submicron particles in ambient air and health risk posed by particle bound metals. Chemosphere 146:582–590

Kelly FJ, Fussell JC (2012) Size, source and chemical composition as determinants of toxicity attributable to ambient particulate matter. Atmos Environ 60:504–526

Kim KH, Kabir E, Kabir S (2015) A review on the human health impact of airborne particulate matter. Environ Int 74:136–143

Klepeis NE (2006) Modeling human exposure to air pollution. In: Human exposure analysis, pp 445–470

Kreyling WG, Semmler-Behnke M, Möller W (2006) Ultrafine particle-lung interactions: does size matter? J Aerosol Med 19(1):74–83

Löndahl J, Pagels J, Boman C, Swietlicki E, Massling A, Rissler J, Blomberg A, Bohgard M, Sandström T (2008) Deposition of biomass combustion aerosol particles in the human respiratory tract. Inhalation toxicology 20(10):923–933

Löndahl J, Massling A, Swietlicki E, Bräuner EV, Ketzel M, Pagels J, Loft S (2009) Experimentally determined human respiratory tract deposition of airborne particles at a busy street. Environ Sci Technol 43(13):4659–4664

Monn C (2001) Exposure assessment of air pollutants: a review on spatial heterogeneity and indoor/outdoor/personal exposure to suspended particulate matter, nitrogen dioxide and ozone. Atmos Environ 35(1):1–32

Nong A, Taylor MD, Clewell HJ, Dorman DC, Andersen ME (2009) Manganese tissue dosimetry in rats and monkeys: accounting for dietary and inhaled Mn with physiologically based pharmacokinetic modeling. Toxicol Sci 108(1):22–34

Oberdörster G, Sharp Z, Atudorei V, Elder A, Gelein R, Kreyling W, Cox C (2004) Translocation of inhaled ultrafine particles to the brain. Inhal Toxicol 16(6–7):437–445

Ott WR (1982) Concepts of human exposure to air pollution. Environ Int 7(3):179–196

Pope CA III, Dockery DW (2006) Health effects of fine particulate air pollution: lines that connect. J Air Waste Manag Assoc 56(6):709–742

Rostami AA (2009) Computational modeling of aerosol deposition in respiratory tract: a review. Inhalation Toxicol 21(4):262–290
Saber EM, Heydari G (2012) Flow patterns and deposition fraction of particles in the range of 0.1–10 μm at trachea and the first third generations under different breathing conditions. Comput Biol Med 42(5):631–638
Salma I, Balásházy I, Winkler-Heil R, Hofmann W, Záray G (2002) Effect of particle mass size distribution on the deposition of aerosols in the human respiratory system. J Aerosol Sci 33 (1):119–132
Salma I, Füri P, Németh Z, Balásházy I, Hofmann W, Farkas Á (2015) Lung burden and deposition distribution of inhaled atmospheric urban ultrafine particles as the first step in their health risk assessment. Atmos Environ 104:39–49
Sánchez-Soberón F, Mari M, Kumar V, Rovira J, Nadal M, Schuhmacher M (2015) An approach to assess the Particulate Matter exposure for the population living around a cement plant: modelling indoor air and particle deposition in the respiratory tract. Environ Res 143:10–18
Sarigiannis DA, Karakitsios SP, Zikopoulos D, Nikolaki S, Kermenidou M (2015) Lung cancer risk from PAHs emitted from biomass combustion. Environ Res 137:147–156
Singh DK, Gupta T (2014) Field performance evaluation during fog-dominated wintertime of a newly developed denuder-equipped PM1 sampler. Environ Sci Pollut Res 21(6):4551–4564
Schikowski T, Sugiri D, Ranft U, Gehring U, Heinrich J, Wichmann HE, Kramer U (2007) Does respiratory health contribute to the effects of long-term air pollution exposure on cardiovascular mortality. Respir Res 8(1):20
Teeguarden JG, Hinderliter PM, Orr G, Thrall BD, Pounds JG (2007) Particokinetics in vitro: dosimetry considerations for in vitro nanoparticle toxicity assessments. Toxicol Sci 95(2):300–312
USEPA (2011) Exposure factors handbook: 2011 edn, National Centre of Environmental Assessment, U.S. Environmental Protection Agency, Washington, D.C., U.S.A. Sept. 2011
Vijayan VK, Kuppurao KV, Venkatesan P, Sankaran K, Prabhakar R (1990) Pulmonary function in healthy young adult Indians in Madras. Thorax 45(8):611–615
Vincent JH (2005) Health-related aerosol measurement: a review of existing sampling criteria and proposals for new ones. J Environ Monit 7(11):1037–1053
Yeh HC, Schum GM (1980) Models of human lung airways and their application to inhaled particle deposition. Bull Math Biol 42(3):461–480

Chapter 5
Bioaerosols Over the Indo-Gangetic Plain: Influence of Biomass Burning Emission and Ambient Meteorology

Prashant Rajput, Amit Singh Chauhan and Tarun Gupta

Abstract Bioaerosols (particles of biological origin) can be produced from living or dead plants and animals. They can potentially serve as the cloud condensation and ice nuclei (CCN and IN). Their role in global carbon cycle further highlights importance of studying their variability to link up with climate relevance parameters. Focusing on tropical region reveals that it holds wealthy number of human population and has massive vegetation cover-area. From Indian region, production estimates of bioaerosols from human population (current: ~1.25 billion; of which over 45% resides in Indo-Gangetic Plain: IGP) and Wildlife Sanctuaries and National Parks (100 in numbers, situated from north to south and east to west) is not known. Most of the forest fires in India occur during March–June (hot and drier season). The detailed information on chemical composition, fingerprinting and radiative forcing from regional forest fires is also lacking. Unlike natural sources (forest cover and fires), the seasonal variability of pollutants emission characteristic and chemical, optical and radiative forcing are relatively well studied from anthropogenic biomass (post-harvest paddy- and wheat-residue and biofuels) burning emission in India. However, the abundance of bioaerosols and their variability over a large stretch of IGP (north-west to north-east) was not well documented. Towards this, we have undertaken a year-long campaign to study and document (first-attempt) bioaerosols variability over a complete annual cycle from central IGP. We observed a parallel enhancement in concentrations of fine-particulate matter ($PM_{2.5}$ in October–November: 158 ± 89 μg m^{-3} as compared to June–September months: 40 ± 18 μg m^{-3}; two-tailed $t = 8.2$, $p < 0.05$) and bioaerosols (particularly Gram-negative bacteria: GNB, a source of endotoxin in ambient air; 186 ± 87 CFU/m^3 during October-November as compared to 114 ± 58 CFU/m^3; $t = 4.0$, $p < 0.05$) with the biomass burning emissions intensification period. The abundance of bioaerosols exhibits influence of ambient meteorology, for example GNB exhibited negative correlations with T, wind speed

---

P. Rajput · A. S. Chauhan (✉) · T. Gupta
Department of Civil Engineering and Atmospheric Particle Technology Lab at Center for Environmental Science and Engineering (CESE), IIT Kanpur, Kanpur 208016, India
e-mail: tarun@iitk.ac.in

T. Gupta et al. (eds.), *Environmental Contaminants*, Energy, Environment, and Sustainability, https://doi.org/10.1007/978-981-10-7332-8_5

and heavy (>4 mm daily) precipitation, whereas it showed positive correlations with RH and low precipitation amount (<4 mm). Studying bioaerosols and establishing its linkage to health and climate appear to be of utmost importance.

Abbreviations

CFU	Colony forming units
EU	Endotoxin units
GNB	Gram-negative bacteria
GPB	Gram-positive bacteria
IGP	Indo-Gangetic Plain
K^{+}_{BB}	Biomass burning derived K^{+}
LPS	Lipopolysaccharides
OC	Organic carbon in particulate phase
RH	Relative humidity (in %)
T	Ambient temperature (in °C)

5.1 Potential Impacts of Assessing and Control of Bioaerosols

(i) Micro-environment of operating rooms could be a high-risk area for both patients and staffs. Thus, managing the air quality and ensuring it free from airborne infectious agents are of utmost importance in such environments. Deployment of filtration equipment (proper air conditioning systems) can reduce airborne concentrations of fungi significantly. It is worthwhile mentioning that airborne bacteria could cause severe infections during surgery. It has been observed that reduction of airborne bacteria in operating rooms is directly linked with the substantial decrease in contamination of wounds.

(ii) In India, there is an urgent need to develop tools to accurately measure bioaerosols, to quickly identify them in order to have a baseline information and establish a link to infections to make appropriate strategy to mitigate and control these especially in the sensitive indoor environments (hospitals, school, shopping malls etc.) and to the influence on climate.

5.2 Concepts of the Synthesis

Biologically produced (from plants/animals) aerosols are widely referred to as bioaerosols (also referred to as primary biological aerosol particles: PBAPs) (Després et al. 2012; Fu et al. 2008; Miyakawa et al. 2015; Pöschl et al. 2010).

Improper sanitation, waste-disposal practices and biomass burning may also result into generation of huge amounts of microbes in the air (Taha et al. 2007). In the atmosphere, they are ubiquitous as bacteria, viruses, fungal spores, bio-debris and pollens (Fišar et al. 1990; Lau et al. 2006; Prussin et al. 2015; Reponen et al. 2007; Sesartic and Dallafior 2011). The size of different types of bioaerosols varies over a large range (Reponen et al. 2001). For example, viruses are generally of the size of less than 300 nm, bacteria are in the range of 0.3–8 μm, fungal spores in the range of 1–30 μm and size of pollens are greater than 17 μm (Gregory 1973; Stanley and Linskins 1974). The size of bio-debris (fragments from plants/animals) may vary from sub-micron to coarser fraction (Graham et al. 2003a, b). It is important to mention here that bioaerosols can adsorb onto existing particles in the ambient atmosphere, and thus, their aerodynamic diameter and residence time can be influenced by the physical characteristics of the suspended particulates (Donaldson et al. 2001; Shaffer and Lighthart 1997). The contribution of bioaerosols in total particulate matter has been assessed previously from different environmental conditions reporting 28% over remote continental, 22% in populated continental and 10% in remote maritime environments (Matthias-Maser et al. 2000). A recent study (Zhu et al. 2016) has studied quantitatively the contribution of PBAP during daytime and night-time in a temperate coniferous forest in Japan (at Wakayama). They have measured biomarkers of PBAP (fungal spores tracers: arabitol, mannitol and trehalose) through solvent extraction followed by derivatization approach. The contribution of fungal spores in organic carbon (OC) was found to be relatively high in their study during night-time (45%) as compared to that in daytime (22%), and they have attributed this observation to nocturnal sporulation under near-saturated RH condition. Thus, the contribution of bioaerosols to OC can be quite significant over/near the forest/polluted region (Huffman et al. 2012, 2013).The spatial distribution of Wildlife Sanctuaries and National Parks in India is shown in Fig. 5.1. They are expected to representing background source of bioaerosols in Indian region and particularly over the Indo-Gangetic Plain.

It has been widely realized that bioaerosols play a very important role in climate change through participation in carbon cycle of the ecosystem and holding potential cloud condensation nuclei as well as ice-nucleation activity (Bauer et al. 2003; Fröhlich-Nowoisky et al. 2009, 2016; Garcia et al. 2012; Hawkes et al. 2011). In addition, they are widely studied also because of their many allergenic effects on the human health (Annadotter et al. 2005; Balasubramanian et al. 2012; Clark et al. 1983; Domanska and Stroszejn-Mrowca 1994; Semple et al. 2010; Xue et al. 2016). Human exposure to bioaerosols through physical contact or inhalation may lead to adverse health effects like asthma, COPD (chronic obstructive pulmonary disease), whooping cough, and sick building syndrome (GBD 2015). Study to understand their occurrence and effects in indoor and outdoor environments is one of the major thirst areas of research (Iossifova et al. 2007; Kildeso et al. 2003; Lee et al. 2006). However, despite being a very important area of research, bioaerosols have been studied little less over Indian region (Ansari et al. 2015; Kumar and Attri 2016; Kumar et al. 2011; Mamta et al. 2015). Recently, there are studies emerging out from southern part of India documenting the number and mass distribution of

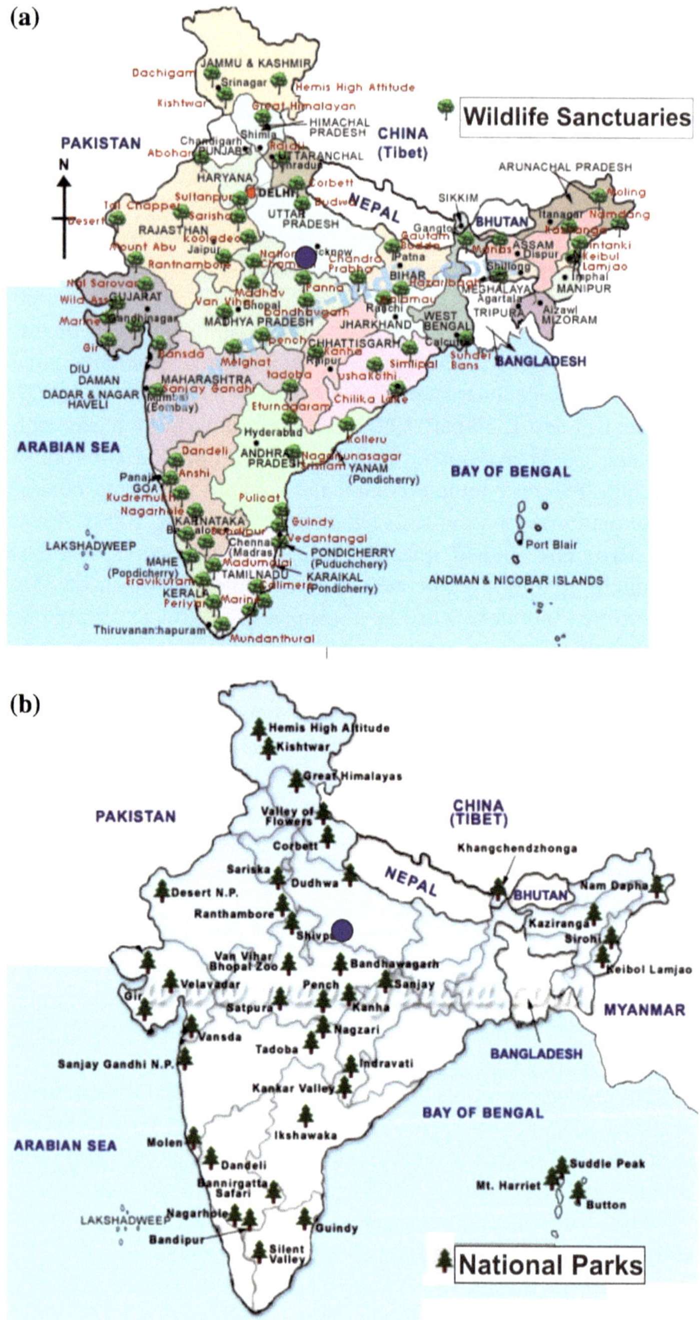

Fig. 5.1 Geographical distribution of: **a** Wildlife sanctuaries and **b** National parks in India (Chauhan 2016)

fluorescent biological aerosol particles (FBAP), a lower limit of PBAP (Valsan et al. 2015, 2016).

Endotoxin is a biologically active LPS (lipopolysaccharide) and a component of outer membrane of the Gram-negative bacteria (GNB) (Clark et al. 1983; Domanska and Stroszejn-Mrowca 1994). They have been studied widely from various environmental conditions (Annadotter et al. 2005; Balasubramanian et al. 2012; Rathnayake et al. 2016a; Semple et al. 2010; Smith et al. 2004; Xue et al. 2016). Epidemiological investigations have suggested a modest effect of endotoxin exposure on morbidity pertaining to asthma (Michel et al. 1996; Salonen et al. 2016). Furthermore, acute exposure of endotoxins to humans can cause blood/lung inflammatory responses (Michel et al. 1992). Towards this, it is worthwhile mentioning that a recent study over eastern part of India (states of Odisha), reported PM_1-bound indoor levels of endotoxin (100–160 EU/m^3; EU: endotoxin units) from biofuel burning (Padhi et al. 2016). They have also reported the estimates on penetration of PM_1-bound endotoxins into alveoli region.

Bioaerosols represent the air suspended particles that are living (Fungi, bacteria and viruses) or have been originated from living organisms (e.g. pollens from plants). Their presence in atmosphere is plausibly a function of dispersal from a site of colonization, survival and/or growth. The health effects of bioaerosols include allergies, infectious diseases and acute toxic effects. Furthermore, cancer in conjunction with the threat of bioterrorism and SARS (severe acute respiratory syndrome) has increased public awareness on the importance of study on bioaerosols. There are numerous technical methods for sampling bioaerosols and can be employed depending on the sensitivity of the method and the concentration of micro-organisms. There have been difficulties and challenges in standardization of sampling methods. The major problems include establishment of a causal relationship arising due to complex composition of bioaerosols and variation in human response as a function of exposure. It has been a widely followed activity to monitor bioaerosols in various micro-environments for epidemiological investigations of infectious diseases. The research on airborne micro-organism's abundance, spread and control represents as a quality control measure on monitoring bio-hazardous and relevance to their impact on climate (as CCN: cloud condensation nuclei and IN: ice nuclei). In many developing countries including India, there is a very little awareness on the indoor air quality, contamination of mould and potential factors for transmission of infections (ranging from mild influenza to deadly tuberculosis).

There is an urgent need to assess indoor air quality, develop tools to accurately measure these bioaerosols, cater techniques to quickly identify them and explore link to infections, and finally to make appropriate strategy to mitigate and control these especially in the sensitive indoor environments (hospitals, schools, shopping malls etc.). The present synthesis involves extensive literature review to screen through available techniques for bioaerosol monitoring, its adaptation and modification for Indian condition followed by its use in different indoor and outdoor micro-environments to investigate the types and concentration levels of viable (living) bioaerosols.

5.3 State of the Art and Progress Envisaged

Details on the state of the art and envisioned progress relevant to the bioaerosols sampling, analysis and health effects are described below. It also points out the technological baseline on different concerned aspects, for example technologies that have been brought into operation over a period of time to collect and study the bioaerosols. Finally, the ongoing research aims to advance the state of the knowledge on bioaerosols.

(A) **Health effects of bioaerosols**

Exposure to high concentrations (or unfamiliar forms) of bioaerosols could lead to biological hazards to humans. The three broad classes of diseases associated with the bioaerosols exposure are respiratory diseases, infectious diseases and cancer. It is worthwhile mentioning that current knowledge on the risk to cancer from exposure to biological agents is limited.

5.3.1 Allergies

Various kinds of bioaerosols are responsible for irritation and allergic responses and few common ones include pollens like ragweed, insect or their body parts like dust mites, cockroach body parts. The major concern among these allergens is on the pollens. Pollen grains represent as the male gametophyte in sexual reproduction of flowering plants (angiosperms) and conifers (gymnosperms). Pollination refers to a process of transfer of pollen grains from male to reproductive structure of female. It can be accomplished via three routes: vectors-wind, water or animals. The flowers of these plants often do not have petals, and thus, anthers (pollen sacs) are directly exposed to air. Hence, pollens from anemophilous plants are most abundant in ambient atmosphere and have tremendous influence in terms of human exposure and related seasonal allergies. Pollen grains are nearly spherical in shape, at least under hydrated condition, with a rigid cell wall structure formed up of a complex polysaccharide substance known as sporopollenin.

5.3.2 Infections

Infectious diseases basically arise from bacteria, fungi, viruses, protozoa and helminthes. It involves transmission of infectious agents in air from a reservoir to a susceptible host.

(i) **Bacteria-induced diseases**: Various diseases due to bacteria such as tuberculosis and legionellosis are of more concern as far as public health is concerned owing to their pretty fast response against low infectious dose. Most important ones have been discussed as follows.

(a) **Anthrax:** The transmission occurs due to inhalation of the *Bacillus anthracis* spores and the outbreaks are usually linked with occupational exposure-based bioterrorism.
(b) **Illness by endotoxin:** Endotoxin is the lipopolysaccharide (toxin) component of Gram-negative bacterial cell wall. These are potent pyrogens, capable of causing fever at very low concentrations. High exposure to endotoxins could lead to nausea and diarrhoea.
(c) **Legionellosis:** *Legionella pneumophila* causes human legionellosis and nosocomial pneumonia in adults following occupational/non-occupational exposures. Certain active aerosolize processes, for example aeration of contaminated waters, cause legionellae to become airborne. There occurrence has been noticed in various water environments including man-designed water systems, biofilms in the cooling towers, A/C (air conditioners).
(d) **Tuberculosis:** The transmission of tubercle bacilli occurs upon inhalation of aerosolized bacilli in droplet of expectorated sputum-positive from tuberculosis patients during ejection of cough, sneezing and occasionally while talking. There have been outbreaks of multi-drug resistant tuberculosis in the UK that have highlighted the potential for transmission within the hospital environment.

(ii) **Fungal diseases:** Fungi are saprophytic parasitic organisms that occupy a kingdom of their own. They can cause the aerobic decay of plant materials and are ubiquitous in air, often as the predominant component of bioaerosols. The fungal cell is eukaryotic and embodies a well-developed membrane system. The rigid cell wall is made up of β-glucans and acetyl glucosamine polymers (chitin). Their wall often contains waxes and has extra cellular polysaccharides coating. Fungi can be identified by the method and nature of spore production. Lichens can be formed from the fungi and algae following a symbiotic relationship. Most of the spores and pollens have density near to unity, and thus, their aerodynamic diameter primarily depends on its shape as well as its size. It is worthwhile mentioning that many spores are hygroscopic, and therefore, their aerodynamic diameter increases as a function of increasing humidity. Mycotoxins are secondary metabolites of some fungal species such as *Fusarium, Aspergillus, Penicillium and Trichoderma.* Fungi causing respiratory infections and allergenic reactions to humans include *Penicillium, Cladosporium, Acremonium, Paecilomyces, Aspergillus and Mucor*. Most infections (commonest being *Aspergillosis*) occur in immune-compromised hosts upon inhalation of fungal spores or toxins produced by them. Symptoms include watering eyes, persistent cold, prolonged muscular cramp and joints pain. *Histoplasma, Coccidioides* and *Blastomyces* grow in soil or may be carried by bats and birds and are linked with the exposure to airborne/animal-borne contamination. The volatile organic products released from fungal metabolism have ability to induce irritation in eyes and in the upper respiratory tract. *Aspergillus* species that can grow in indoor environments include *Aspergillus fumigatus* and *Aspergillus ftavus* and can cause nosocomial infections, allergic broncho-pulmonary aspergillosis (ABPA) and sinusitis.

Mycotoxins-induced illness: Mycotoxins can get absorbed on the intestinal lining, airways and skin. *Aspergillus*, *Fusarium* and *Stachybotrys* act as aero-allergens and also act as a source for mycotoxins. A case report from the USA described upper respiratory tract irritation and rash in a family living in a Chicago home with a heavy growth of *Stachybotrys atra* producing trichothecene mycotoxins. The chemical structure of trichothecene is shown in Fig. 5.2.

The symptoms diminish as a function of substantial reduction in the amount of mould. Other adverse health effects include pre-term births or late abortions in farm women exposed to mycotoxins with immunotoxic and hormone-like effects.

(iii) **Cancer:** It is widely believed that occupational carcinogens of biological nature are the mycotoxins. Aflatoxin from *Aspergillus flavus* is capable of causing liver cancer whereas Ochratoxin Asia is a plausible human carcinogen. Exposure to aflatoxin and ochratoxin usually occurs by the ingestion. However, it can also occur by inhalation in industries of peanut processing, livestock feed processing or when grain dust exposure occurs. Studies have found association between exposure to wood dust and cancer in specific part of body. For example, sinonasal cancer has been found to be prominent in the people working on furniture making or doing wood-related jobs like in sawmills.

(B) **Factors affecting the fate and transport of bioaerosols**

Transport and the ultimate settling of bioaerosols are influenced by its physical properties and the environmental parameters that it encounters. Important physical characteristics include shape, density and size of droplets/particles, whereas the environmental factors include relative humidity, temperature and magnitude of air currents. These parameters basically determine the capacity of bioaerosols to remain airborne. Bioaerosols produced from liquid suspensions (or bursting of bubbles) undergo desiccation, whereas those generated during particle emanation may partially rehydrate under ambient atmospheric conditions. The presence of

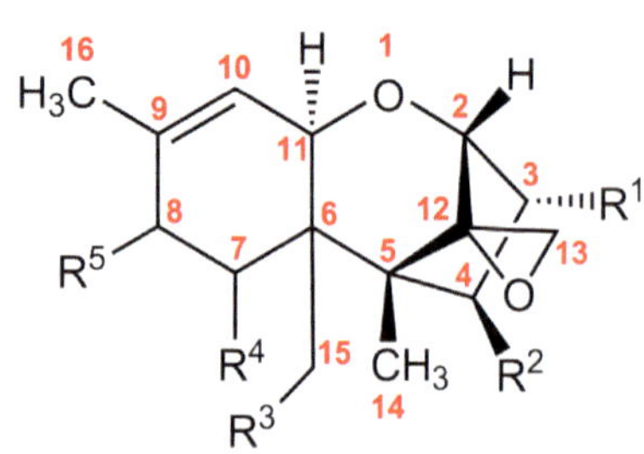

Fig. 5.2 Chemical structure of trichothecene (Appell and Bosma 2015; Chauhan 2016)

moulds refers to a problem with high humidity or water penetration. Typical concentrations of viable bacteria and fungi from air systems and indoor surfaces are given in Table 5.1.

(C) **Sampling methods of bioaerosols**

There are different techniques for collecting bioaerosols. Major characteristics of bioaerosols sampling techniques have been enumerated as follows:

I. **Gravitational samplers (e.g. settle plates)**

- Collection medium includes: coated microscopic slides, agar medium plates etc.
- Particles collection by passive sampler (non-volumetric) is based on gravity settling principle.
- Collection efficiency of particles under turbulent air flow is poor.
- Can suffer with particles overloading particularly for larger particles.

II. **Inertial bioaerosols samplers**

- Facilitates size-segregated particles sampling.
- Relies on inertia that helps particles to deviate from airflow streamlines.
- It includes impactor, sieves and stacked sieves.

III. **Spore traps**

- Firstly designed for collection of pollens and fungal spores. For example, Air-o-cell, Burkhard, Hirst and Allergenco.
- Particles are basically allowed to impact onto coated glass slide/adhesive surface.
- Air sampling is preferred at low volume (10–20 L/min).
- Direct analysis after sample collection is feasible.
- This approach of sampling can mask some of the species, and viability tests are not possible.

IV. **Impaction-based samplers (Impactors)**

- Particles are allowed to impact on slide/agar plates.
- Used at air flows of 10–30 L/minute.

Table 5.1 Typical concentrations of bioaerosols in air conditioning systems and on indoor surfaces

Category	Activity type	Bacteria	Fungi
		(CFU/m^3)	
Air conditions	HVAC	10–10^4	10–10^7
Indoor surfaces	Ceilings and walls	10–10^3	10–10^4
	Carpet	10^3–10^6	10^2–10^5
	House plants	10–10^4	10^2–10^5
	Operating room	10–10^2	10–10^2

HVAC Heating, ventilation and air conditioning

- Bouncing effect of smaller particles can be an issue.
- Different types of impactors are as follows:

  (a) Single or multistage samplers (e.g. Anderson).
  (b) Sieve and stacked sieve samplers (e.g. SAS).
  (c) Rotary arm samplers (e.g. Rotorod, Mesosystems BT550).
  (d) Agar samplers.

The impaction mechanism of aerosols and inertial bioaerosol sampler is shown in Fig. 5.3.

V. **Impingers**

- Particles in the air are removed by impingement when air is allowed to pass through liquid (e.g. water and oil; Fig. 5.4).
- Operated at a flow rate of 0.1–15 L/min (e.g. 12.5 LPM for AGI 30).
- It allows for the dilution.
- There are several challenges and issues involved: particle bounce, pass through, bubbling of liquid and loss of viability features.
- Collection efficiency decreases with particles size (decreases significantly for the particles >10 μm).

(D) **Analytical techniques for qualitative and quantitative determination of bioaerosols**

(i) ***Microscopy***

It is simply described as the use of microscope to magnify an object. It was invented by *Antonie van Leeuwenhoek*. In aeroallergen studies, suitable magnification can vary from low (e.g. 3.5X), for viewing fungal fruiting bodies with a stereo microscope, to very high (e.g. 10,000X), for assessing surface of pollen grain

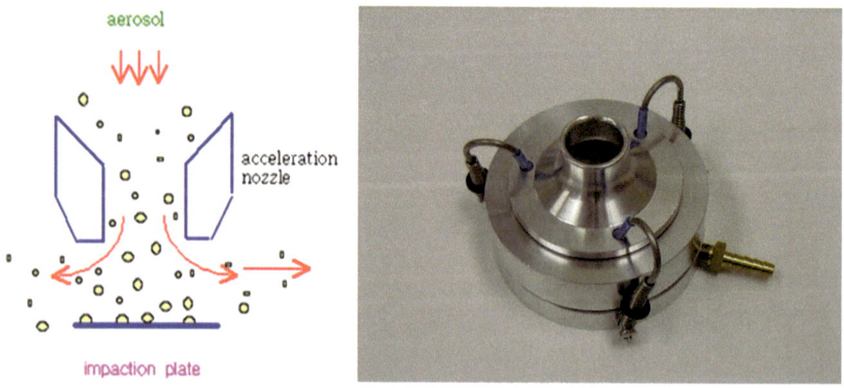

Fig. 5.3 Aerosol impaction mechanism (shown on left) and an inertial-based bioaerosol sampler (on right, MSP Inc., USA; on right)

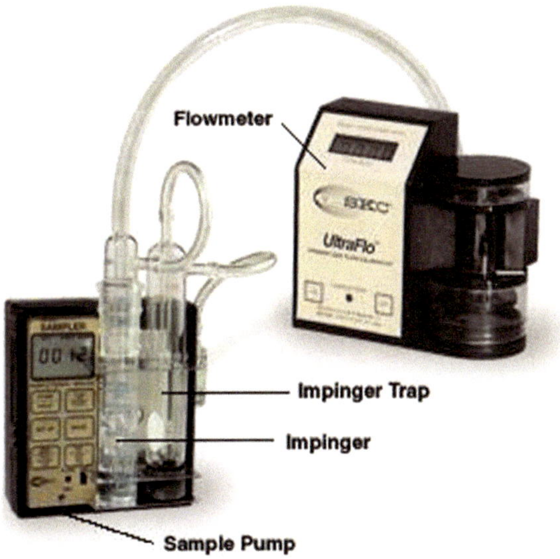

Fig. 5.4 Liquid-impinger-based bioaerosol sampler (SKC, USA)

from electron microscope. Although commonly used to identify fungal spores and pollens in air samples, microscopy also facilitates identifying fungal colonies in culture and spores associated with dust or other source samples. Most common airborne pollen grains can be studied with an optimal magnification for sample scanning and identification of 300–400X. Counting of many fungal spores and most pollen grains can be accomplished with a microscope having magnification at least a 100X oil immersion as well as 40–10X objectives. Bacteria are rarely recognizable in particulate air samples unless proper staining techniques are used.

Types of Microscopy: bright-field microscopy, light microscopy, dark-field microscopy, fluorescence microscopy, phase-contrast microscopy, electron microscopy: transmission electron microscope (TEM) and scanning electron microscope (SEM). Some of the widely used microscopic techniques in bioaerosols research are shown in Fig. 5.5.

(ii) ***Bioaerosols collection/sampling and culture***

A bioaerosol sampler based on the concept of inertial impactor and specific agar plate as a collection substrate and culture medium has been designed and developed at IIT Kanpur, India (Fig. 5.6). Schematic below depicts a typical experimental set-up which is usually employed for characterizing an inertial-based sampler. Apart from the dry aerosol, we also used PSL (polystyrene latex) particles tagged with active groups to determine and optimize the impactor cut-point. Most of the relevant bioaerosols lie in the range of 1–10 μm. So a logical cut-point will be 10 μm. The calibrated airflow rate of this sampler is 12 LPM.

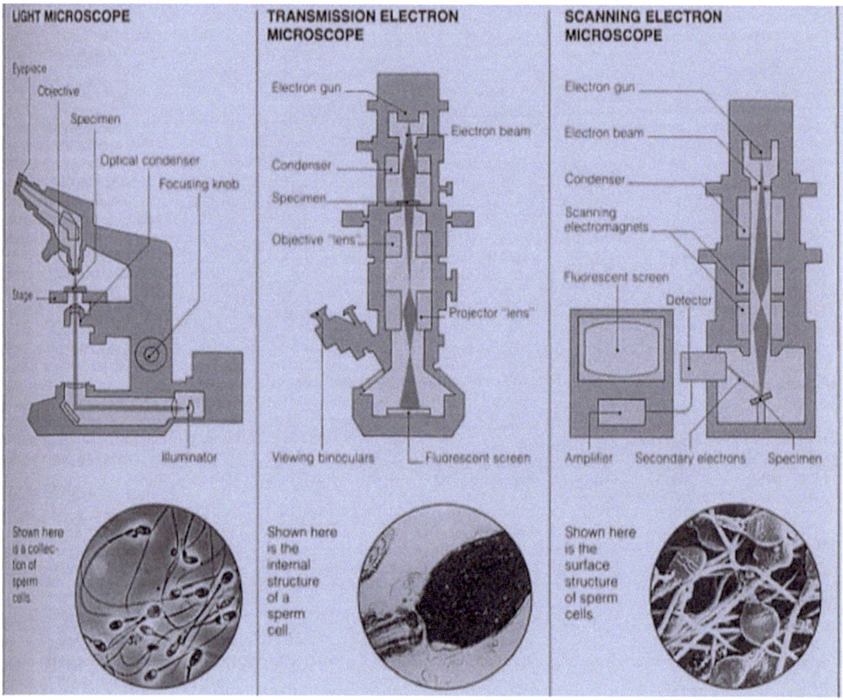

Fig. 5.5 Different types of microscopic techniques used in bioaerosols research (Chauhan 2016)

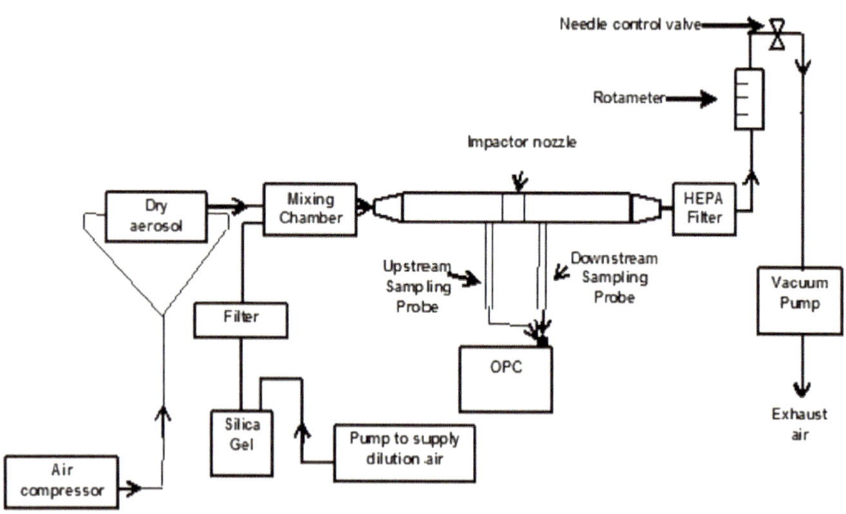

Fig. 5.6 Experimental set-up of bioaerosol impactor sampler characterization (indigenously designed and developed at IIT Kanpur)

For fabrication of the developed bioaerosols sampler, brass has been used due to its advantage in machining, long-term durability, stability and inert characteristics. Chrome plating was further done to shield it with an inert and corrosion-free layer. The sampling device has basically three zones. Top of the sampling device is provided with the rain cover.

Zone 1 Facilitates collection of bigger particles and spores settled due to gravity.

Zone 2 Consists of two stages with cut-off points of PM_{10} and $PM_{0.6}$ achieved at an airflow rate of 12 LPM. The first stage is PM_{10} stage. It consists of a nozzle plate containing four round nozzles and an impaction plate. The impaction plate can accommodate four 35 mm petri dishes in the respective slots below the nozzles. Second stage is $PM_{0.6}$ stage. It consists of one nozzle plate with single round nozzle and one corresponding impaction plate. This impaction plate can house one 35 mm petri dish.

Zone 3 It has provision to house one 47 mm filter. This zone can collect $PM_{0.6}$ particulate matter on filter paper that can be subjected to further gravimetric and chemical analysis.

The sharp cut-off points at 0.6–10 μm make this device suitable for collecting the particles from 0.6 μm (stage 2) to 10 μm (stage 1; Fig. 5.7). Impaction plate at first stage collects the particle with size greater than 10 μm. Particles of size less than 10 μm travel down towards the second stage. The impaction plate on the second stage collects the particles of size 0.6–10 μm.

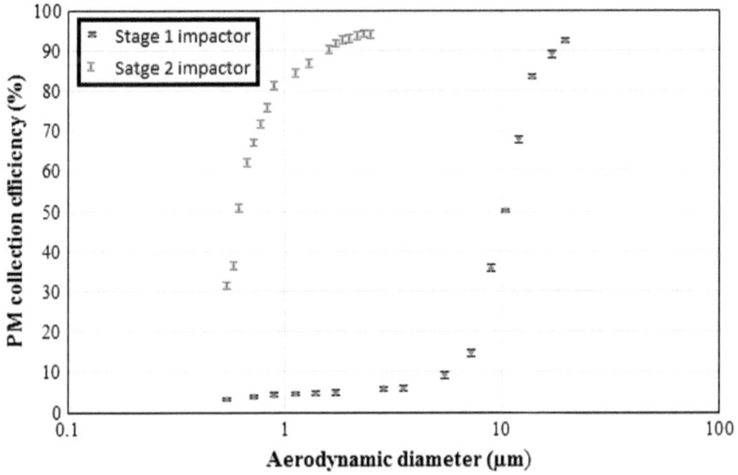

Fig. 5.7 Particulate matter collection efficiency curve of two stage impactor sampler at cut-points of 0.6 (stage 2) and 10 μm (stage 1), adopted from (Gupta and Chauhan 2014a)

(iii) ***Bioassay***

This technique is most commonly used to analyse endotoxin components of the aerosols. Endotoxin is pro-inflammatory substance present in Gram-negative bacteria (GNB). They have been noticed to associate with workplace illnesses and suspected of playing a role in the development of non-specific building-related symptoms, widely referred to as the sick building syndrome (SBS). At times they are also responsible for severity of asthma. Endotoxin is shed off from the outer membrane of GNB as membrane fragments while growing or dying GNB. When endotoxin is purified, it consists of family of proteins called as lipopolysaccharides (LPS). LPS is composed of lipids and carbohydrates; the lipid portion is basically responsible for the toxicity (Fig. 5.8).

Endotoxin is an integral part of the GNB, and therefore, their occurrence is proportional to the occurrence of GNB. Their abundance can be influenced by environmental conditions such as substrate availability, humidity and temperature. Since these conditions are favourable outdoor, GNB and therefore endotoxin are ubiquitous outdoors. High levels have been detected in numerous settings especially where organic dust is present such as in agricultural and related industries. Filter media is usually used to collect endotoxin aerosols because they allow for long collection times and are easy to practically use. At times, all glass impingers

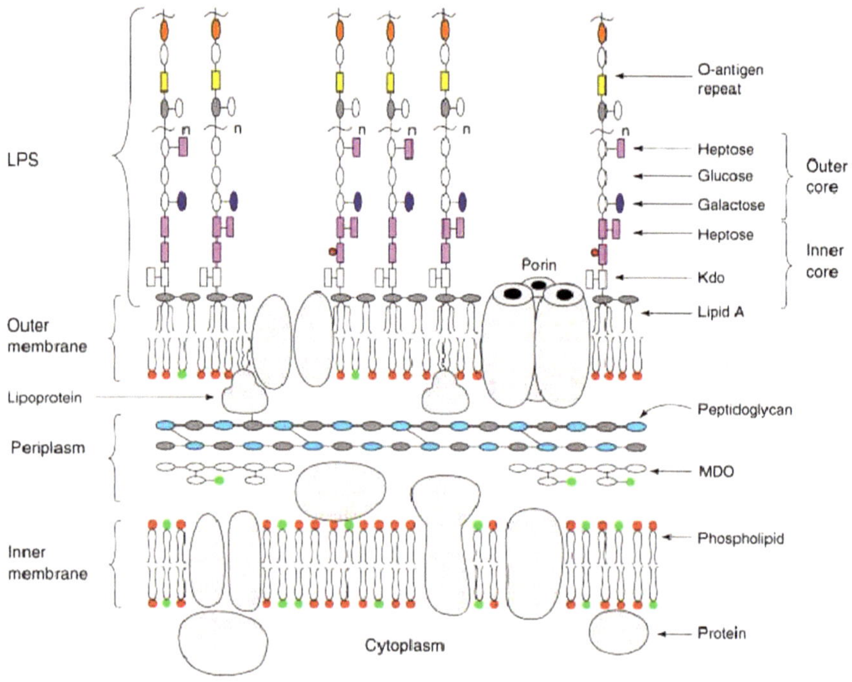

Fig. 5.8 Structure of cell wall of Gram-negative bacteria (Brown et al. 2015; Chauhan 2016)

are used but they may underestimate endotoxin because of their low collection efficiency for sub-micron particles which contain large amount of endotoxin.

The detection of the endotoxin is based on the use of a Limulus Amoebocytes Lysate (LAL) assay performed with the LAL reagent: aqueous extract of circulating amoebocyte of horseshoe crab (Limulus polyphemus). LAL assay is based on the observation of a gel clot formation when an endotoxin comes in contact with clot table protein from circulating amoebocytes of Limulus. For LAL assay, endotoxin is extracted from filters with an aqueous extraction medium. Most laboratories use pathogen-free water, while some use buffers as tris and phosphate triethylamine (pH: 7.5) or dispersing agents such as tween −20, and it was reported that endotoxin activity was seven times higher in tween medium as compared to that in any pathogen-free water.

The most common method of filter extraction is sonication and rocking in an extraction medium. Generally, environmental samples for endotoxin analysis should not be frozen especially after their extraction. Alternatively, the most widely preferred *in vitro* method is the LAL assay owing to high sensitivity. However, this assay does not mimic for the in vivo pyrogenic activity. Following this assay, major and minor pyrogens, endotoxin (C pathway), and 1, 3–glucans β (G pathway) of fungi can be measured. However, this assay gives negative result for the pyrogenic substances from Gram-positive bacteria (GPB). Depending on the chemical and physiochemical structure of the endotoxin, the reactivity of the LAL reagent also differs. The LAL assay is a comparative toxicity bioassay and not an analytical assay. This means the measured endotoxin levels can be altered by factors other than the actual LPS concentration. Due to this very reason, the interpretation of the results becomes difficult especially when comparing observation from different filter media.

Detection technique: LAL reagent is added to the sample (in a pre-cleaned/pyrogen-free test tube). The sample is then incubated at 37 °C temperature for nearly 1 h. The tube is then gently inverted. The formation of a gel or clot confirms the positive result. LAL assay can be applied following the endpoint method or kinetic reaction. A variety of LAL assays are gel clot, chromogenic measurements and turbidimetry. The chromogenic assay depends on a chromogen, which changes its color in the presence of endotoxin (in sample). The chromogen release is a function of the concentration of endotoxin. Optical measurement device operating at wavelength pertaining to chromogen signal is required. LPS can also be detected in environmental samples via chemical methods. These methods are based on assessing chemical markers of LPS. Of these markers, 3-hydroxy fatty acids are most commonly utilized for characterization of endotoxin and GNB. Gas chromatography technique is the most followed method to do this characterization.

5.4 Case Study on Bioaerosols Assessment Over Central Indo-Gangetic Plain

i. *Site description*

The study site at Kanpur (Urban area: 26.30 °N; 80.14 °E; 142 m above mean sea level) is situated in central part of IGP (Chauhan 2016; Gupta and Mandaria 2013; Kumar and Gupta 2015); IGP is stretched from north-west to north-east region in India (Chakraborty et al. 2017; Rajput et al. 2011b, 2013, 2016a; Singh and Gupta 2016b). This region holds ~40% of the south Asia's population and produces over 85% of the rice-wheat (Gupta et al. 2004; Rajput and Sarin 2014; Singh et al. 2014a). Nearly 20 million hectares of agricultural land area is located in NW part of IGP (states of Punjab, Haryana and western part of Uttar Pradesh). Due to crop rotation activity, a conspicuous seasonal and annual feature, farmers burn 100s of million tons of paddy residues (during October–November) and wheat residues (during April–May) (Gupta et al. 2004; Momin et al. 1999; Punia et al. 2008; Rajput et al. 2014a, b, c). Under prevailing NW winds, our sampling site is strategically located downwind of the major agricultural fields in IGP (Rajput et al. 2011a, 2015; Singh and Gupta 2015, 2016a). Thus, the sampling location is influenced by massive biomass burning activities (Kumar et al. 2017; Rajput et al. 2016b). The region experiences usually ~1000 mm annual precipitation with harsh summers and cold winters associated with fog events (Rajput et al. 2016a; Singh et al. 2014b). However, year 2015 (annual rainfall of 375 mm) was influenced due to El Niño (Rajeev et al. 2016).

ii. *Sampling and measurements*

Measurements of viable bioaerosols (n = 130) and ambient particle number concentrations have been performed for 1 year (12:30–1:30 h) from June 2015–May 2016 at CESE building (Center for Environmental Sciences and Engineering) in the campus of Indian Institute of Technology Kanpur. Using a single-stage impactor (aerodynamic diameter >0.6 μm) sampler (flow rate: 12 LPM) (Gupta and Chauhan 2014b), viable bioaerosols were collected at ~1.5 m from ground level and cultured in petri dishes equipped with specific nutrient agar mediums. We have collected (three days a week) and cultured GPB in Mannitol Salt Agar Broth (MSAB), GNB in MacConkey and Fungi in Sabouraud Dextrose Agar medium. Briefly, petri dishes (n = 3 for each day sampling) equipped with specific nutrient agar mediums were placed in the air sampler and the collection was subjected to 4 min for each three types of mediums sequentially. Soon after the collection, these bioaerosols in separate petri dishes were incubated at 35 °C for microbial culture in our lab (Atmospheric Particle Technology Lab, in CESE). Subsequently, their enumeration (counting) through magnifying lens was performed into a bio-safety cabinet. Well established protocol has been followed to proper sterilize the sampler as well as nutrient media every time prior to sampling. The observations of colony counting for GPB, GNB and Fungi for different incubation periods are shown in

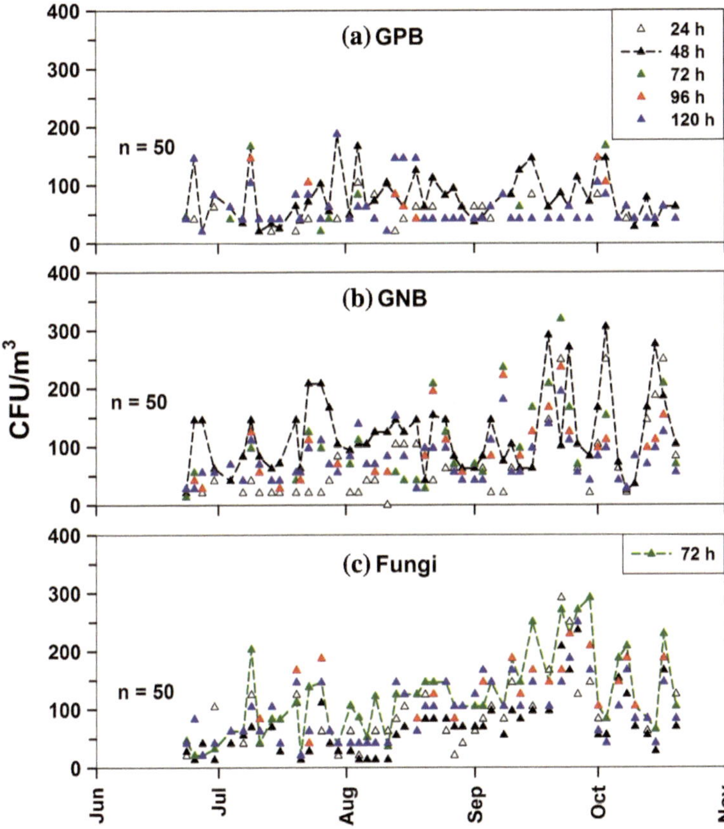

Fig. 5.9 Experiment on optimum incubation period for bioaerosols culture (48 h for both GPB and GNB, whereas about 72 h is required for Fungi growth and counting) (Rajput et al. 2017)

Fig. 5.9. The results shown in Fig. 5.6 suggest that 48 h of incubation is the optimum culture period for GPB and GNB, whereas 72 h is optimum period for Fungi culture.

The images of specific agar mediums prior and post to bioaerosols sampling and incubation are shown in Fig. 5.10.

Furthermore, we have collected PM_{10} (particles with aerodynamic diameter ≤ 10 μm) samples (n = 130) using a high-volume air sampler (Envirotech, India, flow rate: ~1 m^3/min), onto combusted quartz filter substrate for 30 min covering bioaerosols sampling and PNC measurements. Owing to low loading of aerosols onto filter substrate in 30 min sampling-time, gravimetric determination of PM_{10} mass was not carried out. However, our main purpose of collecting PM_{10} samples was to determine the concentrations of organic carbon (OC). OC

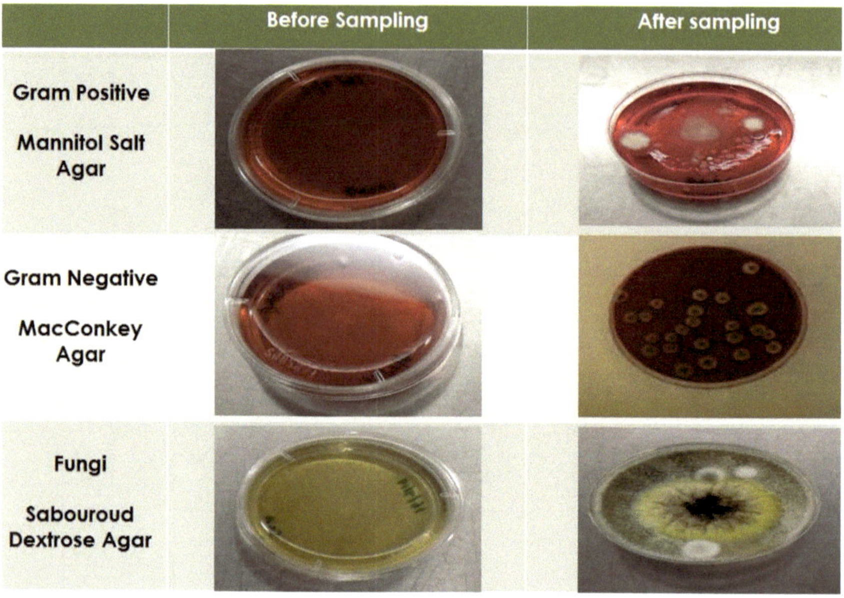

Fig. 5.10 Morphology of GPB, GNB and Fungi colonies in air samples from Kanpur location (IGP)

concentration (n = 130) in each sample has been measured on EC-OC analyser (Sunset lab) with NIOSH (National Institute for Occupational Safety and Health) protocol (Birch and Cary 1996).

5.5 Results and Discussion

i. *Meteorological parameters*

Relevant meteorological parameters including temperature (T), relatively humidity (RH), wind speed and rainfall during the study period (~1300 h) are shown in Fig. 5.11.

It is important to mention here that wind direction assessed from air-mass back trajectories (AMBTs; Fig. 5.12) shows their most plausible origin in conjunction with overall transport of air-mass, whereas the winds measured at a site represent the striking intensity (wind speed) and direction.

ii. *Seasonal variability of viable bioaerosols colonies*

The data set of OC and bioaerosols is given in Table 5.2.

Total viable bioaerosols (Σviable bioaerosols = GPB + GNB + Fungi) concentration averages at 312 ± 118 CFU/m^3 in monsoon, 421 ± 114 CFU/m^3 in

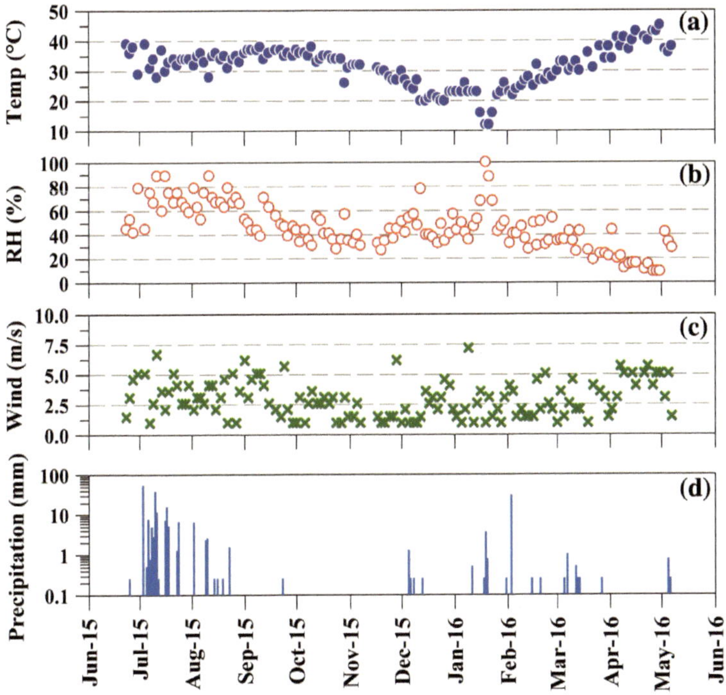

Fig. 5.11 Temporal variability record of meteorological parameters: **a** ambient temperature, **b** relative humidity, **c** wind speed and **d** daily rainfall

post-monsoon, 486 ± 141 CFU/m^3 in winter and 223 ± 61 CFU/m^3 in pre-monsoon season, at Kanpur in IGP (Fig. 5.13). Thus, maximum concentration of Σviable bioaerosols was observed during wintertime followed by post-monsoon, monsoon and pre-monsoon. We reiterate that in wintertime, emissions from fossil-fuel combustion and biofuel burning in conjunction with low temperature and shallower boundary layer height are vital parameters co-governing the atmospheric concentrations of PM and viable bioaerosols.

Furthermore, we have also assessed the relative contribution of GPB, GNB and Fungi of the Σviable bioaerosols (Fig. 5.14). Accordingly, GPB has highest fraction of 32% in wintertime. GNB has highest fraction of 42% during the post-monsoon. This is also reflected in GPB/GNB ratio: 0.85 ± 0.55 in monsoon, 0.73 ± 0.44 in post-monsoon, 0.82 ± 0.17 in wintertime and 0.84 ± 0.37 in pre-monsoon. Summing up, GPB/GNB average ratio is >0.80 in all seasons, exception being the post-monsoon period wherein this ratio averages at 0.73. Relatively lower ratio of GPB/GNB further revisits the observation that post-harvest PRB emissions are associated with elevated concentrations of GNB. However, Fungi have higher fractions during monsoon (37%) and pre-monsoon (39%) with maximum concentration of 292 CFU/m^3 during the monsoon (Fig. 5.13). Bacteria/Fungi average

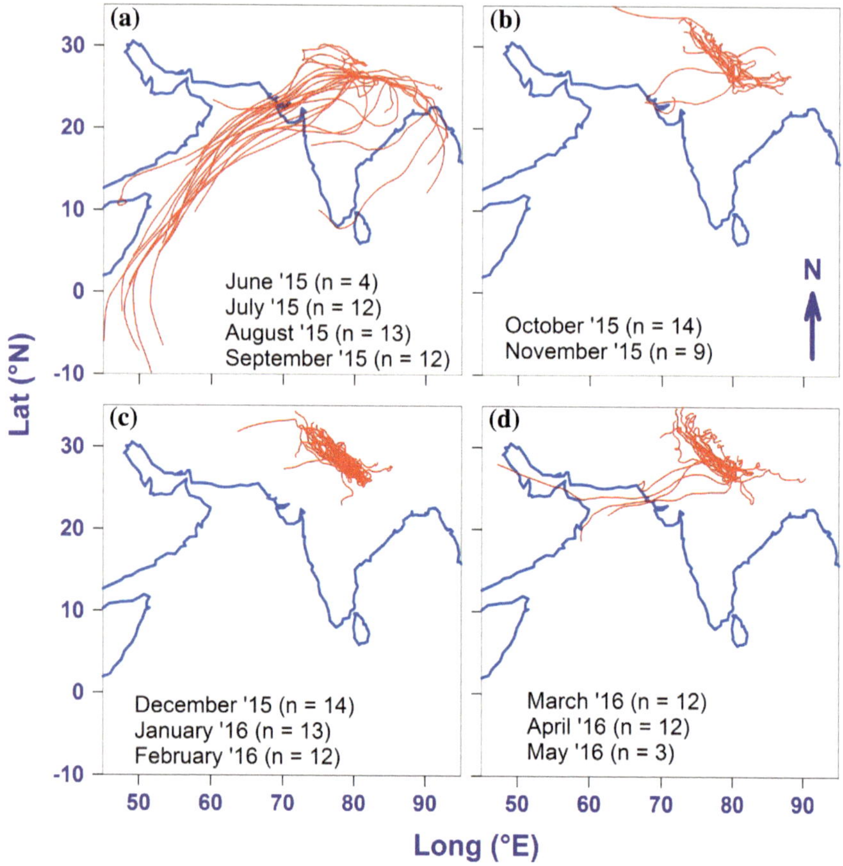

Fig. 5.12 Hysplit-based air-mass back trajectories (7 d; 100 m above ground level) during study period (n = 130) in: **a** monsoon (June–September 2015), **b** post-monsoon (October–November 2015), **c** winter (December 2015–February 2016) and **d** pre-monsoon (March–May 2016)

ratio is ≥ 2.5 from monsoon through wintertime, whereas it decreases to 1.7 during the pre-monsoon (Table 5.2). Dry weather condition (low RH and high temperature) prevailing in pre-monsoon is attributable to lower abundance of bacteria.

5.5.1 *Correlation Analyses of Viable Bioaerosols (GPB, GNB and Fungi) with Organic Carbon (OC) and Meteorological Parameters*

Assessing inter-relationship of bioaerosols with prevailing meteorology is very important to understand the feedback between ecosystem and meteorology (Jones

Table 5.2 Atmospheric concentrations of assessed species [min–max (Avg. ± 1σ)] during different seasons in Indo-Gangetic Plain (IGP; at Kanpur)

Parameters	Monsoon	Post-monsoon	Winter	Pre-monsoon
	(June–September)	(October–November)	(December–February)	(March–May)
	(*n* = 41)	(*n* = 23)	(*n* = 39)	(*n* = 27)
[a]OC	0.3–42.1 (21.2 ± 8.4)	11.8–47.6 (29.6 ± 10.3)	18.9–51.8 (34.7 ± 8.5)	9.6–36.2 (20.5 ± 5.4)
[b]GPB	21–188 (80 ± 41)	28–166 (112 ± 44)	63–272 (157 ± 57)	21–125 (61 ± 25)
[b]GNB	21–292 (114 ± 58)	25–352 (186 ± 87)	63–325 (199 ± 76)	42–146 (77 ± 26)
[b]Fungi	21–292 (118 ± 71)	65–229 (122 ± 43)	63–242 (130 ± 35)	42–146 (85 ± 25)

[a]particle mass concentration (μg/m^3); [b]bioaerosols abundance in colony forming units (CFU/m^3)

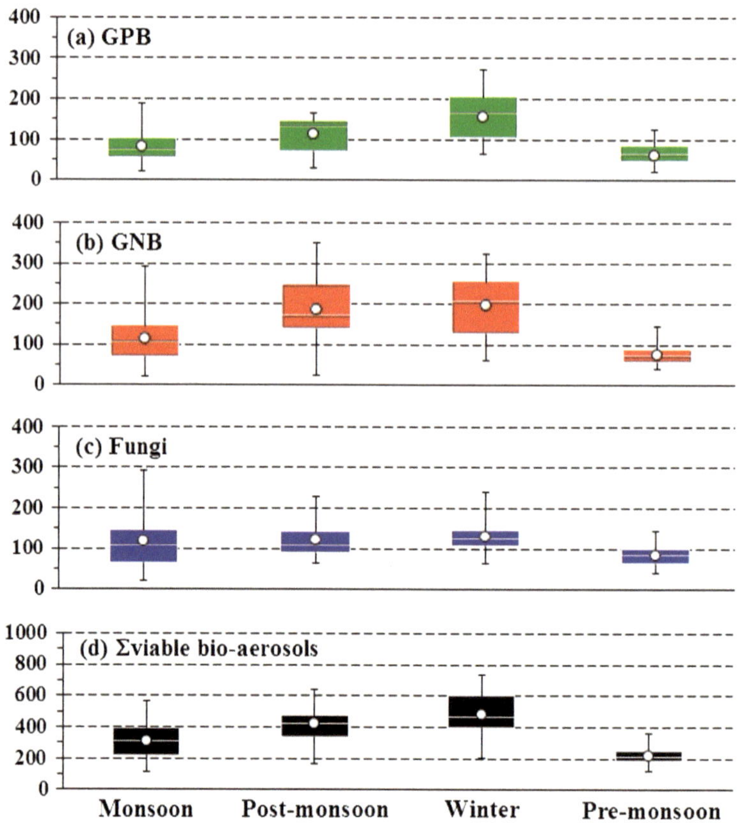

Fig. 5.13 Seasonal variability of: **a** Gram-positive bacteria, **b** Gram-negative bacteria, **c** Fungi and **d** all viable bioaerosols (= GPB + GNB + Fungi) in IGP (at Kanpur)(Rajput et al. 2017)

and Harrison 2004; Murata and Zhang 2016; Rathnayake et al. 2016b; Ziemba et al. 2016). In order to identify potential predictors (factors) influencing the viable bioaerosols concentration (GPB, GNB and Fungi) over IGP, a correlation matrix based on linear regression analysis has been generated utilizing concentrations of these bioaerosols and OC along with several meteorological parameters [temperature (T), relative humidity (%RH), wind speed (WS) and daily rainfall; Table 5.3]. The uncertainty level on interpreting results on correlation analysis is less than 5% ($p < 0.05$). It is evident from the figure that GPB exhibits a significant positive linear relation with GNB ($r = 0.68$; $n = 130$), suggesting plausibility on their co-genetic sources and/or their viability under identical ambient conditions (Table 5.3). A positive linear correlation also exists between GPB and Fungi, but not to a significant level. GPB correlates positively with OC concentrations ($r = 0.60$) and negatively with ambient T ($r = -0.66$). The other meteorological parameters viz. RH, wind speed and rainfall do not show any trend on a significant level. Interestingly, close observations of influence of rainfall on ambient levels of

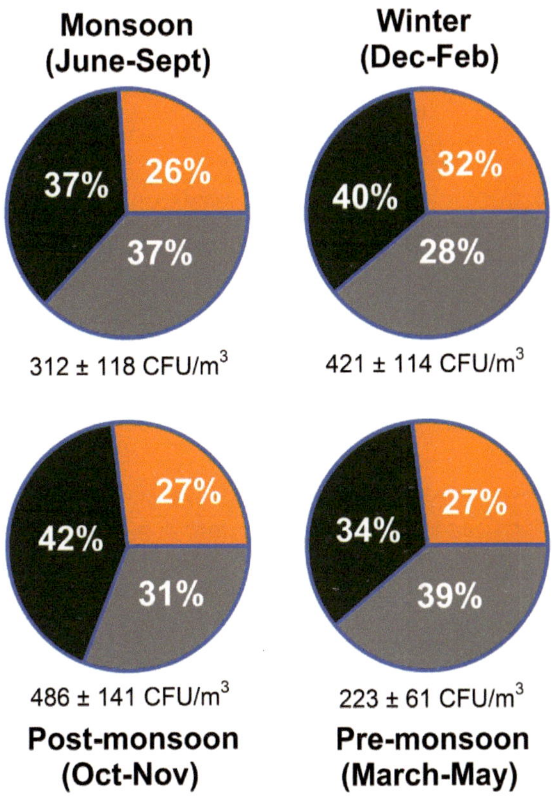

Fig. 5.14 Seasonal scenario on fractional contribution of: GPB (orange color), GNB (dark green) and Fungi (grey) colonies in IGP (at Kanpur). Total concentration of viable bioaerosols colonies are also mentioned (Avg. ± SD) (Rajput et al. 2017)

GPB reveals that rainfall greater than 4 mm relates to lowering of GPB concentration. In contrast, daily rainfall ≈ 1–4 mm appears to positively correlate with GPB concentrations (Table 5.3). Similar to GPB, GNB also exhibits a positive trend with Fungi, but not to a significant level. Likewise, GNB correlates positively with OC concentrations ($r = 0.87$) and negatively with ambient T ($r = -0.64$). Other observations through linear correlation analysis of GNB with meteorological parameters viz. RH, wind speed and rainfall are near similar to that discussed above for GPB, and hence not repeated again. Fungi colony concentrations do not correlate significantly with any herein reported parameter (Table 5.3). Influence of daily rainfall ≈ 1–4 mm appears to positively correlate with Fungi concentrations; higher precipitation decreases levels of Fungi along with other atmospheric pollutants. Massive emissions from open field crop residue and biofuel burning in the IGP have been documented previously (Gustafsson et al. 2009; Momin et al. 1999; Rajput et al. 2014b, 2016c; Ram et al. 2010; Venkataraman et al. 2005). As aforementioned, the air-masses during the post-monsoon and winter season exhibit influence of transport from upwind IGP. Temporal variations in FFA and viable bioaerosols abundance with a positive linear relationship among GPB, GNB and OC suggest their co-genetic source and/or common atmospheric transport process.

Table 5.3 Correlation analysis of viable bioaerosols (GPB, GNB and Fungi) with organic carbon (OC) and meteorological parameters (T, RH, daily rainfall and wind speed: WS)

	G+	*G−*	*Fungi*	*OC*	$\mathbf{K^+_{BB}}$	*Temp*	*RH%*	*WS*	*Rainfall*
G+	1.00								
G−	**0.68**	1.00							
Fungi	0.24	0.26	1.00						
OC	**0.60**	**0.87**	0.29	1.00					
$\mathbf{K^+_{BB}}$	**0.52**	**0.72**	0.26	**0.87**	1.00				
Temp	**−0.70**	**−0.64**	−0.23	**−0.63**	−0.48	1.00			
RH%	0.20	0.20	0.11	0.14	0.08	**−0.50**	1.00		
WS	−0.24	−0.24	−0.23	−0.21	−0.08	0.25	−0.15	1.00	
Rainfall	−0.29	−0.10	−0.07	−0.14	−0.22	−0.06	0.56	0.08	1.00

Significant correlation ($p < 0.05$) is highlighted

These inter-relationships highlight the significant role of biomass burning emissions (PRB and biofuels) in contributing to bioaerosols.

5.6 Conclusions

A year-long measurement of viable bioaerosols has been conducted from central IGP to assess their abundance and temporal/seasonal variability (Rajput et al. 2017). As far as viable bioaerosols are concerned, the highest concentration of GPB was recorded during December–January (Avg.: 189 CFU/m^3), GNB during November (Avg.: 244 CFU/m^3) and Fungi in September (Avg.: 188 CFU/m^3). A significant positive linear correlation-ship for GPB and GNB with OC ($p < 0.05$) in conjunction with their variability pattern indicates their co-genetic source. Fungi also exhibit a positive trend with OC, but not to a significant level ($p > 0.05$). We have attributed major source of bioaerosols to be associated with massive emissions from PRB and biofuel burning in the IGP. Ambient temperature shows a negative impact on the abundance of GPB and GNB, whereas RH and wind speed do not exhibit any pronounced effects. Low precipitation (1–4 mm) relates to higher concentrations of bioaerosols particularly the Fungi. This study provides field-based data on bioaerosols and ambient particulate matter with a consideration of their potential role in influencing climate and human health.

Acknowledgements Authors thank internal funding support received from IIT Kanpur to conduct this study. PR is thankful to Council of Scientific and Industrial Research (India) for providing CSIR-SRA Fellowship. We thank anonymous reviewer for providing fruitful comments and suggestions that helped in improving the content of this chapter.

References

Annadotter H, Cronberg G, Nystrand R, Rylander R (2005) Endotoxins from cyanobacteria and gram-negative bacteria as the cause of an acute influenza-like reaction after inhalation of aerosols. EcoHealth 2:209–221

Ansari TU, Valsan AE, Ojha N, Ravikrishna R, Narasimhan B, Gunthe SS (2015) Model simulations of fungal spore distribution over the Indian region. Atmos Environ 122:552–560

Appell M, Bosma WB (2015) Assessment of the electronic structure and properties of trichothecene toxins using density functional theory. J Hazard Mater 288:113–123

Balasubramanian R, Nainar P, Rajasekar A (2012) Airborne bacteria, fungi, and endotoxin levels in residential microenvironments: a case study. Aerobiologia 28:375–390

Bauer H, Giebl H, Hitzenberger R, Kasper-Giebl A, Reischl G, Zibuschka F, Puxbaum H (2003) Airborne bacteria as cloud condensation nuclei. J Geophys Res: Atmos 108:4658

Birch ME, Cary RA (1996) Elemental Carbon-Based Method for Monitoring Occupational Exposures to Particulate Diesel Exhaust. Aerosol Sci Technol 25:221–241

Brown L, Wolf JM, Prados-Rosales R, Casadevall A (2015) Through the wall: extracellular vesicles in Gram-positive bacteria, mycobacteria and fungi. Nat Rev Microbiol 13:620–630

Chakraborty A, Rajeev P, Rajput P, Gupta T (2017) Water soluble organic aerosols in indo gangetic plain (IGP): insights from aerosol mass spectrometry. Sci Total Environ 599:1573–1582

Chauhan AS (2016) Year long study to understand the role and significance of bioaerosols in defining respiratory health of community residing in a wellplanned academic campus and nearby. (Ph.D. Thesis, Awarded from Indian Institute of Technology Kanpur)

Clark CS, Rylander R, Larsson L (1983) Levels of gram-negative bacteria, Aspergillus fumigatus, dust, and endotoxin at compost plants. Appl Environ Microbiol 45:1501–1505

Després VR, Huffman JA, Burrows SM, Hoose C, Safatov AS, Buryak G, Fröhlich-Nowoisky J, Elbert W, Andreae MO, Pöschl U, Jaenicke R (2012) Primary biological aerosol particles in the atmosphere: a review. Tellus-B64

Domanska A, Stroszejn-Mrowca G (1994) Endotoxin in the occupational environment of bakers: method of detection. Int J Occup Med Environ Health 7:125–134

Donaldson AI, Alexandersen S, Sorensen JH, Mikkelsen T (2001) Relative risks of the uncontrollable (airborne) spread of FMD by different species. Vet Rec 148:602–604

Fišar Z, Hýsek J, Binek B (1990) Quantification of airborne microorganisms and investigation of their interactions with non-living particles. Int J Biometeorol 34:189–193

Fröhlich-Nowoisky J, Pickersgill DA, Després VR, Pöschl U (2009) High diversity of fungi in air particulate matter. Proc Natl Acad Sci USA 106:12814–12819

Fröhlich-Nowoisky J, Kampf CJ, Weber B, Huffman JA, Pöhlker C, Andreae MO, Lang-Yona N, Burrows SM, Gunthe SS, Elbert W, Su H, Hoor P, Thines E, Hoffmann T, Després VR, Pöschl U (2016) Bioaerosols in the Earth system: Climate, health, and ecosystem interactions. Atmos Res 182:346–376

Fu P, Kawamura K, Okuzawa K, Aggarwal SG, Wang G, Kanaya Y, Wang Z (2008) Organic molecular compositions and temporal variations of summertime mountain aerosols over Mt. Tai, North China Plain. J Geophys Res: Atmos 113:D19107

Garcia E, Hill TCJ, Prenni AJ, DeMott PJ, Franc GD, Kreidenweis SM (2012) Biogenic ice nuclei in boundary layer air over two U.S. High Plains agricultural regions. J Geophys Res: Atmos 117:D18209

GBD (2015) Institute for Health Metrics and Evaluation (IHME), GBD Compare, Seattle, WA: IHME, University of Washington, 2015 (Accessed [30-09-2016])

Graham B, Guyon P, Maenhaut W, Taylor PE, Ebert M, Matthias-Maser S, Mayol-Bracero OL, Godoi RHM, Artaxo P, Meixner FX, Moura MAL, Rocha CHEDA, Grieken RV, Glovsky MM, Flagan RC, Andreae MO (2003a) Composition and diurnal variability of the natural Amazonian aerosol. J Geophys Res: Atmos 108:4765

Graham B, Guyon P, Taylor PE, Artaxo P, Maenhaut W, Glovsky MM, Flagan RC, Andreae MO (2003b) Organic compounds present in the natural Amazonian aerosol: characterization by gas chromatography–mass spectrometry. J Geophys Res: Atmos 108:4766

Gregory PH (1973) The microbiology of the atmosphere. 2nd ed. Leonard Hall

Gupta T, Chauhan A (2014a) Comprehensive air sampling device application for indian patent. 1474/DEL/2014 filed on 03.06.2014

Gupta T, Chauhan AS (2014b) comprehensive air sampling device application for indian patent. 1474/DEL/2014 filed on 03.06.2014

Gupta T, Mandaria A (2013) Sources of submicron aerosol during fog dominated wintertime at kanpur. Environ Sci Pollut Res. https://doi.org/10.1007/s11356-11013-11580-11356

Gupta PK, Sahai S, Singh N, Dixit CK, Singh DP, Sharma C, Tiwari MK, Gupta RK, Garg SC (2004) Residue burning in rice-wheat cropping system: causes and implications. Curr Sci 87:1713–1717

Gustafsson Ö, Kruså M, Zencak Z, Sheesley RJ, Granat L, Engström E, Praveen PS, Rao PSP, Leck C, Rodhe H (2009) brown clouds over south asia: biomass or fossil fuel combustion? Science 323:495–498

Hawkes CV, Kivlin SN, Rocca JD, Huguet V, Thomsen MA, Suttle KB (2011) Fungal community responses to precipitation. Glob Change Biol 17:1637–1645

Huffman JA, Sinha B, Garland RM, Snee-Pollmann A, Gunthe SS, Artaxo P, Martin ST, Andreae MO, Pöschl U (2012) Size distributions and temporal variations of biological aerosol particles in the Amazon rainforest characterized by microscopy and real-time UV-APS fluorescence techniques during AMAZE-08. Atmos Chem Phys 12:11997–12019

Huffman JA, Prenni AJ, DeMott PJ, Pöhlker C, Mason RH, Robinson NH, Fröhlich-Nowoisky J, Tobo Y, Després VR, Garcia E, Gochis DJ, Harris E, Müller-Germann I, Ruzene C, Schmer B, Sinha B, Day DA, Andreae MO, Jimenez JL, Gallagher M, Kreidenweis SM, Bertram AK, Pöschl U (2013) High concentrations of biological aerosol particles and ice nuclei during and after rain. Atmos Chem Phys 13:6151–6164

Iossifova YY, Reponen T, Bernstein DI, Levin L, Kalra H, Campo P, Villareal M, Lockey J, Hershey GK, LeMasters G (2007) House dust (1–3)-beta-D-glucan and wheezing in infants. Allergy 62:504–513

Jones AM, Harrison RM (2004) The effects of meteorological factors on atmospheric bioaerosol concentrations—a review. Sci Total Environ 326:151–180

Kildeso J, Wurtz H, Nielsen KF, Kruse P, Wilkins K, Thrane U, Gravesen S, Nielsen PA, Schneider T (2003) Determination of fungal spore release from wet building materials. Indoor Air 13:148–155

Kumar A, Attri AK (2016) Characterization of fungal spores in ambient particulate matter: a study from the Himalayan region. Atmos Environ 142:182–193

Kumar A, Gupta T (2015) Development and laboratory performance evaluation of a variable configuration $PM_1/PM_{2.5}$ impaction-based sampler. Aerosol Air Qual Res 15:768–775

Kumar P, Mahor P, Goel AK, Kamboj DV, Kumar O (2011) Aero-microbiological study on distribution pattern of bacteria and fungi during weekdays at two different locations in urban atmosphere of Gwalior, Central India. Scientific Research and Essays 6:5435–5441

Kumar V, Goel A, Rajput P (2017) Compositional and surface characterization of HULIS by UV-Vis, FTIR, NMR and XPS: wintertime study in Northern India. Atmos Environ 164: 468–475

Lau APS, Lee AKY, Chan CK, Fang M (2006) Ergosterol as a biomarker for the quantification of the fungal biomass in atmospheric aerosols. Atmos Environ 40:249–259

Lee T, Grinshpun SA, Kim KY, Iossifova Y, Adhikari A, Reponen T (2006) Relationship between indoor and outdoor airborne fungal spores, pollen, and (1 $\rightarrow$ 3)-β-D-glucan in homes without visible mold growth. Aerobiologia 22:227–235

Mamta Shrivastava JN, Satsangi GP, Kumar R (2015) Assessment of bioaerosol pollution over Indo-Gangetic plain. Environ Sci Pollut Res 22:6004–6009

Matthias-Maser S, Bogs B, Jaenicke R (2000) The size distribution of primary biological aerosol particles in cloud water on the mountain Kleiner Feldberg/Taunus (FRG). Atmos Res 54:1–13

Michel O, Ginanni R, Sergysels R (1992) Relation between the bronchial obstructive response to inhaled lipopolysaccharide and bronchial responsiveness to histamine. Thorax 47:288–291

Michel O, Kips J, Duchateau J, Vertongen F, Robert L, Collet H, Pauwels R, Sergysels R (1996) Severity of asthma is related to endotoxin in house dust. Am J Respir Crit Care Med 154: 1641–1646

Miyakawa T, Kanaya Y, Taketani F, Tabaru M, Sugimoto N, Ozawa Y, Takegawa N (2015) Ground-based measurement of fluorescent aerosol particles in Tokyo in the spring of 2013: Potential impacts of nonbiological materials on autofluorescence measurements of airborne particles. J Geophys Res: Atmos 120:1171–1185

Momin GA, Rao PSP, Safai PD, Ali K, Naik MS, Pillai AG (1999) Atmospheric aerosol characteristic studies at Pune and Thiruvananthapuram during INDOEX programme-1998. Curr Sci 76:985–989

Murata K, Zhang D (2016) Concentration of bacterial aerosols in response to synoptic weather and land-sea breeze at a seaside site downwind of the Asian continent. J Geophys Res: Atmos 121:11, 636–611,647

Padhi BK, Adhikari A, Satapathy P, Patra AK, Chandel D, Panigrahi P (2016) Predictors and respiratory depositions of airborne endotoxin in homes using biomass fuels and LPG gas for cooking. J Expo Sci Environ Epidemiol

Pöschl U, Martin ST, Sinha B, Chen Q, Gunthe SS, Huffman JA, Borrmann S, Farmer DK, Garland RM, Helas G, Jimenez JL, King SM, Manzi A, Mikhailov E, Pauliquevis T, Petters MD, Prenni AJ, Roldin P, Rose D, Schneider J, Su H, Zorn SR, Artaxo P, Andreae MO (2010) Rainforest aerosols as biogenic nuclei of clouds and precipitation in the amazon. Science 329:1513–1516

Prussin AJ, Garcia EB, Marr LC (2015) Total concentrations of virus and bacteria in indoor and outdoor air. Environ Sci Technol Lett 2:84–88

Punia M, Nautiyal VP, Kant Y (2008) Identifying biomass burned patches of agricultural residue using satellite remote sensing data. Curr Sci 94:1185–1190

Rajeev P, Rajput P, Gupta T (2016) Chemical characteristics of aerosol and rain water during an El-Niño and PDO influenced Indian summer monsoon. Atmos Environ 145:192–200

Rajput P, Sarin MM (2014) Polar and non-polar organic aerosols from large-scale agricultural-waste burning emissions in Northern India: Implications to organic mass-to-organic carbon ratio. Chemosphere 103:74–79

Rajput P, Sarin MM, Rengarajan R (2011a) High-precision GC-MS analysis of atmospheric polycyclic aromatic hydrocarbons (PAHs) and isomer ratios from biomass burning emissions. J Environ Prot 2:445–453

Rajput P, Sarin MM, Rengarajan R, Singh D (2011b) Atmospheric polycyclic aromatic hydrocarbons (PAHs) from post-harvest biomass burning emissions in the Indo-Gangetic Plain: isomer ratios and temporal trends. Atmos Environ 45:6732–6740

Rajput P, Sarin MM, Kundu SS (2013) Atmospheric particulate matter ($PM_{2.5}$), EC, OC, WSOC and PAHs from NE-Himalaya: abundances and chemical characteristics. Atmos Pollut Res 4:214–221

Rajput P, Sarin MM, Sharma D, Singh D (2014a) Atmospheric polycyclic aromatic hydrocarbons and isomer ratios as tracers of biomass burning emissions in Northern India. Environ Sci Pollut Res 21:5724–5729

Rajput P, Sarin MM, Sharma D, Singh D (2014b) Characteristics and emission budget of carbonaceous species from post-harvest agricultural-waste burning in source region of the Indo-Gangetic Plain. Tellus-B. https://doi.org/10.3402/tellusb.v66.21026

Rajput P, Sarin MM, Sharma D, Singh D (2014c) Organic aerosols and inorganic species from post-harvest agricultural-waste burning emissions over northern India: impact on mass absorption efficiency of elemental carbon. Environ Sci Proces Impacts 16:2371–2379

Rajput P, Mandaria A, Kachawa L, Singh DK, Singh AK, Gupta T (2015) Wintertime source-apportionment of PM_1 from Kanpur in the Indo-Gangetic plain. Clim Change 1: 503–507

Rajput P, Gupta T, Kumar A (2016a) The diurnal variability of sulfate and nitrate aerosols during wintertime in the Indo-Gangetic Plain: implications for heterogeneous phase chemistry. RSC Adv 6:89879–89887

Rajput P, Mandaria A, Kachawa L, Singh DK, Singh AK, Gupta T (2016b) Chemical characterization and source-apportionment of PM_1 during massive loading at an urban location in Indo-Gangetic Plain: impact of local sources and long-range transport. Tellus-Bdx. https://doi.org/10.3402/tellusb.v68.30659

Rajput P, Mandaria A, Kachawa L, Singh DK, Singh AK, Gupta T (2016c) Chemical characterization and source-apportionment of PM_1 during massive loading at an urban location in Indo-Gangetic Plain: impact of local sources and long-range transport. Tellus-B 68:30659

Rajput P, Anjum MH, Gupta T (2017) One year record of bioaerosols and particles concentration in Indo-Gangetic Plain: Implications of biomass burning emissions to high-level of endotoxin exposure. Environ Pollut 224:98–106

Ram K, Sarin MM, Hegde P (2010) Long-term record of aerosol optical properties and chemical composition from a high-altitude site (Manora Peak) in Central Himalaya. Atmos Chem Phys 10:11791–11803

Rathnayake CM, Metwali N, Baker Z, Jayarathne T, Kostle PA, Thorne PS, O'Shaughnessy PT, Stone EA (2016a) Urban enhancement of PM_{10} bioaerosol tracers relative to background locations in the Midwestern United States. J Geophys Res: Atmos 121:5071–5089

Rathnayake CM, Metwali N, Jayarathne T, Kettler J, Huang Y, Thorne PS, O'Shaughnessy PT, Stone EA (2016b) Influence of Rain on the Abundance and Size Distribution of Bioaerosols. Atmos Chem Phys Discuss 2016:1–29

Reponen T, Grinshpun SA, Conwell KL, Wiest J, Anderson M (2001) Aerodynamic versus physical size of spores: Measurement and implication for respiratory deposition. Grana 40:119–125

Reponen T, Seo S-C, Grimsley F, Lee T, Crawford C, Grinshpun SA (2007) fungal fragments in moldy houses: a field study in homes in new orleans and southern ohio. Atmos Environ (Oxford, England: 1994) 41:8140–8149

Salonen H, Duchaine C, Létourneau V, Mazaheri M, Laitinen S, Clifford S, Mikkola R, Lappalainen S, Reijula K, Morawska L (2016) Endotoxin levels and contribution factors of endotoxins in resident, school, and office environments—A review. Atmos Environ 142: 360–369

Semple S, Devakumar D, Fullerton DG, Thorne PS, Metwali N, Costello A, Gordon SB, Manandhar DS, Ayres JG (2010) Airborne endotoxin concentrations in homes burning biomass fuel. Environ Health Persp 118:988–991

Sesartic A, Dallafior TN (2011) Global fungal spore emissions, review and synthesis of literature data. Biogeosciences 8:1181–1192

Shaffer BT, Lighthart B (1997) Survey of culturable airborne bacteria at four diverse locations in oregon: urban, rural, forest, and coastal. Microb Ecol 34:167–177

Singh DK, Gupta T (2015) Speciation of atmospheric polycyclic aromatic hydrocarbons (PAHs) present during fog time collected submicron particles. Environ Sci Pollut Res. https://doi.org/10.1007/s11356-11015-14413-y

Singh DK, Gupta T (2016a) Effect through inhalation on human health of PM_1 bound polycyclic aromatic hydrocarbons collected from foggy days in northern part of India. J Hazard Mater 306:257–268

Singh DK, Gupta T (2016b) Role of transition metals with water soluble organic carbon in the formation of secondary organic aerosol and metallo-organics in PM_1 sampled during post monsoon and pre-winter time. J Aerosol Sci 94:56–69

Singh DK, Lakshay Gupta T (2014) Field performance evaluation during fog-dominated wintertime of a newly developed denuder-equipped PM1 sampler. Environ Sci Pollut Res 21:4551–4564

Singh A, Rajput P, Sharma D, Sarin MM, Singh D (2014) Black carbon and elemental carbon from postharvest agricultural-waste burning emissions in the Indo-Gangetic Plain. Adv Meteorol 2014:10

Smith JA, Blanchette RA, Newcombe G (2004) Molecular and morphological characterization of the willow rust fungus, Melampsora epitea, from arctic and temperate hosts in North America. Mycologia 96:1330–1338

Stanley R, Linskins H (1974) Pollen: biology, chemistry and management. Springer-Verlag, Berlin

Taha MPM, Drew GH, Tamer A, Hewings G, Jordinson GM, Longhurst PJ, Pollard SJT (2007) Improving bioaerosol exposure assessments of composting facilities—Comparative modelling of emissions from different compost ages and processing activities. Atmos Environ 41: 4504–4519

Valsan AE, Priyamvada H, Ravikrishna R, Després VR, Biju CV, Sahu LK, Kumar A, Verma RS, Philip L, Gunthe SS (2015) Morphological characteristics of bioaerosols from contrasting locations in southern tropical India—a case study. Atmos Environ 122:321–331

Valsan AE, Ravikrishna R, Biju CV, Pöhlker C, Després VR, Huffman JA, Pöschl U, Gunthe SS (2016) Fluorescent biological aerosol particle measurements at a tropical high-altitude site in southern India during the southwest monsoon season. Atmos Chem Phys 16:9805–9830

Venkataraman C, Habib G, Eiguren-Fernandez A, Miguel AH, Friedlander SK (2005) Residential biofuels in south asia: carbonaceous aerosol emissions and climate impacts. Science 307: 1454–1456

Xue J, Zhang J, Xu B, Xie J, Wu W, Lu Y (2016) Endotoxins: the critical risk factor in reclaimed water via inhalation exposure. Environ Sci Technol 50:11957–11964

Zhu C, Kawamura K, Fukuda Y, Mochida M, Iwamoto Y (2016) Fungal spores overwhelm biogenic organic aerosols in a midlatitudinal forest. Atmos Chem Phys 16:7497–7506

Ziemba LD, Beyersdorf AJ, Chen G, Corr CA, Crumeyrolle SN, Diskin G, Hudgins C, Martin R, Mikoviny T, Moore R, Shook M, Thornhill KL, Winstead EL, Wisthaler A, Anderson BE (2016) Airborne observations of bioaerosol over the Southeast United States using a wideband integrated bioaerosol sensor. J Geophys Res: Atmos 121:8506–8524

Chapter 6
Particulate Matter Dispersion in Indian Non-coal Opencast Mines

Sneha Gautam and Aditya Kumar Patra

Abstract Exposure of mine workers to particulate matter (PM) in opencast mines is of major concern because of associated adverse health impacts. Dispersion studies of PM generated during opencast mining until now have mostly been confined to estimation of emissions from individual mining operations as well as total emission from an opencast mine. No literature could be found on PM dispersion inside the mine since its generation until it escapes the mine. However, this is important particularly for deep mines where mine workings are confined to bottom benches and emission from it passes across all benches before it reaches surface, thus affecting the exposure level of workers at higher benches. The proposed research works have been conducted in three non-coal opencast mines in India: (i) Kiriburu Iron Ore Mine (KIOM), (ii) Meghahatuburu Iron Ore Mine (MIOM), and (iii) Malanjkhand Copper Project (MCP). Mining at KIOM and MIOM was carried out at 9–10 m depths. At MCP, mining activities at 168–180 m depth were the source of PM. The aims were to investigate: (i) vertical PM concentration profiles at different depths during mining operations, (ii) particle travel time and its relationship with mine depth, (iii) particle retention time and fraction at workplace, (iv) particle mass balance in terms of the dispersion and settling proportion, and (v) the relationship of particle concentration with depth. For a deep mine, this is of significant interest. The authors propose more studies in different seasons of the year in deeper mines and mines with varying pit geometry for better understanding of the influence of depth of mine. To assess role of particle density on PM dispersion, similar studies should be carried out in coal mines.

Keywords Opencast mine · Particulate matter · Dispersion · Travel time
Retention time · Retention fraction · Empirical equation

S. Gautam
Marwadi University, Rajkot 360003, GJ, India

S. Gautam · A. K. Patra (✉)
Indian Institute of Technology, Kharagpur 721302, WB, India
e-mail: patrakaditya@gmail.com

T. Gupta et al. (eds.), *Environmental Contaminants*, Energy, Environment, and Sustainability, https://doi.org/10.1007/978-981-10-7332-8_6

6.1 Introduction

Opencast mining consists of removal of soil and rock materials, known as overburden, from surface downward to access the mineral to be extracted. This mining method is applicable to mineral deposits located at shallow depth. Opencast mines consist of different unit operations/activities, i.e., overburden removal, drilling, blasting, loading, transportation, unloading, and several other ancillary activities. Each of these activities is a source of particulate matter (PM). PM generated during these activities is directly emitted to the atmosphere. The degradation of air quality is a major problem in surface mining areas (Gautam et al. 2015, 2016b; Patra et al. 2016). It is of more concern because PM is very harmful to human health (Gautam et al. 2016a; Zhang et al. 2013). Inhalation of PM results in increased mortality, morbidity, cardiovascular diseases, and other lung diseases (Dockery et al. 1993; Nel 2005). Particles that are harmful to health are PM_{10} (PM having aerodynamic diameter below 10 μm), and this includes $PM_{2.5}$ (PM having aerodynamic diameter below 2.5 μm) and PM_1 (PM having aerodynamic diameter below 1.0 μm). Epidemiological studies have indicated the relationship between adverse health effects and $PM_{2.5}$ (Dockery et al. 1993; Atkinson et al. 2014). Evidence has shown that mortality, mobility, cardiovascular diseases, etc., are triggered by fine particles (i.e., $PM_{0.1}$, PM_1, and $PM_{2.5}$) (Dockery et al. 1993). In order that the mine workers have less exposure to PM, the PM generated during different mining operations need to be dispersed from the workplace as quickly as possible.

6.1.1 PM Dispersion: Role of Wind Flow and Pit Geometry

Wind flow is the prime mover of the PM from one place to the other. Wind flow across and inside the opencast mine depends primarily on wind speed (WS) and mine geometry. Three parameters that determine the overall geometry of the pit are length of the pit (L), depth of the pit (D), and width of pit (W). On the basis of length-to-depth ratio (also known "aspect ratio"), Larcheveque et al. (2003) classified the pit geometry into the following categories: (a) $L/D > 1$: Shallow pit; (b) $L/D < 1$: Deep pit; (c) $L/D < 10$: Open pit; (d) $L/D > 13$: Closed pit; (e) $L/W < 1$: Two-dimensional pit; and (f) $L/W > 1$: Three-dimensional pit. Depending on the aspect ratio, Chowdhary (1977) have reported that wind velocity will be more in shallow mines than in deep mines. The WS gradually increases with the slope of the open pit (Ngo and Lechford 2008). The deeper parts of the mine will have poor ventilation, and the shallow parts will have better ventilation. The flow type will change from unsteady to laminar as depth increases. In case of shallow pits, air flow expands gradually, and it covers the whole of the mine area without leaving any part of mine with very little compression of air. In case of deep pits, air gets compressed as it enters to the deeper parts, and expansion of air is also restricted.

Pit slope guides the air into the pit. This also results in increase of WS as the wind enters deep into the pit. However, the profile of benches creates a somewhat obstructed flow of wind, especially the layer close to the bench slope, i.e., WS is high along the bench slope and WS is low as it meets the high wall. Local deflection of wind path due to bench geometry causes reduction of speed leading to recirculation, besides changing the direction of flow (Peng and Lu 1995; Grainger and Merony 1993).

One of the earliest studies on PM movement inside the mine was carried out using tracer (Richardson 1926). Peng and Lu (1995) used smoke to simulate the dispersion of PM inside the mines. They released smoke from one point in pit bottom and measured the pollution concentration at different places inside the mine by photoelectric sensor. Simulation of PM inside the mines has also been studied by using the computational fluid dynamics model (CFD) (Loomans and Lemaire 2002). Larcheveque et al. (2003) and Chowdhary (1977) have discussed the variation of wind profile on the basis of aspect ratio (ratio of length to depth of the mine). All the above studies tried to simulate the wind flow pattern with the assumption that the PM dispersion will follow it. No field study on real particle movement has been conducted to verify these.

6.1.2 Pit Retention Studies

Pit retentions studies provide an estimate of the proportion of the particles generated in the mine that does not escape from the mine and therefore contributes to the enhanced PM concentration inside the mine. Wings (1981) proposed an expression in terms of particle deposition velocity, vertical diffusivity, and pit depth to estimate fractions of the total emissions that escape the mine.

$$\epsilon = \frac{1}{1 + \left(\frac{V_d}{K_z}\right)H}$$

where

ϵ is the mass fraction of total emissions
V_d is the particle deposition velocity (ms^{-1})
K_z is the vertical diffusivity (m^2s^{-1})
H is the pit depth (m)

Expression of pit retention by Fabrick (1982) included parameters like WS at the pit top, pit width, and particle deposition velocity.

$$\epsilon = 1 - V_d\left[\frac{C}{u}\left(\frac{1}{2} + \ln\frac{w}{4}\right)\right]$$

where

ϵ is the escape fraction
V_d is the particle deposition velocity (ms^{-1})
u is the WS (ms^{-1})
w is the pit width (m)
C is an empirical dimensionless constant with a value of ~ 7

Retention fractions for PM from 10 μm to 95 μm varied from 0.14 to 0.73 (Winges 1981) and 0.23 to 1.0 (Fabrick 1982), respectively. The weighted average values of pit retention were 0.50 (Winges 1981) and 0.84 (Fabrick 1982). No recent studies on pit retention could be found in literature.

6.1.3 Variation of PM Exposure in an Opencast Mine

Workers work at different places depending on the mining operation he is associated with. A person engaged in excavation and loading the ore usually works one or more benches below the bench where the overburden excavation is in progress. A drill operator engaged in drilling blast holes works both in overburden and ore benches. The dispersion of pollutant from these two operations at different locations will not be the same because of variation of pit geometry and micro-meteorology (mainly wind direction (WD) and WS) inside the mine. Dumpers plying on haul roads are another source of pollution where location of pollution source changes over time. The cumulative effect of these sources gets manifested in terms of enhanced concentration of pollution inside the mine. Therefore, at different locations and mining operations inside the mine, mine workers are exposed to different levels of PM (Naidoo et al. 2006).

PM generated at the lower benches of a mine travels across all the higher benches before it escapes the mine. Therefore, an activity at deeper parts of the mine contributes to the enhanced concentration at higher benches. Time taken by the PM generated at a certain depth to escape the mine is therefore important. If the time taken for the PM to escape the mine is very long, it indicates that workers inside the mine are exposed to elevated concentration for a longer duration. On the other hand, if the PM escapes the mine in a short time, its adverse effect on health will be less due to less exposure duration. Additionally, different size of particles does not follow same speed and proportion, while traveling from the lower part of the mine to upper horizons. Because of higher settling velocity, coarse particle settles on the lower horizons faster than fine particle. Simultaneously, the concentration of coarse particles may get depleted from the plume as it travels upper horizons. On the other hand, fine particle with lower settling velocity takes more time to escape from the mine. These are the causes of higher exposure to particle of mine workers. Therefore, the movement and dispersion of fine-sized particle in deeper part of the mine need to be studied.

No study on the literature could be found that takes into account the variation of particle size and WS and relative location between the mining operation and workers at different parts of the mine to understand the occupational exposure of the miners. No on-field measurement in literature exists that reports actual measurements inside the mine to address this. With opencast mines going deeper in future, this type of study of dust dispersion inside the mine needs attention.

6.2 In-Pit PM Dispersion Study

The study was conducted in three opencast mines of India, namely Kiriburu Iron Ore Mine (KIOM) and Meghahatuburu Iron Ore Mine (MIOM) of Steel Authority of India Limited (SAIL) and Malanjkhand Copper Project (MCP) of Hindustan Copper Limited (HCL). The experiments were conducted between May and October 2013. KIOM and MIOM are located in one deposit that is divided into three blocks, namely South, Central, and North block (Fig. 6.1). KIOM and MIOM are located in South and North block, respectively. The MCP is the single largest copper deposit in India constituting nearly 80% of the country's reserve. In MCP, the surface and bottom are at 580–376 mRL (mRL refers to depth/height above datum/mean sea level), respectively.

The proposed pit depth is 204 m, and it makes MCP one of the deep non-coal opencast mines in India. The entire MCP is divided into three broad sections, namely the North pit, Central pit, and South pit. The selected study area was Central

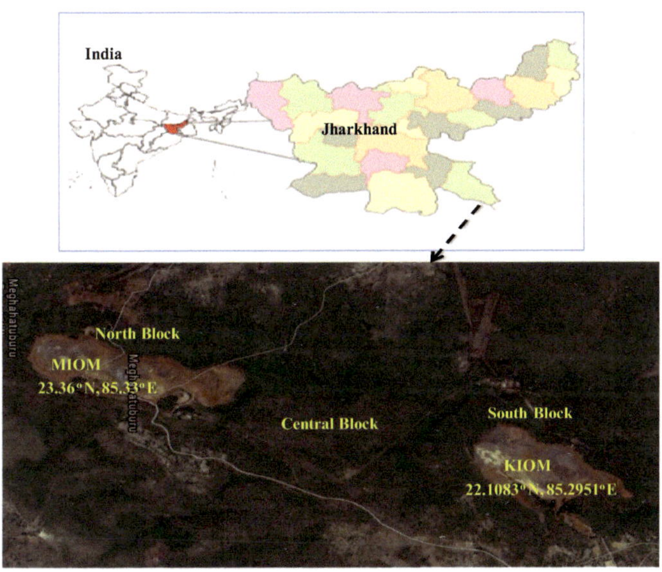

Fig. 6.1 Study site: KIOM and MIOM

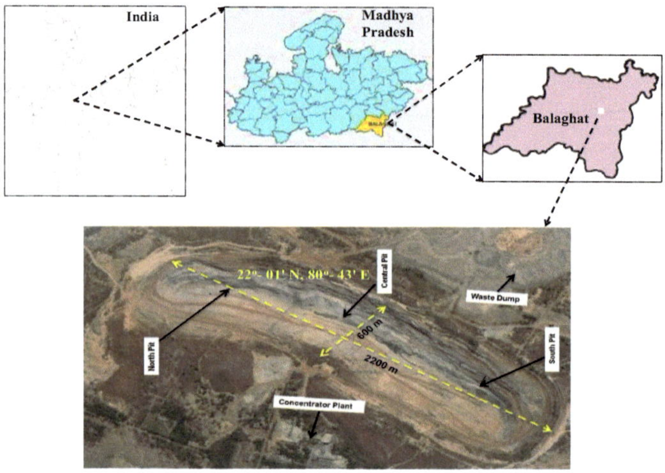

Fig. 6.2 Study site: MCP

pit, according to running mining operations (Fig. 6.2). The study involved the measurement of PM concentration and meteorological parameter in opencast mines. During May 21, 2013 to October 24, 2013, real-time measurement campaigns were done by an aerosol spectrometer (Model 1.108, Grimm, GRIMM Aerosol Technik GmbH & Co. KG, Germany) (Grimm 2010) in which particles were measured at 1-min interval. Moreover, meteorological parameters are recorded by using portable weather stations (Spectrum Technologies, Inc., Model Watchdog 2000) at 1-min interval. The equipment with full weather sensing capabilities has a combination of internal and external sensors to record WD, WS, relative humidity (RH), and temperature (T), (Spectrum 2010).

At KIOM, the source (mining operations) was located at the first and second benches. The PM concentration was measured on the surface near the pit boundary. At MIOM, the source was located at the second bench, and the PM concentration was measured at the first bench. The objective was to determine the travel time and concentration of different sizes of PM as it travels from the source to the receptor. The Central pit of MCP is 180 m deep with surface RL 580 m and the pit bottom RL 400 m. Sampling was carried out on benches having 400, 412, 448, and 460 mRL and also at the pit boundary on surface at 580 mRL. At the Central pit, drilling of blast holes by the drill machines, loading of blasted ore by a shovel, and transport of ore to the concentrator plant by the dumpers at 400–412 mRL constituted the sources of PM emission. Therefore, PM concentration at these two benches provided the estimate of increased concentration during mining operations. 580 mRL refers to the surface and therefore measurement at this location indicated the PM level at the pit boundary. Between 412 and 580 mRL, 13 benches, each 12 m high, are present. Measurements were taken at 2 out of these 13 benches (448–460 mRL) which were accessible at the time of the study. PM concentration and meteorological parameters data were recorded by using GRIMM and portable weather station at each selected locations inside the mine.

6.3 Results

6.3.1 Dependence of PM Concentration on Meteorological Parameters

As expected, significant negative correlations between the WS and RH and between RH and T were obtained at 0.05 level. With the increase of WS, more particles from the source inside the mine reach the Grimm aerosol spectrometer placed at the periphery of the mine. Further, WS has more influence on the dispersion of coarse particles than fines because fine particles can remain airborne in a very low WS. Coarser particles need more WS to remain suspended in air. Therefore, at KIOM, a positive significant correlation was observed between WS and PM ($PM_{2.5-10}$, $PM_{1-2.5}$) concentration at 0.01 level, and no correlation was obtained between WS and PM_1. No correlation between WS and PM concentration was obtained at MIOM. This could presumably be due to low WS (75th percentile $\leq 2.5\ ms^{-1}$ and very low wind speed (WS < 1 ms^{-1} prevailed for 67–92% of the time) prevailing at MIOM. In both the mines, PM sizes were significantly correlated (Table 6.1).

Table 6.1 Correlation matrix for WS, RH, T, and PM concentration

KIOM (n = 1173)	WS	RH	T	$PM_{2.5-10}$	$PM_{1-2.5}$	PM_1
WS	1	−0.51[b]	0.51[b]	0.18[b]	0.19[b]	−0.02
RH		1	−0.97[b]	−0.05	−0.10[b]	0.39[b]
T			1	0.07[a]	0.11[b]	−0.30[b]
$PM_{2.5\_10}$				1	0.97[b]	0.76[b]
$PM_{1\_2.5}$					1	0.78[b]
PM_1						1
MIOM (n = 1113)	WS	RH	T	$PM_{2.5-10}$	$PM_{1-2.5}$	PM_1
WS	1	−0.23[b]	0.16[b]	0.03	0.05	0.01
RH		1	−0.91[b]	0.02	0.02	0.03
T			1	0.02	0.03	−0.02
$PM_{2.5\_10}$				1	0.95[b]	0.36[b]
$PM_{1\_2.5}$					1	0.40[b]
PM_1						1

[a]Correlation is significant at the 0.05 level
[b]Correlation is significant at the 0.01 level

6.3.2 Particle Travel Time

Particle travel time is the time taken by a particle to travel from source (workplace) to a surface (different horizons of interest). It is estimated as the time interval between the start of the mining operations and first observation of enhanced particle concentration (Fig. 6.3; time in figures and tables is mentioned in hh:mm or h:mm in 24-hour format). The particle travel times between source bench (i.e., 412 mRL) and higher benches/horizons (i.e., 448, 460, and 580 mRL) are presented in Tables 6.2 and 6.3. Estimated travel times (i.e., nearly one hour and seven min for MCP (168 m) and KIOM/MIOM (9–10 m), respectively) indicate that with increase in depth, the particle travel time increases significantly. The findings suggest that it is very important occupational exposure parameter that should be considered while assessing the health risk of the miners.

Correlation analysis shows significant relationship ($p = 0.034$) at 0.05 level. An empirical relationship ($R = 0.75$) between particle travel time and depth of mine has been developed through regression analysis by using IBM SPSS statistical package (version 20).

$$T = 0.21d + 13.20$$

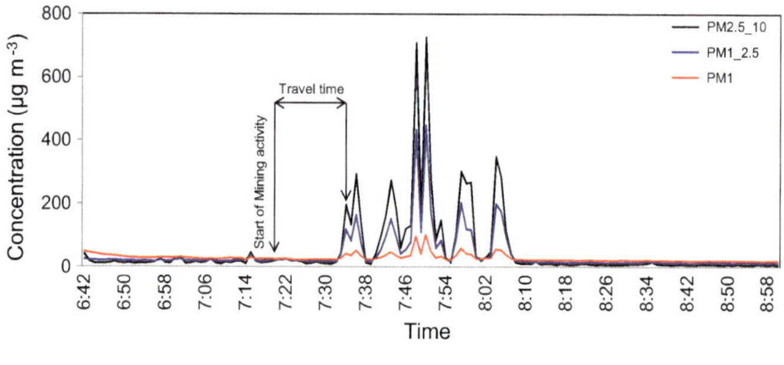

Fig. 6.3 Schematic diagram showing particle travel time

Table 6.2 Travel time estimates at MCP

Source-receptor location (mRL)	Relative depth of source (m)	Travel time (min)	
		First measurement	Second measurement
412–580	168	39	54
412–460	48	30	35
412–448	36	18	13

Note The source is located at 412 mRL

Table 6.3 Travel time estimates at KIOM and MIOM

KIOM	First phase of working: 21.05.2013 (7:30–8:06); 22.05.2013 (7:24–7:44); 23.05.2013 (7:20–7:58)						
Date	Time of start of first phase work	Time when first elevated concentration is observed			Time taken for the particle to reach the surface		
		$PM_{2.5-10}$	$PM_{1-2.5}$	PM_1	$PM_{2.5-10}$ (min)	$PM_{1-2.5}$ (min)	PM_1 (min)
21/05/13	7:30	7:34	7:34	7:34	4	4	4
22/05/13	7:24	7:29	7:29	7:29	5	5	5
23/05/13	7:20	7:22	7:22	7:22	2	2	2
KIOM	Second phase of working: 21.05.2013 (9:18–12:35); 22.05.2013 (9:44–13:09); 23.05.2013 (9:23–12:22)						
Date	Time of start of second phase work	Time when first elevated concentration is observed			Time taken for the particle to reach the surface		
		$PM_{2.5-10}$	$PM_{1-2.5}$	PM_1	$PM_{2.5-10}$ (min)	$PM_{1-2.5}$ (min)	PM_1 (min)
21/05/13	9:18	9:22	9:22	9:22	4	4	4
22/05/13	9:44	9:48	9:48	9:48	4	4	4
23/05/13	9:23	9:30	9:30	9:30	7	7	7
MIOM	First phase of working: 04.06.2013 (7:53–8:59); 05.06.2013 (6:51–8:45); 06.06.2013 (7:10–8:48)						
Date	Time of start of first phase work	Time when first elevated concentration is observed			Time taken for the particle to travel from second bench to first bench		
		$PM_{2.5-10}$	$PM_{1-2.5}$	PM_1	$PM_{2.5-10}$ (min)	$PM_{1-2.5}$ (min)	PM_1 (min)
4/6/2013	7:53	7:58	7:58	7:58	5	5	5
5/6/2013	6:51	6:55	6:55	6:55	4	4	4
6/6/2013	7:10	7:12	7:12	7:12	2	2	2
MIOM	Second phase of working: 04.06.2013 (10:18–13:03); 05.06.2013 (9:47–13:03); 06.06.2013 (9:43–13:00)						
Date	Time of start of second phase work	Time when first elevated concentration is observed			Time taken for the particle to travel from second bench to first bench		
		$PM_{2.5-10}$	$PM_{1-2.5}$	PM_1	$PM_{2.5-10}$ (min)	$PM_{1-2.5}$ (min)	PM_1 (min)
4/6/2013	10:18	10:35	10:35	10:35	17	17	17
5/6/2013	9:47	9:48	9:48	9:48	1	1	1
6/6/2013	9:43	9:45	9:45	9:45	2	2	2

where

T is the PM travel time (min)
d is the depth of the mine (m)

6.3.3 PM Retention at Workplace

6.3.3.1 Retention Time

Retention time gives a measure of the exposure duration to enhanced PM concentration, at a workplace inside the mine after mining activities stop (Fig. 6.4).

Long retention time indicates more exposure of mine workers to enhanced PM concentration at the workplace. Results show that retention time lies in the range of 10–13 min (Table 6.4). Retention times for different size of particles were the same, indicating no dependence of retention time on particle size. It is expected that retention time will increase with depth of mine because of restricted ventilation/recirculation in deeper parts of the mine. Retention time increases with increase of duration of mining (average duration of second phase of mining was about 3 h while the first phase of mining varied from 30 min to 1 h. Only on June 5, 2013, the duration of the first phase of mining at MIOM was close to 2 h). No retention time for second phase of mining could be calculated because the experiment ended before second phase of mining stopped.

6.3.3.2 Retention Fraction

Retention fraction is the ratio of the amount of PM generated due to a mining activity that remained at the workplace after end of the mining operation to the amount generated while the mining operation was in. This has been estimated using area under the concentration curve (Fig. 6.4).

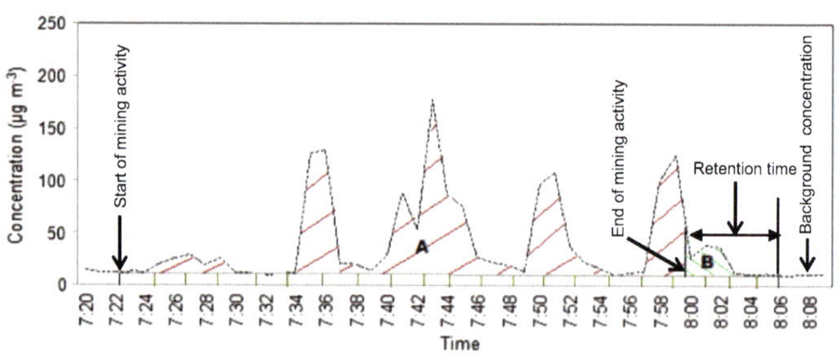

Fig. 6.4 Retention time

Table 6.4 Retention time estimates

KIOM							
Date	Time of stoppage of first phase work	Time when particle concentration reached background concentration			Retention time		
		$PM_{2.5-10}$	$PM_{1-2.5}$	PM_1	$PM_{2.5-10}$ (min)	$PM_{1-2.5}$ (min)	PM_1 (min)
21/05/13	8:06	8:16	8:16	8:16	10	10	10
22/05/13	7:44	7:54	7:54	7:54	10	10	10
23/05/13	7:58	8:09	8:09	8:09	11	11	11
MIOM							
Date	Time of stoppage of first phase work	Time when particle concentration reached background concentration			Retention time		
		$PM_{2.5-10}$	$PM_{1-2.5}$	PM_1	$PM_{2.5-10}$ (min)	$PM_{1-2.5}$ (min)	PM_1 (min)
4/6/2013	8:59	9:12	9:12	9:12	13	13	13
5/6/2013	8:45	8:56	8:56	8:56	11	11	11
6/6/2013	8:48	8:58	8:58	8:58	10	10	10

$$\text{Retention fraction} = \frac{B}{(A+B)}$$

where

A Area under the curve for the mining duration
B Area under the curve for the residual time

A higher retention fraction for fine particles (PM_1 and $PM_{1-2.5}$) in comparison with coarse particle ($PM_{2.5-10}$) is observed (Table 6.5). After mining operations stop, further generation of PM from mining activity ends. Settling of coarse fractions comparatively faster than the fine fractions results in fine particles dominating the mine atmosphere which gives high retention fraction for fine fractions. In MIOM, 76–92% of the time, calm condition was prevailing, and the average WS was less in comparison with WS in KIOM. Therefore, retention fraction was reported higher in MIOM than KIOM, as high WS resulted in better dispersion of PM. Table 6.5 shows that a worker at the workplace in a mine is likely to be exposed to as high as ~10% of the $PM_{2.5-10}$ and 15% of the $PM_{1-2.5}$ and PM_1, making it ~40% PM_{10} after the mining activity stops. As retention time increases with the depth of the mine (discussed in the previous section), retention fraction will also increase. This will potentially result in increased PM exposure of mine workers.

Table 6.5 Retention fraction estimates

Date	KIOM		
	Retention fraction		
	$PM_{2.5-10}$	$PM_{1-2.5}$	PM_1
21/05/13	0.014	0.016	0.045
22/05/13	0.014	0.016	0.045
23/05/13	0.100	0.103	0.128
Date	MIOM		
	Retention fraction		
	$PM_{2.5-10}$	$PM_{1-2.5}$	PM_1
4/6/2013	0.094	0.155	0.155
5/6/2013	0.031	0.044	0.010
6/6/2013	0.009	0.021	0.015

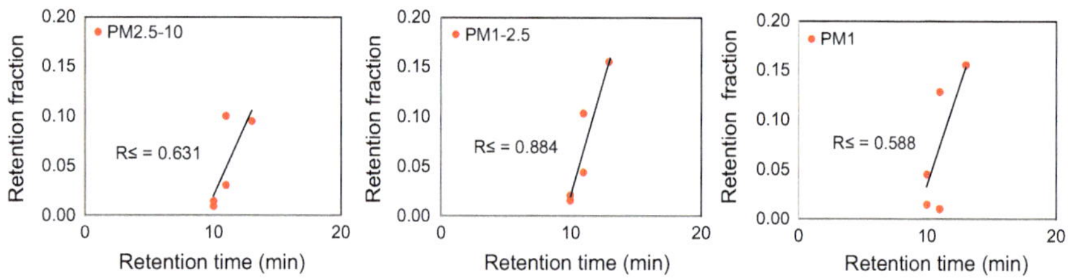

Fig. 6.5 Retention time versus retention fraction

6.3.3.3 Relationship Between Retention Time and Retention Fraction

Irrespective of particle size, retention fraction increases with retention time (Fig. 6.5).

Retention fraction depends on retention time as well as concentration profile of PM during and after the mining operation A moderate-to-strong relationship (R^2 = 0.588–0.884) between retention time and retention fraction for particles of all sizes is observed (Fig. 6.5).

6.3.4 PM Concentration and Depth

Mining operations at 412 mRL has been taken as the source of PM in the mine. From the surface (580 mRL), this is located at a depth of 168 m. Thus, the study is equivalent to a mine where the source is located at 168 m depth, and the PM concentrations are measured at 168 m (412 mRL), 132 m (448 mRL), 120 m (460 mRL), and 0 m (580 mRL) depth.

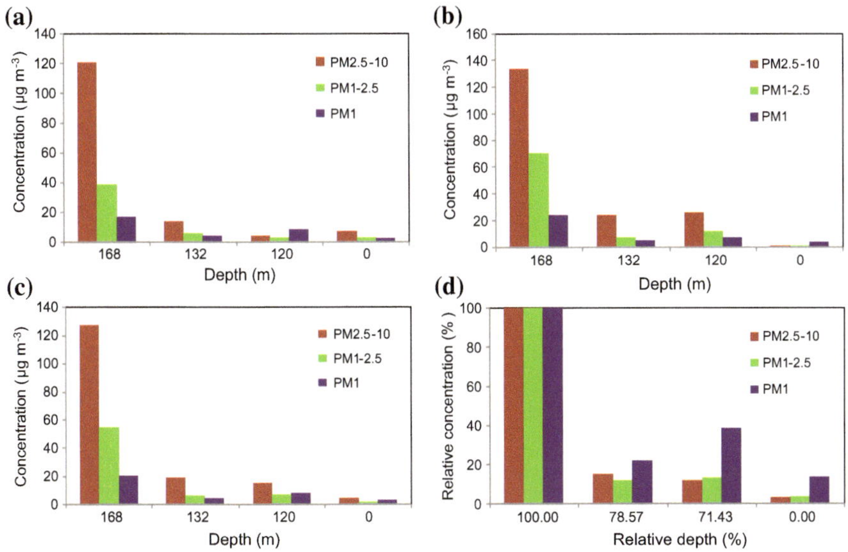

Fig. 6.6 PM concentration at different depths during mining; **a** 1st set; **b** 2nd set; **c** Average of 1st and 2nd sets; **d** Average of 1st and 2nd sets in terms of relative depth and concentration

Table 6.6 Summary of incremental PM concentration at different depths

Depth from the surface (m)	PM Concentration ($\mu g\ m^{-3}$)								
	Day 1 (1st Set)			Day 2 (2nd Set)			Average		
	$PM_{2.5-10}$	$PM_{1-2.5}$	PM_1	$PM_{2.5-10}$	$PM_{1-2.5}$	PM_1	$PM_{2.5-10}$	$PM_{1-2.5}$	PM_1
168	120.6	38.59	16.85	133.44	70.19	23.66	127.02	54.39	20.26
132	13.75	05.88	04.07	23.73	06.74	04.67	18.74	06.31	04.37
120	04.22	02.84	08.36	25.75	11.26	07.10	14.99	07.05	07.73
0	06.99	02.84	02.00	0.96	01.05	03.48	03.98	01.95	02.74

On day 1, incremental concentrations (measured concentration–background concentration) of $PM_{2.5-10}$, $PM_{1-2.5}$, and PM_1 from mining decreased from 120.6 to 6.99 $\mu g\ m^{-3}$, 38.59 to 2.84 $\mu g\ m^{-3}$, and 16.85 to 2.00 $\mu g\ m^{-3}$, respectively, from a depth of 168 m to surface. On day 2, $PM_{2.5-10}$, $PM_{1-2.5}$, and PM_1 concentrations decreased from 133.44 to 0.96 $\mu g\ m^{-3}$, 70.19 to 1.05 $\mu g\ m^{-3}$, and 23.66 to 3.48 $\mu g\ m^{-3}$, respectively (Fig. 6.6; Table 6.6).

In order to establish relationship between the PM concentration and depth, both depth and concentration have been expressed within a range of 0–100%. The depths of the source and surface have been taken as 100% and 0%, respectively. The intermediate depths are expressed as a percentage of source depth. PM concentration at source is taken as 100%, and concentrations at different heights up to the surface are expressed with respect to it (Table 6.7).

Table 6.7 Depths and concentrations at a place relative to source depth and concentration

Relative depth (Depth expressed as a percentage of source depth) (%)	Relative concentration (concentration at a location as a percentage of concentration at source location) (%)								
	Day 1 (1st Set)			Day 2 (2nd Set)			Average		
	$PM_{2.5-10}$	$PM_{1-2.5}$	PM_1	$PM_{2.5-10}$	$PM_{1-2.5}$	PM_1	$PM_{2.5-10}$	$PM_{1-2.5}$	PM_1
100.00	100.00	100.00	100.00	100.00	100.00	100.00	100.00	100.00	100.00
78.57	11.40	15.24	24.15	17.78	09.60	19.74	14.75	11.60	21.57
71.43	03.50	07.36	49.61	19.30	16.04	30.01	11.80	12.96	38.16
0.00	05.80	07.36	11.87	0.72	01.50	14.71	03.13	03.58	13.53

Table 6.8 R^2 and p-value between relative depths and relative concentrations

Particle size	R^2 value	P value
$PM_{2.5-10}$	0.98	0.000
$PM_{1-2.5}$	0.99	0.000
PM_1	0.98	0.001

One-way ANOVA results show a significant relationship between relative depth and relative concentration (Table 6.8).

Therefore, a generalized relationship between PM concentration and depth at MCP has been established by using MATLAB 7.10.0 (R2010a). The following empirical equation shows a strong (R^2 = 0.99, 0.98, and 0.88 for $PM_{2.5-10}$, $PM_{1-2.5}$ and PM_1) relationship between relative PM concentration and relative depth at MCP.

$$C = ae^{bd}$$

where

- C is relative PM concentration (%)
- d is relative depth (%)
- a and b are empirical constants (a is 0.024, 0.015, and 0.823 and b is 0.084, 0.088, and 0.048 for $PM^{2.5-10}$, $PM_{1-2.5}$, and PM_1, respectively).

Quicker settling of coarse particles in comparison with fines results in faster decrease of coarse particle concentration as depth decreases. Therefore, exponential decay constant in the empirical equation for PM_1 (0.823) is an order of magnitude higher than the coefficients of predictor equation for $PM_{2.5-10}$ (0.024) and $PM_{1-2.5}$ (0.015).

6.3.5 *Performance Evaluation of Empirical Equation*

The accuracy of the empirical equation derived from the average of day 1 and day 2 data sets is evaluated against day 1 and day 2 data individually (Fig. 6.7). For particles of all sizes, the predictor equations show good performance. At a depth where source is located (relative depth = 100%), the ratio of the estimated value to measured value is ~1.00, indicating at maximum depth, the estimation from the empirical equation matches with the measurement. At intermediate depths (relative depths of 78.57 and 71.43%), the model overestimates or underestimates the PM concentration about a factor of 2. At the surface, the model underestimated the concentration by a big margin. The best-predicted value for concentration at the surface was 7% of the observed value. While inside the mine, the air flow is channeled following the pit geometry (Peng and Lu 1995); on the surface, no such

Fig. 6.7 Measured and estimated PM concentrations at MCP

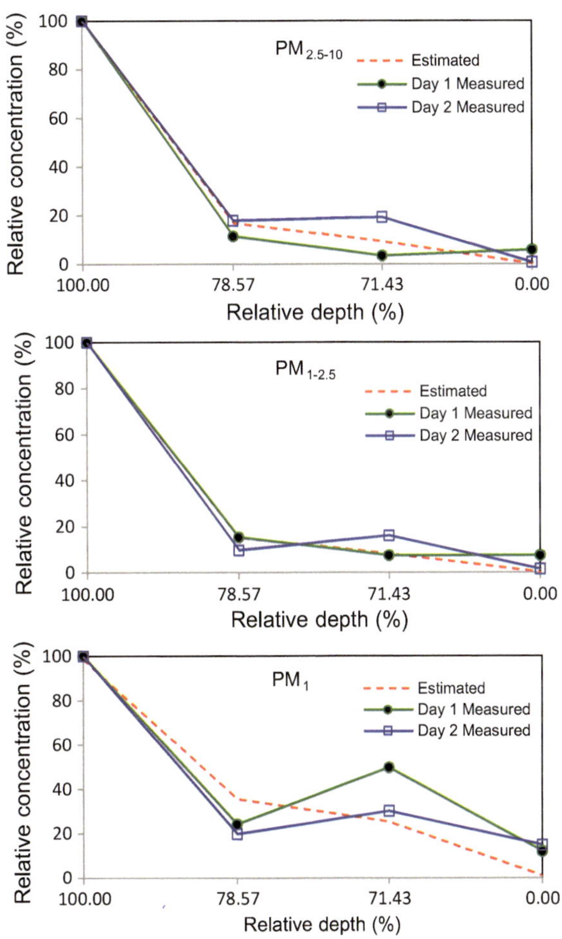

channeling effect of wind flow takes place; and therefore, dilution of PM significantly increases. The empirical equation is not able to account for such phenomenon, and therefore, very low concentration of PM at the surface is estimated. However, in absolute terms the difference in concentration between the measured and estimated PM concentration is limited maximum up to $\sim 10\ \mu g\ m^{-3}$ which is not a significant under-prediction as far as health effects of PM is concerned. This is a realistic accuracy that can be expected from this type of complex source–emission relationship, which is also influenced by pit geometry.

6.4 Discussion

6.4.1 Quantitative Estimate of PM Retention at Workplace: Retention Time and Retention Fraction

Particle retention time was 10–23 min for short duration mining (30 min–2 h) in comparison with 30 min for mining operation of longer duration ($\sim$3 h). Therefore, retention time seems to increase with the increase of mining duration. Therefore, continuous working for a full shift of 8 h will potentially result in a retention time long enough for the PM generated on one shift to affect the workers of the subsequent shift. Retention times for different size of particles were the same, indicating no dependence of retention time on the particle size. Retention fractions of $PM_{2.5-10}$, $PM_{1-2.5}$, and PM_1 varied in the ranges of 0.009–0.100, 0.016–0.155, and 0.010–0.155, respectively. Retention fraction for PM_{10} varied from 7.47% to 33.10% at KIOM and 4.46 to 40.44% at MIOM. The retention fractions estimate of 14% and 23% obtained through modeling by Fabrick (1982) and Winges (1981), respectively, therefore agrees with the findings of this study. The estimates also suggest that workers at the workplace in the mine are likely to be exposed to as high as $\sim$10% of the $PM_{2.5-10}$ and 15% of the $PM_{1-2.5}$ and PM_1, making it $\sim$40% PM_{10} generated during mining for a period of 10–13 min (retention time) after the mining operations end. Retention fraction increases with retention time. However, the rate of increase is more for fine particles. Increase of retention fraction and increase in the rate of retention fraction with decrease in particle size are attributed to faster settling of coarse particles. Therefore, higher retention fraction not only increases the exposures, but also increases more exposure to fine particles that have more potential to enter into the lungs. This is a significant finding, which suggests that workers are exposed to high PM concentration even when no mining activity is in progress. With retention time likely to increase with the depth of mine because of restricted ventilation/recirculation in the deeper parts of the mine, this remains a health risk concern for workers as future opencast mines go deeper.

6.4.2 Temporal Scale of Particle Dispersion: Travel Time and Escape Time

Particle generated at higher depths (168 m at MCP) takes nearly an hour, or even longer, before escaping to the surrounding atmosphere. Particle travel time was found to be independent of the particle size. The three sizes of particle ($PM_{2.5-10}$, $PM_{1-2.5}$, and PM_1) in this study were fine enough to be transported by wind at equal ease. This is presumed to be the reason behind non-dependence of travel time on particle size. Studies carried out for particles larger than this, which have appreciable settling velocities, can provide more information on dependence of particle travel time with depth. An empirical relationship between travel time of particle and depth of mine is developed. The output of this research work is very significant to assess the impact of increased particle concentration on miner at different horizon inside the mine. It is an upcoming challenge to mine planners to come up with new mine geometry that supports easy dispersion (short travel time) of particles generated by mining operations. Studies carried out for particles larger than this, which have appreciable settling velocities, can provide more information on dependence of particle travel time with depth.

6.4.3 Influence of Meteorological Parameter on PM Concentration, Travel Time, and Retention Time

Concentration of coarse particles ($PM_{2.5-10}$, $PM_{1-2.5}$) was increased with WS while no correlations were obtained between PM_1 and WS. Coarse particles ($PM_{2.5-10}$, $PM_{1-2.5}$) need higher WS to remain airborne than that is required for fine particles (PM_1). Therefore, the relationship of WS with coarse particle was significant in KIOM and MCP, and there was no significant relationship between the two at MIOM where average WS was low (75th percentile $\leq$ 2.5 ms^{-1}) and calm condition (WS < 1 ms^{-1}) prevailed for 67–92% of the experiment. No correlation of WS with depth could be obtained. Similar observations of absence of correlation between the WS and depth in opencast mines have been reported in earlier studies (Peng and Lu 1995; Chowdhary 1977).

6.4.4 Proposition of Relationship Between Particle Travel Time and Depth

Particle travel times at MCP were13–18 min, 30–35 min, and 39–54 min for source depths of 36 m, 48 m, and 168 m, respectively. Particle travel time and depth were significantly correlated ($p = 0.045$) at the 0.05 level. An empirical relationship between particle travel time and depth has been proposed as mentioned below.

$$T = 0.19d + 15.80$$

where T is the PM travel time (min) and d is the depth of the source (m)

The empirical equation provides an approach to estimate the particle travel time with depth. This is an important outcome because, with increased depth of opencast working, the corresponding particle travel time can be estimated which will provide information on the particle retention time in the mine. Longer particle travel time will lead to more PM exposure duration of the mine workers.

6.4.5 *Formulation of an Empirical Equation for Variation of Particle Concentration with Depth*

6.4.5.1 When the Source Is Located at Shallow Depth

For a near-surface source location, 72–73% $PM_{2.5-10}$, 53–63% $PM_{1-2.5}$, and 7–15% PM_1 emissions from the mining operations settled in the mine within a short vertical distance (18–20 m) from the source. The higher settling velocity of coarse particles ($PM_{2.5-10}$ and $PM_{1-2.5}$) led to more settling of these particles near the source. Remaining 27–28% $PM_{2.5-10}$, 37–47% $PM_{1-2.5}$, and 85–93% PM_1 either escaped the mine or traveled to the deeper parts of the mine. Concentration of PM at different depths from the source is found to be correlated (R^2 = 0.80−0.99) with depth. For the downward movement of particle in the mine from a source located near the surface, empirical equation has been established showing a generalized relationship between the PM concentration and the depth.

$$C = 100e^{-\beta d}$$

where C is PM concentration (% of source concentration), d is depth (m) from the source, and β is 0.18, 0.17, and 0.09 for $PM_{2.5-10}$, $PM_{1-2.5}$, and PM_1, respectively.

The empirical equation was able to predict the particle concentration between 73–88% of the measured value for shallow (10 m) depth. The prediction accuracy decreased for higher depth (40 m) due to change of settling rate of coarse particles with increase in depth. The rate of settling of $PM_{2.5-10}$ and $PM_{1-2.5}$ gradually decreases as the depth increases due to its decrease in PM plume. PM_1 on the other hand practically does not show differential settling with depth. The equation that was developed from the data up to a depth of 17 m could not account for the variation of settling between 17 m and 40 m for different sizes of particles, and this is suggested to be the reason for the less accuracy of the model prediction for higher depths. Nevertheless, expression of PM concentration–depth relationship gives a new approach that can be applied and further modified with more measurements as an important tool to assess the in-pit exposure of PM.

6.4.5.2 When the Source Is Located at the Bottom of a Deep Opencast Mine

When the source is located at deeper parts of the mine, concentration of PM decreased with increasing height in the mine. For the source depth of 168 m, the concentration of $PM_{2.5-10}$, $PM_{1-2.5}$, and PM_1 at the surface dropped to 3.13%, 3.58%, and 13.53%, respectively, of the concentration at source. An expression showing relationship between relative concentrations (PM concentration at a location in the mine is expressed as a percentage of concentration at the source) and relative depth (Depth of the mine at a location is expressed as a percentage of depth of the source) has been established.

$$C = ae^{bd}$$

where C is relative PM concentration (%), d is relative depth (%), and a and b are the empirical constants (a is 0.024, 0.015, and 0.823 and b is 0.084, 0.088, and 0.048 for $PM_{2.5-10}$, $PM_{1-2.5}$, and PM_1, respectively).

The ANOVA shows a good relationship between the relative depth and the relative concentration (R^2 = 0.98−0.99, $p \leq 0.001$). Quicker settling of coarse particles in comparison with fines results in the faster decrease of the coarse particle concentration with the decrease of the depth. Therefore, coefficient of predictor equation for PM_1 (0.823) is an order of magnitude higher than the coefficients of predictor equations for $PM_{2.5-10}$ (0.024) and $PM_{1-2.5}$ (0.015). The predictor equations established for the upward movement of the particle from the source were able to predict the particle concentration at different benches in the mine by a factor of 2, which is a realistic accuracy that can be expected from this type of complex source–emission relationship, which is also influenced by pit geometry. The equation can thus be applied to estimate PM concentration at a location in the mine as a percentage of PM concentration at the source where the mining operations are in progress.

6.5 Conclusion

Exposure to enhance PM concentration has been the main reason for increased mortality of workers in opencast mines. The particle sizes of interest are $PM_{2.5-10}$, $PM_{1-2.5}$, and PM_1, all of which are important for health. The enhanced understanding of particle dispersion in the mine in terms of identification and evaluation of particle dispersion parameters and development of relationship among them have been the major outcomes of the research. The following conclusions are drawn from this study.

- At the KIOM and MIOM, the particle travel time was recorded up to 17 min for working depth of 9–10 m. At higher depth (MCP), particle took approximately an hour or even longer to come from source to surface (168 m). The findings give a better understanding of the miners' exposure to particle as longer travel time inside the mine is associated with more exposure on miners. Developing a suitable mine geometry for easy dispersion of generated particles from mining operations that makes travel time short should be focussed.

- No correlation between travel time and WS is obtained. Particle travel time was found to be independent of the particle size. The sizes of particle ($PM_{2.5-10}$, $PM_{1-2.5}$, and PM_1) in this study were fine enough to be transported by wind at equal ease. This is presumed to be the reason behind non-dependence of travel time on particle size.
- Particle travel time and depth are significantly correlated. An empirical relationship between particle travel time and depth has been proposed that can be used to estimate the time taken by the PM from mining operations to reach at a location in the mine. Therefore, the exposure duration of the PM inside the mine can be estimated. This can be used as a tool to predict the mine workers' exposure from a source located far away from their workplace and located at a greater depth.
- Particle retention time at the workplace varied from 10 to 13 min. Retention fraction varied between 7 and 40%, and it increased with retention time. Retention time has an inverse relationship with particle size, and this is more pronounced at higher WS that results in effective segregation of PM dispersion. Workers are exposed to the elevated particulate pollution, as high as ~40%, when no mining operation is in progress. Retention time is likely to increase with depth of mine, and therefore, this remains a health risk concern for workers in deep opencast mines.
- An empirical equation has been developed giving relationship between the PM concentration and the depth when the source is located at higher elevation. This can be applied for the estimation of PM concentration in the mine when the mining operation is located in benches near to the surface.
- Relative to source concentration at a depth of 168 m, the concentration of $PM_{2.5-10}$, $PM_{1-2.5}$, and PM_1 dropped to 3.13%, 3.58%, and 13.53%, respectively, at the surface, suggesting very little coarse particles escape the mine.
- Empirical relationship has been established between particle concentration and depth when source is located at the deeper parts of the mine. This can be used as a tool to predict the mine workers' exposure to PM emitted from the workplace located at a greater depth.

The research has led to the development an approach to estimate the PM concentration at different parts of the mine due to mining operations in the mine. The exposure level of workers who are working away from the source can thus be estimated by this approach. For a deep mine, this is of significant interest because PM generated at the bottom benches could take very long time to escape the mine boundary, and during this period, workers at different benches would be exposed to elevated level of PM concentration.

References

Atkinson RW, Kang S, Anderson HR, Mills IC, Walton HA (2014) Epidemiological time series studies of PM2.5 and daily mortality and hospital admissions: a systematic review and meta analysis. Thorax. doi.org/:10.1136/thoraxjnl-2013-204492

Chowdhary KK (1977) An experimental and theoretical investigation of turbulent separated reattached and redeveloped flows with transverse rectangular cavities. Ph.D. thesis, Indian Institute of Technology Delhi, New Delhi

Dockery DW, Pope CA, Xu XP, Spengler JD, Ware JH, Fay ME, Ferris BG, Speizer FE (1993) An association between air-pollution and mortality in six United States Cities. N Engl J Med 329:1753–1759

Fabrick AJ (1982) Technical note: calculation of the effective emissions from mine pit operations by incorporation particulate deposition in the excavated pit. Environmental Research Associates, Chapel Hill

Gautam S, Patra AK (2015) Dispersion of particulate matter generated at higher depths in opencast mines. Environ Technol Innov 3:11–27

Gautam S, Kumar P, Patra AK (2016a) Occupational exposure to particulate matter in three Indian opencast mines. Air Qual Atmos Health 9(2):143–158

Gautam S, Prasad N, Patra AK, Prusty BK, Singh P, Pipal AS, Saini R (2016b) Characterization of PM2.5 generated from opencast coal mining operations: A case study of SonepurBazari opencast project of India. Environ Technol Innov 6:1–10

Grainger C, Meroney RN (1993) Dispersion in an open-cut coal mine in stably stratified flow. Bound-Layer Meteorol 63:117–140

Grimm (2010) Operation manual of Portable Laser Aerosol spectrometer and dust monitor (Model 1.108/1.109). GRIMM Aerosol Technik GmbH & Co. KG, Ainring, Germany

Larcheveque L, Sagaut P, Mary I, Labbe O, Comte P (2003) Large-eddy simulation of a compressible flow past a deep cavity. Phy Fluids 15(1):193–210

Loomans M, Lemaire T (2002) Particle concentration calculations using CFD a comparison. In: Proceedings of Indoor Air, Monterey, California, USA

Naidoo R, Seixas N, Robins T (2006) Estimation of respirable dust exposure among coal miners in South Africa. J Occup Environ Hyg 3:293–300

Nel A (2005) Atmosphere, Air pollution—related illness: effects of particles. Sci 308:804–806

Ngo T, Letchford C (2008) A Comparison of topographic effects on gust wind speed. J Wind Englnd Aerodyn 96:2273–2293

Patra AK, Gautam S, Kumar P (2016) Emissions and human health impact of particulate matter from opencast mining operation—A review. Environ Technol Innov 5:233–249

Peng X, Lu GR (1995) Physical modeling of natural wind and its guide in a large open pit mine. J Wind Englnd Aerodyn 54–55:473–481

Richardson LF (1926) Atmospheric diffusion shown on a distance neighbour graph. Proc R Soc Lond Ser A 110:709–737

Spectrum (2010) Operation manual of Watchdog 2000 series Portable Weather Station. Spectrum Technologies, Inc., USA

Winges KD (1981) Description of the ERTEC mining air quality (EMAQ) model. ERTEC Northwest Inc, Seattle

Zhang X, Chen W, Ma C, Zhan S (2013) Modeling particulate matter emissions during mineral loading process under weak wind simulation. Sci Tot Environ 449:168–173

Chapter 7
Measurement and Removal of Endosulfan from Contaminated Environmental Matrices

Ansaf V. Karim, Swatantra Pratap Singh and Amritanshu Shriwastav

Abstract Endosulfan is a highly toxic chlorinated organic compound that has widely been used throughout the world as a pesticide. Commercial endosulfan is a mixture of two stereoisomers: α-endosulfan and β-endosulfan, in a ratio of 7:3. Due to its recalcitrant nature, it is highly persistent in the environment and has been included in the list of persistent organic pollutants (POPs) by Stockholm Convention in 2011. Further, high residual levels of endosulfan and their metabolites have gradually built up in contaminated environmental matrices at the point of application in all three phases of soil, water, and air. In addition, numerous studies have reported that due to its high chemical stability, semi-volatile, and hydrophobic nature, endosulfan contamination is frequently found in the environment at considerable distances from the point of its original application. Due to the high toxicity to humans, its application has been discontinued in most countries including India. However, the issues of legacy contamination still persist. The partitioning of endosulfan and its metabolites complicates their accurate quantification in the environment and requires complex analytical procedure as well as sophisticated instrumentation. This chapter provides a detailed review of available detection protocols for endosulfan, their advantages, and limitations. Further, the chapter also reviews the existing efforts for the removal of this persistent compound from a variety of environmental matrices.

A. V. Karim · A. Shriwastav (✉)
Centre for Environmental Science and Engineering, IIT Bombay,
Mumbai 400076, India
e-mail: amritan@iitb.ac.in

S. P. Singh
Zuckerberg Institute for Water Research, Ben-Gurion
University of the Negev, Beersheba, Israel

T. Gupta et al. (eds.), *Environmental Contaminants*, Energy, Environment, and Sustainability, https://doi.org/10.1007/978-981-10-7332-8_7

7.1 Introduction

7.1.1 History

Endosulfan is a broad-spectrum organochlorine pesticide belonging to the group of cyclodienes and characterized by the cyclic structure, presence of chlorine atoms, high toxicity, and low aqueous solubility. Also, endosulfan consists of two stereoisomers, α-endosulfan, and β-endosulfan as shown in Fig. 7.1, which exist approximately in a ratio of 2:1–7:3, and both possessing similar insecticidal properties (Kumar and Philip 2006b). It was developed in the early 1950s in USA by Farbwerke Hoechst A.G. under the registered trademark, Thiodan and used as an insecticide. It is a derivative of hexachlorocyclopentadiene and produced by Diels–Alder reaction with butenediol, followed by cyclization with thionyl chloride. It has served as a cost-effective crop protection tool throughout the world, especially in the developing countries. The global consumption of this pesticide was about 9000 tonnes/year in the early 1980s, which increased to 12,800 tonnes/year in 1990s. In 2010, the worldwide production of endosulfan was estimated in the range of 18,000–20,000 tonnes/year, 50–70% of which was produced in India (Janssen 2011).

The environmental effects of endosulfan became clear when its toxicity was linked to the widespread health issues among the exposed population, and awareness for environment increased. Most of the cases on toxicity of endosulfan were reported from India and USA (Randhawa et al. 2007; Weber et al. 2010). Due to its recalcitrant nature, it is highly persistent in the environment and has been included in the list of persistent organic pollutants (POPs) by Stockholm Convention in 2011. In the present scenario, most of the countries, including India, have prohibited the use of endosulfan. A large number of alternatives for endosulfan, such as Thiamethoxam, Coregen, Imidacloprid, have been identified, most of which are much more expensive than endosulfan. However, even after the global ban of the pesticide in 2009 in almost 89 countries, it was still used in limited amount for some of the crops in some countries, viz. India, China etc., because of its high efficiency for pest prevention and killing (Singh and Singh 2014). Also, despite its discontinued application, the legacy contamination of the environmental matrices with endosulfan and metabolites is still a major environmental concern because of their persisting nature.

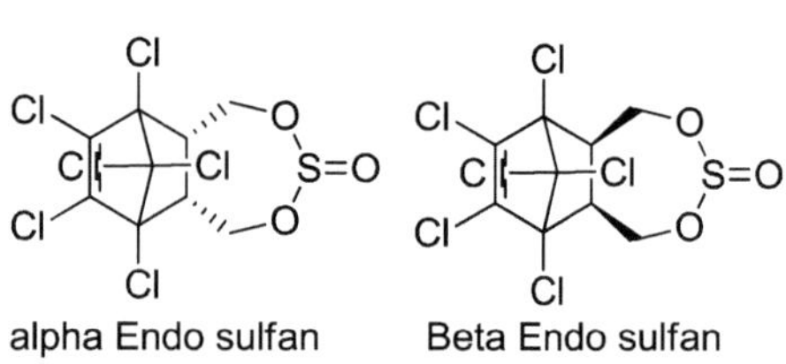

Fig. 7.1 Stereoisomers of endosulfan

7.1.2 Fate and Transport of Endosulfan in the Environment

Endosulfan has been introduced into the environment during its application as a pesticide, after which, its fate depends upon the medium it gets exposed to. Transport of the pollutant in the environment is mainly by adsorption/desorption, volatilization, spray drift, runoff, and abiotic/biotic degradation (Kumar and Philip 2006a; Rauf et al. 2012). The low water solubility and semi-volatile nature of the compound result in the transport and accumulation of the pesticide and its metabolites into the soil, which is controlled by several factors, such as partitioning coefficient, infiltration rate, precipitation, and soil structure. (Dores et al. 2016). These are further transported to the surface water bodies by the surface runoff. Further, depending on the soil characteristics, endosulfan and metabolites are leached into the groundwater. Due to its semi-volatile nature, endosulfan is also transported globally through atmosphere to remote locations with no application history, viz. Arctic (Weber et al. 2010). Such transports are also dependent on the temperature. Also, indirect transport, as wind-induced transport of contaminated dry soil as dust, is a major contributor of the regional and global-scale redistribution (Shen et al. 2005).

The pesticide residue levels in soil have a positive correlation with the clay content in the soil. The presence of clay immobilizes the pesticide, which holds the nutrients and organic molecules to their surfaces (Harikumar et al. 2014). Endosulfan molecules have greater tendency of binding to carbon in soil, which results in its greater persistence in soil and sediments than water samples. The sorption kinetics is fast, and sorption equilibrium is achieved within 3–6 h for endosulfan onto the soil. Sorption of this pollutant governs the leachability, mobility, and bioavailability of the pesticide in soil. Sorption depends upon organic matter, soil type, particle size, clay content, cation exchange capacity, conductivity, and pH of the soils and sediments (Tiwari and Guha 2012).

In normal conditions, the half-life of endosulfan isomers in water is about 35–150 days (Jayaprabha and Suresh 2016). However, these are frequently detected in soil for longer periods with half-life of more than 60–800 days depending on prevalent environmental conditions (Rivero et al. 2012). The half-life of the isomers of endosulfan is lower in tropical region while compared to temperate region due to their volatilization (Dores et al. 2016). α-endosulfan is more volatile and dissipative, while the β-isomer is generally more adsorptive and persistent (Wan et al. 2005).

7.1.3 Toxicity

US Environmental Protection Agency (US EPA) stated that exposure to endosulfan can result in both acute and chronic risks in terrestrial and aquatic environments. According to the US EPA, endosulfan concentrations above 0.22 μg/L (acute) and

0.056 μg/L (chronic) will have an adverse impact on the health of aquatic organisms (Mersie et al. 2003). It is classified in US EPA toxicity class I-b (highly hazardous), while the European Union also rated it as highly hazardous. The permissible limit of endosulfan in drinking water is 0.4 μg/L according to Bureau of Indian Standards, 2012 (Jayaprabha and Suresh 2016). The increased consumption of pesticide for pest control and crop diseases resulted in damaging the ecosystem. Bio-accumulation or bio-concentration of the pollutant in an organism can occur in different ways, either by eating food contaminated with it, direct contact with endosulfan contaminated water, by skin contact with contaminated soil or by breathing the dust while spraying, etc.

The major route for endosulfan absorption is through dermal or respiratory intake, as it has the potential to evaporate and travel long distances in the atmosphere (Venugopal and Sumalatha 2011). Endosulfan contaminated food and water cause serious endocrine disruption, central nervous system disorders, as well as blood chemistry, parathyroid gland, and reproductive system disorders. The population which is potentially at greater risks comprises of unborn and elderly people, and people with liver, kidney, or neurological diseases (Lu et al. 2000; Saiyed et al. 2003).

Human exposure can cause severe health issues as shown in Table 7.1, such as liver and kidney damage, nervous disorders, reproductive and developmental effects (Li et al. 2009; Thangadurai and Suresh 2014). Exposure to very large amount for short duration can lead to the acute toxicity toward the nervous system, blood, liver, kidneys, and which can further lead to death. Neurotoxicity is the primary effect observed in human beings. Further, acute exposure in humans and animals can have neurological effects, such as hyperactivity, decreased respiration, dyspnoea,

Table 7.1 Toxicity effects of endosulfan on humans

Compound	Toxicity effects on human	Location	Reference
Endosulfan	Male reproductive development	Kerala, India	Saiyed et al. (2003)
α-endosulfan and β-endosulfan	Genotoxicity on human HepG2 cells (liver tissue)	Saitama, Japan	Lu et al. (2000)
Endosulfan	Endocrine disruptor and a genotoxin	Granada	Fernandez et al. (2007)
Endosulfan	Effects on immune, reproductive, endocrine, and nervous systems of human beings	Kerala, India	Jayaprabha and Suresh (2016)
Endosulfan residues such as endosulfan sulfate, endosulfan ether	Women of reproductive age affected because chemicals can get mobilized during pregnancy and lactation	Spain	Silva and Beauvais (2010)
Endosulfan	Reproductive health events including infertility, precocious puberty, abortion, intra-uterine death	Kerala, India	Government of Kerala, 2001

salivation. It can cause breast cancer and also interferes in male hormones estrogenic and antiandrogenic activities (Westbom et al. 2008). Laboratory studies have reported that it can be mutagenic and genotoxic and can cause birth defects in animals and humans on exposure (Siddique et al. 2003; Li et al. 2009).

The toxicity of α-endosulfan is more to insects and mammals than β-endosulfan, because of its physicochemical properties (Wan et al. 2005). It is extremely toxic to birds, bees, earthworms and had caused massive killing of fish (Greve and Wit 1971; Verma et al. 2006). The residual concentration of endosulfan in soil and water is a common problem which causes several health abnormalities in fish and aquatic invertebrates due to its low water solubility, and high acute oral, and inhalation toxicity effect in human beings. The release of endosulfan sulfate, one of the more toxic metabolites of endosulfan, into the water can result in its adsorption onto the sediments, and which may further bio-accumulate in aquatic organisms.

7.2 Quantification of Endosulfan and Metabolites

7.2.1 *Extraction of Endosulfan and Metabolites from Contaminated Environmental Matrices*

Extraction of the compound for accurate quantification depends upon the phase where it is present. During the extraction of a semi-volatile compound like endosulfan from aqueous or solid phase, the chemical properties of the compound, such as vapor pressure, molecular weight, solubility, hydrophobicity, are considered. The determination of pesticide residues in environmental matrices primarily depends upon the accuracy of their extraction and enrichment. The most common methods of extraction of pesticides from environmental matrices include liquid–liquid extraction (LLE) and solid-phase extraction (SPE) (EPA Method 8081). The major steps involved in the extraction of endosulfan from aqueous or solid (soil) matrix are shown in Fig. 7.2.

7.2.1.1 Extraction from Aqueous Phase

The solvent system for extraction involves two immiscible liquids. In liquid–liquid extraction (LLE), one phase is usually aqueous and the other is an organic solvent immiscible in water. The extraction requires the analyte to have favorable solubility in the organic solvent. The extraction of endosulfan and metabolites usually requires organic solvents, such as n-hexane, ethyl acetate, acetone, dichloromethane. n-hexane alone or mixed with acetonitrile is a common extractant for endosulfan.

Further, though the liquids are immiscible, they are often somewhat soluble in each other and would be mutually saturated while mixing, which allows them to

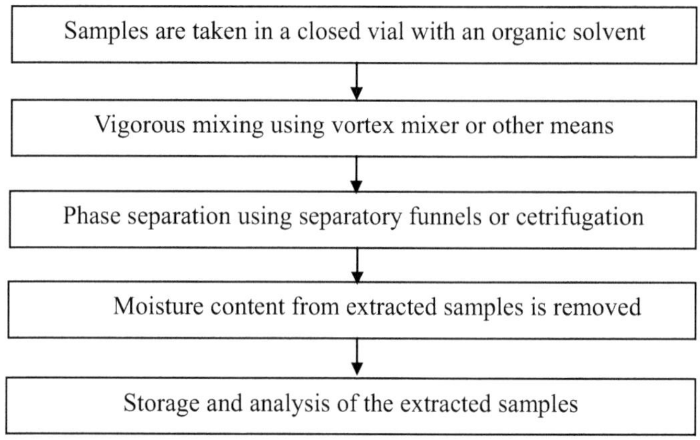

Fig. 7.2 Flow diagram of steps required for extraction of endosulfan from different phases

form a dispersion. Extraction is usually accomplished by mixing the mixture using either a separatory funnel or a vortex mixer to enable extraction of these compounds from aqueous phase to the organic solvent, and then the distinct layers of these two liquids are separated after mixing. The separation of two liquids after the process is usually done in a separatory funnel or with centrifugation. The residual moisture content inside the extracted samples in organic solvents is removed using inorganic solids, viz. anhydrous sodium sulfate. The organic solvent with extracted compounds is evaporated using a rotary evaporator. The extracted compounds are again dissolved in the organic solvent containing internal standard and stored for analysis (El-Gawad 2016).

Tiwari and Guha (2012) extracted endosulfan from aqueous phase using three-step liquid–liquid extraction method. Samples were mixed with an organic solvent (ethyl acetate) inside a tightly capped vial and mixed in a rotary shaker for 30 min. The contents were centrifuged at 3000 g for phase separation, and the pooled extracts after three successive extractions to achieve sixfold dilutions were stored for analysis. Memon et al. (2011) used ethyl acetate/n-hexane for extracting endosulfan from water samples inside a separatory funnel. The upper organic layer in the funnel was separated after mixing, and moisture content was removed using anhydrous sodium sulfate. Samples were stored for analysis after filtering through Whatman filter paper (0.45 μm). Mishra and Patel (2008) extracted endosulfan from aqueous solution in a separatory funnel using n-hexane as the solvent by liquid–liquid separation method. For low concentration of pesticides, the extracts were condensed in a rotary evaporator and stored for analysis.

Although LLE is an effective and fully developed technique, its application is limited due to the more time consumption, cost, and difficulty in handling large volume of solvents (Martínez Vidal et al. 2009). Hence, SPE is preferred as it reduces the use of solvents and can combine extraction and clean up in a single step. In SPE, the pesticide residues present in an aqueous solution are extracted using a cartridge column (usually C18) using a metering pump which controls the flow and aspirated through the cartridge under vacuum (Pablos-Espada et al. 1999). Arrebola et al. (2001) conditioned the column using a mixture of serum, internal standard, and methanol before the extraction process to avoid the dryness and interferences which decrease the column efficiency. To overcome the drawbacks associated with LLE, several other extraction procedures are also developed. The common methods are solid-phase micro-Extraction (SPME), pressurized liquid extraction (PLE), supercritical fluid extraction (SFE), matrix solid-phase dispersion (MSPD), and microwave-assisted extraction (MAE) (Martínez Vidal et al. 2009; Wille et al. 2012). SPME is normally an automated extraction process in which the sampling handling issues and solvent consumption are very less with higher sensitivity for trace analysis which is directly coupled to GC and MS. These methods are highly efficient, which consumes very less amount of solvent and the average extraction times are less compared to other methods.

7.2.1.2 Extraction from Soil

Extraction of endosulfan from soil was done after shaking the mixture of soil samples and organic solvent (mixture of hexane and acetone) in mechanical shaker and then filtering using Whatman filter paper (Goswami and Singh 2009). The solvent extracts were redissolved in organic solvent after being dried in rotary evaporator and stored in vials for analysis. In SPE of pesticides, C-18 cartridges conditioned with solvents, such as methanol and water, were used. Samples were pre-concentrated under vacuum. Mixed extracting solvent of n-hexane and acetone having higher polar nature was used for extraction of endosulfan from soil. The extraction was carried out following the standard methods with some modification by addition of an inorganic salt NaCl resulting in salt-induced phase separation (Harikumar et al. 2014). Coextractives in pooled extracts were removed in an alumina column and the cleaned-up extracts were stored in airtight vials until analysis.

Ultrasonic extraction from homogenized soil samples can be done in glass with extraction solvent. Samples are manually agitated initially and exposed to ultrasonic waves for 10 min at 25 °C. Extracts are passed through a glass fiber filter packed Buchner funnel where extracts are further evaporated and stored for analysis. Li et al. (2009) extracted endosulfan residues from an aqueous solution of bacterial assay using acetonitrile as solvent. An equal volume of solvent was taken and shaken for 1 h in a reciprocating shaker followed by centrifugation to separate the supernatant. Kumar et al. (2008) extracted the pesticide residues from soil samples using hexane: acetone as solvent in an ultrasonic water bath. Endosulfan present in the supernatant was evaporated and redissolved in acetone for quantification. The extraction efficiency of the process varied in the range of 85–90%.

Selective pressurized liquid extraction (SPLE) technique can also be used for extraction of endosulfan from soil samples which allows simultaneous extraction and cleanup. It is a fast and low solvent-consuming process in which an adsorbent, usually Florisil, is introduced as the adsorbent inside the PLE cell downstream the sample to remove interfering compounds (Westbom et al. 2008). The packing of the extraction cell consists of glass fiber filters at the top, followed by a soil sample blended with diatomaceous earth, activated Florisil, sodium sulfate for in-cell cleanup and two glass fiber filters at the bottom inside a stainless steel packing (Hussen et al. 2007).

Awasthi et al. (2000) extracted endosulfan from bacterial incubated soil samples by first drying the sample in oven. The dried soil was transferred to a test tube, and ethyl acetate was used as the solvent for extraction. One g of dried soil was mixed with 3 mL of ethyl acetate and extracted by vortexing. Ethyl acetate layer was decanted, and the process was repeated for three times. The moisture content in pooled ethyl acetate fraction was removed by passing it through anhydrous sodium sulfate. The efficiency of the process is ranged between 85 ± 2%. Soil samples used for microbial degradation can also be extracted using a Teflon homogenizer. Shivaramaiah and Kennedy (2006) added hydrochloric acid into the aqueous sample, followed by extraction by vortexing with extracting solvent, and centrifugation at 2000 rpm. Organic layer was separated, and the moisture content was removed using anhydrous sodium sulfate as the drying agent prior to storage.

Acetonitrile was used to extract endosulfan and its metabolites from bacterial enriched medium. The extracted samples were filtered through Gelman Acrodisc syringe filter of 0.2 μm (Hussain et al. 2007). Kumar and Philip (2006a) extracted endosulfan inside a standard separatory funnel with Teflon stopper using n-hexane as the solvent. The water layer was decanted after vigorous shaking of the mixture to extract the supernatant, which was further dehydrated using sodium sulfate. Extraction of endosulfan containing soil sample was carried out inside an orbital shaker at 150 rpm for 4 h followed by centrifugation at 5000 rpm for 15 min (Kumar and Philip 2006b). Extraction of endosulfan after bioremediation from a bacterial culture is usually done by adding equal volume of hexane to the culture and removing the moisture content by anhydrous sodium sulfate.

7.2.2 Quantification of Endosulfan and Metabolites

The accurate quantification of extracted endosulfan and metabolites is still a complex process and relies on complex analytical procedure as well as sophisticated instrumentation. Chromatography is the most widely utilized process for their determination, in which the relative affinity of different compounds in a sample to some stationary phase is utilized for their separation and subsequent quantification. Gas chromatography (GC) and high-performance liquid chromatography (HPLC) are two such techniques which have found increasing application for the quantification of these compounds. The inclusion of mass spectrometry, viz. GC-MS, has further resulted in identifying the components at molecular level with higher accuracy.

The semi-volatile nature, as well as thermal stability of this pesticide and metabolites, makes these suitable for GC analysis. GC equipped with Ni electron-capture detector (ECD) or hydrogen-flame ionization detector (FID) with N_2 as the carrier gas is widely utilized for endosulfan determination in the extracted samples (Kong et al. 2013). The type of carrier gas used for the analysis depends on the detector. Westbom et al. (2008) utilized GC-MS equipped with a capillary column with helium as the carrier gas for analyzing the metabolites. The flow rate of the carrier gas was usually within the range of 1–1.5 mL/min. Also, the temperature of the injector and detector were generally maintained between 200–300 °C. GC/ECD equipped with an on-column (sample is introduced directly into the column with a thin injection needle), and split/split less capillary injection system with Perkin Elmer PE 35 capillary column was used for the analysis of endosulfan (Kumar and Philip 2006a, b). The injection of the samples was made in split mode with a split ratio of 1:10, and N_2 at a flow rate of 30 mL/min was used as the carrier gas (Kumar and Philip 2007). Tiwari and Guha (2013a, b) established the superior selectivity of tandem mass spectrometry GC-MS/MS to GC-MS for better accuracy in identification and quantification of this pesticide and intermediates. Various compounds have also been utilized as internal standards for accurate quantification. Westbom et al. (2008) used a mixture of tetrachloro-m-xylene (retention time reference) and decachlorobiphenyl (quantification reference) prepared in n-hexane and used as internal standard. Table 7.2 summarizes the details of the GC analysis of Endosulfan by various studies.

In HPLC analysis, the extracted sample having mixture of target compounds is separated into components by allowing it to pass through a stationary solid phase, usually in a column, under high pressure of up to 400 atm using mobile liquid phase. The components of the sample get separated by the column packing due to various chemical or physical interactions, and separated components are then detected at the exit of the column by an external detector. The common mobile phase for the analysis is a mixture of methanol or acetonitrile and water.

The amount of endosulfan present in extracted samples was analyzed using an HPLC equipped with a reverse-phase column and UV detector with methanol as the mobile phase (Yonli et al. 2012). Endosulfan residues extracted from soil samples were detected by HPLC equipped with Zorbax SB-C18 column and acetonitrile: water (80:20, v/v) as the mobile phase (Li et al. 2009). Extracted samples from soil after biodegradation studies filtered through Gelman GHP Acrodisc 0.2 µm syringe filter were analyzed with HPLC using an ODS-Hypersil column as stationary phase and acetonitrile: water (70:30) as the mobile phase. Hussain et al. (2007) analyzed the extracted samples from a bacterial enriched medium with HPLC using ODS-Hypersil column as the stationary phase and acetonitrile: water (70:30, v/v) as the mobile phase. The injection volume was 20 µL, and the solutes were identified using UV-VIS detector at 214 nm. Table 7.3 summarizes the details of the HPLC analysis of Endosulfan by various studies.

In addition to the chromatography, Venugopal and Sumalatha (2011) had used spectrophotometric method for the analysis of endosulfan present in aqueous solution at 605 nm. The method was based on the release of sulfur dioxide from

Table 7.2 Quantification of endosulfan using GC

Type of column	Carrier gas and flow rate	Type of detector	Analysis range	Internal standard	Temperature	Reference
Perkin Elmer-35 capillary column	Nitrogen, 30 mL/min	GD/ECD	0–50 (mg/L)		Initial oven temperature of 120 °C (1 min), 30 °C/min from 120 to 180 °C, 20 °C/min to 240 °C (3 min) Injector—260 °C, Detector—300 °C	Kumar and Philip (2006b)
Shimadzu 2010-DB-1701 capillary column	Nitrogen, 1 mL/min	GC/ECD, Nickel	0–50 (mg/L)		Initial oven temperature of 80 °C (0.5 min), 35 °C/min from 80 to 160 °C (1 min), 8 °C/min from 160 to 200 °C (1 min) 3 °C/min from 200 to 240 °C (6 min) Injector—240 °C, Detector—300 °C	Deng et al. (2016)
Agilent 6890—HP-5 capillary column	Nitrogen, 1.5 mL/min	GC/ECD, Nickel	0–100 (mg/L)		Initial oven temperature of 150 °C (4 min) 6 °C/min from 150 to 290 °C (4 min) Injector—250 °C, Detector—300 °C	Thangadurai and Suresh (2014)
Perkin Elmer-silica capillary column MXT-5	Nitrogen 1 mL/min	GC/ECD, Nickel	0–1.229 (μmol/L)	2,4,5,6-tetrachloro-m-xylene	Initial oven temperature of 90 °C (7 min), 10 °C/min from 90 to 120 °C, 15 °C/min from 120 to 270 °C (5 min) Injector—250 °C Detector—375 °C	Tiwari and Guha (2012)
Agilent 6980 HP-5 capillary column	Nitrogen	GC/ECD, Nickel	0–7.5 (mg/L)		Initial oven temperature of 150 °C (2 min), 10 °C/min from 150 to 250 °C (2 min) Injector—200 °C Detector—290 °C	Begum et al. (2014)
Capillary column Mega 68		GC/ECD		Chlorpyrifos-methyl	Initial temperature of 100 °C (3 min), 10 °C/min from 100 to 180 °C (15 min), 5 °C/min from 180 to 270 °C (10 min) Injector—280 °C Detector—280 °C	Rivero et al. (2012)

Table 7.3 Quantification of endosulfan using HPLC

Type of column	Mobile phase	Type of detector	Analysis range	Flow rate	Reference
Zorbax SB-C18 column	Acetonitrile: water (80:20, v/v)	PDA detector with gradient UV–vis detection	0–100 mg/L	1 mL/min	Li et al. (2009)
Separation column was Pecosphere 5CR C18	Water: acetonitrile (35:65, v/v)	UV/Vis detector		1 mL/min	Singh and Singh(2014)
Agilent eclipse PAH 5 μm column	Acetonitrile: water (70:30, v/v)	PDA detector	1–10 mg/L	1 mL/min	Sivagami et al. (2016)
Reverse-phase C18 column	Methanol: water (70:30, v/v)	UV/Vis detector		1 mL/min	Randhawa et al. (2007)

endosulfan by alcoholic potassium hydrolysis. The released gas is passed through the reagents hydrogen peroxide and diphenyl amine resulting in violet color formation, which was measured at 605 nm.

7.3 Removal of Endosulfan from Contaminated Environmental Matrices

Control measures to achieve the detoxification of the compound rely on the chemical and physical properties of endosulfan, such as solubility, volatility, environmental stability, biodegradability, and the nature of the contaminated matrix. Also, the potential of a process for the extent of mineralization of the endosulfan and the formation of intermediate metabolites is crucial for its effectiveness.

7.3.1 Biological Processes for Endosulfan Degradation

Biodegradation of pesticide in soil depends upon factors, such as availability of pesticide or metabolite to the microorganisms, sustainable population of these microorganisms, survival and/or proliferation of pesticide-degrading microorganisms at contaminated site, and physiological status of the microorganisms. For successful achievement of biodegradation of the pollutant, the creation of unique microhabitats for desired microbes is required, which may depend on factors such as pH, temperature, nutrients, and amount of pesticides. The microbial degradation of endosulfan is generally a slow process.

In biodegradation, endosulfan can be degraded by oxidation to form toxic metabolite endosulfan sulfate via sulfate group attacking or hydrolysis to metabolize endosulfan and produce predominantly less toxic endosulfan diol (Kong et al. 2013). The formation of endosulfan sulfate was through microbial transformation, while hydrolysis is dominant at alkaline pH. The water-soluble endosulfan diol can be further degraded to less toxic metabolites, such as endosulfan ether, endosulfan hydroether, endosulfan lactone, and endosulfan hydroxycarboxylate (Goswami et al. 2009). Endosulfan sulfate is the most stable transformation product which is frequently found in water. Due to the formation of highly toxic metabolites, viz. endosulfan sulfate and poor removal efficiency, biodegradation of endosulfan is less effective. Biodegradation of endosulfan can be either aerobic or anaerobic. The formation of endosulfan sulfate and endosulfan diol was the rate-limiting steps in aerobic process, while conversion to endosulfan diol was the rate-limiting step in anaerobic process (Tiwari and Guha 2013b). The removal of sulfur group from the compound decreases the vertebrate toxicity of endosulfan, which further results in detoxification of the compound. Endosulfan in aqueous solution can be degraded by

Table 7.4 Biodegradation of endosulfan

Microbes used	Type of reactor and removal phase	Removal achieved	Major metabolites	Experimental duration	Reference
Agrobacterium tumefaciens	Batch, liquid phase	99%	No intermediates	160 h	Thangadurai and Suresh (2014)
Aspergillus sydoni	Batch, soil	95% of α-isomer and 97% of β-isomer	Endosulfan sulfate, endosulfan ether, endosulfan lactone	18 days	Goswami et al. (2009)
Achromobacter xylosoxidans	Batch, soil and aqueous phase	80–90% degradation depending upon the culture medium	Endosulfan ether and endosulfan diol	8 days in aqueous phase and 30 days in soil	Li et al. (2009)
Staphylococcus sp., *Bacillus circulans*–I, and *Bacillus circulans*–II	Batch, soil	71.82 in aerobic and 76.04 in facultative condition	No intermediates	28 days	Kumar and Philip (2006c)
Aspergillus niger	Batch, aqueous	100%	Endosulfan diol, endosulfan sulfate, and an unidentified metabolite	12 days	Bhalerao and Puranik (2007)
Alcaligenes faecalis JBW4	Batch, aqueous	87.5 of α-isomer and 83.9% of β-isomer	Endosulfan diol and endosulfan lactone	5 days	Kong et al. (2013)

certain bacteria, such as *Agrobacterium tumefaciens*, *Alcaligenes faecalis*, by using it as a carbon and energy source (Thangadurai and Suresh 2014; Kong et al. 2013). Pure as well as mixed bacterial cultures (*Staphylococcus* species., *Bacillus circulans* I, *Bacillus circulans* II, and methanogens) have been reported to degrade both the isomers of endosulfan without producing any toxic intermediates in both aqueous and soil medium(Kumar and Philip 2006c; Tiwari and Guha 2013a). Due to the limited bioavailability of endosulfan in soil, the rate of degradation is slow compared to aqueous phase. Since endosulfan contains only six potential reducing electrons, it is a poor biological energy source. Even though biological treatments are slow in nature, these are more preferable in comparison with physicochemical methods, since the later produce harmful metabolites and involve high treatment cost.

The degradation rate by microorganism is affected by factors such as non-optimal temperature, higher moisture content, and pH variation among others. Bacteria and fungi are commonly used for bioremediation, and the production of different metabolites is species specific. Rot fungi having considerably higher tolerance to toxic pollutants can decompose the pollutant by the production of extracellular laccases, oxidases, and hydrolase. Fungi grow in a heterogeneous media and degrade the endosulfan through complex biochemical pathways (Rivero et al. 2012). Faster degradation of β-endosulfan in soil has been observed, which could be due to higher partition coefficient of β-isomer (1.8 times) than that of α-isomer, or due to the conversion of β-isomer to α-isomer in the presence of bacteria (Tiwari and Guha 2013b). Table 7.4 summarizes some major findings about the biodegradation of endosulfan in the literature.

7.3.2 PhysicoChemical Processes for Endosulfan Degradation

Incineration of endosulfan at higher temperature in the presence of hydrochloric acid as the scrubber was the disposal method followed initially. The residue or sludge generated during the process is disposed of in a designated landfill site. Underground disposal of the pesticide and disposal at deeper wells located far below freshwater aquifers were also followed as disposal techniques. The hydrophobic nature of the pollutant and affinity to sediments favors adsorption mechanism as an effective method for the removal of endosulfan from aqueous matrix (Yedla and Diksiht 2008). The common adsorbents used for the removal of endosulfan are activated carbon, zeolites, natural organic substances, wheat straw, peach-nut shells carbon slurry, salwood charcoal, etc. (Mishra and Patel 2008; Memon et al. 2009). Most of the studies have reported that 90–95% removal was achieved using adsorption. Zeolites are commonly used to remove organochlorinated pesticides due to their low cost and regeneration ability (Yonli et al. 2012).

Reductive transformation of Endosulfan and its metabolites were attained by treating with micro- or nano-sized metals and bimetallic systems. Zero-valent iron (ZVI) reduces the endosulfan either via beta-elimination or elimination of chlorine molecules sequentially from the compound resulting in dechlorination (Singh and Bose 2017). Begum and Gautam (2011) dechlorinated endosulfan in aqueous phase using $Mg^0/ZnCl_2$ bimetallic system into hydrocarbon skeletons. The solubility of the compound in reaction mixture was increased by adding acetone which accelerated the mass transfer reaction. Zero-valent zinc (Zn^0) in powdered form in acidic conditions degraded endosulfan in soil and water by reducing the four chlorine atoms in the compound (Cong et al. 2015). Zero-valent magnesium (Mg^0) immobilized on Palladium catalyst transformed endosulfan into harmless hydrocarbon product suggesting the complete dechlorination and desulfurization of endosulfan (Thangadurai and Suresh 2013). Singh and Bose (2017) inferred that nano-ZVI and (Mg^0) dechlorinated the compound through sequential electron transfer at the metallic surface.

Photodegradation of endosulfan by UV irradiation yields endosulfan diol as the major product, while minor amounts of endosulfan ether, endosulfan lactone, and hydroxy endosulfan ether are also produced. The insecticidal activity of endosulfan is reduced after photolysis due to the loss of activity with the conversion of active substances to polar compounds whose penetration into insects is hindered. Photolysis of endosulfan solution only results in metabolites formation, while the complete degradation is not achieved (Amin et al. 2014). The removal rate is less for degradation with sunlight as the absorption spectra of the pollutant does not overlap with the incident light. The degradation in the presence of TiO_2 as catalyst yields better removal than direct sunlight. Photolysis, with addition of H_2O_2 as the oxidant, enhanced the reduction rate due to the availability of more hydroxyl radicals (Amin et al. 2014). Photodegradation of endosulfan in soil is restricted to a depth of 0.2–0.4 m and 2 mm in top soil owing to the strong shielding action of soil. Also, the hydrophobic nature of the pollutant hinders the degradation rate due to its sorption on soil. The presence of inorganic ions, such as carbonates, bicarbonates, nitrates, chloride, can inhibit photolytic degradation of endosulfan. Further, radical-based removal of endosulfan using advanced oxidation processes is usually achieved by hydrogen abstraction and electron transfer (Shah et al. 2013). During the degradation, electron abstraction occurs which produces radical—cation intermediate, and which is further oxidized into endosulfan sulfate through hydroxylation. Chemical oxidation results in the formation of toxic metabolites which are more persistent than the parent isomers. Table 7.5 summarizes some major findings about the physicochemical removal of endosulfan in the literature.

Table 7.5 Endosulfan degradation by physicochemical processes

Process	Type of reactor and removing agent	Removal achieved	Major metabolites	Experimental duration	Reference
Adsorption	Batch, on wood charcoal in aqueous solution	90%	No studies on intermediate identification	5 h	Yedla and Diksiht(2008)
Adsorption	Batch and column, on carbon slurry	81.55% in column studies	No studies on intermediate identification	90 min	Gupta and Ali (2008)
Adsorption	Batch, sand along with activated charcoal and salwood charcoal	94% for activated charcoal and 87% for sal wood charcoal	No studies on intermediate identification	24 h	Mishra and Patel (2008)
Photocatalytic	Batch, soil aqueous phase system	100%	Endosulfan sulfate	240 min	Xiong et al. (2015)
Fenton's process	Batch, aqueous solution	83%	No intermediates identified	1 h	Begum et al. (2014)
Ozonation	Batch, aqueous solution	89–93%	No intermediates identified	1 h	Begum and Gautam (2012)
ZVI reduction	Batch in borosilicate glass vials of 40 mL, aqueous solution	100% more rate with nano-ZVI than micro ZVI	No intermediates identified		Singh and Bose (2016)

7.4 Conclusion

Endosulfan is a broad-spectrum organochlorine pesticide with high toxicity and bio-accumulation potential. The residual concentrations of endosulfan in soil and water is a common problem, which causes several health abnormalities in fish and aquatic invertebrates, and also results in high acute oral and inhalation toxicity effect in human beings. Exposure to endosulfan and its metabolites can result in endocrine disruption, neurological effects, and severe health issues in humans and animals. The persistence of the compound in the soil results in bio-accumulation of the pollutant in an organism either by direct contact with skin, eating food contaminated with it or by breathing the dust while spraying, etc.

Extraction of endosulfan from aqueous or solid phase depends upon the chemical properties of the compound, such as vapor pressure, molecular weight, solubility, hydrophobicity. The most common methods of extraction of pesticides from environmental matrices include liquid–liquid extraction (LLE) and solid-phase extraction (SPE). Extraction is usually achieved by mixing the sample in a closed vial, and the distinct layer is separated and stored for analysis. The sensitivity for determination of pesticide residues in environmental matrices primarily depends upon the accuracy in extraction and quantification of extracted samples are usually done using sophisticated instruments, such as GC and HPLC. Removal of the pollutant from the environmental matrices is usually achieved by incineration, biodegradation, adsorption, ozonation, chemical oxidation, advanced oxidation process, etc. The efficiency of different processes depends on the concentration of the pollutant and the nature of the contaminated matrix and the formation of intermediate metabolites. In addition, the toxicity of the intermediate metabolites also plays a major role in the effectiveness of a treatment process.

References

Amin MM, Jaberian B, Bina B, et al (2014) Advanced oxidation of the endosulfan and profenofos in aqueous solution using UV/H2O2 process. https://doi.org/10.14456/ea.2014.9

Arrebola FJ, Martínez Vidal JL, Fernández-Gutiérrez A (2001) Analysis of endosulfan and its metabolites in human serum using gas chromatography-tandem mass spectrometry. J Chromatogr Sci 39:177–182

Awasthi N, Ahuja R, Kumar A (2000) Factors influencing the degradation of soil applied endosulfan isomers. Soil Biol Biochem 32:1697–1705

Begum A, Gautam SK (2011) Dechlorination of endocrine disrupting chemicals using $Mg^0/ZnCl_2$ bimetallic system. Water Res 45:2383–2391. https://doi.org/10.1016/j.watres.2011.01.017

Begum A, Gautam SK (2012) Endosulfan and lindane degradation using ozonation. Environ Technol 33:943–949. https://doi.org/10.1080/09593330.2011.603752

Begum A, Agnihotri P, Mahindrakar AB, Gautam SK (2014) Degradation of endosulfan and lindane using Fenton's reagent. Appl Water Sci 207–215. https://doi.org/10.1007/s13201-014-0237-z

Bhalerao TS, Puranik PR (2007) Biodegradation of organochlorine pesticide, endosulfan, by a fungal soil isolate, *Aspergillus niger*. Int Biodeterior Biodegrad 59:315–321. https://doi.org/10.1016/j.ibiod.2006.09.002

Cong L, Guo J, Liu J et al (2015) Rapid degradation of endosulfan by zero-valent zinc in water and soil. J Environ Manage 150:451–455. https://doi.org/10.1016/j.jenvman.2014.12.028

Deng F, Xiong B, Chen B et al (2016) Microbial degradation of endosulfan in contaminated soil with the elution of surfactants. Environ Sci Pollut Res 23:13268–13275. https://doi.org/10.1007/s11356-016-6523-6

Dores EFGC, Spadotto CA, Weber OLS et al (2016) Environmental behavior of chlorpyrifos and endosulfan in a tropical soil in central Brazil. J Agric Food Chem 64:3942–3948. https://doi.org/10.1021/acs.jafc.5b04508

El-Gawad HA (2016) Validation method of organochlorine pesticides residues in water using gas chromatography–quadruple mass. Water Sci 30:96–107. https://doi.org/10.1016/j.wsj.2016.10.001

Fernandez MF, Olmos B, Granada A, López-Espinosa MJ, Molina-Molina JM, Fernandez JM, Olea N (2007) Human exposure to endocrine-disrupting chemicals and prenatal risk factors for cryptorchidism and hypospadias: a nested case-control study. Environ Health Perspect (SUPPL1) 115:8–14. https://doi.org/10.1289/ehp.9351

Goswami S, Singh DK (2009) Biodegradation of a and b endosulfan in broth medium and soil microcosm by bacterial strain Bordetella sp. B9. Bull Environ Contam Toxicol 20:199–207. https://doi.org/10.1007/s00128-013-1168-3

Goswami S, Vig K, Singh DK (2009) Biodegradation of α and β endosulfan by Aspergillus sydoni. Chemosphere 75:883–888. https://doi.org/10.1016/j.chemosphere.2009.01.057

Greve PA, Wit WL (1971) Endosulfan in the Rhine River. J (Water Pollut Control Fed 43:2338–2343. https://doi.org/10.1177/03063127067078012

Gupta VK, Ali I (2008) Removal of endosulfan and methoxychlor from water on carbon slurry. Environ Sci Technol 42:766–770. https://doi.org/10.1021/es7025032

Harikumar PS, Jesitha K, Megha T, Kokkal K (2014) Persistence of endosulfan in selected areas of Kasaragod district, Kerala. Curr Sci 106:1421–1429

Hussain S, Arshad M, Saleem M, Khalid A (2007) Biodegradation of α- and β-endosulfan by soil bacteria. Biodegradation 18:731–740. https://doi.org/10.1007/s10532-007-9102-1

Hussen A, Westbom R, Megersa N et al (2007) Selective pressurized liquid extraction for multi-residue analysis of organochlorine pesticides in soil. J Chromatogr A 1152:247–253. https://doi.org/10.1016/j.chroma.2007.02.076

Janssen MPM (2011) Endosulfan. A closer look at the arguments against a worldwide phase out, National Institute for Public Health and Environment, Ministry of Health Welfare and Sport, RIVM Letter report, Netherlands

Jayaprabha KN, Suresh KK (2016) Endosulfan contamination in water: a review on to an efficient method for its removal. J Chem Chem Sci 6:182–191

Kong L, Zhu S, Zhu L et al (2013) Biodegradation of organochlorine pesticide endosulfan by bacterial strain *Alcaligenes faecalis* JBW4. J Environ Sci (China) 25:2257–2264. https://doi.org/10.1016/S1001-0742(12)60288-5

Kumar M, Philip L (2006a) Adsorption and desorption characteristics of hydrophobic pesticide endosulfan in four Indian soils. Chemosphere 62:1064–1077. https://doi.org/10.1016/j.chemosphere.2005.05.009

Kumar M, Philip L (2006b) Enrichment and isolation of a mixed bacterial culture for complete mineralization of endosulfan. J Environ Sci Health B 41:81–96. https://doi.org/10.1080/03601230500234935

Kumar M, Philip L (2006c) Bioremediation of endosulfan contaminated soil and water-Optimization of operating conditions in laboratory scale reactors. J Hazard Mater 136:354–364. https://doi.org/10.1016/j.jhazmat.2005.12.023

Kumar M, Philip L (2007) Biodegradation of endosulfan-contaminated soil in a pilot-scale reactor-bioaugmented with mixed bacterial culture. J Environ Sci Health B 42:707–715. https://doi.org/10.1080/10934520701304344

Kumar M, Lakshmi CV, Khanna S (2008) Biodegradation and bioremediation of endosulfan contaminated soil. Bioresour Technol 99:3116–3122. https://doi.org/10.1016/j.biortech.2007.05.057

Li W, Dai Y, Xue B et al (2009) Biodegradation and detoxification of endosulfan in aqueous medium and soil by Achromobacter xylosoxidans strain CS5. J Hazard Mater 167:209–216. https://doi.org/10.1016/j.jhazmat.2008.12.111

Lu Y, Morimoto K, Takeshita T et al (2000) Genotoxic effects of alpha-endosulfan and beta-endosulfan on human HepG2 cells. Environ Health Perspect 108:559–561. https://doi.org/10.2307/3454619

Martínez Vidal JL, Plaza-Bolaños P, Romero-González R, Garrido Frenich A (2009) Determination of pesticide transformation products: a review of extraction and detection methods. J Chromatogr A 1216:6767–6788. https://doi.org/10.1016/j.chroma.2009.08.013

Memon GZ, Bhanger MI, Akhtar M (2009) Peach-nut shells-an effective and low cost adsorbent for the removal of endosulfan from aqueous solutions. Pak J Analyt Env Chem 10(1):14–18

Memon FN, Memon S, Memon S, Memon N (2011) Synthesis and application of a new calix[4] arene based impregnated resin for the removal of endosulfan from an aqueous environment. J Chem Eng Data 56:3336–3345. https://doi.org/10.1021/je2002732

Mersie W, Seybold CA, McNamee C, Lawson MA (2003) Abating endosulfan from runoff using vegetative filter strips: the importance of plant species and flow rate. Agric Ecosyst Environ 97:215–223. https://doi.org/10.1016/S0167-8809(03)00035-5

Mishra PC, Patel RK (2008) Removal of endosulfan by sal wood charcoal. J Hazard Mater 152:730–736. https://doi.org/10.1016/j.jhazmat.2007.07.091

Pablos-Espada MC, Arrebola-Liébanas FJ, Garrido-frenich A, Martínez-Vidal JL (1999) Analysis of pesticides in water samples using GC-ECD and GC-MS/MS techniques. Int J Environ Anal Chem 75:165–179. https://doi.org/10.1080/03067319908047309

Randhawa MA, Anjum FM, Asi MR et al (2007) Removal of endosulfan residues from vegetables by household processing. J Sci Ind Res (India) 66:849–852

Rauf N, Tahir SS, Kang JH, Chang YS (2012) Equilibrium, thermodynamics and kinetics studies for the removal of alpha and beta endosulfan by adsorption onto bentonite clay. Chem Eng J 192:369–376. https://doi.org/10.1016/j.cej.2012.03.047

Rivero A, Niell S, Cesio V et al (2012) Analytical methodology for the study of endosulfan bioremediation under controlled conditions with white rot fungi. J Chromatogr B Anal Technol Biomed Life Sci 907:168–172. https://doi.org/10.1016/j.jchromb.2012.09.010

Saiyed H, Dewan A, Bhatnagar V et al (2003) Effect of endosulfan on male reproductive development. Environ Health Perspect 111:1958–1962. https://doi.org/10.1289/ehp.6271

Shah NS, He X, Khan HM et al (2013) Efficient removal of endosulfan from aqueous solution by UV-C/peroxides: a comparative study. J Hazard Mater 263:584–592. https://doi.org/10.1016/j.jhazmat.2013.10.019

Shen L, Wania F, Lei YD et al (2005) Atmospheric distribution and long-range transport behavior of organochlorine pesticides in North America. Environ Sci Technol 39:409–420. https://doi.org/10.1021/es049489c

Shivaramaiah HM, Kennedy IR (2006) Biodegradation of endosulfan by a soil bacterium. J Environ Sci Health B 41:895–905. https://doi.org/10.1080/03601230600806004

Siddique T, Okeke BC, Arshad M, Frankenberger WT (2003) Biodegradation kinetics of endosulfan by *Fusarium ventricosum* and a Pandoraea species. J Agric Food Chem 51:8015–8019. https://doi.org/10.1021/jf030503z

Silva MH, Beauvais SL (2010) Human health risk assessment of endosulfan. I: toxicology and hazard identification. Regul Toxicol Pharmacol 56:4–17. https://doi.org/10.1016/j.yrtph.2009.08.013

Singh SP, Bose P (2016) Degradation kinetics of endosulfan isomers by micron- and nano-sized zero valent iron particles (MZVI and NZVI). J Chem Technol Biotechnol 91:2313–2321. https://doi.org/10.1002/jctb.4818

Singh SP, Bose P (2017) Reductive dechlorination of endosulfan isomers and its metabolites by zero-valent metals: reaction mechanism and degradation products. RSC Adv 7:27668–27677. https://doi.org/10.1039/c7ra02430d

Singh M, Singh DK (2014) Biodegradation of endosulfan in broth medium and in soil microcosm by Klebsiella sp. M3. Bull Environ Contam Toxicol 92:237–242. https://doi.org/10.1007/s00128-013-1168-3

Sivagami K, Vikraman B, Krishna RR, Swaminathan T (2016) Chlorpyrifos and endosulfan degradation studies in an annular slurry photo reactor. Ecotoxicol Environ Saf 134:327–331. https://doi.org/10.1016/j.ecoenv.2015.08.015

Thangadurai P, Suresh S (2013) Reductive transformation of endosulfan in aqueous phase using magnesium-palladium bimetallic systems: a comparative study. J Hazard Mater 246–247:245–256. https://doi.org/10.1016/j.jhazmat.2012.12.031

Thangadurai P, Suresh S (2014) Biodegradation of endosulfan by soil bacterial cultures. Int Biodeterior Biodegrad 94:38–47. https://doi.org/10.1016/j.ibiod.2014.06.017

Tiwari MK, Guha S (2012) Role of soil organic matter on the sorption and cosorption of endosulfan and chlorpyrifos on agricultural soils. J Environ Eng 138:426–435. https://doi.org/10.1061/(ASCE)EE.1943-7870.0000490

Tiwari MK, Guha S (2013a) Simultaneous analysis of endosulfan, chlorpyrifos, and their metabolites in natural soil and water samples using gas chromatography-tandem mass spectrometry. Environ Monit Assess 185:8451–8463. https://doi.org/10.1007/s10661-013-3186-3

Tiwari MK, Guha S (2013b) Kinetics of the biodegradation pathway of endosulfan in the aerobic and anaerobic environments. Chemosphere 93:567–573. https://doi.org/10.1016/j.chemosphere.2013.07.005

Venugopal NVS, Sumalatha B (2011) Spectrophotometric determination of endosulfan in environmental samples. Int Proc Chem Biol Environ Eng 6:195–197

Verma K, Agrawal N, Farooq M et al (2006) Endosulfan degradation by a Rhodococcus strain isolated from earthworm gut. Ecotoxicol Environ Saf 64:377–381. https://doi.org/10.1016/j.ecoenv.2005.05.014

Wan MT, Kuo J-N, Buday C et al (2005) Toxicity of alpha-, beta-, (alpha + beta)-endosulfan and their formulated and degradation products to *Daphnia magna*, *Hyalella azteca*, *Oncorhynchus mykiss*, *Oncorhynchus kisutch*, and biological implications in streams. Environ Toxicol Chem 24:1146–1154

Weber J, Halsall CJ, Muir D et al (2010) Endosulfan, a global pesticide: a review of its fate in the environment and occurrence in the Arctic. Sci Total Environ 408:2966–2984. https://doi.org/10.1016/j.scitotenv.2009.10.077

Westbom R, Hussen A, Megersa N et al (2008) Assessment of organochlorine pesticide pollution in Upper Awash Ethiopian state farm soils using selective pressurised liquid extraction. Chemosphere 72:1181–1187. https://doi.org/10.1016/j.chemosphere.2008.03.041

Wille K, De Brabander HF, Vanhaecke L et al (2012) Coupled chromatographic and mass-spectrometric techniques for the analysis of emerging pollutants in the aquatic environment. TrAC Trends Anal Chem 35:87–108. https://doi.org/10.1016/j.trac.2011.12.003

Xiong B, Zhou A, Zheng G et al (2015) Photocatalytic degradation of endosulfan in contaminated soil with the elution of surfactants. J Soils Sediments 15:1909–1918. https://doi.org/10.1007/s11356-016-6523-6

Yedla S, Diksiht AK (2008) Removal of Endosulfan from water using wood charcoal-adsorption and desorption. J Environ Eng 134:102–109. https://doi.org/10.1061/(Asce)0733-9372

Yonli AH, Batonneau-Gener I, Koulidiati J (2012) Adsorptive removal of α-endosulfan from water by hydrophobic zeolites. An isothermal study. J Hazard Mater 203–204:357–362. https://doi.org/10.1016/j.jhazmat.2011.12.042

Part II
Modelling

Chapter 8
Groundwater Contamination Problems and Numerical Simulation

T. I. Eldho and B. Swathi

Abstract In the last few decades, due to the overuse of groundwater resources and human mismanagement, groundwater quality has been severely affected. If the quality is not safe, even if it is abundant, it is not useful. This chapter essentially deals with the nature and movement of contaminants in groundwater due to the transport mechanisms such as advection, dispersion, retardation, and other chemical reactions and its numerical modeling. The numerical modeling refers to the process of simulation where approximate solutions to the governing equation of contaminant transport are obtained. The numerical modeling of groundwater contamination had undergone much advancement over the past half century. The driving factors behind this continuous progress are due to the alarming change in groundwater quality due to different anthropogenic reasons leading to the need for more accurate and efficient groundwater remediation and management models. Further, due to the rapid development in computational techniques and powerful data processing systems, groundwater contamination modeling is made more easily available to hydrogeologists and engineers. This chapter gives an insight into the groundwater contamination process, from discussing contaminant transport mechanisms to challenges faced in its numerical modeling. The details of the governing partial differential equations and the evolution of the methodology in obtaining these approximate solutions using different numerical techniques are presented. The process of numerical simulation is demonstrated with the help of example problems, and the results are discussed and compared with solutions from different numerical models. The chapter highlights the need for the qualitative and quantitative reasoning behind the simulation of any contaminant transport problem and how the numerical modeling helps in identifying the best way of solving it. Finally, a field case study is given in which the numerical simulation provides an integrated

T. I. Eldho (✉) · B. Swathi
Department of Civil Engineering, IIT Bombay, Mumbai 400076, India
e-mail: eldho@civil.iitb.ac.in

B. Swathi
e-mail: boddulaswathi@gmail.com

T. Gupta et al. (eds.), *Environmental Contaminants*, Energy, Environment, and Sustainability, https://doi.org/10.1007/978-981-10-7332-8_8

framework wherein the conceptual principles are applied to the observed data and investigative and predictive calculations are done which helps in developing management and protection policies.

Keywords Groundwater contamination · Transport process · Simulation Numerical modeling

8.1 Introduction

For the existence of life on earth, water is essential and it is available as surface water and groundwater. Groundwater is more reliable as its availability does not depend directly on the annual rainfall, it is free from enormous losses, does not require extra land and is assumed to be purer than surface water. As groundwater is a scarce resource, it should be managed properly for sustainable use. With increasing demand and excessive pumping for domestic and agricultural purposes, groundwater levels have declined in many regions leading in decreased well yields, increased pumping costs, water quality deterioration, stream and wetland desiccation, land subsidence, and other adverse environmental and social impacts (Freeze and Cherry 1979). Hence, there is an urgent need to protect the shrinking groundwater resources and manage it efficiently to ensure sustainable development.

In the last few decades, groundwater contamination has become a major problem in many parts of the world making it unusable for domestic or agricultural purposes (Bedient et al. 1999). As groundwater exists in the subsurface, it is very complex to understand the groundwater hydrodynamics and contaminant movement. In the past, huge efforts were put by many researchers to understand the groundwater flow and transport mechanism. Due to the complexity of the problems, groundwater simulation using computer models is essential to understand the flow and contaminant transport mechanism, for further management of the resources. The main focus of this chapter is on the groundwater contamination problems and its numerical simulation.

8.2 Groundwater Contamination Problems

The main reasons for groundwater contamination are indiscriminate disposal of industrial waste, extensive use of chemicals in agriculture such as fertilizers and pesticides, and a host of other human interventions (Freeze and Cherry 1979). The surface contamination after affecting soils and surface water further extends to the groundwater system as well. In majority of the cases, the groundwater contamination can be detected through quality assessment studies in terms of the amount of dissolved or suspended contaminants, other than color, odor, and palatability.

8.2.1 Sources of Groundwater Contamination

The contamination of groundwater can be from various sources as shown in Fig. 8.1. A brief discussion on various sources of groundwater contaminations is given below (Fried 1975; Bedient et al. 1999; Eldho 2001).

The groundwater contamination sources can be from natural or human-made sources. The natural sources can include sea water intrusion, landslides, and breaking down of natural minerals existing in ground strata. The human-made sources are plenty ranging from domestic sources like leakages from septic tanks and sewers; percolation of chemical fertilizers and pesticides used in agricultural activities; urban sources like leakage from landfills where leachate coming out generally seeps into the groundwater; to industrial sources like deep well disposals (Eldho 2001).

Industries disposing only inadequate treated solid or liquid wastes to landfills or surface water bodies and drains have caused severe groundwater contamination. Further accidental spills or leaks from tanks and pipelines of petroleum products and other chemicals pose a serious risk to groundwater quality. Mining industries are also one of the major sources of contamination (Downing 1998). Also, when the groundwater is overexploited, minerals such as fluorides and arsenic may leach to the groundwater and cause many health hazards (Freeze and Cheery 1979).

8.2.2 Challenges in Groundwater Contamination Problems

In groundwater systems, whenever a contaminant enters, it is carried by the flow of the groundwater (Freeze and Cherry 1979). Due to complexity of the problem, to understand the nature of contamination and spread at a particular location,

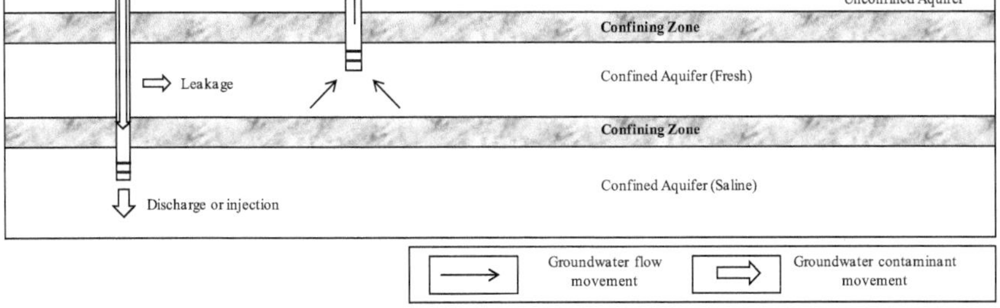

Fig. 8.1 Possible sources of groundwater contamination

appropriate field studies and computer simulations are essential. Mostly nowadays, the scope of groundwater contamination investigation is to identify the presence and extent of contamination in various forms, likely direction and rate of its movement, containment of the contaminant plume, proper preventive measures/ remediation, and management measures for sustainable groundwater management (Eldho 2001).

As groundwater flow and transport processes are highly complex in nature, it is important to study the movement of these contaminants in an aquifer to predict their migration. For these studies, modeling plays a pivotal role. Such studies help end users/decision makers in evolving useful guidelines for future planning of waste disposal operations and for controlling the existing contaminant plumes and can help planners in working out necessary remedial and preventive measures.

8.2.3 Numerical Modeling of Groundwater and Transport

Mathematical models of groundwater flow and contaminant transport are properly conceptualized version of a complex porous media or aquifer system which approximates the flow and transport phenomena (Bear and Verrujit 1987). The mathematical models are represented in terms of governing partial differential equations with appropriate initial and boundary conditions, which are solved by numerical models. The approximations in the numerical models are introduced through set of assumptions for the geometry of the domain which is assumed as a continuum, smoothened heterogeneities, the nature and properties of the porous medium and the fluid, and the type of the flow regime. In groundwater modeling, the continuum is viewed as a network of several representative elementary volumes, each representing a portion of the entire volume of porous media or an aquifer with fluid and porous media properties taken over it and assigned to the nodes of superimposed grid/mesh used for the discretization of the domain (Bear 1979). The numerical models are solved with the help of a computer and hence called as computer models. These computer models greatly help in analyzing the implications of proposed excitations such as pumping or recharge on the response of the aquifer systems. The computer models are much easy to develop than the conventional physical or analog models. The computer models can also handle and process large amount of field information for regional/large-scale problems (Wang and Anderson 1982).

8.3 Groundwater Transport Simulation—Theoretical Developments

For understanding the groundwater contamination process and simulation, it is very essential to understand the mechanisms related to groundwater flow and transport process. Generally, the groundwater flow in the saturated porous media occurs in all directions and is controlled by the hydraulic gradient/head and the hydraulic conductivity (K) or transmissivity (T), which expresses the ability of the soil to transport water (Bear 1972). The variations in hydraulic head/gradient are the driving force which causes water to move from one location to another, according to Darcy's law (Bear 1979).

The groundwater contaminant transport in porous media is governed by many processes including advection, dispersion, retardation, and reaction (Freeze and Cherry 1979). The advection is the process by which solutes are transported by the bulk motion of the flowing groundwater (Freeze and Cherry 1979). Owing to advection, non-reactive solutes are carried at an average rate equal to the average linear velocity, $\bar{v}$, of the water. So, $\bar{v} = \frac{v}{\theta}$, v being the flow velocity and θ the porosity. The hydrodynamic dispersion is a process of the solutes to spread out from the path that it would be expected to follow according to the advective hydraulics of the flow system (Freeze and Cherry 1979). It causes dilution of the solute because of mechanical mixing during fluid advection and because of molecular diffusion due to the thermal-kinetic energy of the solute particles. Dispersion equation in general is derived using Fick's law (Anderson 1984). In groundwater contaminant transport, there are a number of chemical and physical processes which retard, that is, delay the movement of constituents in groundwater. The general mechanisms that can retard the movement of chemical constituents in groundwater can be adsorption and absorption (called as sorption), dilution, filtration. When reactions of the contaminant are considered, as it travels through the porous media, it is bound to undergo weak or strong reactions. These can be due to the interaction between contaminant and soil or contaminant and pore water or due to decay or biodegradation or transformation of the contaminant under various conditions. The incorporation of these mechanisms into models has been described in many references such as Bear (1972), Fried (1975), Freeze and Cherry (1979), Anderson and Woessner (1992), Bedient et al. (1999).

8.3.1 Groundwater Contaminant Transport Modeling

Numerical modeling techniques are often used for the predictions of the groundwater flow and transport of contaminants in subsurface media wherein the processes under simulation are modeled by a set of governing equations, which are solved using numerical methods. Generally, numerical models of groundwater flow and contaminant transport are designed to allow a modeler to simulate the groundwater conditions in an aquifer based on measurable aquifer and contaminant characteristics (Istok 1989). The reliability of modeling results depends upon a good understanding of the fundamental processes governing the flow and contaminant transport in the media and the various assumptions used in the modeling (Wang and Anderson 1982).

8.3.1.1 Governing Equations for Groundwater Flow and Transport

The governing groundwater flow equations for saturated porous media are derived based on Darcy's law and mass balance approach (Bear 1972). The solution of groundwater flow equations provides the head distribution throughout the domain. Using these values, the velocities are calculated using the Darcy's law (Bear 1979). The general solute-transport equation in groundwater relates concentration changes with reference to various transport processes such as advection, hydrodynamic dispersion, retardation, and chemical reactions with respect to initial and boundary conditions. For the solution of contaminant transport problems, the velocities of the flow field are essential and hence, the solute-transport equation must be solved in conjunction with the groundwater flow equations, as coupled problem.

The governing partial differential equations describing the flow in a two-dimensional inhomogeneous, anisotropic confined and unconfined aquifer are given as (Bear 1979; Wang and Anderson 1982):

$$\frac{\partial}{\partial x}\left[T_x\frac{\partial h}{\partial x}\right]+\frac{\partial}{\partial y}\left[T_y\frac{\partial h}{\partial y}\right]=S\frac{\partial h}{\partial t}+Q_w\delta(x-x_i)(y-y_i)-q \tag{8.1}$$

$$\frac{\partial}{\partial x}\left[K_xh\frac{\partial h}{\partial x}\right]+\frac{\partial}{\partial y}\left[K_yh\frac{\partial h}{\partial y}\right]=S_y\frac{\partial h}{\partial t}+Q_w\delta(x-x_i)(y-y_i)-q \tag{8.2}$$

where $h(x,y,t)$ = piezometric head [L]; $T_i(x,y)$ = anisotropic transmissivity [L^2T^{-1}]; $K_i(x,y)$ = anisotropic hydraulic conductivity [LT^{-1}]; $S(x,y)$ = storage coefficient; $S_y(x,y)$ = specific yield; Q_w = source or sink function; ($-Q_w$ = source, Q_w = sink) [LT^{-1}]; δ = Dirac delta function; x_i, y_i = pumping or recharge well location; $q(x,y,t)$ = vertical inflow rate [LT^{-1}]; x, y = horizontal space variables [L]; and t = time [T].

The seepage velocity necessary to the solution of the solute-transport model is computed using Darcy's law (Bear 1972) and can be written as: $V_i=-\frac{K_i}{\theta}\frac{\partial h}{\partial x_i}$, $i=x,y$; where $V_i(x,y)$ = seepage velocity in i-direction [LT^{-1}]; and θ = porosity [−].

The governing partial differential equation for transport of a single chemical constituent in saturated porous media in two dimensions, considering advection, dispersion, and fluid sources/sinks is given as (Freeze and Cherry 1979; Wang and Anderson 1982; Zheng and Bennett 1995):

$$R\frac{\partial C}{\partial t}=-\frac{\partial}{\partial x}(V_xC)-\frac{\partial}{\partial y}(V_yC)+\frac{\partial}{\partial x}\left(D_x\frac{\partial C}{\partial x}\right)+\frac{\partial}{\partial y}\left(D_y\frac{\partial C}{\partial y}\right)-\frac{c'w}{\theta b}-R\lambda C-\frac{q_wC}{\theta} \tag{8.3}$$

where $C(x, y, t)$ = solute concentration [ML^{-3}]; D_x, D_y = components of dispersion coefficient tensor [L^2T^{-1}]; λ = the reaction rate constant [T^{-1}]; w = the elemental recharge rate [LT^{-1}] with solute concentration c'; b = aquifer thickness under the element [L]; $R = 1 + \frac{\rho_b K_d}{\theta}$ = retardation factor[-] with ρ_b = media bulk density [ML^{-3}], and K_d = sorption coefficient [L^3M^{-1}]; and q_w = specific pumping rate from a source [LT^{-1}].

For transient flow and transport analysis, the following initial conditions are used:

$$h(x, y, 0) = h_0(x, y); C(x, y, 0) = f \quad x, y \in \Omega \tag{8.4}$$

The flow and transport equations should be solved with appropriate boundary conditions. Generally, the boundary conditions can be prescribed variable (head or concentration) or gradient of the variable or flux. The boundary conditions can be written as:

$$h(x, y, t) = h_1(x, y, t); C(x, y, t) = g_1 \quad x, y \in \Gamma_1 \tag{8.5a}$$

$T\frac{\partial h}{\partial n} = q_1(x, y, t)$ for confined aquifer; $K h \frac{\partial h}{\partial x} = q_2(x, y, t)$ for unconfined aquifer;

$$\left(D_x \frac{\partial C}{\partial x}\right) n_x + \left(D_y \frac{\partial C}{\partial y}\right) n_y = g_2 \quad x, y \in \Gamma_2 \tag{8.5b}$$

where Ω = flow region; $\Gamma = \Gamma_1 \cup \Gamma_2$ = region boundary; $\frac{\partial}{\partial n}$ = normal derivative; $h_0(x, y)$ = initial head in the flow domain [L]; $h_1(x, y, t)$ = known head value at the boundary section Γ_1[L]; f is a given function in Ω, g_1 and g_2 are given functions along boundary sections Γ_1 and Γ_2; and $n_x n_y$ are the components of the unit outward normal vector to the boundary section Γ_2.

For the simulation of groundwater flow and contaminant transport, the above flow and transport equations are to be solved with the initial and boundary conditions appropriately, either through analytical or numerical methods. In most of the cases, analytical solutions are not possible and are available only for simplified hypothetical problems. Hence, numerical methods-based computational models are mostly preferred and commonly used (Wang and Anderson 1982).

8.3.2 Numerical Methods For Groundwater Transport Modeling

Due to the complexity of groundwater transport problems such as irregularity of domain boundaries, heterogeneity, nonlinearity, irregular source/sink functions, especially for the solution of field problems, it is impossible to derive analytical/exact solutions. Numerical techniques are usually used for solving these problems. In numerical simulation, various numerical methods are employed to solve governing equations and transform the mathematical model into a numerical one. The governing equations appearing in the mathematical model are represented by their numerical counterparts and solved using appropriate initial and boundary conditions.

In the last five decades, a variety of numerical methods such as finite difference method (FDM), method of characteristics (MOC), finite volume method (FVM), finite element method (FEM), boundary element method (BEM), analytical element method (AEM), meshless method (Li and Gu 2005) have been developed for the groundwater system simulation. Depending on the problem to be solved, each of these methods has its own advantages and disadvantages. The choice depends upon the complexity of the problem, data availability, computational facilities, and user's familiarity with the method. However, for groundwater simulation, FDM and FEM are the most popular techniques. Recently, meshless methods are also used for groundwater simulation (Mategaonkar and Eldho 2012; Swathi and Eldho 2017). Brief descriptions of these techniques are given below.

8.3.2.1 Finite Difference Method (FDM)

Using FDM, the governing equations are approximated by standard backward, forward, or central difference schemes (Wang and Anderson 1982). Initially, the problem domain is discretized to a grid. In the grid, the nodes can be considered either at the center of grid cells or at the intersection. Then, the governing equation is approximated using suitable scheme of difference at each node, and system of equations is formed. Here, the unknown variables are represented by a set of values at points on a grid of intersecting lines. The gradient of the function is represented by differences in the values at neighboring points, and FDM version of the equation is formed. This system of equations, after application of the boundary conditions, can be solved using direct or iterative solution techniques, to get the unknown groundwater head or concentration. Considering homogeneous isotropic confined aquifer in 2D, Eq. (8.1) can be written as (Bear 1979),

$$\frac{\partial^2 h}{\partial x^2} + \frac{\partial^2 h}{\partial y^2} = \frac{S}{T}\frac{\partial h}{\partial t} - \frac{R(x,y,t)}{T} \tag{8.6}$$

where S is the storage coefficient, T is the transmissivity coefficient, R is recharge or pumping. Referring to grid given in Fig. 8.2, corresponding FDM form in explicit form can be written as (Wang and Anderson 1982):

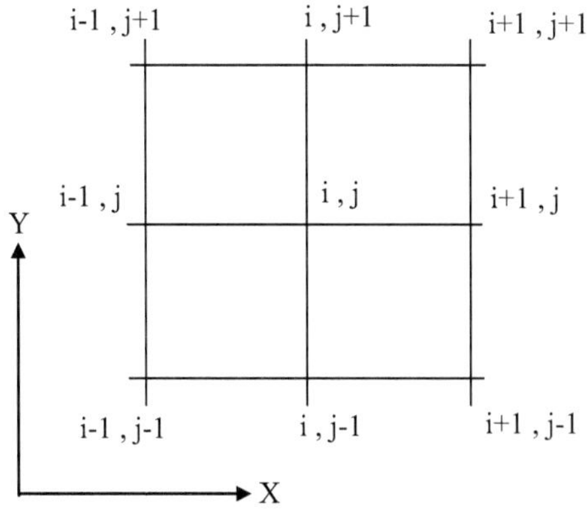

Fig. 8.2 A typical grid used in FDM

$$\frac{h^n_{I+1,J} - 2h^n_{I,J} + h^n_{I-1,J}}{(\Delta x)^2} + \frac{h^n_{I,J+1} - 2h^n_{I,J} + h^n_{I,J+1}}{(\Delta y)^2} = \left(\frac{S}{T}\right)\frac{h^{n+1}_{I,J} - h^n_{I,J}}{(\Delta t)} - \frac{R^n_{I,J}}{T} \tag{8.7}$$

The corresponding FDM form in implicit form can be written as (Wang and Anderson 1982):

$$\frac{h^{n+1}_{I+1,J} - 2h^{n+1}_{I,J} + h^{n+1}_{I-1,J}}{\Delta x^2} + \frac{h^{n+1}_{I,J+1} - 2h^{n+1}_{I,J} + h^{n+1}_{I,J-1}}{\Delta y^2} = \frac{S}{T}\frac{h^{n+1}_{I,J} - h^n_{I,J}}{\Delta t} - \frac{R^{n+1}_{I,J}}{T} \tag{8.8}$$

Here n indicates previous time step and $n + 1$, current time step. In order to apply explicit or implicit scheme, the stability conditions are to be taken into consideration. It is also reported that numerical stability problems arise while simulating solute-transport equation using FDM (Bear 1979). Detailed formulation of FDM for groundwater systems can be found in Wang and Anderson (1982), Bear (1979), Rastogi (2010), etc.

8.3.2.2 Finite Element Method (FEM)

FEM is one of the widely used numerical methods for solving differential and partial differential equations (Pinder and Gray 1977; Desai et al. 2011). The first step in FEM is discretizing the domain to a mesh with element shape of our choice. Then, interpolation or shape functions are defined in a piece-wise fashion over the defined individual elements to represent the unknown variables. One of the most commonly used FEM techniques is Galerkin's approach (Desai et al. 2011) in which the resulting error or residuals due to the trial solution chosen for the governing equation are weighted by a weighting function and set equal to zero over the entire domain. Using the elemental equations, matrices are assembled for the global domain, forming a system of algebraic equations. These set of equations are solved by iterative or direct techniques after applications of boundary conditions (Desai et al. 2011). While FEM is used for approximation in space, generally FDM is used for time discretization (Desai et al. 2011).

Using Galerkin's finite element method (Desai et al. 2011) and using simple 2D triangular element (Fig. 8.3) for the approximation of flow Eq. (8.1), the first step is to define a trial solution.

$$\hat{h}(x, y, t) = \sum_{L=1}^{NP} h_L(t) N_L(x, y) \tag{8.9}$$

where h_L is the unknown head, N_L is the known basis function at node *L*, and *NP* is the total number of nodes in the problem domain. A set of simultaneous equations is obtained when residuals weighted by each of the basis function are forced to be zero and integrated over the entire domain Ω (Desai et al. 2011). Thus, Eq. (8.1) can be written as,

$$\iint_{\Omega.} \left[\frac{\partial}{\partial x}\left(T_x \frac{\partial \hat{h}}{\partial x} \right) + \frac{\partial}{\partial y}\left(T_y \frac{\partial \hat{h}}{\partial y} \right) - Q_w + q - S \frac{\partial \hat{h}}{\partial t} \right] N_L(x, y) \mathrm{d}x\mathrm{d}y = 0 \tag{8.10}$$

Equation (8.10) can further be written as the summation of individual elements as,

$$\begin{aligned} &\sum_e \iint \left(T_x^e \frac{\partial \widehat{h^e}}{\partial x} \left\{ \frac{\partial N_L^e}{\partial x} \right\} + T_y^e \frac{\partial \widehat{h^e}}{\partial y} \left\{ \frac{\partial N_L^e}{\partial y} \right\} \right) \mathrm{d}x\mathrm{d}y + \sum_e \iint \left(S \frac{\partial \widehat{h^e}}{\partial t} \right) \{N_L^e\} \mathrm{d}x\mathrm{d}y \\ &= \sum_e \iint (Q_w) \{N_L^e\} \mathrm{d}x\mathrm{d}y - \sum_e \iint (q) \{N_L^e\} \mathrm{d}x\mathrm{d}y \end{aligned} \tag{8.11}$$

where $\{N_L^e\} = \begin{Bmatrix} N_i \\ N_j \\ N_k \end{Bmatrix}$.

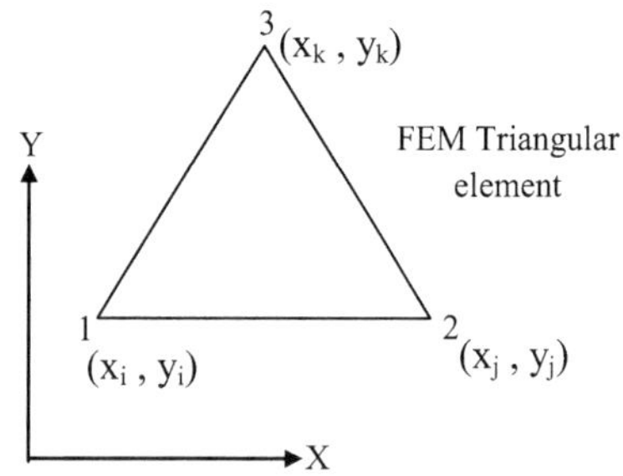

Fig. 8.3 A typical linear triangular element used in FEM

For an element, Eq. (8.11) can be written in matrix form as

$$[G^e]\{h_I^e\} + [P^e]\left\{\frac{\partial h_I^e}{\partial t}\right\} = \{F^e\} \tag{8.12}$$

where $I = i, j, k$ are three nodes of triangular elements and G, P, F are the element matrices known as conductance, storage matrices, and recharge vectors, respectively. Summation of elemental matrix Eq. (8.12) for all the elements lying within the flow region gives the global matrix as

$$[G]\{h_I\} + [P]\left\{\frac{\partial h_I}{\partial t}\right\} = \{F\} \tag{8.13}$$

Applying the implicit finite difference scheme for the $\frac{\partial h_I}{\partial t}$, term in time domain for Eq. (8.13) gives,

$$[G]\{h_I\}_{t+\Delta t} + [P]\left\{\frac{h_{t+\Delta t} - h_t}{\Delta t}\right\} = \{F\} \tag{8.14}$$

The subscripts t and $t + \Delta t$ represent the groundwater head values at earlier and present time steps. By rearranging the terms of Eq. (8.14), the general form of the equation can be given as (Istok 1989; Desai et al. 2011):

$$[[P] + \omega\Delta t[G]]\{h\}_{t+\Delta t} = [[P] - (1-\omega)\Delta t[G]]\{h\}_t + \Delta t(1-\omega)\{F\}_t + \omega \tag{8.15}$$

where Δt = time step size, $\{h\}_t$ and $\{h\}_{t+\Delta t}$ are groundwater head vectors at the time t and $t + \Delta t$, respectively, ω = Relaxation factor which depends on the type of finite difference scheme used. For fully explicit scheme $\omega = 0$; Crank–Nicolson scheme $\omega = 0.5$; fully implicit scheme $\omega = 1$.

Similarly applying the Galerkin's finite element method for the groundwater solute-transport Eq. (8.3), we get (Desai et al. 2011)

$$\iint_{\Omega.} \left[R\frac{\partial C}{\partial t} - \frac{\partial}{\partial x}\left(D_{xx}\frac{\partial C}{\partial x}\right) - \frac{\partial}{\partial y}\left(D_{yy}\frac{\partial C}{\partial y}\right) + \frac{\partial}{\partial x}(v_x C) + \frac{\partial}{\partial y}(v_y C) + \frac{c'W}{nb} + R\lambda c\right] N_L(x,y)\,\mathrm{d}x\mathrm{d}y = 0 \tag{8.16}$$

Writing the above equation in the matrix form and applying finite difference scheme over the time domain will give the following equation:

$$[[S] + \omega\Delta t[D]]\{C\}_{t+\Delta t} = [[S] - (1-\omega)\Delta t[D]]\{C\}_t + \Delta t(1-\omega)\{F\}_t + \omega\{F\}_{t+\Delta t} \tag{8.17}$$

where $[S]$ = element sorption matrix, $[D]$ = element advection–dispersion matrix, ω = relaxation factor, $\{F\}$ = flux matrix, and t and $t + \Delta t$ = beginning and ending time steps.

Detailed formulation of FEM for groundwater flow and transport can be found in Wang and Anderson (1982), Istok (1989), Rastogi (2010), Desai et al. (2011), etc.

8.3.2.3 Meshless Methods

Meshless methods are the numerical techniques wherein the elements or meshes are replaced by only nodes (Li and Mulay 2013). Meshless methods use a set of nodes scattered within the problem domain as well as on the boundaries of the domain to represent the problem domain and its boundaries (Fig. 8.4). These nodes do not require prior information for the approximation of the unknown field variables (Swathi and Eldho 2014). Many meshless methods have found good applications and show very good potential to become powerful numerical tools (Liu and Gu 2005).

While solving Eq. (8.1) using a meshless technique, meshless local Petrov–Galerkin (MLPG) as described in Swathi and Eldho (2014, 2017), the first step is to define the trial solution for the unknown variable.

$$\hat{h}(x, y, t) = \sum_{i=1}^{N} \varnothing_i(x, y)\widehat{h_i} \tag{8.18}$$

where h_i is the unknown head, $\varnothing_i(x, y)$ is the shape function at node i, and n is the total number of nodes in the support domain (Fig. 8.4). The details of the shape function and the MLPG methodology can be referred from Atluri (2004) and Swathi and Eldho (2014, 2017). The dependent variable h is computed from Eq. (8.18) and also its derivatives (in case of 2D w.r.t x and y) as below.

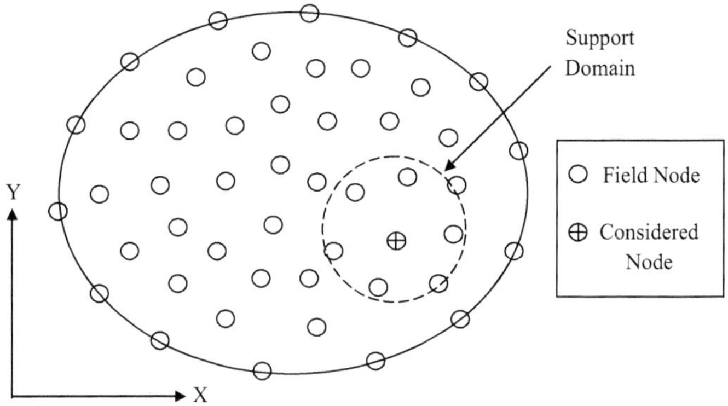

Fig. 8.4 Typically used local support domain in meshless method to construct shape function

$$\frac{\partial h(\mathbf{r})}{\partial x} = \sum_{i=1}^{N} \frac{\partial \overline{\varnothing_i}(\mathbf{r})}{\partial x} \widehat{h_i} \text{ and } \frac{\partial h(\mathbf{r})}{\partial y} = \sum_{i=1}^{N} \frac{\partial \overline{\varnothing_i}(\mathbf{r})}{\partial y} \widehat{h_i} \tag{8.19}$$

For the time derivative term, Crank–Nicolson time-stepping scheme with weighting factor $\theta = 0.5$ is considered and the system of equations can be written in the matrix form as

$$\mathbf{K}\left\{\widehat{\boldsymbol{h}}_{t+1}\right\} = \boldsymbol{f} \tag{8.20}$$

where

$$\mathbf{K} = \left(T \int\limits_{\partial\Omega_{s,j}} \left(\frac{\partial \phi_i(\mathbf{r})}{\partial x} n_x + \frac{\partial \phi_i(\mathbf{r})}{\partial y} n_y \right) \mathrm{d}\Omega_j \right) \theta - \int\limits_{\Omega_s} \frac{S}{\Delta t} \mathrm{d}\Omega \tag{8.21}$$

$$\begin{aligned} \boldsymbol{f} &= Q_j - f_j A_\Omega \\ &- \left(\int\limits_{\Omega_s} \frac{S}{\Delta t} \mathrm{d}\Omega + \left(T_j \int\limits_{\partial\Omega_{s,j}} \left(\frac{\partial \phi_i(\mathbf{r})}{\partial x} \widehat{h}_t n_x + \frac{\partial \phi_i(\mathbf{r})}{\partial y} \widehat{h}_t n_y \right) \mathrm{d}\Omega_j \right) (1-\theta) \right) \left\{\widehat{h}_t\right\} \end{aligned} \tag{8.22}$$

Now, we can approximate the solution by just computing the matrix **K** and the vector $\boldsymbol{f}$, knowing $\left\{\hat{\boldsymbol{h}}_t\right\}$. Then, just iteratively, the right-hand side of Eq. (8.20) acts as the new $\boldsymbol{f}$ vector. By a solution, we get a vector $\hat{\boldsymbol{h}}_{t+1}$, which is used as a value at right-hand side, in the next iteration. Equation (8.20) is solved using Gauss–Jordan method.

In a similar way, the groundwater contaminant transport equation (Eq. 8.3) is approximated using MLPG with the proposed unknown variable as (Swathi and Eldho 2017):

$$c(\mathbf{r}) = \sum_{i=1}^{n} \overline{\varnothing_i}(\mathbf{r}) \widehat{c_i} \tag{8.23}$$

where $\hat{c}_i$ is the so-called *fictitious* value of c at node i. From Eq. (8.3), the system of equations can be written in the matrix form as below:

$$\bar{\mathbf{K}}\{\hat{\boldsymbol{c}}_{t+1}\} = \bar{\boldsymbol{f}} \tag{8.24}$$

where

$$\bar{\mathbf{K}} = \left\{ \int\limits_{\Omega_x} \frac{R}{\Delta t} \mathrm{d}\Omega_x - \theta \left(\int\limits_{\partial\Omega_x} D_i \left(\frac{\partial \overline{\varnothing_i}(\boldsymbol{r})}{\partial x} n_x + \frac{\partial \overline{\varnothing_i}(\boldsymbol{r})}{\partial y} n_y \right) \mathrm{d}\Omega_x + \left(\int\limits_{\Omega_x} v_x \frac{\partial \overline{\varnothing_i}(\boldsymbol{r})}{\partial x} + \int\limits_{\Omega_s} v_y \frac{\partial \overline{\varnothing_i}(\boldsymbol{r})}{\partial y} \right) \mathrm{d}\Omega \right) \right\} \tag{8.25}$$

$$\bar{f} = \left\{ \int\limits_{\Omega_x} \frac{R}{\Delta t} \mathrm{d}\Omega_x + (1-\theta) \left(\int\limits_{\partial\Omega_x} D_j \left(\frac{\partial \overline{\varnothing_i}(\boldsymbol{r})}{\partial x} n_x + \frac{\partial \overline{\varnothing_i}(\boldsymbol{r})}{\partial y} n_y \right) \mathrm{d}\Omega_j - \left(\int\limits_{\Omega_x} v_x \frac{\partial \overline{\varnothing_i}(\boldsymbol{r})}{\partial x} + \int\limits_{\Omega_x} v_y \frac{\partial \overline{\varnothing_i}(\boldsymbol{r})}{\partial y} \right) \mathrm{d}\Omega \right) \right\} \{\hat{\boldsymbol{c}}_t\} + \frac{c'w'}{nb} \tag{8.26}$$

The solution is computed by solving Eq. (8.24) using Gauss–Jordan method. Finally, we get the approximate values, $\boldsymbol{c}$ at different time intervals.

Detailed description on application of different meshless methods to groundwater flow and transport can be found in Li et al (2003), Kumar and Dodagoudar (2008), Mategaonkar and Eldho (2012), Kovářík and Muzik (2013), Swathi and Eldho (2014, 2017), Guneshwor et al. (2016) etc.

8.4 Numerical Modeling

8.4.1 Modeling Procedure

For groundwater flow and transport simulation, a systematic approach is to be followed. Figure 8.5 gives the modeling procedure for the application of groundwater modeling.

The first step is to develop a conceptual model based on the field data available and decide on the assumptions which simplifies and best describes the aquifer system. The corresponding governing equations with the appropriate boundary and initial conditions are finalized. Further, the governing equations are converted to system of algebraic equations using a selected numerical approach through the computer code developed (Wang and Anderson 1982). The developed numerical model for verification is compared either with the available analytical solutions or the collected field data. In case of field problems, the fitness of the model is improved by calibration process. The appropriate aquifer parameters such as hydraulic conductivity, storage coefficient, and dispersivity are adjusted till the results from the models fall within certain range of fitness criteria (Willis and Yeh 1987). Later, the numerical model can be used to simulate the desired aquifer system scenario.

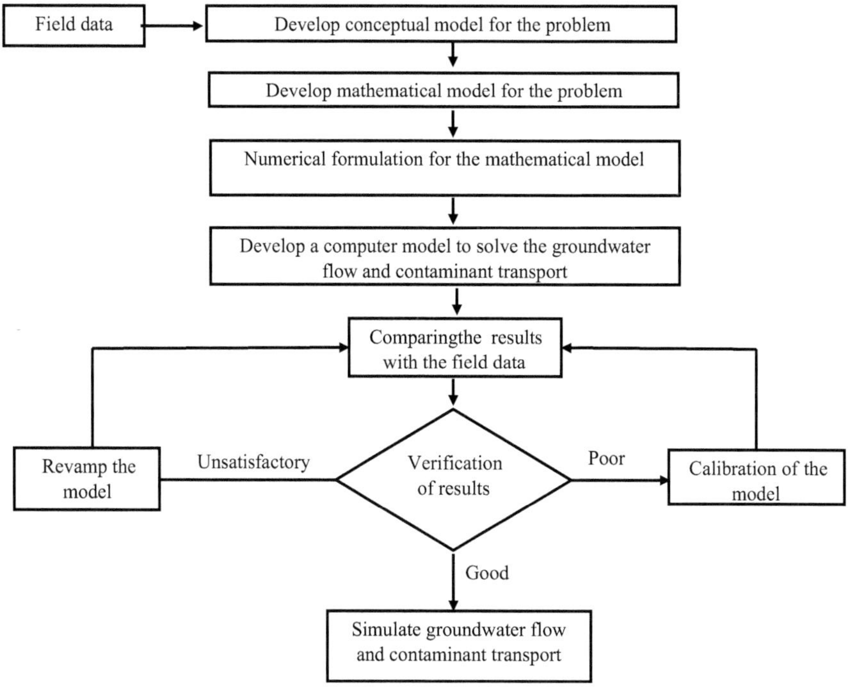

Fig. 8.5 Numerical modeling procedure for groundwater flow and transport

8.5 Case Studies

The application of numerical modeling for groundwater flow and transport problems is demonstrated here with the help of two case studies.

8.5.1 Case Study 1—Hypothetical Problem

Here a hypothetical confined aquifer (Sharief et al. 2008) is considered. The aquifer domain is of 1800 m × 1000 m in area as shown in Fig. 8.6. The hydrogeologic parameters taken for the aquifer are as given in Table 8.1. The entire aquifer has storativity of 0.0004, and a pond is considered to be in Zone A with rate of seepage of 0.009 m/d. Zone A and Zone C are assumed to be recharged between nodes 6–24 and 42–60 at a rate of 0.00024 and 0.00012 m/d, respectively, as seen from Fig. 8.7. Three pumping wells are located at nodes 26, 29, and 46 with water discharging at a rate of 200, 500, and 700 m^3/d, respectively. One recharge well at node 17 with rate of 800 m^3/d is also assumed.

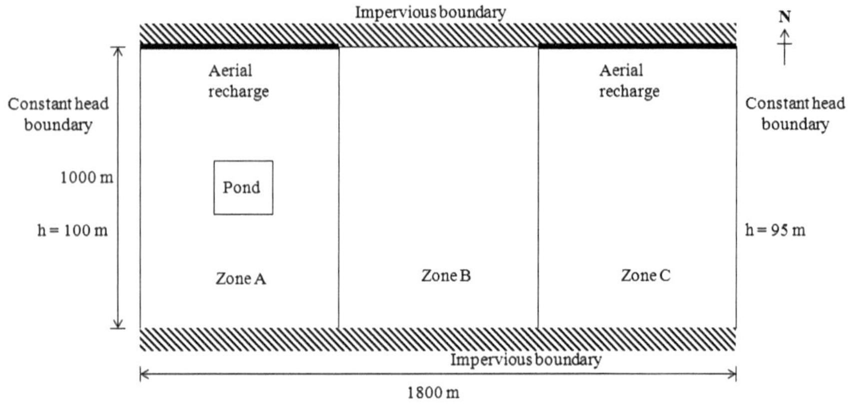

Fig. 8.6 Hypothetical confined aquifer configuration

Table 8.1 Hydrogeologic data used for the coupled flow and transport model

Properties	Zone A	Zone B	Zone C
Transmissivity T_x (m^2/d)	500	400	250
Transmissivity T_y (m^2/d)	300	250	200
Porosity	0.20	0.25	0.15
Longitudinal dispersivity (m)	150	75	50
Transverse dispersivity (m)	12.5	7.5	5.0

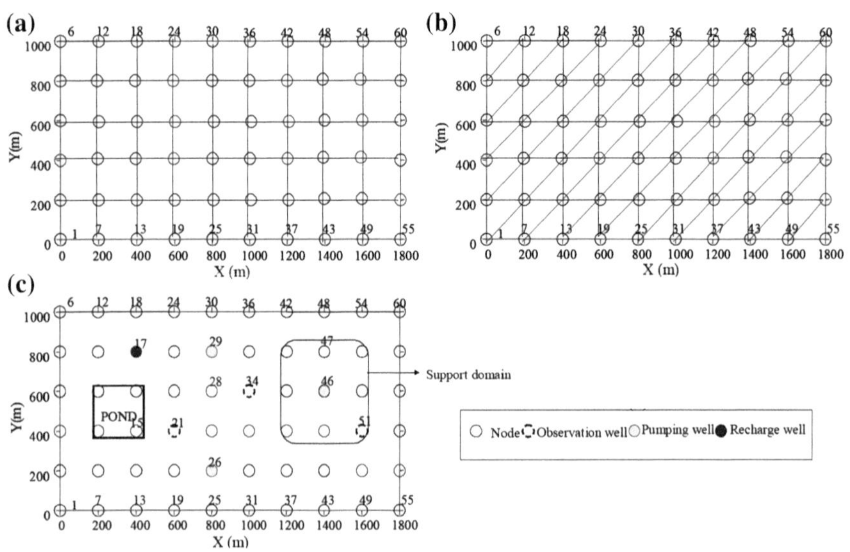

Fig. 8.7 FDM grid, FEM mesh, and nodal arrangement in MLPG for the study area

The contaminant is assumed to be total dissolved solids (TDS), and the contamination of the aquifer system is due to the seepage of the contaminated water from the pond and due to injection of water through recharge well at the rate 4000 and 1500 ppm/day, respectively. Initially, the entire aquifer is assumed to be uncontaminated, i.e., c = 0. To simulate the contaminant plume, the flow model is coupled with the transport model through Darcy's law.

The flow model has constant head conditions on its western (h = 100 m) and eastern (h = 95 m) boundaries and no flow boundaries in northern and southern directions. For transport model, only the eastern boundary is kept open and all other boundaries are assumed to be impervious.

The aquifer is simulated for an assumed period of 10,000 days to study the extent of contamination. The problem is simulated using three different numerical approaches, namely FDM, FEM, and MLPG. For FDM model, implicit time scheme and forward difference in space derivative approach with 6 × 10 regularly distributed nodes forming a uniform grid as shown in Fig. 8.7a are used. In case of FEM model, Galerkin's approach with Crank–Nicolson time scheme having 6 × 10 regularly distributed nodes and 90 triangular elements as shown in Fig. 8.7b is implemented. Further in MLPG model, only 6 × 10 regularly distributed nodes are used as shown in Fig. 8.7c. The time step of 1 day is chosen for both flow and transport models.

To monitor and compare the results from different models, three observation wells at nodes 21, 34, and 51 are considered.

The numerical models are initialized based on the steady-state head in the aquifer wherein no pumping is assumed. Through the entire simulation period of 10,000 days, it is considered that there is continuous contamination. The groundwater flow and contaminant concentration distribution are simulated using FDM, FEM, and MLPG models. The head and concentration distribution after 10,000 days of simulation from MLPG model are shown in Figs. 8.8 and 8.9, respectively.

The head and concentration values at the observation wells through the simulation using three different numerical models are given in Tables 8.2 and 8.3. Good agreement is observed between the results from different models. As the aquifer considered is simpler and smaller in size, the computational efforts required in modeling through either of the models are almost same. The application of different methodology for a given problem is solely dependent on the user's discretion and the computational facilities available.

8.5.2 Case Study 2—Field Problem

As can be seen in the previous case study, the application of grid-/mesh-based techniques such as FDM and FEM is quite cumbersome. It is quite burdensome and time consuming to construct a mesh or grid with elements of proper shapes for field problems. Even though some automatic mesh generators have been developed,

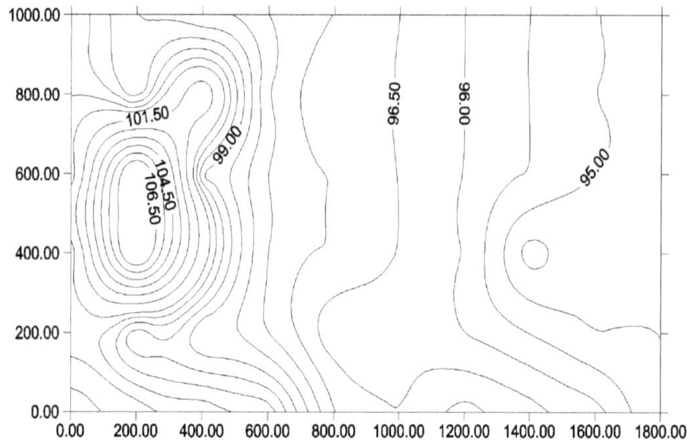

Fig. 8.8 Head distribution after 10,000 days from MLPG model

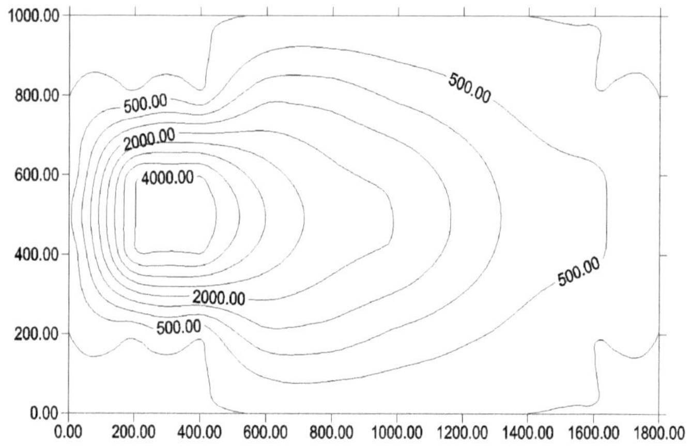

Fig. 8.9 Concentration distribution after 10,000 days from MLPG model

Table 8.2 Comparison of head distribution in m by FDM, FEM, MLPG model simulation for hypothetical problem

Node No.	FDM	FEM	MLPG
21	98.18	98.2	98.34
34	97.12	97.1	97.22
51	95.26	95.25	95.54

manual work is still needed to complete a proper mesh for each problem. Thus, the motivation in application of MLPG model is to cut down modeling costs in industrial-level applications by avoiding the labor-intensive step of mesh generation, and this is demonstrated with the help of an example here.

Table 8.3 Comparison of concentration distribution in ppm by FDM, FEM, MLPG model simulation for hypothetical problem

Node No.	FDM	FEM	MLPG
21	2696.13	2698.37	2697.36
34	1984.32	1985.16	1983.18
51	598.26	600.87	600.12

In this case study, the MLPG model for 2D coupled groundwater flow and transport is applied to a real field aquifer near Patancheru, Hyderabad, India. Many chemical and pharmaceutical industries have been established during 1970–1980s in the Patancheru Industrial Development Area near Hyderabad city in Medak district, Telangana, India (Rao et al. 2001). The treated and untreated effluents from the industries were being discharged in two streams: Nakkavagu and Peddavagu. Groundwater contamination has been reported since 1985, and the groundwater level and water quality monitoring study was carried out during 1997 and 1998 in Patancheru and its surrounding areas, covering an area of approximately 120 km^2 (Dhar et al. 1998; Rao et al. 2001) (Fig. 8.10). The streams which are the surface water bodies were found to be contaminated with TDS of concentration between 3000 and 5000 ppm. The surface water streams, all along its course, were acting as source of groundwater contamination. The groundwater contaminant concentrations of 1000–3500 ppm were assessed for 20-year period (1977–1997) as reported by Rao et al. (2001).

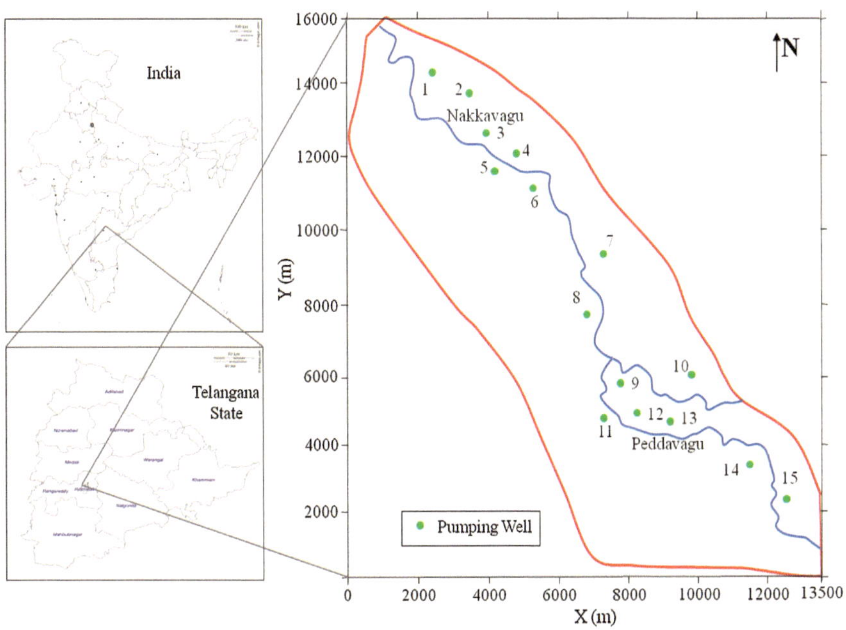

Fig. 8.10 Location map of the study area

The following information is used initially for developing a conceptual model of the given aquifer system (Fig. 8.10). Both Nakkavagu and Peddavagu streams have groundwater flow from north to southeastern direction. As the groundwater divide coincides approximately with the watershed boundary, no flow across the boundaries is assumed for the aquifer system. The entire aquifer domain is receiving groundwater recharge due to monsoon rainfall. The continuous operation of pumping wells causes fluctuating groundwater levels resulting in seepage of water from the streams to the groundwater regime. The contamination in the aquifer is due to the seepage of contaminated water from the two streams.

Based on the conceptual model, the numerical model using MLPG approach is simulated. A uniformly distributed set of nodes with nodal spacing of $\Delta x = \Delta y =$ 250 m is considered in the domain as shown in Fig. 8.11. The water level data of July 1997 is considered for initialization of groundwater flow model. The unconfined aquifer has the thickness of 15 m with groundwater recharge at the rate of 110 mm per year and permeability of 62.5 m/day, which considered as constant throughout the aquifer (Dhar et al. 1998; Rao et al. 2001). There exist 206 boring wells in the problem domain as shown in Fig. 8.12 (Dhar et al. 1998). The bore wells are assumed to be working at a rate of 1.1 L/s (95 m^3/d) working for over only 60% of day. The seepage from the water streams is simulated by giving additional recharge input to the model.

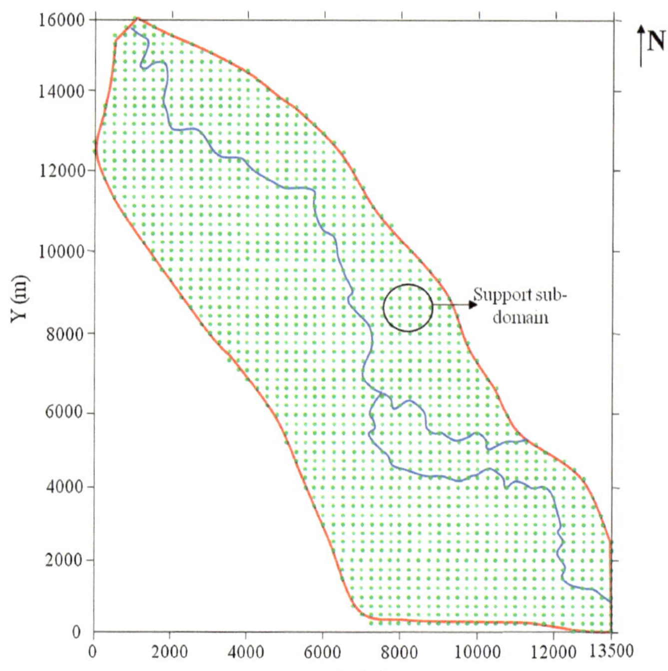

Fig. 8.11 Problem domain with nodal distribution

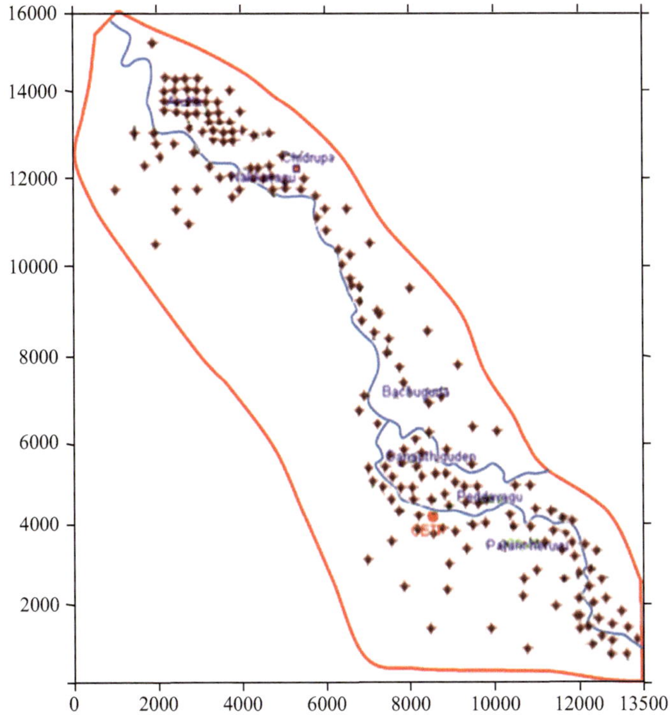

Fig. 8.12 Problem domain with bore wells Adapted from Dhar et al. (1998)

For calibration of the model, a steady-state analysis is carried out. For verification, the recharges are given to see whether the desired head values at all the nodes are obtained; if so, the model is calibrated. During calibration of the model, the hydrogeologic parameters are adjusted within a range of 20% from their field-estimated values. This matching is done by minimizing the sum of some function of the differences between the observed and predicted (calculated from MLPG model) groundwater head values.

Further, the following error measure is defined for measuring the accuracy of the predictions from the MLPG model: Root-mean-square error in groundwater head predictions is given below:

$$\text{RMS} = \sqrt{\frac{\sum_{j=1}^{N}\left(h_j^{\text{Pred}} - h_j^{\text{Obs}}\right)^2}{N}} \tag{8.27}$$

The RMS is a measure of the effectiveness of the MLPG model in simulation. The calibrated initial groundwater levels from the model are shown in Fig. 8.13 with RMS error of 0.832. These values are considered as the initial groundwater levels for the transient simulation by MLPG model for the year 1997 (Fig. 8.13).

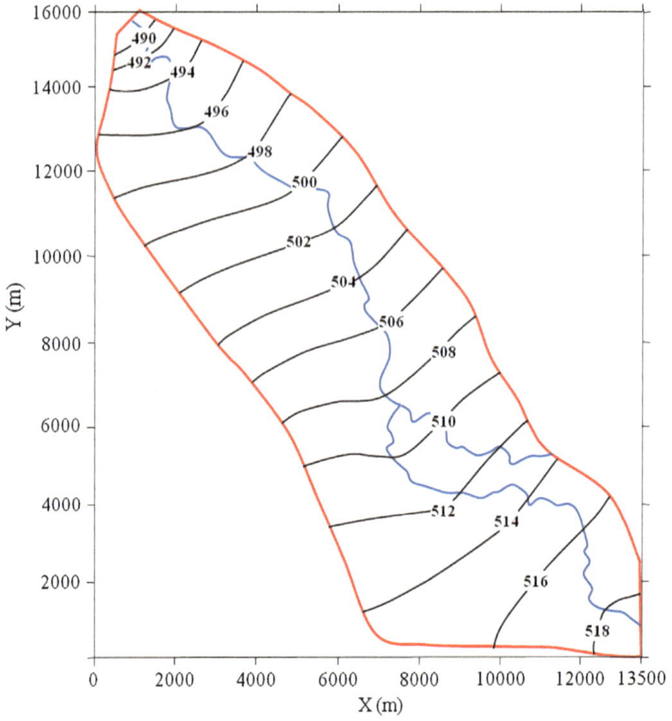

Fig. 8.13 Initial groundwater head distribution (m), 1997

The velocity vectors at the steady state for the considered data are shown in Fig. 8.14. The calculated heads from MLPG model are compared with the observed heads values as shown in Fig. 8.15 considering 15 observation nodes.

For the contaminant transport modeling, the values of dispersivity in longitudinal and transverse directions are taken as 50 and 5 m, respectively (Dhar et al. 1998; Rao et al. 2001). A constant TDS concentration at different nodes on the Nakkavagu was assigned varying from 3500 ppm near the Patancheru to 1000 ppm along downstream of the Nakkavagu. For calibration of the model, a steady-state analysis is carried out coupled with flow model. The initial contaminant concentration levels for the year 1997 are shown in Fig. 8.16.

The initial groundwater head levels of the MLPG model are set for 1997, and the model is run to predict the head levels in 2017 (at the end of 20 years). The groundwater levels due to boundary conditions as discussed above are used in the numerical simulation. It is assumed that there is no change in the average groundwater recharge and pumping wells are working at the same rate. The time step $\Delta t = 10$ days is considered, and the head distribution after 20 years of simulation is obtained, for year 2017, as shown in Fig. 8.17. From Figs. 8.13 and 8.17, it can be seen that the aquifer system may have reached steady state in the course of 20-year simulations as not much difference in the head distribution is observed. For

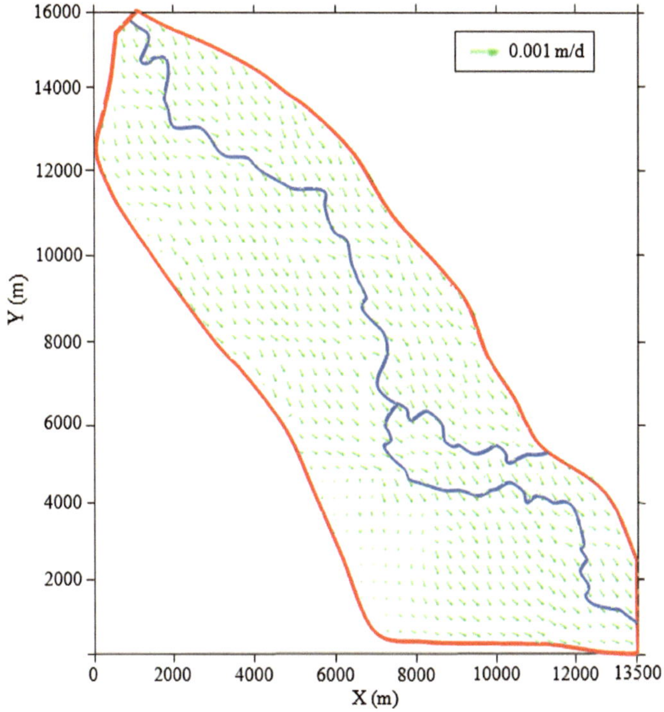

Fig. 8.14 Velocity vector distribution for steady-state condition, 1997

Fig. 8.15 Comparison of calculated versus observed groundwater heads, 1997 (m)

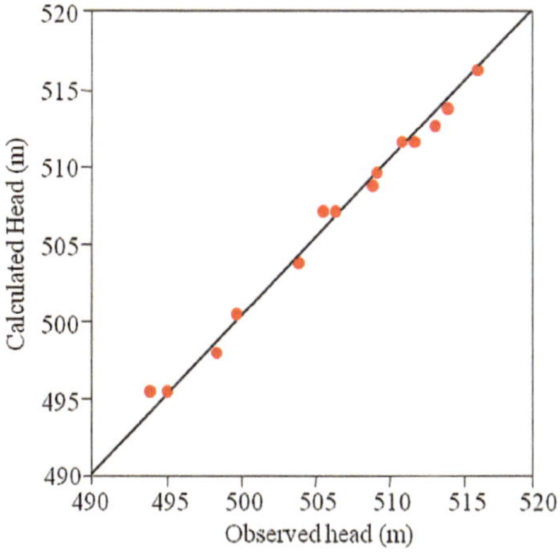

transport simulation, it is assumed that no further contamination is taking place in the system, other than the existing contamination. The transport simulation is done for 20 years, and the concentration distribution is shown in Fig. 8.18. From the

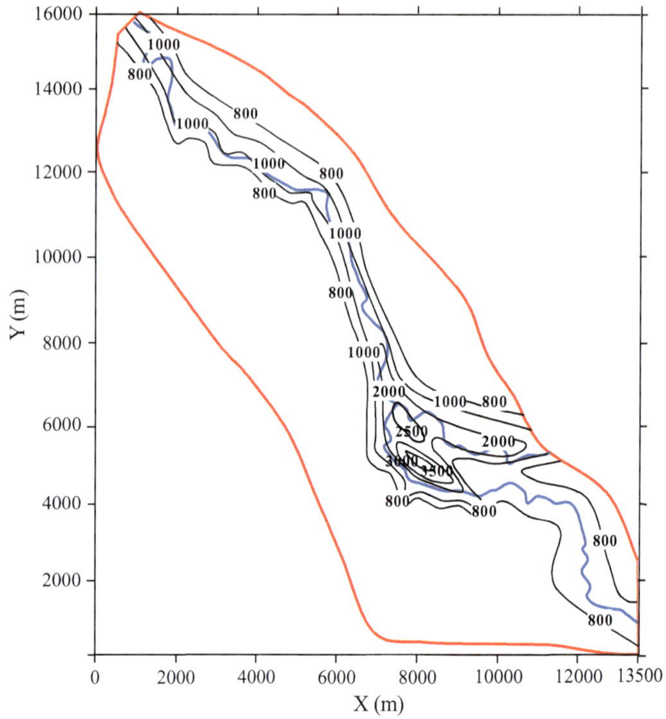

Fig. 8.16 Initial contaminant concentration (ppm) distribution, 1997

Figs. 8.16 and 8.18, it can be observed that the maximum TDS contaminant concentration in the aquifer domain has decreased from 3500 to 2800 ppm, which is about 20% decrease in the maximum concentration of TDS. This is attributed due to the natural attenuation of the contaminant transport process mechanism in the aquifer due to mainly dilution and advection.

Based on the results, the MLPG coupled groundwater flow and transport model prove to be an effective and efficient simulation model. As the contamination levels are high in the aquifer, the groundwater system can be recommended for remediation. Due to the presence of TDS contamination, pump and treat remediation system can be adopted. The MLPG simulation model can be further coupled with an optimization model to form a simulation–optimization model. It can be used to investigate and predict the responses of the groundwater aquifer systems under various remediation scenarios.

Additional case studies are demonstrated in Sharief et al. (2012), Mategaonkar and Eldho (2012), Guneshwor et al. (2016), Swathi and Eldho (2017), etc., wherein different simulation models are used in solving the groundwater contamination problems.

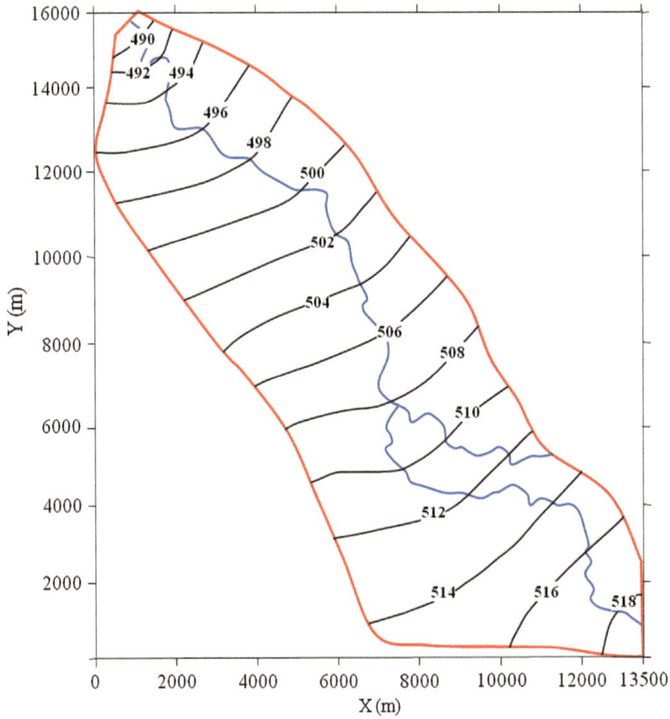

Fig. 8.17 Head distribution (m) in the aquifer, 2017 after 20 years

8.6 Summary and Conclusion

Groundwater is a major source of water, and its sustainable development is very important. In the recent times, the groundwater contamination has become a major problem in many parts of the world. In this chapter, the groundwater contamination issues and its numerical simulation are discussed in detail. The importance and necessity of numerical modeling of the contaminant transport problems have been emphasized. The theoretical background of groundwater flow and transport and its governing equations are presented, and numerical procedures followed using different approaches for solving these equations are explained. Through two case studies, the numerical modeling approach has been demonstrated. In the first example, the contamination is occurring due to the seepage from a polluted pond. Over the simulation period, the movement of the contaminant plume can be calculated and observed such that appropriate action can be taken. This example has been simulated using three different modeling approaches, FDM, FEM, and MLPG. The results from these different models are in good agreement with each other; however, the manual effort in the labor-intensive step of mesh generation is cut down in case of MLPG model. Further, the contaminant transport problem in case

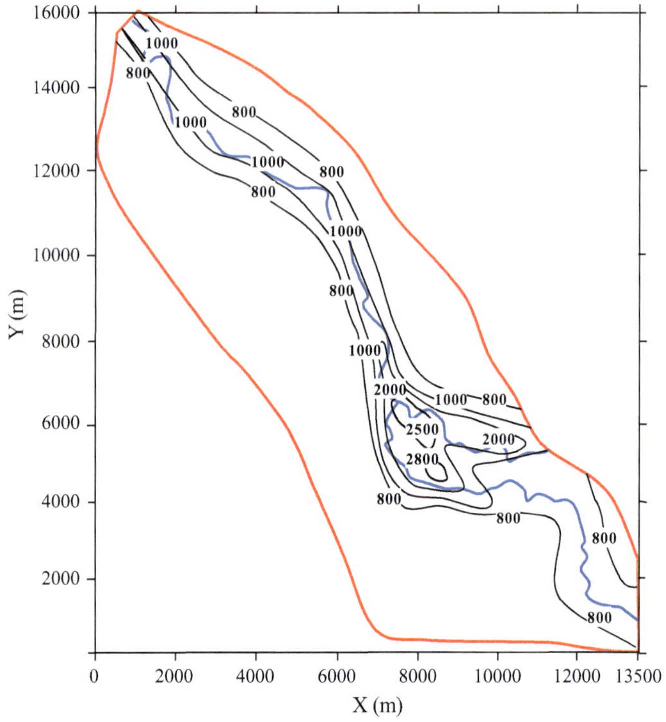

Fig. 8.18 Contaminant concentration distribution (ppm) in the aquifer, 2017 after 20 years

of real field study is simulated using MLPG model. The contamination in the aquifer is due to the seepage of contaminated water from the bed of streams existing in the problem domain. The MLPG model is calibrated based on the observed data from 1997, and the model is used to simulate the conditions in 2017. From the results, it can be seen that around 20% of the maximum contaminant concentration has been reduced. This can be attributed to the natural attenuation of the aquifer system. It can be concluded that if further contamination is prevented, the groundwater aquifer systems can self-clean to certain extent, though it takes a long time. However, based on the type of contaminant and recovery time period set up by governing bodies, different kinds of groundwater remediation systems can be recommended.

As the roles of simulation models in case of groundwater contamination problems are very clear, it is imperative to use numerical models for recreating present conditions and to predict the future variations under different conditions. These simulations help in deciding on optimal management or remediation policies during the stage of planning and governance by end users. This chapter provides the basic understanding of groundwater contamination mechanism and its numerical simulation and can be a reference for the researchers and policymakers.

References

Anderson MP (1984) Movement of contaminants in groundwater: groundwater transport–advection and dispersion, groundwater contamination. National Research Council, National Academy Press, Washington D.C

Anderson MP, Woessner WW (1992) Applied groundwater modeling—simulation of flow and transport. Academic Press Inc., San Diego, California

Atluri SN (2004) The Meshless method (MLPG) for domain and BIE discretizations. Tech Science Press, Forsyth

Bear J (1972) Dynamics of fluids in porous media. Dover Books, New York

Bear J (1979) Hydraulics of groundwater. McGraw Hill Publishing, New York

Bear J, Verruijt A (1987) Modeling groundwater flow and pollution. D. Reidel Publishing Co., Dordrecht

Bedient PB, Rifai HS, Newell CJ (1999) Groundwater contamination-transport and remediation, 2nd edn. Prentice Hall, New Jersey, USA

Desai YM, Eldho TI, Shah AH (2011) Finite element method with applications in engineering. Pearson Education, New Delhi

Dhar RL, Rao VVSG, Subrahmanayam K, Yadaiah P, Govil PK, Gyaneswara Rao T, Keshavkrishna A (1998) Study of Groundwater Pollution in Patancheru and Bolaram Industrial Development Areas. Medak district, Andhra Pradesh. Technical Report: NGRI-99-GW- 252:127

Downing RA (1998) Groundwater—Our hidden asset. British Geological Survey Monograph

Eldho TI (2001) Groundwater contamination—the challenge of pollution control and protection. J Indian Water Works Assoc 33(2):171–180

Freeze RA, Cherry JA (1979) Groundwater. Prentice Hall, Englewood Cliffs

Fried JJ (1975) Groundwater pollution. Elsevier, New York

Guneshwor SL, Eldho TI, Kumar AV (2016) Coupled groundwater flow and contaminant transport simulation in a confined aquifer using meshfree radial point collocation method (RPCM). Eng Anal Bound Elem 66:20–33

Istok J (1989) Groundwater modeling by the finite element method. American Geophysical union, Washington

Kovářík K, Mužík J (2013) A Meshless solution for two dimensional density-driven groundwater flow. Eng Anal Bound Elem 37(2):187–196

Kumar P, Dodagoudar GR (2008) Two-dimensional modeling of contaminant transport through saturated porous media using the radial point interpolation method (RPIM). J Hydrol 16:1497–1505

Li H, Mulay SS (2013) Meshless methods and their numerical properties. CRC, Boca Raton

Li J, Chen Y, Pepper D (2003) Radial basis function method for 1-D and 2-D groundwater. Comput Mech 32:10–15

Liu GR, Gu YT (2005) An introduction to meshfree methods and their programming. Springer, Berlin

Mategaonkar M, Eldho TI (2012) Groundwater remediation optimization using a point collocation method & particle swarm optimization. Environ Model Softw 32:37–48

Pinder GF, Gray WG (1977) Finite element simulation in surface and subsurface hydrology. Academic press, New York

Rao VVSG, Dhar RL, Subrahmanyam K (2001) Assessment of contaminant migration in groundwater from an industrial development area, Medak district, Andhra Pradesh, India. Water Air Soil Pollu 128(3):369–389

Rastogi AK (2010) Numerical Groundwater Hydrology. Penram Int. Publishing, New Delhi

Sharief SMV, Eldho TI, Rastogi AK (2008) Optimal pumping policy for aquifer decontamination by pump and treat method using genetic algorithm. ISH J Hydraul Eng 14(2):1–17

Sharief SMV, Eldho TI, Rastogi AK (2012) Optimal groundwater remediation by pump and treat using FEN and EGA based Simulation—Optimization Model. J Hazard Toxic Radioact Waste, ASCE 16(2):106–117

Swathi B, Eldho TI (2014) Groundwater flow simulation in unconfined aquifers using Meshless Local Petrov-Galerkin (MLPG) method. Eng Anal Bound Elem 48:43–52

Swathi B, Eldho TI (2017) A moving least squares based Meshless Local Petrov-Galerkin method for the simulation of contaminant transport in porous media. Eng Anal Bound Elem 78:8–19

Wang H, Anderson MP (1982) Introduction to groundwater modeling finite difference and finite element methods. W. H. Freeman and Company, New York

Willis R, Yeh WW-G (1987) Groundwater systems planning and management. Prentice Hall, Inc. Englewood Cliffs, New Jersey

Zheng C, Bennett GD (1995) Applied contaminant transport modeling—theory and practice. Van Nostrand Reinhold, New York

Chapter 9
A Systematic Development of Uncertainty Modeling in the Assessment of Health Risk to Environmental Contaminants

Harshit Mishra, Subhankar Karmakar and Rakesh Kumar

Abstract The objective of science-based risk assessment is to protect public health by providing profound decisions. Health risk analysis involves various uncertainties and highly variable parameters like multiple routes (ingestion, dermal, and inhalation), complex environmental contaminants, various pathways, and different exposure to population; which makes the risk estimation procedure extremely challenging and rigorous. The uncertainties in risk assessment majorly result from two reasons, firstly, the lack of knowledge of input variable (mostly random), and secondly, data obtained from an expert judgment or subjective interpretation of available information (non-random). The NRC (1994) states that to ignore the uncertainty in any step of risk assessment process is almost as likely as to leave critical parts of the process has been left incompletely examined and therefore increase the probability of generating a risk estimate that is incorrect, incomplete, or misleading. Each step of the risk assessment process involves various assumptions, both quantitative and qualitative, must be evaluated through uncertainty analysis. However, it is necessary that risk process of evaluation must treats uncertainty and variability scientifically and robustly. Moreover, addressing uncertainties in health risk assessment is a critical issue while evaluating the effects of environmental contaminants on public health. The uncertainty propagation in health risk can be assessed and quantified using probability theory, possibility theory, or a combination of both. This chapter will systematically

H. Mishra (✉) · S. Karmakar
Centre for Environmental Science and Engineering, Indian Institute of Technology Bombay, Mumbai 400076, India
e-mail: harshit.cese@gmail.com

S. Karmakar
Interdisciplinary Program in Climate Studies, Indian Institute of Technology Bombay (IITB-CECS), Mumbai 400076, India

S. Karmakar
Centre for Urban Science and Engineering, Indian Institute of Technology Bombay, Mumbai 400076, India

R. Kumar
National Environmental Engineering Research Institute (NEERI), Nagpur 440020, India

T. Gupta et al. (eds.), *Environmental Contaminants*, Energy, Environment, and Sustainability, https://doi.org/10.1007/978-981-10-7332-8_9

report the development of various methodologies and frameworks to address the uncertainties that are intrinsic to health risk estimation.

9.1 Introduction

The risk is a composite function of the inherent toxicity of a chemical and the degree of exposure to humans. The toxicity of any chemical and exposure to risk population is independent of each other (Penningroth 2010). A risk assessor performs the risk analyzes by identifying the toxic effect associated with a specific chemical and evaluates the harm caused to environment and humans. The preliminary risk assessment (RA) is broadly classified into two independent approaches; firstly, to analyze the toxicity of the chemical of concern and secondly to estimate the degree of exposure to humans. However, a comprehensive risk analysis is a science-based tool and requires the knowledge from multiple disciplines, which ultimately helps the risk analyst in discerning specific public health threats from any contaminant.

The involvement of enormous amount of data and complexity in risk procedures makes the RA extremely rigorous and uncertain. The National Research Council (NRC) in 1983, published a report entitled as "*assessment in the federal government: managing the process*" and proposed RA framework for incorporation of toxicology. This report provided a roadmap for the usage of toxicology data to inform decision-making; also the book is later famous as "Red Book" due to the color of its cover. The NRC (1983) reported that RA is a process in which information is analyzed to determine whether an environmental contaminant might cause harm to exposed population or ecosystems. However, the red book committee mentioned explicitly that the RA does not dispel scientific uncertainty and they wrote:

> "The committee believes that the fundamental problem in RA is the sparseness and uncertainty of the scientific knowledge, and this issue has no ready solution. The field has been developing rapidly, and the greatest improvements in RA result from the acquisition of additional and better data, which decreases the need to rely on interference and informed judgment to bridge gaps in knowledge (NRC 1983)."

Health and the environment are intricately interlinked. As a result, need to foster interdisciplinary and cross-domain aspects and integration in the environmental modeling (Reis et al. 2017). However, the application of such models/methods depends on the nature and availability of data (US EPA 1989). The goal of RA is to estimate the severity and likelihood of harm to human health from exposure to a substance (Mishra et al. 2017). The RA process consists of the following four primary steps: hazard identification, exposure assessment, dose-response evaluation, and risk characterization (Fig. 9.1). RA is a science-based tool to help policymakers to discern specific public health threats, to gauge their relative severity, to establish priorities, and to allocate limited resources to address these priorities (Penningroth 2010). This approach also serves as a rational basis for protecting the

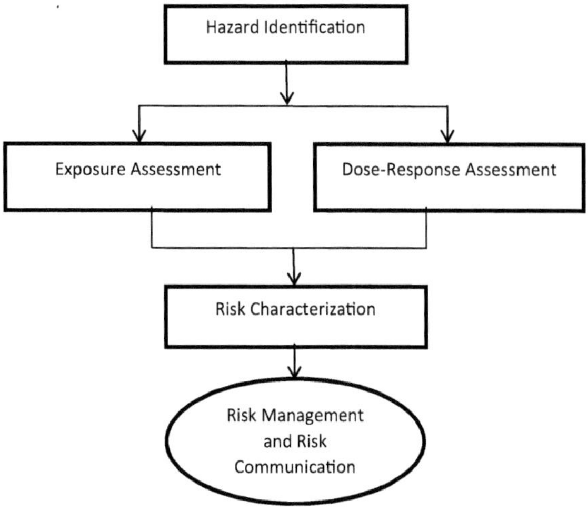

Fig. 9.1 Flow chart of risk assessment (adopted from Masters and Ela 2008)

environment and human health from risks associated with toxic chemicals. However, the quantification of risk is not straight forward in terms of calculating the risk values, as described by NIH. It is a data-intensive and site-specific approach, and it requires sound modeling skills (Mishra et al. 2017). The RA process can assist in drawing a cost-effective compromise between economic and environmental costs, thereby ensuring that the philosophy of sustainable development is followed (Butt et al. 2008).

RA has been used to quantify ecological and human health impacts from a contaminant exposure via multiple exposure routes. The RA process begins by identifying chemicals of concern, potential pathways of exposure, and populations at risk. Hazard identification is the procedure of determining whether or not a particular chemical is linked to any specific health effects such as cancer or congenital disabilities. *Hazard identification* is the initial step of RA and most easily recognized in the actions of regulatory agencies. This step involves characterizing the nature and strength of the evidence of causation and is usually derived from laboratory testing on animals or another test organism, as experiments on humans are restricted. Further, information from short-term in vitro tests and on structural similarity to known chemical hazards may also be considered as a potential method for hazard identification (NRC 1983; Masters and Ela 2008; Penningroth 2010; Dong et al. 2015).

The *dose-response assessment* is the procedure of characterizing the relationship between the dose of a chemical received and its adverse health effect upon exposed populations. The dose-response relationship for any contaminant is derived from toxicology studies that depend on the type of experiment performed, i.e., one-time acute test or a long-term chronic test. A dose-response assessment usually requires extrapolation from high doses on animals to low doses on humans (NRC 1983; Masters and Ela 2008; Penningroth 2010; Dong et al. 2015).

The third step is *exposure assessment*, which involves determining the size and nature of the population exposed to the toxicant and the length of time and toxicant concentration. Exposure assessment is the process of estimating the intensity, frequency, and duration of human exposures to a chemical present in the environment. It also includes the route of exposure, and the size, nature, and classes of the human populations exposed (NRC 1983; Masters and Ela 2008; Penningroth 2010; Dong et al. 2015).

Risk characterization is the combination of the three aforementioned steps (i.e., hazard, dose-response, and exposure), which estimates the risk due to a toxicant or mixture of toxicants for oral, inhalation, and dermal routes. It is an estimate of the degree and severity of public health problem. Risk management and communication is the last step of RA. Risk management ensures that risks are maintained within acceptable levels while risk communication is the dissemination of risk information between regulators and stakeholders (Penningroth 2010).

9.2 Uncertainty Propagation in Risk Assessment

The real-world simulation problems of modern engineering are complex, and they involve many non-deterministic inputs. The non-deterministic nature of inputs into these complex simulation process results in uncertain outputs. Most of the input parameters in the risk model are uncertain, results in the risk outputs as uncertain. Another level of uncertainty exists during the consideration of specific routes (oral, inhalation, and dermal) and pathways (air, water, and soil), which makes the risk uncertain. Thus, using single value estimation for these variables may result in significant error. Each step of the RA process involves various assumptions, both quantitative and qualitative, which must be addressed through uncertainty analysis (Mishra et al. 2017). According to NRC (1994), ignorance of uncertainty in any step of RA leads to an increase in the likelihood of generating an incorrect, incomplete, or misleading risk estimate. Hence, it is important that the risk analyst scientifically treat uncertainty and variability, and ensures that the risk outputs can be confidently translated into management actions (Refsgaard et al. 2007).

The implementation of a robust RA starts by identifying potential types of uncertainties associated with ecological and human health risk assessment (Morgan et al. 1990). Therefore, it is essential that the risk analysts have a clear and defensible typology of uncertainties. Identification of the different kinds of uncertainties that exist in RA is a critical part of uncertainty quantification, which must be addressed using different techniques (Morgan et al. 1990; van der Sluijs et al. 2004; Refsgaard et al. 2007). The uncertainty can be broadly classified into three dimensions, i.e., location, nature, and level of uncertainty (Walker et al. 2003). Figure 9.2 provides broad categorizations of all possible types of uncertainties that may be encountered during risk estimation (Acosta et al. 2010; Arhonditsis et al. 2007; Ascough et al. 2008; Bedford and Cooke 2001; Chen et al. 2007; Croke et al. 2006; Ducey and Larson 1999; Gottschalk et al. 2010; Hromkovic 2005;

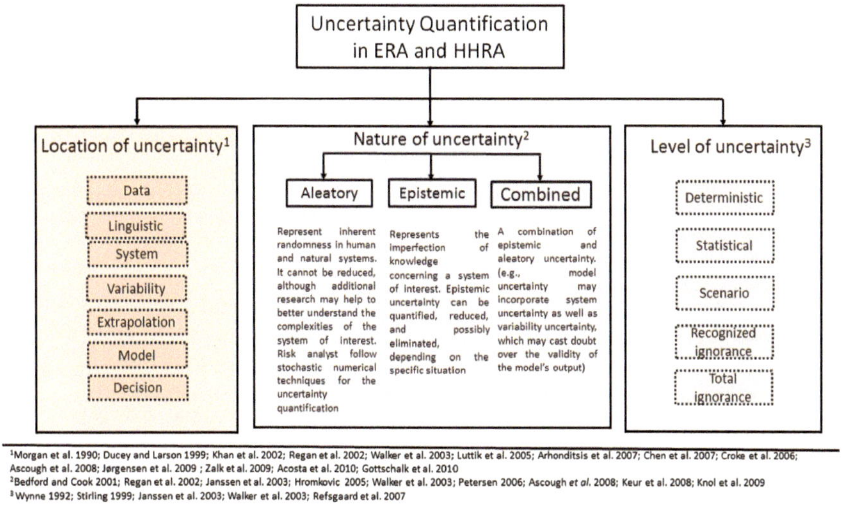

Fig. 9.2 Identifying uncertainties in ecological and human health risk assessments

Janssen et al. 2003; Jørgensen et al. 2009; Khan et al. 2002; Knol et al. 2009; Luttik et al. 2005; Petersen 2006; Regan et al. 2002; Stirling 1999; Keur et al. 2008; Wynne 1992; Zalk et al. 2009).

Addressing uncertainties in ecological and human health, RA is a critical issue when evaluating the effects of contaminants on public health. It has been reported that a range of uncertainties exists through the source-to-outcome continuum, including dose-response, exposure, hazard, and risk characterization. Based on a detailed, the comprehensive literature review, aim and potential sources of uncertainties associated with each step of RA have been summarized in Table 9.1. The Table provides a notion that quantitative RA requires an enormous amount of data to perform ecological and human health risk assessment and subsequently uncertainty modeling. The literature review reports Probability and Fuzzy theories are being used extensively to address uncertainty, and the same have been described in the next section.

9.3 Methods of Addressing Uncertainty

Risk analysis will not remain simple when it involves multiple uncertain variables. Uncertainty is inherent to the RA process and involved in every single step of risk estimation. Typically, two types of uncertainties exist during risk estimation, i.e., random and non-random. The random uncertainty is usually due to the lack of knowledge of input variable, and the non-random is mainly due to the data obtained from an expert judgment or subjective interpretation of available information (Guyonnet et al. 2003). It was found that the uncertainty due to randomness had

Table 9.1 Characterization of key uncertainties in ecological and human health risk assessment

Component	Aim/objective	Uncertainty source
Hazard identification	To determine whether any contaminant can cause adverse health impacts to public health. The quantitative RA involves laboratory and field observations for the determination of contaminant concentration and possible adverse health effects	a. Factor affecting fate and transport of stressor b. Data considerations and requirements c. Appropriateness of assessment endpoints
Dose-response assessment	To develop the relationship between the amount of dose and response by aspecific organism. The aim is to derive the RfD for non-carcinogenic and potency factor for carcinogenic contaminants using extrapolation from high to low doses	a. Dose received by receptor (e.g., extrapolating knowledge for the duration, frequency or intensity of dose) b. Assessment endpoints (e.g., variability in the duration, frequency, intensity of dose) c. Examine assessment endpoints (e.g., variability in receptor development, behavior, survival) d. Distribution assumption e. Extrapolation: extrapolating RfD from PoD, including interspecies, intra-species
Exposure assessment	To estimate the duration, frequency, and magnitude of the exposure to a target population. This step involves the environmental pathways details and the route of exposure and characterization of population	a. Stressor information (e.g., data about composition, distribution, and release) b. Fate and transport information c. Receptor information d. Exposure duration and route e. Model selection f. Exposure profile (e.g., using direct monitoring data)
Risk characterization	To estimate and aggregate the risk from particular contaminant for multiple pathways and routes. This step also describe the nature, likelihood, and magnitude from exposure to chemicals to public health	a. Risk estimation b. Risk aggregation (e.g., combining individual risk estimates to quantify overall risk) c. Assessing the risk (e.g., with regulators or stakeholders) d. Communicating the risk(e.g., to risk professionals, public, and interested parties)

RfD reference dose; *PoD* point of departure

been addressed using probabilistic theory, while subjective uncertainty was characterized through fuzzy set theory. In this regard, the literature further reveals that the practical environmental engineering problems involve both types of uncertainties, which cannot be modeled using probability or fuzzy set theory individually (Singpurwalla and Booker 2004; Kentel and Aral 2004). Therefore, to address the uncertainty in ecological and human health risk assessment, probability theory, fuzzy concept, and a combination of both, i.e., hybrid theory can be applied. The next section demonstrates various approaches for the quantification of uncertainty.

9.3.1 Probabilistic Approach

Probabilistic analyzes, as compared to a deterministic value, are likely to offer more information as it allows a range of input parameters for risk estimation (Finley and Paustenbach 1994). Therefore, probability distribution function rather than single point estimates was used to represent key exposure parameters when sufficient data were available (Buteau and Valcke 2010). In RA, probabilistic analysis has been used extensively in the past as the primary tool for modeling uncertainty during the risk estimation process. Uncertainties in risk estimates may arise from many different sources, such as the measurement of risk parameters, natural variability in the individual responses of various population groups, variability in the environmental concentration of toxicants over time and space, and extrapolations in dose-response models (Kentel and Aral 2004). Therefore, it can be concluded that uncertainty is inherent in the RA process, even when using the most accurate data and sophisticated models (U.S. EPA 2005). However, a robust uncertainty analysis facilitates the risk managers to analyze the risk outcomes appropriately and provides flexibility in suggesting better options for risk management (Kalberlah et al. 2003).

Each step of the RA process involves various assumptions, both quantitative and qualitative, that must be evaluated through uncertainty analysis. Probabilistic risk assessment (PRA) uses the probability theory to quantify the risk levels due to the exposure of toxicants in a population and characterize the uncertainty in risk estimates (Maxwell and Kastenberg 1999). The PRA effectively combines risk characterization with uncertainty analysis by providing the range of possible risk values (Stackelberg and Burmaster 1994). PRA uses Monte Carlo simulation, which uses statistical tools, and currently the most common method for evaluating uncertainty and variability in health RA (Schuhmacher et al. 2001).This approach approximates the probability of outcomes using random variables in a series of multiple trial runs, called simulations. The Monte Carlo simulation illustrated in Fig. 9.3, first selects a random set of input data values drawn from their probability distributions. These values are then used in the risk model to obtain output risk distribution, which reflects the combination of all input distributions. A PRA framework for the estimation of uncertainty associated with human health risk is performed by Mishra et al. (2017) for the oral intake of leachate-contaminated groundwater. The paper proposed an integrated approach for the assessment of uncertainty associated with human health risks from municipal solid waste landfill leachate contamination in the form of a framework. To perform this analysis, a data inventory for the risk estimation has been adopted from Mishra et al. (2016), which provides a set of heavy metal concentrations in the groundwater.

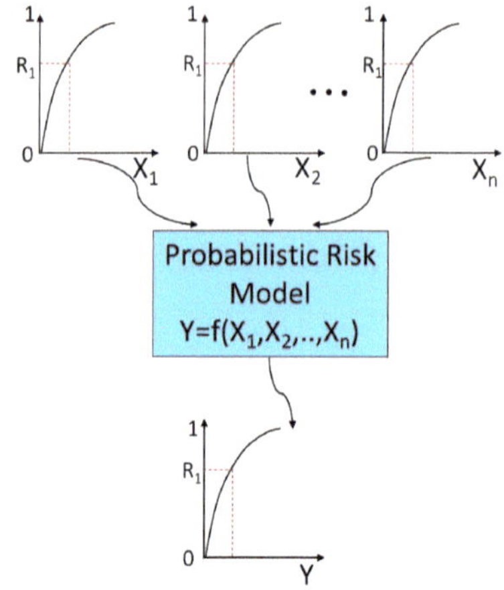

Fig. 9.3 A conceptual diagram for Monte Carlo simulation for evaluating output distribution corresponds to specific reliability level of input distribution

9.3.2 Fuzzy Theory

In many real case scenarios, the RA parameters contain subjective information related to risk and its occurrence, and either expert judgment or subjective interpretation is used to define these parameters. In such cases, the PRA analysis may not be sufficient to represent the actual nature of the parameters' uncertainty. Zadeh (1965) introduced the fuzzy concept to develop a mathematical framework for treating subjective or imprecise information. It is a generalized form of interval analysis, where parameters are defined with lower and upper values (Shakhawat et al. 2006). In the classical set theory, an element is defined as a member of a set, and if the element is in the set, the membership grade is unity while otherwise, it is zero. A fuzzy set is an expansion of the classical set theory in which the element has a certain degree of membership μ(0–1) in the set. Fuzzy set establishes the relationship between uncertain data and the membership function μ, which ranges from 0 to 1 (Eq. 1). The membership function, $\mu(x)$ is defined as the fuzzy subset A in the universe of discourse, x.

$$\mu_A = \begin{cases} 1 \text{ iff} x \in A \\ 0 \text{ iff } x \notin A \\ p;\ 0 < p < 1 \text{ if } x \text{ partially belongs to } A \end{cases} \tag{1}$$

The α cut of a fuzzy set A is a crisp set A_α that contains all the elements in U (Universe of discourse) that have membership values in A greater than or equal to α (Wang 1997). Figure 9.4 shows a parameter P represented as a triangular fuzzy number with the support of A_0. Wider the support of the membership function, higher is the uncertainty. The fuzzy set that contains all elements with a membership of a ε[0, 1] and above is called the α-cut of the membership function. At a resolution level of α, it will have the support of A_α. The higher the value of α, the higher the confidence in the parameter (Li and Yen 1995).

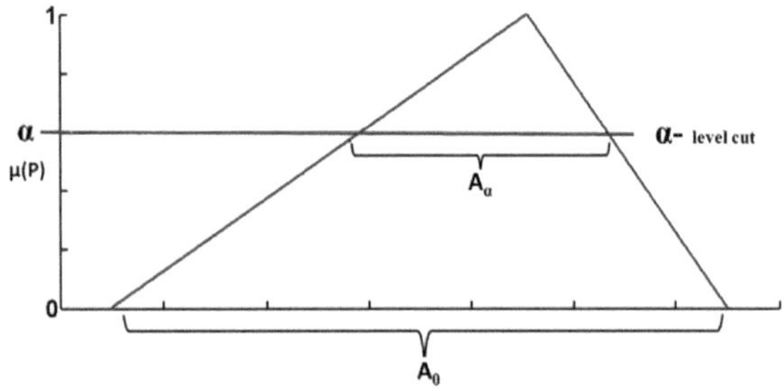

Fig. 9.4 Fuzzy number, its support, and α-cut

The triangular fuzzy membership functions are defined by (*a*, *b*, and *c*), where *a* and *c* represent the minimum and maximum values, respectively, and *b* accounts for the most likely value. For the most likely value, the membership grade is assigned as unity. The triangular membership function for Fig. 9.4 can be represented as:

$$\mu_p(\boldsymbol{x}) = \left\{ \begin{array}{c} \frac{x-a}{b-a}, \boldsymbol{a} \leq \boldsymbol{x} \leq \boldsymbol{b} \\ \frac{x-c}{b-c}, \boldsymbol{b} \leq \boldsymbol{x} \leq \boldsymbol{c} \\ \boldsymbol{0}, \text{otherwise} \end{array} \right\} \tag{2}$$

Fuzzy modeling is a technique that has gained popularity among researchers using RA models, given its efficacy in handling uncertainties caused by subjective interpretations. A collection of the literature on theoretical advances and applications of fuzzy set theory concepts in engineering, economics, medicine, and ecology has been provided by Dubois et al. (2001).

9.3.3 Hybrid Theory

Risk analysis will not remain simple when it involves multiple uncertain variables. Typically, two types of uncertainties exist during risk estimation, i.e., random and non-random. The random uncertainty is usually due to the lack of knowledge of input variable, and the non-random is mainly due to the data obtained from an expert judgment or subjective interpretation of available information (Guyonnet et al. 2003). In the previous studies, the uncertainty due to randomness has been addressed using probabilistic theory, while subjective uncertainty was characterized through fuzzy set theory. In this regard, the literature further reveals that the practical environmental engineering problems involve both types of uncertainties, which cannot be modeled using probability or fuzzy set theory individually (Singpurwalla and Booker 2004; Kentel and Aral 2004). Additionally, it was found that both the procedure, i.e., probability and fuzzy have limitations to address the uncertainty. Firstly, in the application of realistic problems, particularly evaluating the ecological and human health risk estimation; rarely adequate data is available to

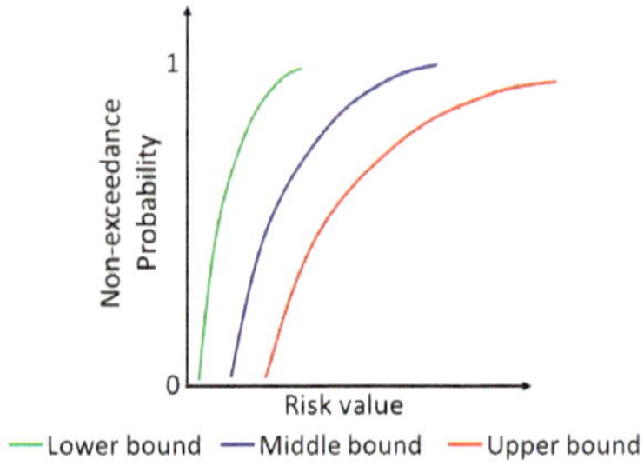

Fig. 9.5 A representation of non-exceedance probability of risk value derived through hybrid approach

estimate, and characterize the probability distribution functions of the input variables. However, PRA typically uses distribution functions of input parameters to quantitatively characterize the variability and uncertainty in the risk output (U.S. EPA 2004) and due to lack of sufficient and reliable data in RA, probability distribution functions of the input variables is often not possible to derive. Secondly, if the available information is in the form of an expert judgment or subjective interpretations of input parameters, then PRA may not be sufficient to represent the actual nature of uncertainty and fuzzy set theory is usually applied. Therefore, an integrated probabilistic-fuzzy (hybrid) risk approach has been required for simultaneous treatment of random and non-random uncertainties associated with input parameters of the risk model.

The procedure for proposed hybrid approach starts with the identification of fuzzy, random variables, or constant in a risk model. The form of information available to the risk analyst determines whether a parameter will be treated as a fuzzy or a random variable. Once all available information is gathered and data evaluated, appropriate probability density functions, and membership functions can be specified for variable and uncertain parameters, respectively. For fuzzy parameters, appropriate membership function must be considered based on the literature review. However, the hybrid framework is not restricted to this membership function choice. The α-cut concept is used to handle fuzzy membership function while reliability level has been used for probability density functions allotted to random variable respectively. Finally, Monte Carlo simulation is used to generate cumulative distribution functions of risk for upper and lower limits of each α-cut and reliability level (Mofarrah and Hussain 2011). Figure 9.5 represents a typical plot of the non-exceedance probability of risk value derived through a hybrid approach.

9.4 Conclusion

The quantification of uncertainties associated with ecological and human health risk assessment is extremely challenging as it involves multiple uncertain variables. The presence of various types of uncertainties (random and non-random) in risk estimations makes the RA procedure difficult. The uncertainties that intrinsically involved in RA can be addressed through probabilistic, fuzzy, or a combination of both. Probability and fuzzy theory are the two main approaches to address the uncertainty, which have an application based on input data inventory. A combination of both probabilistic and fuzzy theory, i.e., the hybrid theory also has been demonstrated to overcome the limitation of simultaneous treatment of random and non-random type of uncertainty in the risk model. However, the present study recommends following an entirely probabilistic approach, which required a continuous and consistent collection of input parameters for health risk model.

References

Acosta H, Wu D, Forrest BM (2010) Fuzzy experts on recreational vessels, a risk modelling approach for marine invasions. Ecol Model 221(5):850–863

Arhonditsis GB, Qian SS, Stow CA, Lamon EC, Reckhow KH (2007) Eutrophication risk assessment using Bayesian calibration of process-based models: application to a mesotrophic lake. Ecol Model 208(2):215–229

Ascough JC, Maier HR, Ravalico JK, Strudley MW (2008) Future research challenges for incorporation of uncertainty in environmental and ecological decision-making. Ecol Model 219 (3):383–399

Bedford T, Cooke RM (2001) Probabilistic risk analysis: foundations and methods. Cambridge University Press, Cambridge, UK

Buteau S, Valcke M (2010) Probabilistic human health risk assessment for quarterly exposure to high chloroform concentrations in drinking-water distribution network of the province of Quebec, Canada. J Toxicol Environ Health Part A 73(23):1626–1644

Butt TE, Lockley E, Oduyemi KO (2008) Risk assessment of landfill disposal sites–state of the art. Waste Manag 28(6):952–964

Chen CF, Ma HW, Reckhow KH (2007) Assessment of water quality management with a systematic qualitative uncertainty analysis. Sci Total Environ 374(1):13–25

Croke BFW, Ticehurst JL, Letcher RA, Norton JP, Newham LTH, Jakeman AJ (2006) Integrated assessment of water resources: Australian experiences. In: Integrated assessment of water resources and global change, pp 351–373. Springer, Netherlands

Dong Z, Liu Y, Duan L, Bekele D, Naidu R (2015) Uncertainties in human health risk assessment of environmental contaminants: a review and perspective. Environ Int 85:120–132

Dubois D, Prade H, Sessa S (2001) Recent Literature. Fuzzy Sets Syst 355–363

Ducey MJ, Larson BC (1999) A fuzzy set approach to the problem of sustainability. For Ecol Manage 115(1):29–40

Finley B, Paustenbach D (1994) The benefits of probabilistic exposure assessment: three case studies involving contaminated air, water, and soil. Risk Anal 14(1):53–73

Gottschalk F, Sonderer T, Scholz RW, Nowack B (2010) Possibilities and limitations of modeling environmental exposure to engineered nanomaterials by probabilistic material flow analysis. Environ Toxicol Chem 29(5):1036–1048

Guyonnet D, Bourgine B, Dubois D, Fargier H, Coλme B, Chilès JP (2003) Hybrid approach for addressing uncertainty in risk assessments. J Environ Eng 129(1):68–78

Hromkovic JI (2005) Zámecniková. Design and analysis of randomized algorithms: introduction to design paradigms. Springer, New York

Janssen P, Petersen A, van der Sluijs J, et al (2003) RIVM/MNP guidance for uncertainty assessment and communication: quickscan hints & actions list. RIVM/MNP Guidance for

Uncertainty Assessment and Communication Series 2. Copernicus Institute for Sustainable Development, Universitat Utrecht und RIVM-MNP, Utrecht, The Netherlands
Jørgensen RB, Hauser T, D'Hertefeldt T, Andersen NS, Hooftman D (2009) The variability of processes involved in transgene dispersal—case studies from Brassica and related genera. Environ Sci Pollut Res 16(4):389–395
Kalberlah F, Schneider K, Schuhmacher-Wolz U (2003) Uncertainty in toxicological risk assessment for non-carcinogenic health effects. Regul Toxicol Pharmacol 37(1):92–104
Kentel E, Aral MM (2004) Probabilistic-fuzzy health risk modeling. Stoch Env Res Risk Assess 18(5):324–338
Khan FI, Sadiq R, Husain T (2002) GreenPro-I: a risk-based life cycle assessment and decision-making methodology for process plant design. Environ Model Softw 17(8):669–692
Knol AB, Petersen AC, Van der Sluijs JP, Lebret E (2009) Dealing with uncertainties in environmental burden of disease assessment. Environmental Health 8(1):21
Li H, Yen VC (1995) Fuzzy sets and fuzzy decision-making. CRC press
Luttik R, Mineau P, Roelofs W (2005) A review of interspecies toxicity extrapolation in birds and mammals and a proposal for long-term toxicity data. Ecotoxicology 14(8):817–832
Masters GM, Ela W (2008) Introduction to environmental engineering and science, vol 3. Prentice Hall, Englewood Cliffs, NJ
Maxwell RM, Kastenberg WE (1999) Stochastic environmental risk analysis: an integrated methodology for predicting cancer risk from contaminated groundwater. Stoch Env Res Risk Assess 13(1):27–47
Mishra H, Rathod M, Karmakar S, Kumar R (2016) A framework for assessment and characterisation of municipal solid waste landfill leachate: an application to the Turbhe landfill, Navi Mumbai, India. Environ Monit Assess 188(6):357(1–26)
Mishra H, Karmakar S, Kumar R, Singh J (2017) A framework for assessing uncertainty associated with human health risks from MSW landfill leachate contamination. Risk Anal 37:1237–1255
Mofarrah A, Husain T (2011) Fuzzy based health risk assessment of heavy metals introduced into the marine environment. Water Qual Exposure Health 3(1):25–36
Morgan MG, Henrion M, Small M (1990). Uncertainty: a guide to dealing with uncertainty in quantitative risk and policy analysis. Cambridge university press
National research Council (1983) Risk assessment in the federal government: managing the process. National Academies Press
National Research Council (1994) Science and judgment in risk assessment. National Academies Press
Penningroth S (2010) Essentials of toxic chemical risk: science and society. CRC Press
Petersen AC (2006) Simulating nature: a philosophical study of computer-simulation uncertainties and their role in climate science and policy advice. CRC Press
Refsgaard JC, van der Sluijs JP, Højberg AL, Vanrolleghem PA (2007) Uncertainty in the environmental modelling process–a framework and guidance. Environ Model Softw 22 (11):1543–1556
Regan HM, Colyvan M, Burgman MA (2002) A taxonomy and treatment of uncertainty for ecology and conservation biology. Ecol Appl 12(2):618–628
Reis S, Voigt K, Oxley T (2017) Thematic issue on modelling human and ecological health risks
Schuhmacher M, Meneses M, Xifró A, Domingo JL (2001) The use of Monte-Carlo simulation techniques for risk assessment: study of a municipal waste incinerator. Chemosphere 43 (4):787–799
Shakhawat C, Tahir H, Neil B (2006) Fuzzy rule-based modelling for human health risk from naturally occurring radioactive materials in produced water. J Environ Radioact 89(1):1–17
Singpurwalla ND, Booker JM (2004) Membership functions and probability measures of fuzzy sets. J Am Stat Assoc 99(467):867–877
Stackelberg KV, Burmaster DE (1994) A discussion on the use of probabilistic risk assessment in human health impact assessment. Environ Impact Assess Rev 14(5–6):385–401
Stirling A (1999) Risk at a turning point? J Risk Res 1(2):97–109

U.S. EPA (1989) Risk assessment guidance for superfund, volume I, human health evaluation manual (part A). EPA/540/1-89/002. Office of emergency and remedial response, Washington, DC

U.S. EPA (2004) An examination of EPA risk assessment principles and practices. Staff paper prepared for U.S. Environmental Protection Agency by members of the risk assessment task force office of the science advisor. EPA/100/B- 04/001

U.S. EPA (2005) Guidelines for carcinogen risk. U.S. EPA, Washington, DC. EPA/630/P-03/0018

Van der Keur P, Henriksen HJ, Refsgaard JC, Brugnach M, Pahl-Wostl C, Dewulf A, Buiteveld H (2008) Identification of major sources of uncertainty in current IWRM practice. Illustrated for the Rhine Basin. Water Resour Manage 22(11):1677–1708

Van der Sluijs JP, Janssen PHM, Petersen AC, Kloprogge P, Risbey JS, Tuinstra W, Ravetz JR (2004) RIVM/MNP guidance for uncertainty assessment and communication: tool catalogue for uncertainty assessment. Utrecht University (Downloadable from http://www.nusap.net/sections.php

Walker WE, Harremoës P, Rotmans J, van der Sluijs JP, van Asselt MB, Janssen P, Krayer von Krauss MP (2003) Defining uncertainty: a conceptual basis for uncertainty management in model-based decision support. Integrated assessment 4(1):5–17

Wang LX (1997) A Course in fuzzy systems and control. Prentice-Hall PTR, Englewood Cliffs, NJ

Wynne B (1992) Uncertainty and environmental learning: reconceiving science and policy in the preventive paradigm. Glob Environ Change 2(2):111–127

Zadeh LA (1965) Fuzzy sets. Inform Control 8(3):338–353

Zalk DM, Paik SY, Swuste P (2009) Evaluating the control banding nanotool: a qualitative risk assessment method for controlling nanoparticle exposures. J Nanopart Res 11(7):1685

Chapter 10
Stable Carbon Isotope and Bulk Composition of Wintertime Aerosols from Kanpur

Gyanesh Kumar Singh, Debajyoti Paul, Prashant Rajput and Tarun Gupta

Abstract This study assesses stable carbon isotopic composition ($\delta^{13}C$) of total carbon (TC) in ambient aerosols ($PM_{2.5}$) during wintertime (December 2014) from Kanpur (26.30 °N, 80.14 °E) in northern India. Chemical constituents viz organic carbon (OC), elemental carbon (EC) and water-soluble ions in $PM_{2.5}$ have also been measured. Back trajectories of air masses arriving at the sampling site (Centre for Environmental Science and Engineering, IIT Kanpur) have been utilized to infer the air-mass transport. Most of the trajectories showed their origin from northwestern region during the study period. Average $PM_{2.5}$ and TC concentrations were centered around 240 μg m^{-3} and 91 μg m^{-3}, respectively. The OC + EC concentrations averaged at 58 ± 15 μg m^{-3}. Significant linear correlation between OC and EC in conjunction with high OC/EC ratio (9 to 12) suggests dominance of anthropogenic combustion sources of organic aerosols. Concentration of anthropogenic ionic species (SO_4^{2-} + NO_3^- + NH_4^+) averaged at 46.74 μg m^{-3}. The average $\delta^{13}C$ values of TC in the integrated 24-h samples were centered around −25‰. Integrated data analyses of chemical constituents and stable C isotope suggests the influence of mixed emission sources. Future studies are required to better constraint the observations.

Keywords $PM_{2.5}$ · Total carbon · Stable isotopes · Kanpur · Indo-Gangetic plain

---

G. K. Singh · P. Rajput · T. Gupta
Department of Civil Engineering, Indian Institute of Technology Kanpur, Kanpur, India
e-mail: gyanesh@iitk.ac.in

P. Rajput
e-mail: prajput@iitk.ac.in

T. Gupta
e-mail: tarun@iitk.ac.in

D. Paul (✉)
Department of Earth Sciences, Indian Institute of Technology Kanpur, Kanpur, India
e-mail: dpaul@iitk.ac.in

T. Gupta et al. (eds.), *Environmental Contaminants*, Energy, Environment, and Sustainability, https://doi.org/10.1007/978-981-10-7332-8_10

10.1 Introduction

Atmospheric aerosols originate from a wide variety of natural as well as anthropogenic sources (Putaud et al. 2010). High concentration of fine particles can cause an adverse effect on human health (Dockery et al. 1993). Residence time of fine aerosols is quite high as compared to coarser particles. They can exert potential impacts through long-range transport and atmospheric chemistry and further influence radiation budget and climate. Organic matter (OM) and elemental carbon (EC) constitute a dominant fraction of atmospheric fine particles. Complexity in characterizing emission sources of carbonaceous aerosols arise due to atmospheric transformation (Garbaras 2008). Also, the increasing anthropogenic activities in conjunction with varying ventilation coefficient lead to increased polycyclic aromatic hydrocarbons (PAHs) concentration which possess a health hazard risk particularly in an urban environment (Singh et al. 2015; Singh and Gupta 2016a). Biomass burning, industrial emissions, and combustion of fossil fuels are the major sources of anthropogenically derived aerosols. The abundance of organic aerosols is governed by their primary emissions and secondary transformations in ambient atmosphere. However, EC is only derived from primary emissions.

Stable isotopes are used to understand various processes, characterize the sources of pollutants and determine their proportional inputs in the ecosystem. For a given element, an isotope differs from the other due to the difference in number of neutrons. Isotope of an element is said to be stable when the tendency of decay is negligible (i.e., they are not radioactive) and has similar number of neutrons (N) and number of protons (Z). Lighter elements dominate the ecological studies due to bulk presence of them in the biota as well as the mass difference caused due to change in neutrons is greatest for them. Various isotopes are useful in atmospheric research such as carbon, nitrogen, sulfur, oxygen, and lead. High precision stable isotope ratios are important and useful tool to characterize aerosol particulate matter. Application of both the compound-specific and bulk analysis of stable isotope has been reported previously (e.g., Hoefs 1997; Flanagan 2005). Assessment of aerosol aging can also be done using stable carbon isotope ratios (Rudolf 2007; Wang and Kawamura 2006). Isotope analysis in association with the chemical composition can provide a significant information to better characterize the carbonaceous aerosols. This study has been conducted to assess chemical characteristics of ambient aerosols in northern India. Although bulk chemical composition of atmospheric carbonaceous aerosols is relatively well studied over northern India, not much isotope data is available. Because δ^{13}C values of different aerosols sources are distinct, it has been used to trace the origin of pollution (Widory et al. 2004; Widory 2006; Lopez-Veneroni 2009). The δ^{13}C values can render vital information concerning the sources and atmospheric processing of organic carbon species (Fisseha et al. 2009). Chemical processes that occur in the long-range transported aerosols can be very well understood by robust stable isotope measurements.

10.2 Study Area

Study area at Kanpur is located in central part of the Indo-Gangetic plain (IGP). It is the fifth largest city in India based on areal extension and is a prime center for industrial activities. IGP provides shelter to over 60% of the human population in India. Aerosol sampling ($PM_{2.5}$) was carried out in Indian Institute of Technology Kanpur premises at the rooftop of the Centre for Environmental Science and Engineering (CESE) building (26.30 °N; 80.14 °E; 142 m above mean sea level) during winter (December 2014). IIT Kanpur is a research institute located 15 km upwind of Kanpur city. The atmospheric particulate matter concentration of anthropogenic origin in Kanpur is governed by activities of vehicular emissions (petrol and diesel traffic), biomass burning, solid fuel (coal) combustion, and industrial emissions. Major concern in IGP during wintertime is a thick haze cover of pollutants over the entire region and its downwind advection toward the Northern Indian Ocean causing perturbation in radiation balance over the ocean. Large-scale biomass burning emissions and fossil fuel combustion activities (Rajput et al. 2014) contribute to haze episodes over the IGP. Emissions from local sources and those arriving due to long-range transport alter the atmospheric aerosol composition and chemistry over the study region in IGP (Rajput et al. 2016; Rajeev et al. 2016). Variability of aerosol sources in IGP makes the source apportionment process difficult, so to resolve this problem, the sources are needed to be well established with respect to various chemical markers.

10.3 Sample Collection

Aerosol ($PM_{2.5}$) samples were collected continuously for 24 h on pre-combusted 47-mm-diameter quartz-fiber filters at the rooftop of CESE building, IIT Kanpur. These samples were stored at −19°C until chemical and isotopic analysis. Sampling was conducted during winter season in the month of December 2014. Sampling was performed by an air sampler working at a flow rate of 15 L/min (Fig. 10.1). Air sampler designed and developed at IIT Kanpur was used for the sampling of ambient aerosols. The impactor in the sampler is designed to collect ≤ 2.5 μm onto filter, whereas particles greater than 2.5 μm in the airstream are impacted onto vacuum grease (Gupta and Dubey 2011; Kumar and Gupta 2015).

10.4 Chemical Analysis

Measurements of total carbon (TC), organic carbon (OC), elemental carbon (EC), water-soluble organic carbon (WSOC), stable carbon isotope of TC, and inorganic ions in the ambient aerosol samples have been performed. The OC and EC of

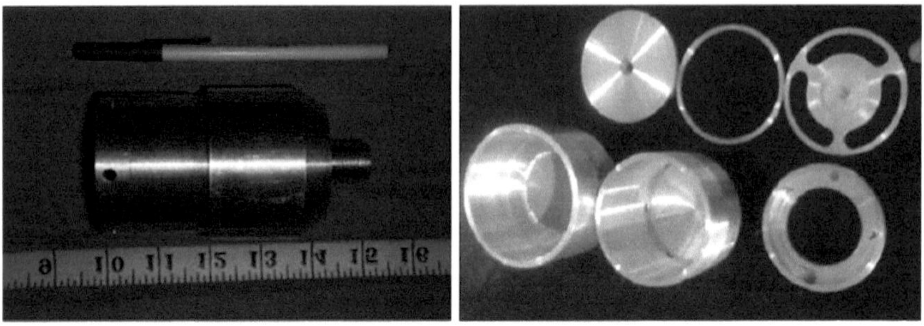

Fig. 10.1 $PM_{2.5}$ air sampler and its internal components (Gupta and Dubey 2011)

samples were measured using National Institute for Occupational Safety and Health (NIOSH) protocol using a thermo-optical carbon analyzer (Sunset Laboratory) (Birch and Cary 1996). For the measurements of WSOC, total organic carbon analyzer (TOC Shimadzu) using a non-dispersive infrared detection (NDIR) technique was utilized. Potassium hydrogen phthalate (KHP) and sodium carbonate-bicarbonate mixture (Na_2CO_3 + $NaHCO_3$) were used for the linear calibration of TC (total carbon) and IC (inorganic carbon) ($R^2 > 0.99$) on TOC analyzer. Analytical uncertainty was well within 5% for the replicate analysis of samples. The ion chromatography (Metrohm compact IC 761) was carried out to quantify ions in samples and the detailed procedure is mentioned in Singh and Gupta (2016b). For determination of stable isotopes, appropriate filter size was taken based on mass-to-signal ratio and placed into pre-cleaned tin cup and closed, which was then inserted into an elemental analyzer (Flash EA 2000, Thermo Scientific) using an autosampler. Carbon isotopic ratio in the samples has been measured on a continuous flow isotope ratio mass spectrometer (EA-IRMS).

Stable isotope ratio mass spectrometer (Thermo Scientific Delta V Plus) used for the measurement is coupled with elemental analyzer and conflo IV universal interface. It is housed in the Advanced Centre for Material Science (ACMS) laboratory, IIT Kanpur. High-precision stable carbon ($\delta^{13}C$), nitrogen ($\delta^{15}N$), oxygen ($\delta^{18}O$), and hydrogen (δ^2H) isotope ratio analysis in various types of organic and inorganic samples such as aerosols, plants, soil, water, carbonate sediments, atmospheric gases can be performed using this assembly. Here, δ represents the isotopic composition of elements and is expressed as given in Eq. (10.1):

$$\delta = [(R_{\text{Sample}}/R_{\text{Standard}} - 1)] \times 1000‰(\text{per mil}) \quad (10.1)$$

where R represents the ratio of heavy to light isotope. Therefore, measure of comparison of heavy to light isotopes in a sample versus a reference standard is represented by δ. Hence, the isotopic composition of carbon can be represented by Eq. (10.2):

$$\delta^{13}C_{sample}(‰, V-PDB) = [\left({}^{13}C/{}^{12}C\right)_{sample}/\left({}^{13}C/{}^{12}C\right)_{V-PDB} - 1] \times 10^3 \quad (10.2)$$

The carbon isotopic data is reported relative to the Vienna Pee Dee Belemnite (PDB) standard. The calibration of EA-IRMS was done using various internal standards such as CH3 (−24.724‰) and IAEA 601 (Benzoic Acid, −28.81‰). For the case of carbon isotope analysis, the uncertainty involved was within 0.2‰.

10.5 Results and Discussion

10.5.1 Source Region of Air Masses

HYbrid Single-Particle Lagrangian Integrated Trajectory (HYSPLIT) model was used to identify the origin of air masses during the sampling period (Stein et al. 2015). Five-day back trajectory at 1000 m above mean sea level (AMSL) was performed for the sampling duration. $\delta^{13}C$ values of aerosol particles transported from longer distances depend upon the transport directions of air masses (Garbaras 2008). The trajectories revealed that northwestern winds were prevailing during the entire campaign as seen in Fig. 10.2. Northwestern India influenced air masses traveling specifically through Punjab, Haryana brings primary pollutants during this time period of the year to the downwind side in Indo-Gangetic plain (IGP) where the study area is located. Post-harvest paddy-residue burning emissions can have a significant impact on the aerosol composition over Northern India during the wintertime (Rajput et al. 2013, 2014, 2016). This region is impacted by large-scale biomass burning during winter which can affect the aerosol composition of locations present at the downwind side in IGP. Obtained back-trajectory results were found to be consistent with the previous studies (Rajput et al. 2016).

10.5.2 $PM_{2.5}$ Aerosol Mass, Total Carbon (TC), Organic Carbon (OC), Elemental Carbon (EC)

Distinct variations in the concentrations of total $PM_{2.5}$ mass and TC were observed during the entire study. Total $PM_{2.5}$ aerosol mass obtained was centered around ~ 240 μg m^{-3}. TC concentrations obtained ranged from 20.43 to 77.23 μg m^{-3}. Massive emissions from Kanpur city where sources are industrial as well as vehicular can lead to higher TC concentrations in ambient aerosols. OC concentration ranged from 20.43 to 77.23 μg m^{-3}, whereas EC concentration ranged from 2.35 to 6.56 μg m^{-3}. OC + EC concentration in the samples averaged at 58 ± 15 μg m^{-3} as seen in Fig. 10.3. As a result of lower mixing height and temperature inversion during wintertime, dilution and dispersion of pollutants are

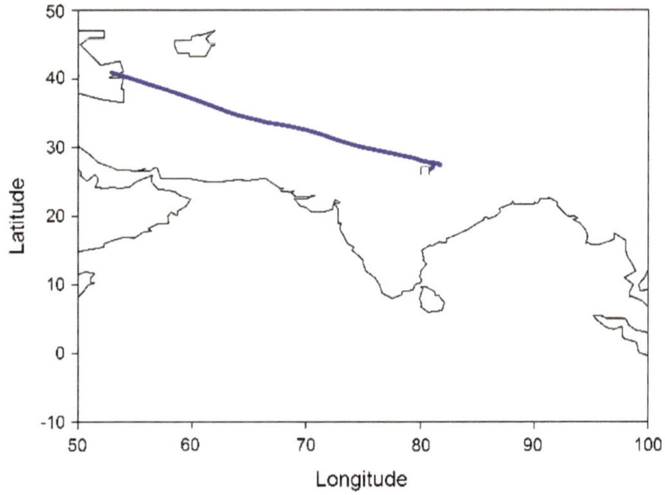

Fig. 10.2 Resultant back trajectory of air masses during the campaign

constrained which causes accumulation of organic compounds such as volatile organic compounds (VOCs) leading to increased levels of OC. Relatively higher OC/EC ratios obtained in this study can be due to aging and atmospheric processing during advective transport of organic compounds or SOA formation (Ram et al. 2010).

Between October and January, i.e., during wintertime, concentrations of OC and EC have been found to be higher in IGP compared to other months (Rajput et al. 2015). The different intensity of emissions sources from biomass burning and/or contributions from secondary organic aerosols (SOA) results in the deviation of concentrations of carbonaceous aerosols (Rajput et al. 2015). The OC/EC ratio is fairly high in these samples, thereby suggesting the dominance of biomass and biofuel combustion dominated emissions and/or secondary organic aerosol

Fig. 10.3 Concentration of EC + OC

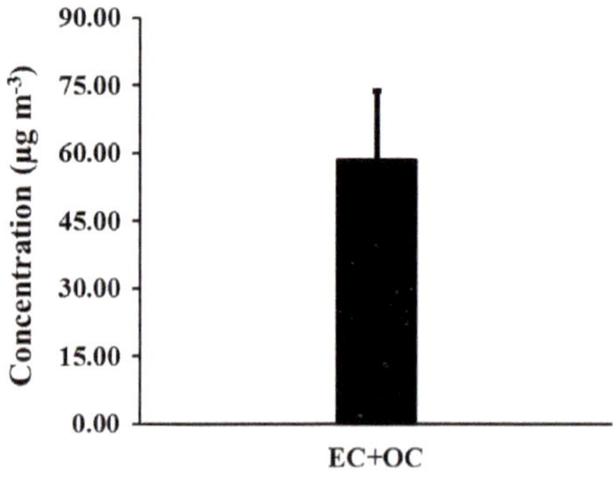

(SOA) formation. Biomass and biofuel emissions have more OC/EC ratios as compared to fossil fuel emissions (Andreae and Merlet 2001; Saarikosi et al. 2008; Zhu et al. 2010).

10.5.3 Anthropogenic Ionic Species

For the entire duration of sampling, percentage contribution of inorganic species to total WSIS (ΣWSIS) has been shown in Fig. 10.4. The major contributors to WSIS mass were anthropogenic ions (NH_4^++ NO_3^- + SO_4^{2-}). The contribution of these species to ΣWSIS was ~ 86%. Temporal variability record of anthropogenic ionic species is shown in Fig. 10.5. SO_4^{2-} contribution was highest followed by NH_4^+ and NO_3^-. Since, SO_4^{2-} is derived from photochemical oxidation of SO_2 that is produced by fossil fuel combustion, high concentration of SO_4^{2-} indicates that air is heavily polluted in Kanpur.

The measured average concentration of anthropogenic ions (i.e., SO_4^{2-} + NO_3^- + NH_4^+) was 46.74 μg m^{-3}. All these species are secondary and form in the atmosphere from their precursors. Secondary aerosol contribution to total PM during winter is quite significant in the IGP (Gupta and Mandariya 2013). During wintertime, there is a significant increase in the concentration of ions like NH_4^+, NO_3^-, and SO_4^{2-} due to high relative humidity (RH), stagnant conditions and various fog/haze events which are the chief contribution of secondary sources. A complete charge balance between anions and cations (Σ^-/Σ^+; R^2 = 0.98) has been observed in this study.

Fig. 10.4 Percentage contribution of inorganic species to ΣWSIS

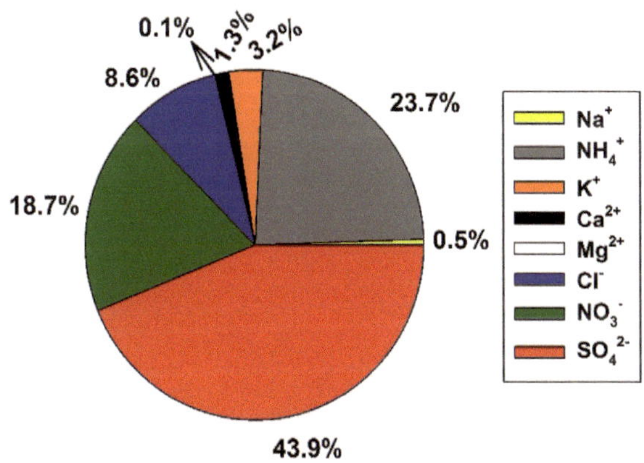

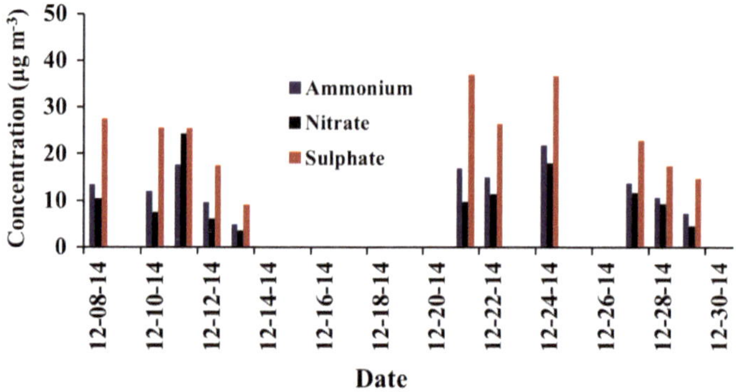

Fig. 10.5 Temporal variability of anthropogenic ions (NH_4^+, NO_3^-, SO_4^{2-}) during the study period at Kanpur

10.5.4 Stable Carbon Isotope Analyses

Temporal variability of δ^{13}C values in atmospheric aerosols has been shown in Fig. 10.6. δ^{13}C of the TC varied from ~ −26.2 to −24.5‰ in aerosol samples. Influence of mixed sources such as vehicle exhaust (−28 to −26‰) (Widory 2006), coal combustion (−24.9 to −21‰) (Widory 2006; Gleason and Kyser 1984), biomass/biofuel combustion emissions (−29.4 to −25.9‰) (Agnihotri et al. 2011), C3 plants (−20 to −32‰) (Smith and Epstein 1971) and C4 plants (−6 to −19‰) (Deines 1980) can give rise to δ^{13}C values obtained in ambient aerosols. The variability in carbon isotope value obtained on different sampling events represents the dominance of different sources. Like for the days, when coal combustion was dominant the values tended to be on the positive side (i.e., closer to −24‰).

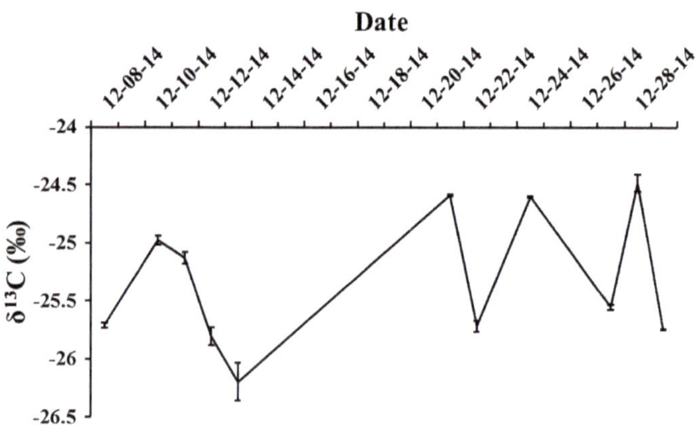

Fig. 10.6 Temporal variability of δ^{13}C in atmospheric aerosol samples

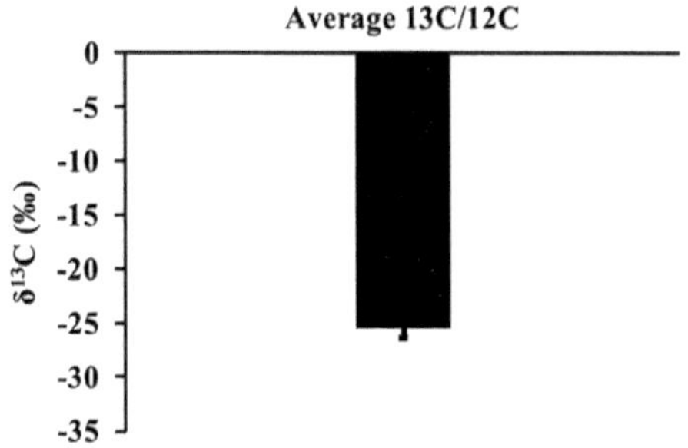

Fig. 10.7 Average $\delta^{13}C$ in the aerosol samples during wintertime at Kanpur

Since Kanpur in the IGP is located far-off from the ocean, therefore, the prominent sources of carbonaceous aerosols can be biogenic emission and/or biomass and fossil fuel combustion. In aerosol samples ($PM_{2.5}$), we have observed the $\delta^{13}C$ of TC centering around −25‰ (Fig. 10.7).

In this study, it is observed that with increase in the carbon content there is an enrichment in the ^{13}C of the TC. $\delta^{13}C$ of the black carbon (BC) do not change significantly from source on transport. However, tremendous interaction of OC with atmospheric oxidants is also recorded in variability of isotope ratio. $\delta^{13}C$ of chemical species can also be affected by the prevalent atmospheric processes. Similarly, fractionation resulted by kinetic isotopic effect due to aging (photochemical) of particulate organics leads to ^{13}C enrichment in the remaining aerosols (Aggarwal and Kawamura 2008; Pavuluri and Kawamura 2012; Bosch et al. 2014; Kirillova et al. 2014). Compared with corresponding gaseous precursors, SOA formation by the photochemical oxidation of VOCs during transport leads to depletion in ^{13}C of TOC (Anderson et al. 2004; Irei et al. 2006, 2011; Bosch et al. 2014; Kirillova et al. 2014). As seen in the previous study (Fisseha et al. 2009), WSOC enriched in ^{13}C influences the $\delta^{13}C$ of TC. Higher fraction of WSOC can cause an increase in $\delta^{13}C$ and vice versa but more studies would be required. We tried to understand the dominant sources of aerosol carbon as source composition (Turekian et al. 1998), particle formation and transport (Cachier et al. 1985) mainly controls the $\delta^{13}C$ of TC in the tropics.

10.6 Summary

This study presents information about $\delta^{13}C$ values in ambient aerosols ($PM_{2.5}$) collected from Kanpur. Back-trajectory analysis along with the determination of TC, OC and EC concentrations and anthropogenic ionic species provided insight into the chemical characteristics of aerosols during wintertime. Information from $\delta^{13}C$ values (and other characteristic ratios like OC/EC, WSOC/OC) in conjunction with air-mass back trajectories were integrated to better infer about sources of ambient aerosols during wintertime over IGP. Integrated analysis of chemical constituents and stable isotopes provides a better constraint on emission characteristics from anthropogenic sources. Higher TC concentrations can be attributed to impact of anthropogenic emission sources of local origin versus long-range transport from upwind IGP. Secondary aerosols contribution to PM mass is quite significant in IGP during wintertime.

Acknowledgements This study has been carried out by the financial support from Indian Institute of Technology Kanpur (India).

References

Aggarwal SG, Kawamura K (2008). Molecular distributions and stable carbon isotopic compositions of dicarboxylic acids and related compounds in aerosols from Sapporo, Japan: Implications for photochemical aging during long-range atmospheric transport. J Geophys Res: Atmos 113(D14)

Agnihotri R, Mandal TK, Karapurkar SG, Naja M, Gadi R, Ahammmed YN, Kumar A, Saud T, Saxena M (2011) Stable carbon and nitrogen isotopic composition of bulk aerosols over India and northern Indian Ocean. Atmos Environ 45(17):2828–2835

Anderson RS, Huang L, Iannone R, Thompson AE, Rudolph J (2004) Carbon kinetic isotope effects in the gas phase reactions of light alkanes and ethene with the OH radical at 296 $\pm$ 4 K. J Phys Chem A 108(52):11537–11544

Andreae MO, Merlet P (2001) Emission of trace gases and aerosols from biomass burning. Global Biogeochem Cycles 15(4):955–966

Birch ME, Cary RA (1996) Elemental carbon-based method for monitoring occupational exposures to particulate diesel exhaust. Aerosol Sci Technol 25:221–241

Bosch C, Andersson A, Kirillova EN, Budhavant K, Tiwari S, Praveen P, Russell LM, Beres ND, Ramanathan V, Gustafsson Ö (2014) Source-diagnostic dual-isotope composition and optical properties of water-soluble organic carbon and elemental carbon in the South Asian outflow intercepted over the Indian Ocean. J Geophys Res: Atmos 119(20)

Cachier H, Buat-Menard P, Fontugne M, Rancher J (1985) Source Terms and Source Strengths of the Carbonaceous Aerosol in the Tropics. J Atmos Chem 3:469–489. https://doi.org/10.1007/BF00053872

Deines P (1980) The isotopic composition of reduced organic carbon. Handb Environ Isot Geochem:329–406

Dockery DW, Pope CA, Xu X, Spengler JD, Ware JH, Fay ME, Ferris BGJ, Speizer FE (1993) An Association between Air Pollution and Mortality in Six U.S. Cities. N Engl J Med 329(24): 1753–1759

Fisseha R, Saurer M, Jaggi M, Siegwolf RTW, Dommen J, Szidat S, Samburova V, Baltensperger U (2009) Determination of primary and secondary sources of organic acids and carbonaceous aerosols using stable carbon isotopes. Atmos Environ 43(2):431–437

Flanagan LB (2005) 1—Introduction: Stable isotopes and earth system science. In: Stable isotopes and biosphere atmosphere interactions. Academic Press, San Diego, pp 1–5

Garbaras A (2008) Tracing of atmospheric aerosol sources using stable carbon isotopes. Lith J Phys Tech Sci 48:259–264

Gleason JD, Kyser TK (1984) Stable isotope compositions of gases and vegetation near naturally burning coal

Gupta T, Dubey S (2011) Field performance evaluation of a newly developed PM2.5 sampler at IIT Kanpur. Sci Total Environ 409(18):3500–3507

Gupta T, Mandariya A (2013) Sources of submicron aerosol during fog-dominated wintertime at Kanpur. Environ Sci Pollut Res 20(8):5615–5629

Hoefs J (1997) Stable isotope geochemistry. Springer

Irei S, Huang L, Collin F, Zhang W, Hastie D, Rudolph J (2006) Flow reactor studies of the stable carbon isotope composition of secondary particulate organic matter generated by OH-radical-induced reactions of toluene. Atmos Environ 40(30):5858–5867
Irei S, Rudolph J, Huang L, Auld J, Hastie D (2011) Stable carbon isotope ratio of secondary particulate organic matter formed by photooxidation of toluene in indoor smog chamber. Atmos Environ 45(4):856–862
Kirillova EN, Andersson A, Tiwari S, Srivastava AK, Bisht DS, Gustafsson O (2014) Water-soluble organic carbon aerosols during a full New Delhi winter: Isotope-based source apportionment and optical properties. J Geophys Res-Atmos 119(6):3476–3485
Kumar A, Gupta T (2015) Development and field evaluation of a multiple slit nozzle-based high volume PM2.5 inertial impactor assembly (HVIA). Aerosol Air Qual Res 15(4):1188–1200 + i
López-Veneroni D (2009) The stable carbon isotope composition of PM2.5 and PM10 in Mexico City Metropolitan Area air. Atmos Environ 43(29):4491–4502
Pavuluri CM, Kawamura K (2012) Evidence for 13-carbon enrichment in oxalic acid via iron catalyzed photolysis in aqueous phase. Geophys Res Lett 39(3)
Putaud J-P, Van Dingenen R, Alastuey A, Bauer H, Birmili W, Cyrys J, Flentje H, Fuzzi S, Gehrig R, Hansson H-C (2010) A European aerosol phenomenology–3: Physical and chemical characteristics of particulate matter from 60 rural, urban, and kerbside sites across Europe. Atmos Environ 44(10):1308–1320
Rajeev P, Rajput P, Gupta T (2016) Chemical characteristics of aerosol and rain water during an El Niño and PDO influenced Indian summer monsoon. Atmos Environ 145:192–200
Rajput P, Sarin M, Kundu SS (2013) Atmospheric particulate matter (PM2.5), EC, OC, WSOC and PAHs from NE–Himalaya: abundances and chemical characteristics. Atmos Pollut Res 4(2):214–221
Rajput P, Sarin MM, Sharma D, Singh D (2014) Organic aerosols and inorganic species from post-harvest agricultural-waste burning emissions over northern India: impact on mass absorption efficiency of elemental carbon. Environ Sci Proces Impacts 16(10):2371–2379
Rajput P, Mandaria A, Kachawa L, Singh DK, Singh AK, Gupta T (2015) Wintertime source-apportionment of PM1 from Kanpur in the Indo-Gangetic plain. Clim Change 1(4): 503–507
Rajput P, Mandaria A, Kachawa L, Singh DK, Singh AK, Gupta T (2016) Chemical characterisation and source apportionment of PM1 during massive loading at an urban location in Indo-Gangetic Plain: impact of local sources and long-range transport. Tellus Ser B 68:30659
Rajput P, Gupta T, Kumar A (2016) The diurnal variability of sulfate and nitrate aerosols during wintertime in the Indo-Gangetic Plain: implications for heterogeneous phase chemistry. RSC Adv 6(92):89879–89887
Ram K, Sarin MM, Hegde P (2010) Long-term record of aerosol optical properties and chemical composition from a high-altitude site (Manora Peak) in Central Himalaya. Atmos Chem Phys 10(23):11791–11803
Rudolph J (2007) Gas chromatography–isotope ratio mass spectrometry. In: Volatile organic compounds in the atmosphere. Blackwell, pp 388–466
Saarikoski S, Timonen H, Saarnio K, Aurela M, Järvi L, Keronen P, Kerminen V, Hillamo R (2008) Sources of organic carbon in fine particulate matter in northern European urban air. Atmos Chem Phys 8(20):6281–6295
Singh DK, Gupta T (2016a) Effect through inhalation on human health of PM1 bound polycyclic aromatic hydrocarbons collected from foggy days in northern part of India. J Hazard Mater 306:257–268
Singh DK, Gupta T (2016b) Role of transition metals with water soluble organic carbon in the formation of secondary organic aerosol and metallo-organics in PM1 sampled during post monsoon and pre-winter time. J Aerosol Sci 94:56–69
Singh DK, Sharma S, Habib G, Gupta T (2015) Speciation of atmospheric polycyclic aromatic hydrocarbons (PAHs) present during fog time collected submicron particles. Environ Sci Pollut Res 22(16):12458–12468

Smith BN, Epstein S (1971) Two categories of 13C/12C ratios for higher plants. Plant Physiol 47(3):380–384

Stein AF, Draxler RR, Rolph GD, Stunder BJB, Cohen MD, Ngan F (2015) NOAA's HYSPLIT atmospheric transport and dispersion modeling system. Bull Am Meteor Soc 96(12): 2059–2077

Turekian VC, Macko S, Ballentine D, Swap RJ, Garstang M (1998). Causes of bulk carbon and nitrogen isotopic fractionations in the products of vegetation burns: Laboratory studies. Chem. Geol. 152:181–192

Wang H, Kawamura K (2006) Stable carbon isotopic composition of low-molecular-weight dicarboxylic acids and keto acids in remote marine aerosols. J Geophys Res: Atmos 111(D7): D07304. https://doi.org/10.1029/2005JD006466

Widory D (2006) Combustibles, fuels and their combustion products: A view through carbon isotopes. Combust Theor Model 10(5):831–841

Widory D, Roy S, Le Moullec Y, Goupil G, Cocherie A, Guerrot C (2004) The origin of atmospheric particles in Paris: a view through carbon and lead isotopes. Atmos Environ 38(7):953–961

Zhu C-S, Chen C-C, Cao J-J, Tsai C-J, Chou CC-K, Liu S-C, Roam G-D (2010) Characterization of carbon fractions for atmospheric fine particles and nanoparticles in a highway tunnel. Atmos Environ 44(23):2668–2673

Chapter 11
Indoor Air Quality Assessment as Related to Household Conditions in Rural Houses During Winter Season

Alfred J. Lawrence and Tahmeena Khan

Abstract Human exposure to indoor air contaminants may be several times higher than outdoor levels because of the majority of their time is spent indoors. People belonging to rural areas in developing countries are usually exposed to high levels of household pollution daily owing to use of biomass for cooking. The situation becomes worsened in winter as the consumption goes up. The unprocessed biofuel is related to asthma, respiratory tract infections, chronic obstructive pulmonary disease (COPD), cataract, and in severe cases lung cancer. The aim of the study was to assess indoor air quality through measurement of CO, CO_2, NO_2, SO_2 and particulate matter (PM_{10} and $PM_{2.5}$) and health symptoms. A questionnaire survey was conducted in 125 households in 15 villages in the vicinity of Lucknow city. After analyzing the survey results, five villages were shortlisted from where fifteen houses were selected for monitoring from November 2013 to February 2014. The survey presented a comprehensive picture of a rural household along with socioeconomic structure. Many other confounding factors including smoking, poor ventilation practices etc. were also identified in the course of study revealing poor household conditions. High particulate levels and health symptoms were likely to be associated mainly to use of crude fuel use.

Keywords Pollutant · Assessment · Ventilation · Questionnaire
Cooking · Rural

11.1 Introduction

Air pollution is one of the leading causes of deaths globally causing about 7 million deaths in 2012 according to the World Health Organization (WHO) findings (WHO 2015a). In the past, air pollution was linked to vehicular exhaust and smoke and

A. J. Lawrence (✉) · T. Khan
Department of Chemistry, Isabella Thoburn College, 7, Faizabad Road, 226007 Lucknow, UP, India
e-mail: Alfred_lawrence@yahoo.com

T. Gupta et al. (eds.), *Environmental Contaminants*, Energy, Environment, and Sustainability, https://doi.org/10.1007/978-981-10-7332-8_11

confined within urban boundaries. Today air pollution has spread like an epidemic and become a global issue of huge concern. The actual exposure to air pollutants should include all microenvironments where a person spends time (Lawrence et al. 2005). Indoor air pollution (IAP) is a direct consequence of outdoor contamination. IAP is one of the major causes of health risks to exposed population. The menace is growing day by day in developing countries. Statistics reveal that in developing countries, health impacts of indoor air pollution far outweigh those of outdoor air pollution (Kankaria et al. 2014). There are many sources like heating, cooking, cleaning, smoking, and building materials which affect the indoor ambience considerably. Around half of the world's population and up to 90% of rural population rely on biomass fuels in the form of wood, crop residue, and animal dung (World Resource Institute 1998). People belonging to low and middle-income countries in Southeast Asia and Western Pacific regions are highly susceptible to the problem. As per the WHO (2015b) 4.3 million deaths were reported due to indoor air pollution in 2012, of which 1.69 million were reported from Southeast Asia only, whereas in high-income countries, 19,000 deaths were reported. Studies on pollution from solid fuels have been conducted in countries including Mexico, the Philippines, China, Zimbabwe, Bangladesh, India, Costa Rica, Bolivia, and Kenya (Zuk et al. 2007; Saksena et al. 2007; Fischer and Koshland 2007; Rumchev et al. 2007; Dasgupta et al 2006; Balakrishnan et al. 2002; Balakrishnan et al. 2004; Park and Lee 2003; Albalak et al. 2001; Boleij et al. 1989). In India, 20–30% population still lives in poverty, and the scenario is more intense in rural regions (Advisory Board on Energy 1984). People in rural areas still lack accessibility to basic energy sources. Access to energy for cooking, heating, transport, and other activities is essential to human health (WHO 2005). Rural energy provision is an essential requirement for rural development (Laxmi et al. 2003). Access and availability to efficient energy sources are still difficult for rural population which makes them vulnerable to various health hazards and wastage of time in fuel gathering, harmful ecological changes, and many other negative outcomes (Laxmi et al. 2003). Smoke originating from incomplete combustion of fuel is a major concern in households which leads to adverse indoor air quality. Inadequate ventilation especially in winter months further exacerbates the situation (Gold 1992). Indoor activity pattern, human occupancy, ventilation, types of stove used for cooking and heating, and tobacco smoke have also been found to influence indoor environment (Baek et al. 1997). These factors are responsible for marked difference between outdoor and indoor concentrations and accountable for innumerous health hazards (Baek et al. 1997). Women and children are more vulnerable as they spend a substantial amount of time indoor. Strong association has been established between smoke exposure and chronic bronchitis in women as they are the primary cooks (Akhtar et al. 2007). Studies in India have reported increased child deaths in solid-fuel consuming households aged between 1 and 4 years. Girls were found to be more affected than boys in the same study (Basani et al. 2010). Tuberculosis in the occupants has been attributed to cooking smoke in the age group of 20 years and above (Pricilla et al. 2011). High odd ratio viz., 2.65 was obtained for women in rural areas in the same study. Results from a study among elderly men and women (age $\geq$60 years) showed higher prevalence of asthma in biomass fuel users (Odd ratio (OR) = 1.59; 95% Confidence interval (CI): 1.30–1.94) as compared to men (OR = 1.83; 95% CI: 1.32–2.53) (Mishra 2003). Prevalence of asthma was greater in women as they are directly exposed to cooking smoke.

11.1.1 Household Fuel and Indoor Air Quality

World Health Organization's report in 2007 (WHO 2007) on national burden of diseases from indoor air pollution establishes the proportionality between indoor contamination owing to usage of solid fuels and health issues including acute and chronic respiratory diseases, tuberculosis, asthma, and cardiovascular diseases, and prenatal health outcomes. Indoor air pollution was responsible for more than 1.6 million annual deaths and 2.7% of global burden of diseases (WHO 2006). In addition, kerosene usage has been linked to be a significant source of black carbon (BC) emission which is tagged as an important pollutant affecting climate change in several parts of the world including South Asia (Lam et al. 2012). It also converts approximately one-tenth of the fuel carbon to particulate matter (PM) (Lam et al. 2012). Smoke and fumes from open fires have been reported to contribute to the death of 1.5 million people per year (World Bank Energy 2014). Biomass fuel has been linked with significantly prolonged nasal mucociliary clearance time (765.8 ± 378.16 s) as compared to clean fuel users (545.4 ± 215.55 s), and reduced peak expiratory flow rate (319.3 l/min) as compared to (371.7 l/min) (Priscilla et al. 2011). A study conducted in South India linked COPD with biomass fuel use (OR: 1.24; 95% CI: 0.36–6.64). Women were more at risk owing to spending more than 2 h a day in cooking (Johnson et al. 2011). Affordability to clean energy sources is crucial for health improvement, especially in developing countries where, there is a prominent gap between social, economic, and developmental conditions which has made access to clean energy sources less accessible (International Energy Agency 2015). The extension of health energy services in remote areas is the need of the hour (International Energy Agency 2013).

11.1.1.1 Main Pollutants in Rural Indoor Environment—Sources and Pathways of Exposure

The nature of contaminants varies in different microenvironments, even within the same microenvironment, varied number of pollutants can be found which can differ in different seasons. It is beyond the scope of this chapter to review every single pollutant. As the focus of this study is to highlight the indoor air pollution due to usage of biomass fuel as the primary factor in rural microenvironment, only those pollutants are discussed in brief which were assessed and associated with most of the respiratory outcomes in the case study under consideration. Health effects associated with them are briefly summarized in Table 11.1.

Table 11.1 Health effects associated with assessed pollutants

Pollutant	Indoor sources	Health effects
CO and CO_2	Kerosene and gas heaters, wood, coal stoves, indoor smoking	Formation of carboxyhemoglobin reduces oxygen intake of blood, headache, shortness in breath, and even death (Braubach et al. 2013)
NO_2	Kerosene, diesel, wood, coal, and outdoor activities infiltration	Chronic Lung diseases, cardiovascular diseases, Hypertension, Skin irritation, eye irritation, and cough (immediate effect) (Lawrence and Taneja 2005)
SO_2	Kerosene, diesel, coal, and outdoor activities infiltration	Respiratory diseases (immediate effect) (Lawrence and Taneja 2005)
RSPM	Wood, peat, biomass, heavy oil, diesel, outdoor activities infiltration, and animal hairs	Affects respiratory tracts and embedded into alveoli (McCracken et al. 2011), carrier of many secondary pollutants and carcinogenic trace elements, lung cancer (cumulative effect) (Kampa and Castanas 2008)

(a) **Carbon monoxide (CO) and Carbon dioxide (CO_2)**—Anthropogenic discharge causes about two-thirds of the carbon monoxide in the atmosphere and natural emission accounts for the rest. Some amount is also produced by human beings body(Alm et al. 1999). Petrol and diesel motor vehicles can also contribute to low amount of CO through infiltration (Kleinman 2009). Combustion sources and inefficient cooking or heating appliances using fossil fuels can lead to production of CO indoors. Production of these gases through combustion may pose danger to the occupants unless they are vented out through proper ventilation means. Carbon dioxide and carbon monoxide are the products of combustion reactions, such as the burning of coal, wood, and natural gas. Incense burning can also be a source of CO (Jetter et al. 2002). It also might be a significant contributor to carbon monoxide production. Carbon monoxide is often produced as a result of incomplete combustion. In presence of adequate oxygen, combustion reactions usually produce carbon dioxide. Human activities and exhalation result in CO_2 emission.

(b) **Nitrogen dioxide (NO_2)**—Combination of oxygen and nitrogen leads to production of oxides of nitrogen. Major indoor sources of NO_2 include tobacco smoke, gas, wood, oil, kerosene, and coal burning appliances such as stoves, ovens, space and water heaters, and fireplaces (Levy et al. 1998; Glorennec et al. 2008). Higher concentrations of NO_2 are obtained in winter season, as there is comparatively lesser mixing in the lower air boundary during winter season leading to higher concentration of NO_2 in winter season.

(c) **Sulphur dioxide (SO_2)**—SO_2 is mainly produced by combustion of fossil fuels. Use of coal for domestic heating purpose is a major source of SO_2 production. Coal and oil contain sulfur ranging between 1 and 5%. On combustion sulfur is converted almost to sulfur dioxide. Diesel generators can also add to SO_2 level.

(d) **Particulate matter**—Particulate matter is responsible for adverse health effects. Studies have found higher particulate concentrations indoor than outdoor (Chao et al. 1998; Jones et al. 2000; Kamens et al. 1991). Increase in particulate concentration can be attributed to indoor activities like tobacco smoking and operation of gas stoves (Spengler et al. 1981). The difference in cooking method between Eastern and Western cultures can have a direct impact on particulate emission (He et al. 2004; Robinson et al. 2006; See and Balasubramanian 2008). The most important sources include cooking, kerosene heating, and fossil fuel burning (Long et al. 2000; Raunemaa et al. 1989). Other sources include household activities like cleaning, dusting, and vacuuming, showering, electric motors, and movement of people (Spengler et al. 1981). Weather also influences the indoor particulate concentration significantly because general ventilation is not enough in winter season, and most of the windows and doors are closed. In addition, rural households burn cow dung cakes and wood sticks to combat cold climate.

11.2 Factors Affecting Indoor Air Quality

Studies on measurement of indoor atmospheric environment have increased during the past few years because of the growing complaints about the IAQ. Quality of the air inside buildings is referred to as IAQ which is represented by concentrations of pollutants and thermal conditions that include temperature and relative humidity conditions that affect the health and performance of occupants. House characteristics including building material, construction techniques, ventilation practices, various indoor processes, activities and outdoor contaminant level influence the indoor air quality to a considerable extent. In a study conducted in Bangladesh, an association was established between wall material of houses with eye irritation, cough and shortness of breath in dwellers (Khalequzzaman et al. 2010) in rural environment. Tin and bamboo were mainly used as wall material in these houses whereas floors were usually made of mud. In another study conducted in rural Rajasthan, symptoms such as cough, phlegm, and blood in sputum were significantly higher among members of households with one room (Laxmi et al. 2003). Climatic conditions and indoor human occupancy can also lead to imbalance and undesired changes in the indoor air quality which can lead to discomfort of the occupants in the form of suffocation, headache, drowsiness, and lack of concentration.

11.3 Important Tools Involved in Air Pollution Studies

During indoor air pollution assessment, it is necessary to collect information on the household characteristics including climatological factors, indoor/outdoor behavioral pattern, surrounding probing for identification of any potential source of pollution. In a nutshell, the important factors under consideration are summarized here.

11.3.1 Questionnaire Survey

The questionnaire tool is used to collect observational data that includes information on various factors like household condition, its environment and ventilation conditions, occupants and their daily activity pattern, and their health status as they perceive. Open or closed structured questionnaire can both be employed for the purposed depending on the depth of information to be extracted. Translation of the questionnaire in the interviewee's friendly language can be helpful in getting them more acquainted.

11.3.1.1 Observational Data Collection and Points to Include

IAQ of a dwelling is by and large influenced by the household conditions and characteristics which include-

1. Average number of people
2. Indoor activity pattern
3. Age group
4. Socioeconomic and educational status
5. Building orientation parameters
   (a) Number of rooms
   (b) Height
   (c) Place of cooking
6. Ventilation means
7. Health records of the occupants.

11.3.2 Meteorological Parameters

Dispersion mechanism of pollutants suggests that their concentration is strongly influenced by meteorological conditions and topographical conditions (Charron and

Harrison 2005). The atmospheric concentration of pollutants also depends upon movement of air and climatic changes. Pollution cloud and its movement are governed by the chemical composition and physical characteristics of the atmosphere which has four major layers-

1. Troposphere
2. Stratosphere
3. Mesosphere
4. Thermosphere

(a) **Atmospheric stability and mixing height**—The atmospheric stability is an important factor for diffusion. Vertical movement of air is fairly influenced by atmospheric stability. The stability is related to change in temperature with height, thermal turbulence which is caused by solar heating and mechanical turbulence. During the temperature inversion, the atmosphere is stable and very little turbulence or mixing takes place. Mixing height is the height above the earth surface to which pollutants will reach by the action of the atmospheric turbulence. It is dependent on direction of wind, its speed and turbulence in atmosphere.

(b) **Atmospheric dispersion**—Dispersion of pollutants signifies how the pollutants are scattered from their source. Dispersion may take place mainly through wind transport mechanism and turbulent diffusion. The variation in wind speed and direction may be indicator of turbulence in atmosphere (Hewson and Gill 1944). Several mathematical equations developed have shown that values of diffusion coefficient vary with meteorological conditions and atmospheric stability. More field studies to ascertain meteorological variables like diffusion coefficient and their variation through the course of season or year, turbulence and lapse rate must be taken up. Dispersion can take place through a) transport of wind b) atmospheric turbulence (Hewson 1956).

(c) **Atmospheric turbulence**—Fluctuations in wind flow which have a frequency of more than 2 cycles/hour signify turbulent behavior. Turbulence is usually indicated by fluctuations of wind speed and direction. It signifies irregular motion of air over short distances in the atmosphere. Under turbulence of wind, diffusion is very fast, and the pollutants diffuse rapidly to cover a larger distance. The degree of turbulence depends upon the roughness of the surface, environmental lapse rate and speed of wind. Turbulence is categorized under following types

1. **Mechanical turbulence**—Occurs when fast moving air passes over the rough surface leading to friction and development of small eddies which move upward and downward within the horizontally moving air (Fig. 11.1). This type of turbulence is easily developed when wind speed is high
2. **Thermal turbulence**—Thermal turbulence takes place when the temperature rises due to solar radiation near the earth's surface.

(d) **Lapse rate**—The rate of decrease in temperature with height is also known as the lapse rate. It is correlated with atmospheric turbulence and diffusion (Fig. 11.1).

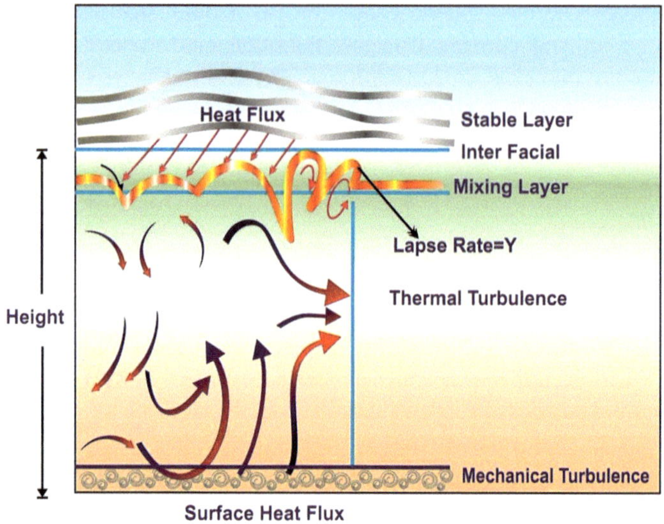

Fig. 11.1 Boundary layer development

(e) **Other factors**—Along with conventional meteorological parameters, atmospheric boundary layer also influences chances, maintenance and vertical dissipation of pollutants (Tang et al. 2016). During heavy pollution, the changes in mixing layer height and temperature inversion layer are important parameters which need to be explored more (Tang et al. 2016).

11.3.3 Selection of Sampling Location

The choice of site selection for measurement depends on the objective of the study. The commonly followed strategy for choosing the sampling locations includes (Maldonado and Woods 1983; Woods et al. 1985)

1. Identification of source of pollution
2. The relative exposure index (REI)
3. The ventilation effectiveness (VE) and
4. Incidence of health complaints.

A large sampling area can be subdivided into different microenvironments which have their typical characteristics. The microenvironments must be decided before sampling. In addition to indoor monitoring, it is also desirable to take outdoor samples for the indoor/outdoor (I/O) analysis. The comparison can help identifying different sources of pollution. An area map of the chosen location may guide to locate probable pollutant sources around the selected location.

11.3.4 Sampling Period, Frequency, and Duration

The sampling period depends upon the objective of the study. The duration should be as such so that the pollutants under consideration are trapped in measurable amount. Twenty-four-hour continuous sampling time provides full day variation pattern, rise and fall in concentration as affected by outdoor and indoor activity pattern. Online samplers are frequently employed to monitor continuous data with specific flow rates. For seasonal changes, it is advisable to take samples in each specific season so as to compare the pollutant level with the climatic changes.

11.3.5 Statistical Tools Used in Air Quality Assessment

In atmospheric pollution research, the following statistical tools are generally applied for evaluation of derived data from observations-

(a) **Regression analysis**—In case of two or more dependent variables in a sample, regression analysis is often used to study the relationship. Regression analysis is applied for a data set describing the relationship between the variables. Linear regressions used when the relationship between variables is close to linear.
(b) **Correlation analysis**—For the quantification of the relationship between two variables, a correlation coefficient is calculated. Correlation coefficient is denoted by (R^2), a larger value of R^2 shows higher magnitude of correlation between two variables. In general, R^2 is from 0 to 1, and r is from −1 to 1. The sign of r indicates whether the two variables increase together or are inversely related. Coefficient of determination, R^2 can predict the indoor level of pollutants as dependent on infiltration from outdoor sources (Kulshreshtha and Khare 2011).
(c) **Principal Component and Cluster Analyses**—For multivariate analysis, Principal Components Analysis (PCA) and Cluster Analysis (CA) are commonly used tools. PCA is a multivariate statistical technique that creates new variables known as principal components (PCs) which are uncorrelated and orthogonal to each other. These PCs represent linear combinations of the original variables (Wang et al. 2004). Cluster Analysis (CA) creates separate clusters so that objects in a same cluster are similar to each other and different from those in other clusters. CA separates data in different sets, hence it is a categorization method (Manly 1994). In a study conducted in coal mining areas of Jharkhand, India assessment of air pollution was made in which multivariate statistical analyses were adopted including PCA and CA to identify the sources of air pollutants in that area. Pollutant dispersion and spatial variations were explained by the use of cluster analysis (Pandey et al. 2014)
(d) ***t*-test and ANOVA**—A *t*-test can be used for estimation of significant difference between two populations. In this research field, student's t-test is usually performed at 95% significance level. The value of $t_{0.05,\ n-1}$ is evaluated from the t-distribution tables and then compared with the t value obtained from

the formula given above. If $|t| < t_{0.05},\ _{n-1}$, then it is concluded that the means of the two populations. In a study conducted in Bangladesh, pollutant concentration was compared between households using biomass or fossil fuels using Student's t-test (Khalequzzaman et al. 2010). The variation in concentrations during cooking and non-cooking time was calculated using two-way analysis of variance (ANOVA) in the same study.

11.4 Objective

11.4.1 Indoor Air Quality Assessment in Rural Houses Around Lucknow District

The study was undertaken to assess the indoor air quality in rural households during winter season from November 2013 to February 2014. According to a recent declaration by the WHO, 13 of the 20 most polluted cities in the world belong to India and majority of them belong to Northern region of the country including Allahabad, Kanpur, and Lucknow, capital city of Uttar Pradesh. Lucknow city is located at 26°51′N and 80°55′E. According to 2011 census, population of the city is 2,815,033, and the present area of Lucknow is envisaged to be 310 sq. km (Lawrence and Fatima 2014).

11.4.1.1 Study Design

A questionnaire-based survey was conducted in fifteen villages surrounding Lucknow city. The study design is being presented schematically (Fig. 11.2). Out of the fifteen villages, Gaura, Malihabad, Arjunganj, Kakori, and Bijnaur were selected for indoor air quality assessment (Fig. 11.3).

11.5 Methodology

11.5.1 Questionnaire Survey

The questionnaire was designed with the consultation of doctors from Department of Pulmonary Medicine, King George's Medical University (KGMU), and Sanjay Gandhi Post Graduate Institute of Medical sciences (SGPGIMS), Lucknow, and translated in Hindi for better understanding of the respondents. Personal observation and focus group discussion methods were also used to collect information. Informed consent was obtained from all subjects. The data were collected to get household conditions, educational attributes, and infrastructural characteristics

15 VILLAGES (125 HOUSES)

(ON THE OUTSKIRTS OF LUCKNOW)

PARAMETERS INCLUDED IN SURVEY CONDUCTED ON HOUSEHOLD CHARACTERISTICS

(HEALTH EFFECTS, TYPE OF FUEL USED, VENTILATION)

↓

5 VILLAGES SELECTED OUT OF 15 ON THE BASIS OF HEALTH EFFECTS REPORTED

↓

3 HOUSES FROM EACH VILLAGE WERE CHOSEN FOR IAQ ASSESSMENT

↓

PARAMETERS – CO, CO_2, SO_2, NO_2, PM_{10}, $PM_{2.5}$

Fig. 11.2 Schematic representation of the study design

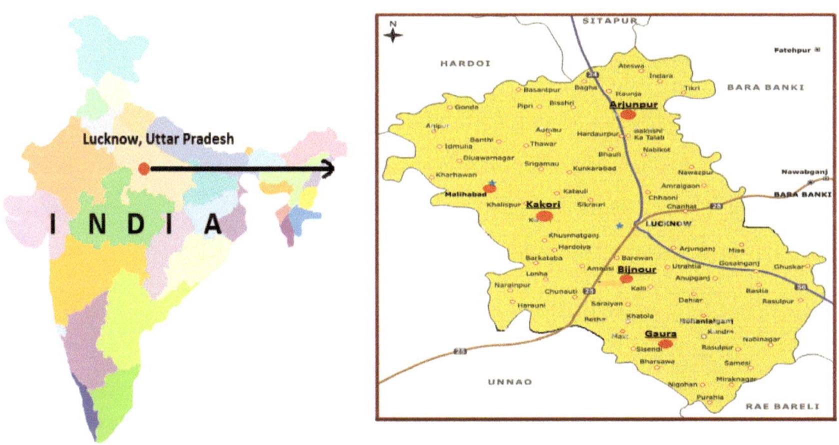

Fig. 11.3 Site map showing selected five villages for IAQ assessment

including house and kitchen type, number of rooms, and ventilation and sanitary conditions. Daily indoor and outdoor activities pattern including cooking, cleaning, and heating were also reported. Questions pertaining to energy sources and consumption were also included depicting the affordability, availability, and accessibility. Another important aspect of the study was to have an overall health impact of indoor air quality. Health symptoms were discussed with medical experts and correlated mainly to indoor contamination.

11.5.1.1 Basis of Site Selection

Out of all the villages, top five reporting health issues were selected for air quality assessment from where three houses each were selected for monitoring. House selection was done taking into account the meteorological data collected from Lucknow weather station.

11.5.2 Meteorological Data

Average temperature, wind speed, and wind direction were collected from Lucknow weather station (Fig. 11.4). A sample questionnaire is given as Fig. 11.5. House selection was done in and against the direction of wind both. While selection of houses was made, it was also noted whether no potent outdoor contamination source was located nearby (Fig. 11.5).

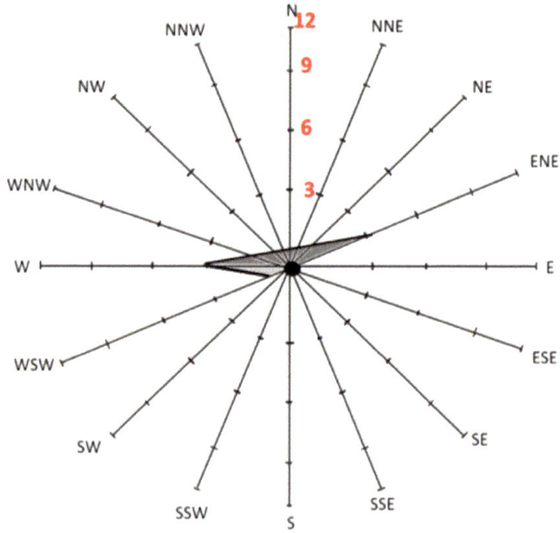

Fig. 11.4 Wind rose diagram of the month of November 2013 (Obtained from Lucknow weather station)

1.Name:

2.Location/Site (Address):

3.Characteristics:
- Area (High/Low Population):
- Building Material:
- Age of House:
- Height of House:
- Number of Rooms:
- Living Room Area:
- Ventilation
  - Acceptable
  - Somewhat Acceptable
  - Somewhat Unacceptable
  - Unacceptable
- Any plant located near or inside your house? Specify

4. Details of Occupants
- Number of Family Members (adults, children and sick)
- Average time you stay indoor?

5. Activity Schedule
- Number of People who smoke/Duration of Smoking.
- Quality of oil used for cooking/ Cooking Hours.
- Material used during Prayer Time/Prayer Time.
- Type of Fuel used/Purpose.
- Heating material used.

Any other remark:
Do you have any other such exposures such as an additional job, hobbies, farming, welding, auto repair etc.? yes/no
If yes please describe:

6. Health Status
6.1 What health complaints you have experienced?
Select any symptoms you have experienced in your home. (This is random list – not all symptoms listed have been noted in houses.)

SYMPTOMS	OCCASIONALLY	FREQUENTLY	NOT RELATED TO HOUSE	NO
Difficulty in Concentrating				
Dry or sore throat				
Dizziness				
Itching				
Heartburn				
Nausea				
Noticeable Odors				
Sinus Congestion				
High stress levels				
Chest tightness				
Eye irritation				
Hyperventilation				
Shortness of breadth				
Headache				
Fatigue/drowsiness				
Temperature too hot				
Temperature too cold				

6.2 Are the symptoms more likely to appear at particular times of the day/year?
6.3 Do these symptoms clear up within 1-2 hours after leaving house? Yes/No
If no, do they clear up overnight or over the weekend? Yes/No
6.4 Have you sought medical attention for your symptoms? Yes/No
If yes, please specify the medicines taken routinely.
6.5 Do you have any allergies or other health problems that may account for any of the listed symptoms? Yes/No. If yes, please describe:
6.6 Can you offer any other comments or observations that may be helpful in determining the environmental condition of your home?

(Signature of Occupant)

Fig. 11.5 An outline of framed questionnaire

11.5.3 Assessment of Air Pollutants—Instrumentation Specifications and Principles

CO, CO_2, SO_2, NO_2, and particulate matter (PM_{10}, $PM_{2.5}$) were assessed with the help of following instruments

11.5.3.1 YES-205 and YES-206 Falcon IAQ Monitor for CO and CO_2 Estimation

The instruments are based on Non-dispersive Infrared absorption method. The main components of the monitors are an infrared source (lamp), a sample chamber or light tube, a light filter, and an infrared detector. The gas in the sample chamber caused absorption of specific wavelengths, and the attenuation of these wavelengths was measured by the detector to determine the gas concentration. The concentration was displayed on a LCD display.

Working principle-Non-dispersive infrared spectroscopic technique—The sampled gas absorbs infrared radiation at a particular frequency. By measuring the amount of absorbed infrared radiation at the necessary frequency, the concentration of the gas component can be determined. Besulfonic acid. The absorbanceer–Lambert's Law is used for operation of NDIR analyzers by measuring the degree of absorption of infrared (IR) radiations when they pass through a column of gas. The fraction of incident radiations absorbed is given by Eq. 11.1,

$$I/Io = \left(1 - e^{-k.c.d.}\right) \tag{11.1}$$

where

I = Radiation energy absorbed
I_0 = Incident radiation energy
K = Characteristic absorption constant for the gas, m^2/gmol
c = Concentration of the gas, gmol/m^3
d = Length of the gas column, m

The method is non-dispersive because the optical filter eliminates all but the wavelength that the selected gas molecules absorb.

11.5.3.2 Handy Sampler Based on Impinger Technique for SO_2 and NO_2 Estimation

The assembly has a small battery operated pump to draw air through a suitable absorption solution contained in an impinge. Two impingers were used to monitor SO_2 and NO_2 separately. Nitrogen dioxide (NO_2) is collected by bubbling air through a solution of sodium hydroxide and sodium arsenite, whereas SO_2 is absorbed by a solution of sodium or potassium tetrachloromercurate (II).

Working principle—West and Gaek method for SO_2 estimation—West and Gaeke method introduced in 1956 is an accurate and sensitive method that is freedom from interferences (West and Gaeke 1956). SO_2 from air is absorbed by a solution of sodium or potassium tetrachloromercurate II to form dichlorosulfito-mercurate (II) complex. The complex is resistant to oxidation by atmospheric oxygen (Nauman et al. 1960). The dichlorosulfitomercurate (II) complex reacts with sulfamic acid (0.6%), pararosaniline and formaldehyde (0.2%) to form colored pararosaniline methyl sulfonic acid. The absorbance of the solution is measured at 560 nm using UV-Vis spectrophotometer.

11.5.3.3 Jacobs and Hochheiser Method for NO_2 Estimation

Determination of Nitrogen dioxide in air was done by Sodium Arsenite Method (Jacobs and Hochheiser 1958). The concentration of nitrite ion (NO_2^-) produced during sampling is determined colorimetrically by reacting the nitrite ion with phosphoric acid, sulfanilamide, and *N*-(1-naphthyl)-ethylenediamine di-hydrochloride (NEDA) and measuring the absorbance of the highly colored azo-dye at 540 nm.

11.5.3.4 APM 550 for Particulate Matter Estimation

The instrument is based on gravimetric method which samples a specific volume of air on a collecting medium at a specific flow rate. The measurement was made at a flow rate of 1 m^3/h controlled by critical orifice. PM_{10} samples were collected using 47 mm diameter, 2 μm pore size PTFE filter and $PM_{2.5}$ samples were collected on PTFE Whatman filter papers with pore size 2 μm, diameter 46.2 mm with supported PP ring. Blank correction was done to avoid the high background values in analysis. The filter paper is weighed before and after the collection of sample, and weight of particles is calculated.

Working principle-Gravimetric estimation—Particulate matter can vary in size from nanometers to micrometers, hence choosing a correct measuring technique is dependent on the size of measured particle. Particles are measured based on their sizes. Primary particles are directly emitted in atmosphere as generated during the combustion process. Gravimetric method of estimation is based on measuring the mass of particles collected on suitable filter media.

The instruments and their specifications are summarized in Table 11.2.

Setting and positioning of the instruments—The instruments must be placed so as to minimize the hindrance in the indoor activities. Ideally, all instruments should be placed centrally the living area at a height of 1–1.5 m above the ground to simulate breathing zone. The placing should also avoid any probable interference

Table 11.2 Specifications of instruments used in the study

Contaminant	Principal of measurement	Instrument used for measurement
CO	Non-Dispersive InfraRed (NDIR)	YES-205 multigas monitor, YES Environment Technologies Inc. Canada
CO_2	Non-Dispersive InfraRed (NDIR)	YES-206 Falcon IAQ monitor, Geo Scientific Ltd., Canada
SO_2	Improved West and Gaeke method	Impinger Technique, Handy Sampler Envirotech India
NO_2	Jacob and Hochheiser modified (NaOH–$NaAsO_2$) method	Impinger Technique, Handy Sampler Envirotech India
$PM_{2.5}$ and PM_{10}	Gravimetric	APM 550, Envirotech, India

caused by re-suspension of particles. The placing of instruments did not interfere with the normal functioning of the household (Kulshreshtha and Khare 2011).

11.5.3.5 Quality Assurance

Instruments were calibrated before and at the end of every monitoring period or seven days. Filter in the wind impactor was changed after 72 h of sampling or when the filter got clogged. Filter was immerged in 3–4 drops of silicon oil at regular intervals. Daily flow rate calculations (gas meter reading/timer reading) of APM550 were made to make sure that the fluctuations in flow rate were within the range.

11.6 Results and Discussion

11.6.1 Socio-demographic Conditions

The socio-demographic conditions of all the 125 surveyed households have been summarized in Table 11.3. Household characteristics revealed high dependence on crude fuel. Cooking was mainly done in closed space with poor ventilation conditions. It has been found in literature that location of kitchen has a significant influence on exposure and concentration of pollutants (Parikh et al. 2001). Chimneys are likely to reduce the level of respirable particles by 80% (Practical Action 2004; McCracken et al. 2007). A study conducted in rural Rajasthan revealed that economic status also influences the choice of fuel (Laxmi et al. 2003). Educational status was fairly low. Occupational status and income of a household also have a significant role in overall living standard. In a study conducted in Pakistan, participants were categorized on the basis of monthly income of less than or equal to or greater than $50 per household and the type of fuel they used (Siddiqui et al. 2009). High unemployment in the surveyed households may be responsible for the unaffordability of clean energy sources. Other than cooking practices and construction material, there are several other factors that have an impact on the quality of indoor air in residential environment. Cleaning, working, and even human occupancy may lead to increase in pollutant concentration (Spengler et al. 1981). Smoking was also identified as another common practice indoors. Smoking has been established as a potent hazard to human health. Sharp elevation in indoor/outdoor ratio of PM_1 reaching to a maximum of 12.95 has been reported in Pakistan (Ahmad et al. 2005). Smoking and passive factors are also being associated with lung cancer in women (Behera and Balamugesh 2005). Environmental tobacco smoke (ETH) is also associated with cough, asthma, wheeze, bronchitis, pneumonia, and deficits in childhood growth (Fox et al. 1990).

Table 11.3 Socio-demographic characteristics and cooking/ventilation practices observed during the study

Characteristics	Category	Percentage
Educational status	Educated	12.2
	Uneducated	87.8
Occupational status	Employed	36.02
	Unemployed	63.98
House type and drainage	Kutcha with poor drainage	26.08
	Pucca with poor drainage	35.27
	Mixed	38.65
Number of rooms	1 room	29.27
	More than 1 room	70.73
Number of occupants in a house	Less than 6	42.35
	More than 6	67.65
Number of children	Less than 6	65.72
	More than 6	34.28
Place of cooking food	Closed room	40.15
	Multipurpose room	30.77
	Open space	27.39
Livestock ownership	Cow	33.67
	Buffalo	16.81
	Others	19.20
	None	18.21
Ventilation in kitchen	Proper	22.26
	Improper	77.74

11.6.1.1 Fuel Inefficiency

Firewood, coal, animal dung, and kerosene were commonly used energy sources mainly because they have no monetary value and are easily available in nature. About 96% of the households used earthen chullah for cooking which are highly polluting. Modern heating appliances were used only be a meager percentage. Only 43.8% of the sampled households availed electricity connection in their homes (Table 11.4).

11.6.2 Assessment of Indoor Air Quality

CO, CO_2, SO_2, NO_2, $PM_{2.5}$, and PM_{10} were measured through 10 h simultaneous monitoring indoor and outdoor. The monitoring period covered all the indoor activities to better understand the variability pattern.

Table 11.4 Household characteristics related to energy usage

Characteristic	Category	Percentage
Stove used for cooking	Traditional Modern	96.62 3.38
Source of energy	Firewood/coal/cow dung Kerosene Gas and Electricity	72.80 16.35 10.85
Appliances used for house heating	Traditional Modern appliances	68.30 31.70
Electricity connection	Yes No	43.80 56.20
Use of electricity in house	For lighting For warming/heating space	76.33 23.67
Willing to change from traditional to modern energy sources	Yes No	58.65 41.35

11.6.2.1 Indoor Air Quality Assessment—Average Concentration of Pollutants

Average concentrations of measured pollutants are given from November 2013 to February 2014 are given in Table 11.5. The particulate concentrations exceeded the WHO 2000 limits (Table 11.5). The indoor concentration of particulate matter was higher than outdoor. Concentrations of NO_2 and SO_2, though high were within the WHO 2000 limits. High CO and CO_2 levels were obtained through December and January. The outdoor concentrations were treated as control. Less usage of clean energy sources may be suggested as potential reason of poor indoor air quality owing to poor affordability and lack of access. The air quality can also be hampered due to several confounders like smoking, use of low-quality cooking oil, ventilation and practices (Goyal and Khare 2009). During winters, the ventilation rate is less.

11.6.2.2 Elevation in Pollutant Concentration During Cooking

Monitoring was done for 10 h at a stretch from 8:00 a.m.to 6:00 p.m. Average 2 h was taken to cook one meal. In all the houses, an average pollutant concentration between non-cooking and cooking hours showed elevated level of particulate and gaseous concentrations (Table 11.6). Even after cooking was completed, smoke from the challahs kept producing making the indoor conditions unfavorable for overall health for as long as 1–1.5 h. Other factors which might contribute significantly to the sudden elevation may be the cumbersome lighting process to light the challah. Use of biomass for cooking caused the particulate concentration to reach from 500 to 2000 mg/m^3, as shown in a study to quantify exposures to particulate matter (Balakrishnan et al. 2002). Women are primary sufferers because of their pivotal role in cooking (Behera et al. 1988). The variation in particulate

Table 11.5 Average concentrations of pollutants

Pollutant	Environment	November	December	January	February
CO	I ± SD	0.2 ± 0.1	0.8 ± 0.3	1.1 ± 0.46	0.2 ± 1
	O ± SD	BDL	0.65 ± 0.1	0.84 ± 0.1	BDL
CO_2	I ± SD	390 ± 26	458 ± 16	508 ± 24	408 ± 12
	O ± SD	368 ± 13	386 ± 11	398 ± 5	374 ± 6
NO_2	I ± SD	0.043 ± 0.03	0.078 ± 0.034	0.008 ± 0.004	0.003 ± 0.002
	O ± SD	0.038 ± 0.18	0.073 ± 0.2	0.007 ± 0.09	0.008 ± 0.9
SO_2	I ± SD	0.03 ± 0.004	0.11 ± 0.01	0.18 ± 0.05	0.009 ± 0.001
	O ± SD	0.023 ± 0.001	0.09 ± 0.023	0.098 ± 0.005	0.0075 ± 0.005
PM_{10*}	I ± SD	109 ± 17	172 ± 40	218 ± 30	107 ± 9
	O ± SD	65 ± 8	83 ± 5	93 ± 14	73 ± 4
$PM_{2.5*}$	I ± SD	71 ± 4	98 ± 12	116 ± 13	74 ± 21
	O ± SD	31 ± 3	48 ± 4	61 ± 7	39 ± 6

Average concentrations in ppm (otherwise mentioned) of pollutants ($TWA^{\#}$) from November 2013 to February 2014
Note #TWA: Time weight Average concentration for a normal 10-hour workday covering major indoor activities
*µg/m^3

Table 11.6 Concentration variation in pollutants between cooking and non-cooking period

Pollutant	Cooking period	Non-cooking period
PM_{10} (µg/m^3)	286	204
$PM_{2.5}$ (µg/m^3)	283	165
SO_2 (ppm)	0.023	0.009
NO_2 (ppm)	0.028	0.013

concentration during cooking in one of the rural sites in Kenya suggested that mean PM10 concentration near the fire was 1250 µg/m^3, though the actual level peaked over 50,000 µg/m^3 (Ezzati et al. 2000).

11.6.2.3 Health Issues Reported by the Respondents

Indoor air pollution causes about 2 million premature deaths per year. During combustion of biomass and fossil fuel, various gases and particles are released in atmosphere degrading the indoor living conditions. These pollutants make the host susceptible to catch respiratory infections by weakening immunity. Respiratory infections have been found to increase occurrences of morbidity and mortality in children. In our findings, inefficient fuel usage was identified as the primary source of contamination. Further, there was inadequate ventilation in sampled houses. The measured particulate concentration was significantly higher in all instances. For health-related issues, the queries were put directly to all adult individuals present during the survey. Headache, sneezing, skin irritation, shortness of breath, cough, dizziness, nausea, eye irritation, and cataract were most prevalent complaints which may be linked to the increased inhalation of RSPM and CO_2 originating from smoke and due to inadequate ventilation (Fig. 11.6). From the questionnaire survey results done in 125 houses in fifteen villages, a simple linear regression analysis was used to study the relationship between the health symptoms and crude fuel usage which is used as a proxy for household air pollution considered in our study. A correlation ($R^2 = 0.71$) was obtained between the two parameters (Fig. 11.7). A similar relationship was studied previously between household air pollution and neonatal mortality conducted in 284 districts of nine states in India (Neogi et al. 2015). Because of the presence of several confounders like quality of cooking oil, ventilation, smoking, incense burning, and animal excreta, interpretation of the correlation results should be taken as suggestive rather than definitive.

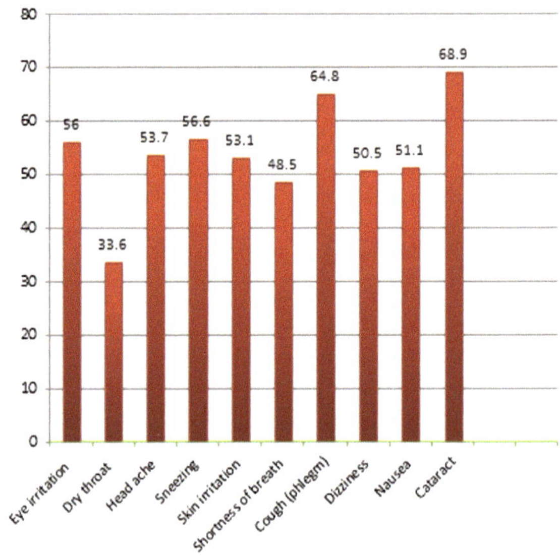

Fig. 11.6 Health problems reported by the respondents

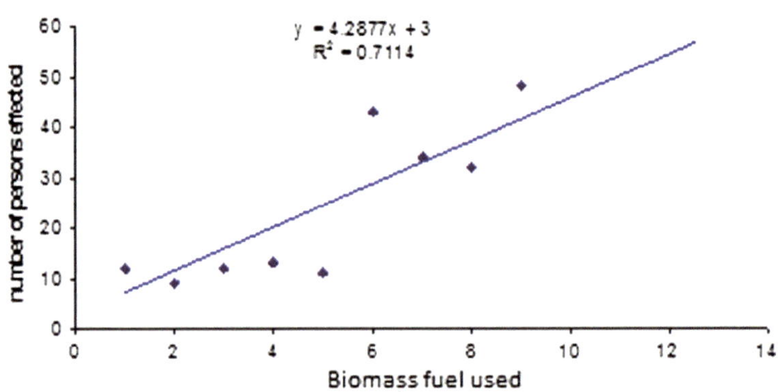

Fig. 11.7 Regression analysis between health symptoms and fuel usage

11.7 Conclusion

Accurate measurement of indoor air quality is a complicated task. The air quality varies in different microenvironments (e.g., schools, restaurants, hospitals, and homes etc.) as related to different sources of pollution. In residential microenvironment, significant differences exist depending on sociological factors. For urban household's, indoor air is hampered through a varied number of sources, whereas for rural houses, pollution generated by combustion processes continues to be the prime source. There is a growing awareness for the significance of indoor environment on health and exposure to indoor air pollution should be given prime importance and priority in policy making (Harrison 2002). The highlighting point of this study is the observation that educational rate in rural areas is still very low which calls for spreading awareness about the serious threat posed to the health. There are societal, customary, and economic factors that influence the decision of people (Sreeramareddy et al. 2011). Flexibility and easy availability to non-polluting energy sources in rural setups should be facilitated. Affordability is another factor of importance when we talk of low-income households. To overcome this hurdle better job prospects for rural population should be created so that they could afford and adopt better and healthy lifestyle. The findings here showed smoky cooking stoves are still very much in use, they should be modified and replaced with the ones which are fuel efficient and smokeless. Many improved cookstove projects are ongoing around the world initiated by local non-governmental organizations and nationwide initiatives under various government-sponsored schemes (WHO 2008; Granderson et al. 2009). Small changes in household habits and structure of housing can have a significant impact for the betterment of indoor environment and inclusive of the health of the occupants as the study presented here shows probable linkage between many socio-demographic variables and respiratory impacts. But at the same time, there are many other factors and reasons, which should be explored in detail.

References

Advisory Board on Energy (1984) Towards a perspective on energy demand and supply in India in 2004/05. Government of India

Ahmad K, Jafary F, Jehan I, Hatcher J, Khan AQ, Chaturvedi N et al (2005) Prevalence and predictors of smoking in Pakistan: Results of the National Health Survey of Pakistan. Euro J Cardio Prev and Rehab 12:203–208

Akhtar T, Uah Z, Khan MH, Nazli R (2007) Chronic bronchitis in women using solid biomass fuel in rural Peshawar, Pakistan. Chest 132:1472–1475

Albalak R, Bruce N, McCracken J, Smith KR, De Gallardo T (2001) Indoor respirable particulate matter concentrations from an open fire, improved cook stove, and LPG/open fire combination in a rural Guatemalan community. Environ Sci Technol 35:2650–2655

Alm S, Jantunen MJ, Vartiainen M (1999) Urban commuter exposure to particle matter and carbon monoxide inside an automobile. J Expo Anal Environ Epidemiol 9:237–244

Baek SO, Kim YS, Perry R (1997) Indoor air quality in homes, offices and restaurants in Korean urban areas indoor/outdoor relationships. Atmos Environ 31:529–544

Balakrishnan K, Parikh J, Sankar S, Padmavathi R, Srividya K, Venugopal V et al (2002) Daily average exposures to respirable particulate matter from combustion of biomass fuels in rural households of southern India. Environ Health Perspect 110:1069–1075

Balakrishnan K, Sambandam S, Ramaswamy P, Mehta S, Smith KR (2004) Exposure assessment for respirable particulates associated with household fuel use in rural districts of Andhra Pradesh. Ind J Expo Anal Environ Epidemiol 14:S14–S25

Bassani DG, Jha P, Dhingra N, Kumar R (2010) Child mortality from solid-fuel use in India: A nationally-representative case-control study. BMC Public Health 10:491

Behera D, Balamugesh T (2005) Indoor air pollution as a risk factor for lung cancer in women. J Assoc Physicians India 53:190–192

Behera D, Dash S, Malik S (1988) Blood carboxyhaemoglobin levels following acute exposure to smoke of biomass fuel. Indian J Med Res 88:522

Boleij JSM, Ruigewaard P, Hoek F, Thairu H, Wafula E, Onyango F et al (1989) Domestic air-pollution from biomass burning in Kenya. Atmos Environ 23:1677–1681

Braubach M, Algoet A, Beaton M, Lauriou S, Heroux ME, Krzyanowski M (2013) Mortality associated with exposure to carbon monoxide in WHO European Member States. Indoor Air J 23:115–125

Chao YH, Tung CW, Burnett J (1998) Influence of different indoor activities on the indoor particulate levels in residential buildings. Indoor Built Environ 7:110–121

Charron A, Harrison RM (2005) Fine ($PM_{2.5}$) and coarse ($PM_{2.5-10}$) particulate matter on a heavily trafficked London highway: Sources and processes. Environ Sci Technol 39:7768–7776

Dasgupta S, Huq M, Khaliquzzaman M, Pandey K, Wheeler D (2006) Indoor air quality for poor families: New evidence from Bangladesh. Indoor Air 16:426–444

Ezzati M, Saleh H, Kammen DM (2000) The contributions of emissions and spatial microenvironments to exposure to indoor air pollution from biomass combustion in Kenya. Environ Health Perspect 108:833–839

Fischer SL, Koshland CP (2007) Daily and peak 1 h indoor air pollution and driving factors in a rural Chinese village. Environ Sci Technol 41:3121–3126

Fox NL, Sexton M, Hebel JR (1990) Prenatal exposure to tobacco: Effects on physical growth at age three. Int J Epidemiology 19:66–71

Glorennec P et al (2008) Is a quantitative risk assessment of air quality in underground parking garages possible? Indoor Air 18:283–292

Gold DR (1992) Indoor air pollution. Clin Chest Med 13:215–229

Goyal R, Khare M (2009) Indoor–outdoor concentrations of RSPM in classroom of a naturally ventilated school building near an urban traffic roadway. Atmos Environ 43:6026–6038

Granderson J, Sandhu JS, Vasquez D, Ramirez E, Smith KR (2009) Fuel use and design analysis of improved wood burning cook stoves in the Guatemalan Highlands. Biomass Bioenerg 33:306–315

Harrison PTC (2002) Indoor air quality guidelines. Occup Environ Med 59:73–74

He LY, Hu M, Huang XF, Yu BD, Zhang YH, Liu DQ (2004) Measurement of emissions of fine particulate organic matter from Chinese cooking. Atmos Environ 38:6557–6564

Hewson EW (1956) Meteorological factors affecting causes and controls of air pollution. J Air Pol Cont Asso 5(4):235–241. https://doi.org/10.1080/00966665.1956.10467718

Hewson EW, Gill GC (1944) Meteorological Investigations in Columbia River Valle. U.S. Bureau of Mines Bull, 453:23–22

International Energy Agency (2015) Chapter 2—Extract: Energy Access, World Energy Outlook, www.worldenergyoutlook.org/resources/energydevelopment

International Energy Agency (2013) Chapter 2—Extract: Modern energy for all, World Energy Outlook. http://www.worldenergyoutlook.org/energydevelopment

Jacobs MB, Hochheiser S (1958) Continuous sampling and ultra–micro determination of nitrogen dioxide in air. Anal Chem 30:426–428

Johnson P, Balakrishnan K, Ramaswamy P, Ghosh S, Sadhasivam M, Abirami O et al (2011) Prevalence of chronic obstructive pulmonary disease in rural women of Tamilnadu: Implications for refining disease burden assessments attributable to household biomass combustion. Global Health Action 4:7226

Jetter J et al (2002) Characterization of emissions from burning incense. Sci Total Environ 295:51–67

Jones NC, Thornton CA, Mark D, Harrison RM (2000) Indoor/outdoor relationships of particulate matter domestic homes with roadside, urban and rural locations. Atmos Environ 34:2603–2612

Kamens R, Lee CT, Weiner R, Leith D (1991) A study to characterize indoor particulates in three non-smoking homes. Atmos Environ 25:939–948

Kampa M, Castanas E (2008) Human health effect of air pollution. Environ Pollut 151(2):269–428

Kankaria A, Nongkynrih B, Gupta SK (2014) Indoor air pollution in India: Implications on health and its control. Indian J Community Med 39:203–207

Khalequzzaman M, Kamijima M, Sakai K, Hoque BA, Nakajima T (2010) Indoor air pollution and the health of children in biomass- and fossil-fuel users of Bangladesh: Situation in two different seasons. Environ Health Prev Med 15:236–243. https://doi.org/10.1007/s12199-009-0133-6

Kleinman MT (2009) Carbon monoxide. In: Lippmann M (ed) Environmental toxicants, human exposures and their health effects. John Wiley and Sons, New Jersey, pp 499–528

Kulshreshtha P, Khare M (2011) Indoor exploratory analysis of gaseous pollutants and respirable particulate matter at residential homes of Delhi, India. Atmospheric Pollution Research 2:337–350

LamNL et al (2012) Household light makes global heat: High black carbon emissions from kerosene wick lamps. Environ Sci. Technol. 46:13531–13538

Laxmi V, Parikh J, Karmakar S, Debrase P (2003) Household energy, women's hardship and health impacts in rural Rajasthan, India: Need for sustainable energy solutions. Energy Sustain Dev 7:50–68

Lawrence A, Fatima N (2014) Urban air pollution and its assessment in Lucknow city-the second largest city of north India. Sci. Tot Environ 488–489(1):447–455

Lawrence AJ, Masih A, Taneja A (2005) Indoor/outdoor relationships of carbon monoxide and oxides of nitrogen in domestic homes with roadside, urban and rural locations in a central Indian region. Indoor Air 15:76–82

Lawrence A, Taneja A (2005) An Investigation of Indoor Air Quality in Rural Residential Homes in India—A Case Study. Indoor Built Environ 14:321–329

Levy JI et al (1998) Determinants of nitrogen dioxide concentrations in indoor ice skating rinks. Am J Public Health 88:1781–1786

Long CM, Suh HH, Koutrakis P (2000) Characterization of indoor particle sources using continuous mass and size monitors. J Air Waste Manag Assoc 50:1236–1250

Maldonado EAB, Woods JE (1983) A method to select Locations for indoor air quality sampling. Building Environ 18:171–180

Manly BFJ (1994) Multivariate statistical methods: A primer. Chapman and Hall/CRC, London, pp 129–133

McCracken JP, Smith KR, Siscovick DS, Sheppard L, Shepherd K, Sullivan JH, Anderson GL, Kaufman JD (2007) Long-term exposure to air pollution and incidence of cardiovascular events in women. N Engl J Med 356:447–458

McCracken JP, Smith KR, Stone P, Díaz A, Arana B, Schwartz J (2011) Intervention to lower household wood smoke exposure in Guatemala reduces ST-segment depression on electrocardiograms. Environ Health Perspect 119:1562–1568

Mishra V (2003) Effect of indoor air pollution from biomass combustion on prevalence of asthma in elderly. Environ Health Perspect 111:71–78

Neogi SB, Pandey S, Sharma J, Choksi M, Chauhan M, Zodpey S, Paul VK (2015) Association between household air pollution and neonatal mortality: an analysis of Annual Health Survey results, India. WHO South East Asia J Public Health 4(1):30–37

Pandey B, Agrawal M, Singh S (2014) Assessment of air pollution around coal mining area: Emphasizing on spatial distributions, seasonal variations and heavy metals, using cluster and principal component analysis. Atmos. Pol. Res. 5:79–86

Parikh J, Balakrishna K, Laxmi V, Biswas H (2001) Exposure from cooking with biofuels: Pollution monitoring, and analysis from rural Tamil Nadu, India. Energy 26:949–962
Park E, Lee K (2003) Particulate exposure and size distribution from wood burning stoves in Costa Rica. Indoor Air 13:253–259
Practical-Action. (2004) Smoke—The killer in the kitchen. London: ITDG Publishing
Priscilla J, Padmavathi R, Ghosh S, Paul P, Ramadoss S, Balakrishnan K et al (2011) Evaluation of mucociliary clearance among women using biomass and clean fuel in a periurban area of Chennai: A preliminary study. Lung India 28:30–33
Priscilla D, Hamidin A, Azhar MZ, Noorjan KON, Salmiah MS, Bahariah K (2011) Quality of life among patients with hematological cancer in a Malaysian hospital. Med J Malaysia 66:117–120
Raunemaa T, Kulmala M, Saari H, Olin M, Kulmala MH (1989) Indoor air aerosol model—Transport indoors and deposition of fine and coarse particles. Aerosol Science and Technology 11:11–25
Robinson AL, Subramanian R, Donahue NM, Bernardo-Bricker A, Rogge WF (2006) Source apportionment of molecular markers and organic aerosol. 3. Food cooking emissions. Environ Sci Technol 40:7820–7827
Rumchev K, Spickett JT, Brown HL, Mkhweli B (2007) Indoor air pollution from biomass combustion and respiratory symptoms of women and children in a Zimbabwean village. Indoor Air 17:468–474
Saksena S, Subida R, Buttner L, Ahmed L (2007) Indoor air pollution in coastal houses of southern Philippines. Indoor and Built Environment 16:159–168
See SW, Balasubramanian R (2008) Chemical characteristics of fine particles emitted from different gas cooking methods. Atmos Environ 42:8852–8862
Siddiqui AR, Lee K, Bernett D, Zyang X, Brown KH, Bhutta ZA, Gold EB (2009) Indoor carbon monoxide and PM 2.5 concentrations by cooking fuels in Pakistan. Indoor Air 19:75–82
Spengler JD, Dockery DW, Turner WA, Wolfson JM, Ferris BG (1981) Long-term measurements of respirable sulphates and particles inside and outside homes. Atmos Environ 15:23–30
Sreeramareddy CT, Shidhaye RR, Sathiakumar N (2011) Association between biomass fuel use and maternal report of child size at birth—An analysis of 2005–06 India Demographic Health Survey data. BMC Public Health 11:403
Tang G, Zhang J, Zhu X, Song T, Münkel C, Hu B, Schäfer K, Liu Z, Zhang J, Wang L et al (2016) Mixing layer height and its implications for air pollution over Beijing. China Atmos Chem Phys 16:2459–2475
Wang SW, Xiao F (2004) AHU sensor fault diagnosis using principal component analysis method. Energy Build 36:147–160
West PW, Gaeke GC (1956) Anal Chem 28:1816–1956
WHO (2005) Energy and Health. http://www.who.int/indoorair/publications/energyhealthbrochure.pdf?ua=1
WHO (2006) Fuel for life: Household energy and health. WHO, Geneva
WHO (2007) Indoor air pollution: National burden of disease estimates. Geneva: WHO
WHO (2008) Evaluating household energy and health interventions: A catalogue of methods. Geneva: WHO
WHO (2015a) 7 million premature deaths annually linked to air pollution. http://www.who.int/mediacentre/news/release/2014airpollution/inf. Accessed 22nd February 2015
Woods JE, Krafthefer BC, Janssen JE (1985) Solutions to indoor air quality problems in light housing, presented at Energy 85. Washington DC, March 1985
World Resource Institute (1998) UNEP, UNDP, World Bank. 1998–99 World resources: A guide to the global environment. Oxford: Oxford University Press
Zuk M, Rojas L, Blanco S, Serrano P, Cruz J, Angeles F et al (2007) The impact of improved wood-burning stoves on fine particulate matter concentrations in rural Mexican homes. J of Exp Sci and Environ Epid 17:224–232

Chapter 12
Effects of Biomass Burning Emissions on Air Quality Over the Continental USA: A Three-Year Comprehensive Evaluation Accounting for Sensitivities Due to Boundary Conditions and Plume Rise Height

Anirban Roy, Yunsoo Choi, Amir Hossein Souri, Wonbae Jeon, Lijun Diao, Shuai Pan and David Westenbarger

Abstract We report a comprehensive evaluation of the impacts of biomass burning on regional ozone and fine particulate matter ($PM_{2.5}$) over the continental USA, southern Canada, and northern Mexico during 2012–2014 using the Community Multiscale Air Quality (CMAQ) chemical transport model. Inputs included the Fire INventory from National Center for Atmospheric Research (FINN) for fire emissions, Biogenic Emission Inventory System (BEIS) for biogenics, the US Environmental Protection Agency (USEPA)'s National Emissions Inventory of

A. Roy · Y. Choi (✉) · A. H. Souri · L. Diao
Department of Earth and Atmospheric Sciences,
University of Houston, Houston, TX 77004, USA
e-mail: ychoi23@central.uh.edu

A. Roy
e-mail: aaroy@central.uh.edu

A. H. Souri
e-mail: asouri@central.uh.edu

L. Diao
e-mail: ljdiao@gmail.com

W. Jeon
Pusan National University, Pusan, South Korea
e-mail: wbjeon@pusan.ac.kr

D. Westenbarger
Office of Air, Texas Commission on Environmental Quality,
Park 35 Circle, PO Box 12100, Austin, TX 78753, USA
e-mail: david.westenbarger@tceq.texas.gov

S. Pan
Cornell University, Ithaca, NY 14850, USA
e-mail: sp694@cornell.edu

T. Gupta et al. (eds.), *Environmental Contaminants*, Energy, Environment, and Sustainability, https://doi.org/10.1007/978-981-10-7332-8_12

2011 (NEI2011) for anthropogenic sources, and Weather Research and Forecasting (WRF) model fields for meteorology. In situ data were taken from the Texas Commission on Environmental Quality (TCEQ)'s Continuous Ambient Monitoring Stations (CAMS) and the USEPA's Air Quality System (AQS) networks. This study has marked improvements over the previous biomass burning evaluations, which are as follows: (a) a significantly longer simulation episode; (b) use of 3-D dynamic boundary conditions; (c) grid nudging to improve meteorological fields; and (d) physically representative fire plume rise model. Observations showed ozone hot spots of 60–70 parts per billion (ppb) across the Western Mountain region and California. The model was able to reproduce these only in 2012, underpredicting in California otherwise. Monthly mean biomass impacts of 2–3 ppb, averaged over daylight hours (6:00–18:00 CST), were predicted for California and Idaho in 2012 and 2013. The largest impacts were predicted for summer 2013, adding 3 ppb in northern Mexico and southeastern Canada, and 1 ppb in Florida, New Mexico, and Colorado. For April 2014, the model predicted 1–2 ppb disparities in ozone over the southern USA; a 1–2 ppb impact in southeastern Oregon, northwestern Nevada, and southern Idaho during July 2014; and in August, up to 3 ppb changes in western California, Central Oregon, Idaho, southwestern Canada, and southern Georgia. The model was unable to accurately capture the high $PM_{2.5}$ concentrations across the domain. Large monthly mean fire impacts of up to 10 μg m^{-3} were predicted for southeastern Canada in July 2012 and June and July 2013, and for Alabama, Georgia, Idaho, and southwestern Canada for October 2013. In June 2014, the model significantly underpredicted when the biomass impact was minimal, indicating that uncertainty in biomass emissions was not the probable cause for model-measurement error.

12.1 Introduction

Several metropolitan areas in the USA are currently above or close to the 2008 Federal National Ambient Air Quality Standards (NAAQS) for eight-hour ozone of 75 parts per billion (ppb). In 2014, the US Environmental Protection Agency (EPA) further proposed to tighten the standard to 65–70 ppb, putting additional areas in danger of non-attainment (79 Federal Register 242, December 17, 2014, pp. 75234–75411). Previous studies have indicated that surface ozone (O_3) and fine particulate matter ($PM_{2.5}$) concentrations can be influenced by emissions from wildfires burning in far upwind areas (e.g., Bertschi and Jaffe 2005; Fiore et al. 2014; Jaffe et al. 2004; Jiang et al. 2012; Lapina et al. 2006; McKeen et al. 2002; Morris et al. 2006; Pfister et al. 2008a, b; Singh et al. 1996; Sinha et al. 2004; Val Martin et al. 2006). Wildfire emissions can include ozone and ozone precursors such as oxides of nitrogen (NO_x) which often travel as peroxyacetyl nitrate (PAN) before converting back into NO_x; carbon monoxide (CO), fine particulate matter ($PM_{2.5}$), organic and elemental carbon (OC and EC), and formaldehyde (HCHO). Some of these compounds can undergo photochemical reactions and be transported very long distances at high altitudes before descending into the mixing layer and impacting surface air quality.

This study focused on photochemical transport modeling over North America for the ozone seasons (April–October) of 2012–2014 to study the impacts of biomass burning emissions on some of the above mentioned pollutants. In order to do this, we considered two scenarios, with and without biomass emissions. We conducted model-measurement comparisons and estimated the monthly impacts of biomass burning, touching upon pertinent issues such as effects of dynamic boundary conditions and differing plume rise height modeling approaches on model-measurement comparisons.

12.2 Observations, Methodology, and Model Setup

12.2.1 Observations

The study uses surface observations from multiple sources. Each of these is described in the following sections, along with the procedure to extract the data and make it compatible with the model format.

12.2.2 Surface Data

Surface observational data for species and meteorological variables were retrieved over the continental USA. We used hourly surface O_3, CO, and $PM_{2.5}$ concentrations from the US Environmental Protection Agency (USEPA)'s Air Quality System (AQS) network (http://aqsdr1.epa.gov/aqsweb/aqstmp/airdata/download_files.html). The number of ozone sites over this network varies by month of the year, but on average, 1300 stations were available during 2012–2014. Additionally, the network provides hourly $PM_{2.5}$ concentrations at more than 400 sites. The network is more concentrated over the eastern and western regions of the USA, where population density is higher. AQS data were archived using a MATLAB code which reads and decodes the raw format file.

Surface measurements were also taken from the Continuous Ambient Monitoring Stations (CAMS) network operated by the Texas Commission on Environmental Quality (TCEQ). Measured parameters differ from station to station, with station density in southeastern Texas being relatively high. The locations of sites measuring ozone and nitrogen oxides (NO_x) are shown in Fig. 12.1. Comparison of CAMS data with model predictions is relatively straightforward. The first step is to extract surface-layer model variables. The CMAQ model outputs binary netCDF files, and we have developed in-house scripts to extract data for any

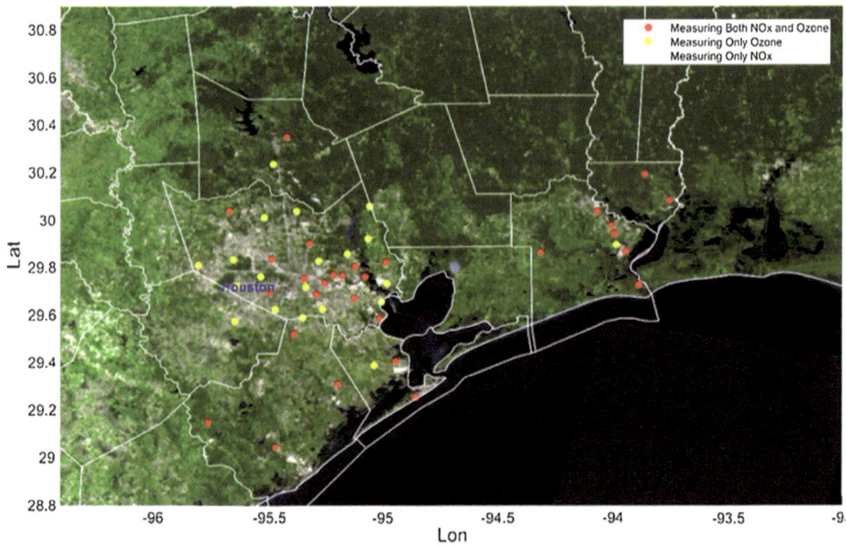

Fig. 12.1 MODIS true color image showing the location and status of the CAMS ozone and NO_x monitoring stations in southeast Texas

variable and layer. In order to compare a model value with an observation from a given site, we first use the latitude and longitude of a site to determine its cell location in the model grid and then extract the corresponding value from model output. The temporal frequency of model output and CAMS are both hourly, making direct comparison easy.

12.2.3 Methodologies Used for Modeling

12.2.3.1 The Chemical Transport Model, CMAQ

This study used the USEPA's Community Multiscale Air Quality (CMAQ) model (v5.0.2) (Byun and Schere 2006), with a 12-km spatial resolution including 459 × 299 grid cells in the *x*- and *y*-directions, respectively. Gas-phase chemistry (including chlorine) was simulated using the Carbon Bond 5 (CB05) mechanism (Yarwood et al. 2005), while aerosol chemistry used the sixth-generation aerosol mechanism (AERO6) using updated sea salt and aqueous/cloud chemistry. Inorganic aerosol thermodynamics was simulated using the ISORROPIA II module (Fountoukis and Nenes 2007). Horizontal advection was modeled using the Yamartino scheme and vertical advection on WRF-derived omega values. Vertical mixing and diffusion were modeled by the asymmetric convective model (ACM2), which simulates vertical motions caused by shear-driven turbulence (e.g., topography friction) and large-scale eddies (Pleim 2007). The model also accounts for metal catalyzed aqueous-phase sulfur oxidation (Alexander et al. 2009). Species removal through dry deposition was modeled based on the approach by Venkatram and Pleim (1999), who used an updated version of the deposition equation that is consistent with mass conservation, as opposed to the standard electrical analogy approach.

12.2.3.2 The Meteorological Model, WRF

This study used meteorological fields simulated using the Weather Research and Forecasting (WRF) model (v3.7). Here, major WRF physics options included the Lin microphysics scheme, RRTMG long-wave radiation, unified Noah land-surface modeling, the Kain–Fritsch cumulus cloud, and the Yonsei University boundary layer schemes. The model domain uses a Lambert conformal projection centered at 33°N and 97°W with 470 horizontal and 310 lateral grid cells having a 12-km spatial resolution. The vertical grid includes 41 pressure-sigma levels extending from the surface up to 32 km. The complexity of many physical processes is still not fully understood in meteorological models, causing poor model performance in representing energy and transport patterns. Therefore, we used a grid nudging technique adopted from Jeon et al. (2015), which partially reduces model error by nudging the modeled values toward lateral boundary conditions. The initial and lateral boundary conditions were retrieved from the National Centers for Environmental Prediction (NCEP) North American Regional Reanalysis (NARR) with a 32-km spatial resolution and 3-hr frequency.

12.2.3.3 The Emissions Inventory Model, SMOKE

The USEPA's National Emissions Inventory of 2011 (NEI-2011) was used for anthropogenic sources, processed by the Sparse Matrix Operator Kernel Emissions (SMOKE) model (Houyoux et al. 2000) representing all emission sectors (e.g., area, point, agricultural NH_3) except mobile sources. Mobile emissions were modeled using the Motor Vehicle Emissions Simulator (MOVES) (USEPA 2014). The SMOKE model performs spatial and temporal allocation and speciates VOC and $PM_{2.5}$ source-specific emissions. The USEPA's Biogenic Emission Inventory System (BEIS) version 3.14 was used to generate emissions of VOCs and NO_x from vegetation and soils. We chose BEIS over MEGAN since previous studies (e.g., Kota et al. 2015; Warneke et al. 2010) indicated that CMAQ modeling using MEGAN overpredicted concentrations of several biogenic VOCs such as isoprene.

The FINN Inventory for Fire Emissions

The Fire INventory from NCAR (FINN) (v1.5) (Wiedinmyer et al. 2011) was used to provide daily emissions of trace gases and particles from biomass burning at a spatial resolution of 1 km. These emissions are a function of burnt area (from MODIS), emission factors, biomass loading, and fractional area burned (from the MODIS vegetation product and lookup tables). The FINN is not a standard product released with SMOKE or CMAQ and hence not speciated for a standard chemistry mechanism options such as CB05. Therefore, we mapped the FINN MOZART-4 mechanism to CMAQ CB05/AERO6 (Choi et al. 2016) as per Tai et al. (2008). Here, primary $PM_{2.5}$ emissions are split into 18 species: organic carbon (OC), elemental carbon (EC), sulfate (SO_4^{-2}), nitrate (NO_3^-), water (H_2O), sodium (Na^+), chloride (Cl^-), ammonium (NH_4^+), selected trace elements (Al, Ca^{2+}, Fe, Si, Ti, Mg^{2+}, K^+, Mn), non-carbon organic matter (NCOM), and unspeciated fine PM (PM_{OTHR}). Primary unspeciated coarse particulate matter (PM) is named PMC (Table 12.1).

Plume Rise Approaches

Plume top and bottom define a vertical zone within which the smoke begins to transport away from the source (Raffuse et al. 2012). Plume rise height defines the terminal height to which the smoke plume rises. This parameter determines smoke transport distance and direction and can be associated with significant downwind enhancement of species such as ozone (Leung et al. 2007). This study compares plume rise algorithms from two models. The first one is the Western Regional Air Partnership's (WRAP) plume model, which parameterizes plume rise as a function of other physical quantities such as acreage and buoyancy. The other one is the Weather Research and Forecasting Chemistry's (WRF-Chem) online plume model, which employs a detailed physical approach.

Plume Rise Estimation Using the WRAP Model

The WRAP scheme (Tai et al. 2008; WRAP 2005) is an empirical parameterization that estimates the plume top and bottom as a function of the time of day and fire size. Fires are categorized into five classes (Table 12.2) based on burned virtual acreage (WRAP 2005).

$$\text{Acreage}_{\text{virtual}} = \text{Acreage}_{\text{actual}} \cdot \sqrt{\frac{\text{Fuel Loading}}{\text{Normalizer}}} \tag{12.1}$$

where $\text{Acreage}_{\text{actual}}$ is the actual fire size in acres and Normalizer equals 13.8 tons per acre.

Table 12.1 Mapping of MOZART-4 to CB05 species

CB05	MOZART-4
NO	NO
NO_2	NO_2
CO	CO
FORM	CH_2O
ALD2	CH_3CHO
ALDX	GLYALD
ETOH	C_2H_5OH
MEOH	CH_3OH
ETHA	C_2H_6
PAR	C_3H_6 + 1.7*BIGENE + 5.0*BIGALK + 1.5*C_3H_8 +3.0*CH_3COCH_3 + 4.0*MEK + 1.0*C_2H_2 + 3.0*HYAC
ETH	C_2H_4
OLE	C_3H_6 + BIGENE
ISOP	ISOP
TERP	$C_{10}H_{16}$
TOL	0.3*TOLUENE
XYL	0.1*TOLUENE
SO_2	SO_2
NH_3	NH_3
BENZENE	0.6*TOLUENE
NR	C_2H_2 + 1.5*C_3H_8 + 0.5*TOLUENE + 0.3*BIGENE
CH_4	CH_4
POC	OC
PEC	BC
PM_{OTHR}	PM_{25}–OC–BC
PMC	PM_{10}–PM_{25}

Table 12.2 Fire size classes

Class	1	2	3	4	5
Virtual acres (a)	0–10	10 < a < 100	100 < a < 1000	1000 < a < 5000	≥ 5000

Fuel loadings for North America were converted from Wiedinmyer et al. (2011). These values are 114 tons $acre^{-1}$ and 43 tons $acre^{-1}$ for tropical and extratropical forests respectively, while savanna and grassland both have a value of 4 tons $acre^{-1}$. Using these numbers provided a more accurate representation of fuel loading for the inventory. Hourly plume top and bottom are calculated as follows:

$$
\begin{aligned}
p_{\mathrm{top}_{\mathrm{hour}}} &= (\mathrm{BE}_{\mathrm{hour}})^2 \cdot (\mathrm{BE}_{\mathrm{size}})^2 \cdot p_{\mathrm{top}_{\max}} \\
p_{\mathrm{bot}_{\mathrm{hour}}} &= (\mathrm{BE}_{\mathrm{hour}})^2 \cdot (\mathrm{BE}_{\mathrm{size}})^2 \cdot p_{\mathrm{bot}_{\max}}
\end{aligned}
\tag{12.2}
$$

where $\mathrm{BE}_{\mathrm{hour}}$ and $\mathrm{BE}_{\mathrm{size}}$ are buoyancy efficiencies as a function of hour of day and fire size, respectively. The values and formulations follow in the tables and equations below, taken from WRAP (2005) (Tables 12.3 and 12.4).

Plume Rise Estimation Using the 1-D WRF-Chem Plume Rise Model

One major drawback of the WRAP scheme is that it releases all nighttime fire emissions into the first model layer, which may lead to overestimation of ground-level biomass burning emissions' impacts near the source. Hence, we evaluate the 1-D plume rise model developed by Freitas et al. (2007), which is an alternative approach with detailed physics treatment. The model explicitly treats pyro-convection by solving balance equations for momentum (vertical motion), energy (the first law of thermodynamics), and mass balance of water vapor (continuity) iteratively. The model obtains environmental input conditions from the host model and returns estimated plume heights. As described by Freitas et al. (2007), the water balance can be written as

$$
\frac{\partial y}{\partial t} + w\frac{\partial w}{\partial z} = \frac{1}{1+\gamma} gB - \frac{2\alpha}{R_p} w^2 + \frac{\partial}{\partial z}\left(K_m \frac{\partial w}{\partial z}\right) \tag{12.3}
$$

$$
\frac{\partial T}{\partial t} + w\frac{\partial T}{\partial z} = -w\frac{g}{c_p} - \frac{2\alpha}{R_p}|w|(T - T_e) + \frac{\partial}{\partial z}\left(K_T \frac{\partial T}{\partial z}\right) + \left(\frac{\partial T}{\partial z}\right)_{\text{microphysics}} \tag{12.4}
$$

Table 12.3 Buoyancy efficiencies as a function of hour of day

Hour	1	2	3	4	5	6	7	8	9	10	11	12
$\mathrm{BE}_{\mathrm{hour}}$	0.03	0.03	0.03	0.03	0.03	0.03	0.03	0.03	0.06	0.1	0.2	0.4
Hour	13	14	15	16	17	18	19	20	21	22	23	24
$\mathrm{BE}_{\mathrm{hour}}$	0.7	0.8	0.9	0.95	0.99	0.8	0.7	0.4	0.06	0.03	0.03	0.03

Table 12.4 Fire-related parameters as a function of fire size classes

Class	1	2	3	4	5
Virtual acres (a)	0–10	10 < a < 100	100 < a < 1000	1000 < a < 5000	≥ 5000
$\mathrm{BE}_{\mathrm{size}}$	0.4	0.6	0.75	0.85	0.90
$p_{\mathrm{top}_{\max}}$ (m)	160	2400	6400	7200	8000
$p_{\mathrm{bot}_{\max}}$ (m)	0	900	2200	3000	3000

$$\frac{\partial r_v}{\partial t} + w\frac{\partial r_v}{\partial z} = -\frac{2\alpha}{R_p}|w|(r_v - r_{v_e}) + \frac{\partial}{\partial z}\left(K_T\frac{\partial r_v}{\partial z}\right) + \left(\frac{\partial r_v}{\partial z}\right)_{\text{microphysics}} \tag{12.5}$$

$$\frac{\partial r_c}{\partial t} + w\frac{\partial r_c}{\partial z} = -\frac{2\alpha}{R_p}|w|r_c + \frac{\partial}{\partial z}\left(K_T\frac{\partial r_c}{\partial z}\right) + \left(\frac{\partial r_c}{\partial z}\right)_{\text{microphysics}} \tag{12.6}$$

$$\frac{\partial r_{\text{ice,rain}}}{\partial t} + w\frac{\partial r_{\text{ice,rain}}}{\partial t} = -\frac{2\alpha}{R_p}|w|r_{\text{ice,rain}} + \frac{\partial}{\partial z}\left(K_T\frac{\partial r_{\text{ice,rain}}}{\partial z}\right) + \left(\frac{\partial r_{\text{ice,rain}}}{\partial z}\right)_{\text{microphysics}} + \text{sedim}_{\text{ice,rain}} \tag{12.7}$$

where w, T, r_v, r_c, r_{rain}, and r_{ice} are the vertical velocity, air temperature, water vapor, and cloud, rain, and ice mixing ratios respectively, associated with in-cloud air parcels. In Eq. (12.3), γ is a term introduced to offset exclusion of non-hydrostatic pressure perturbations as explained by Simpson and Wiggert (1969), taken to be 0.5 in this study. The acceleration due to gravity is denoted by g, and B is the buoyancy term due to temperature difference between the in-cloud air parcel and its surrounding environment. It also includes the additional downward drag by condensate water (due to the higher density and hence specific weight of the condensed liquid phase). In Eq. (12.7), the term $r_{\text{ice,rain}}$ refers to the combined mixing ratios of ice and rain as resulting from entrainment, turbulent transport, and microphysical processes such as coagulation, growth on salt particles and auto-conversion processes as indicated by Simpson and Wiggert (1969). Freitas et al. (2007) assumed entrainment to be directly proportional to the vertical velocity in the cloud. The entrainment coefficient is calculated as $\frac{2\alpha}{R_p}$, where R_p is the plume radius and $\alpha = 0.1$. The lower boundary condition is based on a virtual source of buoyancy placed below the model surface (Turner 1973; Latham 1994). The buoyancy generated by this source is obtained from the convective energy flux E and the plume radius:

$$F = \frac{gR}{c_p p_e} ER_p^2 \tag{12.8}$$

where R is the ideal gas constant and p_e is the surface pressure. Once the buoyancy flux is determined, it provides the vertical velocity (w_0) and temperature excess ($T_0 - T_{e,0}$) for surface air parcels according to Morton et al. (1956) and Latham (1994). The index e refers to the surrounding environment, while c_p denotes specific heat at constant pressure. Eddy coefficients for diffusivity of momentum and heat are represented by K_m ψ and K_T, respectively. These coefficients are calculated using the approach of Smagorinsky (1963) and include corrections for the influence of the Brunt–Vaisala frequency (Hill 1974) and Richardson number (Lilly 1962).

$$w_0 = \frac{5}{6\alpha}\left(\frac{0.9\alpha F}{z_v}\right)^{1/3} \tag{12.9}$$

$$\frac{\Delta\rho_0}{\Delta_{\rho e,0}} = \frac{5}{6\alpha}\frac{F}{g}\frac{z_v^{-5/3}}{(0.9\alpha F)^{1/3}} \tag{12.10}$$

$$T_0 = \frac{T_{e,0}}{1 - \frac{\Delta\rho_0}{\Delta_{\rho e,0}}} \tag{12.11}$$

Figure 12.2 plots the modeled plume rise height from both approaches. Results indicate that the WRAP approach is unable to capture the diurnal variation of the plume rise height, showing no residual height during nighttime. It is understood that physically, fire emissions rise a finite, nonzero height during the night. The 1-D WRF-Chem plume rise approach captures this physical trend quite well.

Vertical Allocation of Fire Emissions

The vertical distribution of FINN emissions is calculated on both pressure and smoldering smoke effect. The computed vertical layer fraction (LFRAC) for one model grid cell is shown in Fig. 12.3. The fraction of emissions on the surface (Lay1F) is small (0.01) without considering smoke smoldering. Pouliot et al. (2005) estimated the smoldering effect as follows:

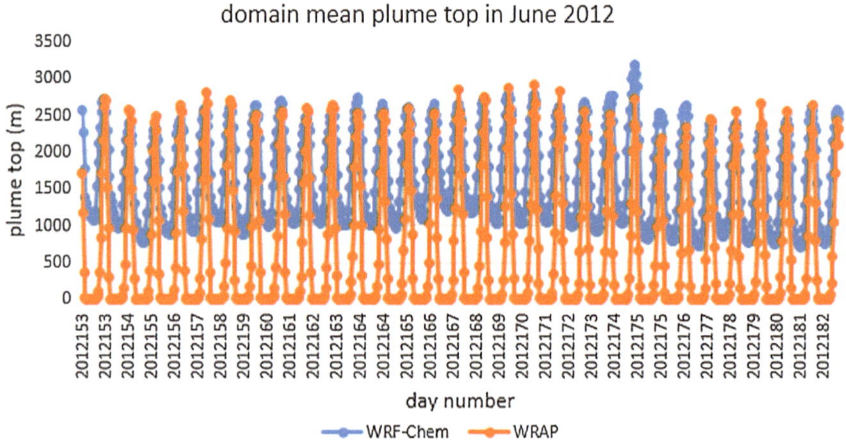

Fig. 12.2 Diurnal variation of domain-average plume rise height using the WRAP and 1-D WRF-Chem plume rise approaches for June 2012

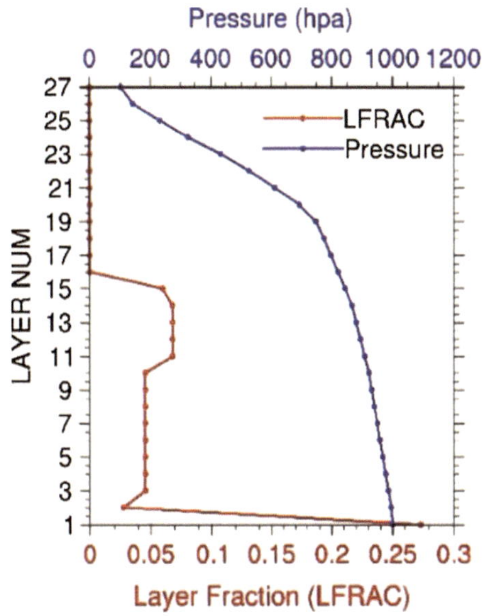

Fig. 12.3 An example of vertical layer fraction (LFRAC) for one model grid cell with nonzero FINN emissions

$$BE_{size} = 0.0703 \times \ln(acres) + 0.3 \tag{12.12}$$

$$S_{fract} = 1 - BE_{size} \tag{12.13}$$

where BE_{size} is buoyancy efficiency as a function of fire size, acres is fire size in the unit of acres, and S_{fract} is the smoldering fraction. In the above example, the new Lay1F is 0.27, which is consistent with the values from Tai et al. (2008).

12.2.3.4 Dynamic Chemical Boundary Conditions

One of the drawbacks of the standard CMAQ model is that it uses temporally static boundary conditions, implying modeled species concentrations over boundary grid cells do not exhibit any diurnal variation. This could potentially bias model-measurement comparisons, especially at grid cells near the lateral and upper boundary. In order to mitigate these potential errors, we used input boundary conditions simulated by a global chemical transport model, GEOS-Chem (Bey et al. 2001), to generate 3-D gridded species concentrations over a lateral grid, where the $2^{o} \times 2.5^{o}$ GEOS-Chem spatial resolution is gridded to the new CMAQ chemical boundary conditions over the continental USA. Additionally, GEOS-Chem has 47 vertical layers extending from the surface to about 80 km, generating vertical boundary conditions over a larger height. This could help simulate the long-range transport of ozone from wildfires and capture their stratospheric impact on surface ozone.

12.3 Model Evaluation

In this section, we evaluate the performance of the CMAQ model and the impact of biomass burning over different regions of the USA. We divided our simulation into monthly and seasonal periods (spring: April–May; summer: June–August; and fall: September–October) for each year. The ozone season does not include winter months. The following maps report model simulated monthly mean surface concentrations of monthly average ozone. Figures 12.7, 12.8 and 12.9 represent ozone time series, while Fig. 12.5 represents $PM_{2.5}$ time series in Texas (Figs. 12.4, 12.5, 12.6, 12.7, 12.8, 12.9, 12.10, 12.21, 12.22, 12.23, 12.24, 12.25, 12.26) and $PM_{2.5}$ (Figs. 12.14, 12.15, 12.16, 12.17, 12.18, 12.19, 12.20) along with the portion of those simulated concentrations that are due to biomass burning emissions. For each spatial plot, the base case represents simulations with biomass emissions included, while the impacts represent the difference between this case and that with biomass emissions turned off.

12.3.1 Ozone Model-Measurement Comparisons and Biomass Burning Impacts in 2012

A comparison for spring 2012 is plotted in Fig. 12.4, where the panels (a) and (b) represent April 2012. Broadly, the model does a good job of capturing ozone

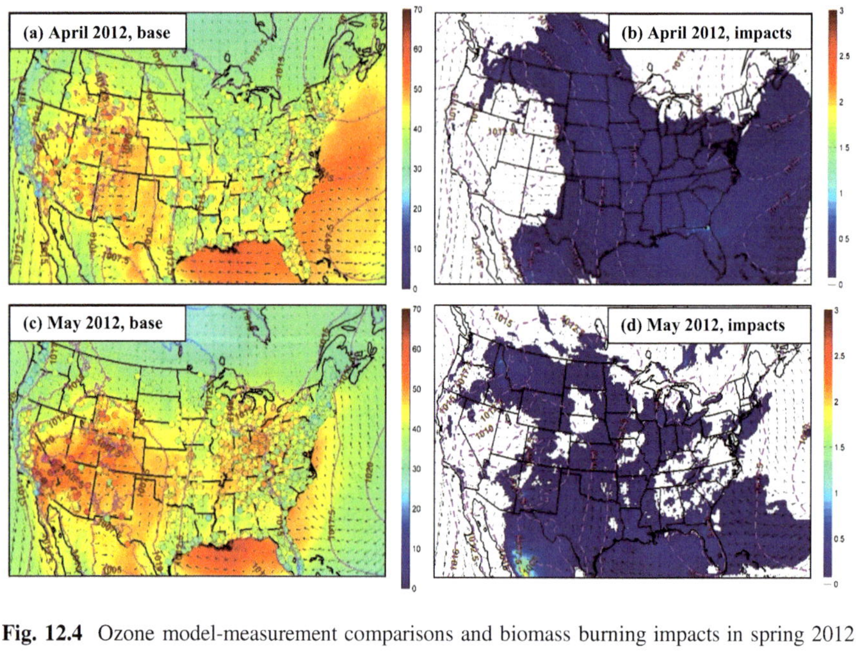

Fig. 12.4 Ozone model-measurement comparisons and biomass burning impacts in spring 2012

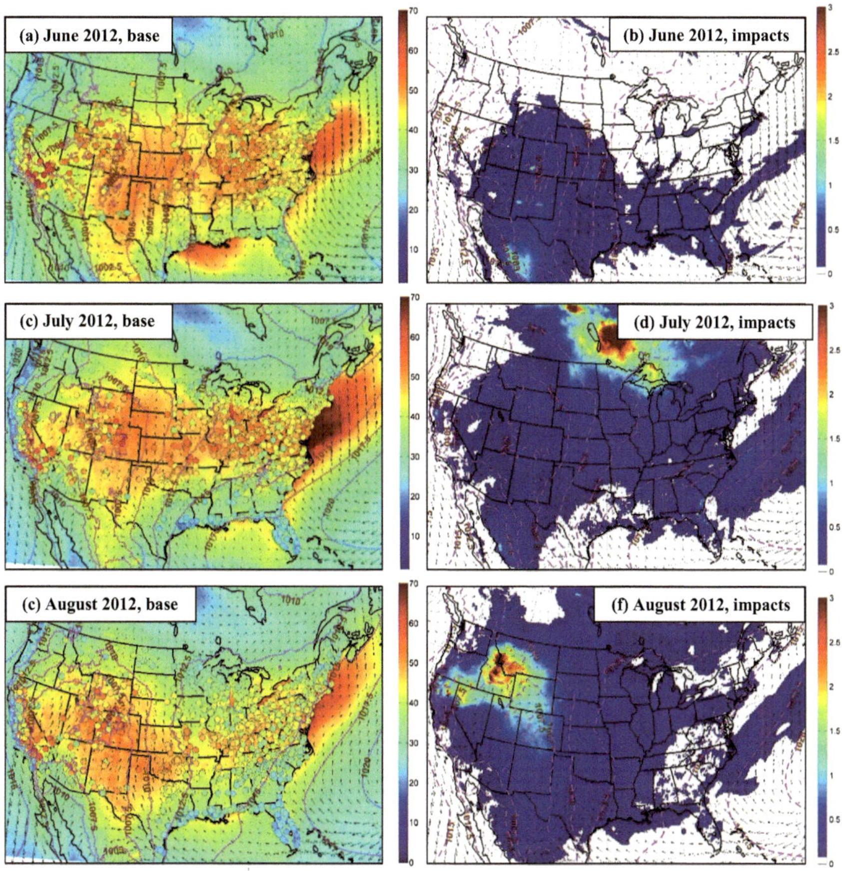

Fig. 12.5 Ozone model-measurement comparisons and biomass burning impacts in summer 2012

hot spots over the Western Mountains region. Panel 4(b) predicts a monthly mean biomass impact of 0.5–1 ppb across the eastern USA. The model-measurement comparison for ozone in May 2012 is plotted in panel 4 (c). In situ data show several ozone hot spots exceeding 70 ppb over the Western Mountains region and southern California. Additionally, there are 60 ppb hot spots over Ohio and Pennsylvania in the northeast. The model does a good job of capturing these hot spots. The biomass impact plot in 4(d) indicates a difference of ~0.5–1 ppb in monthly mean concentrations predicted all across the USA except for the West Coast and a disparity of 1–2 ppb across northern Mexico.

Figure 12.5 plots the ozone results for summer, with panel 5(a) representing June 2012. There are several 60–70 ppb hot spots in the Western Mountains, Midwest, East North Central, and Northeastern regions, which the model captures quite well. However, it is unable to capture hot spots higher than 70 ppb in Central California. A broadly similar story is told by the ozone comparison for July 2012 in

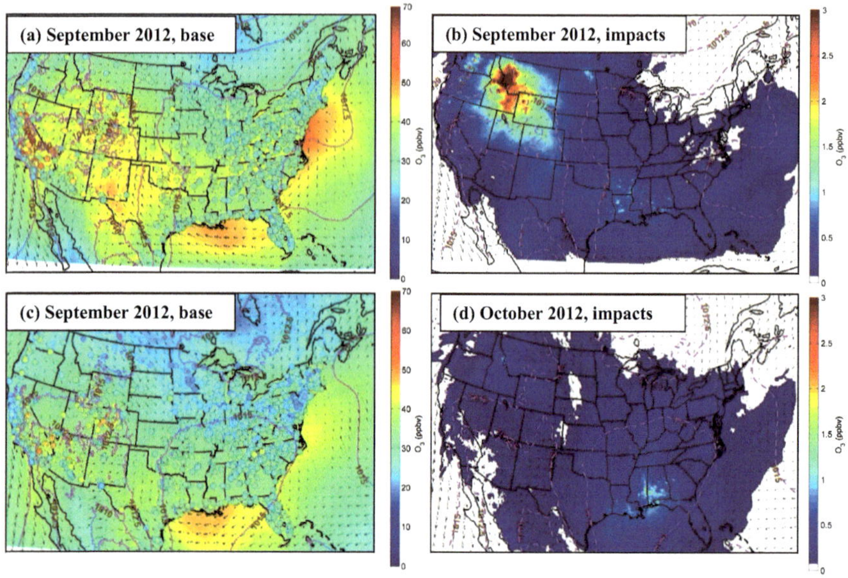

Fig. 12.6 Ozone model-measurement comparison and biomass burning impacts in fall 2012

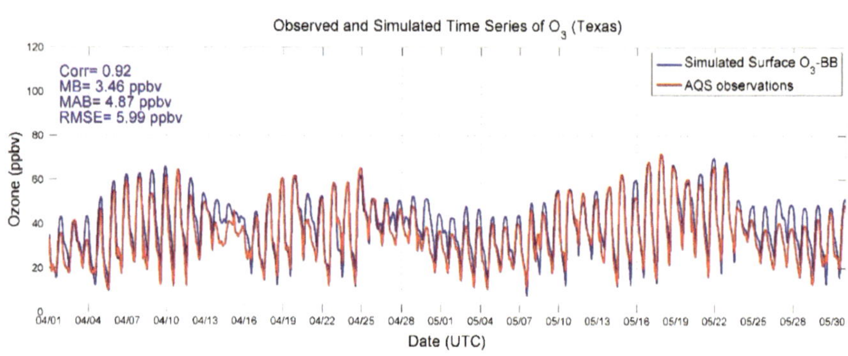

Fig. 12.7 Ozone time series for model-measurement comparison for Texas in spring 2012

panel 5(c). The biomass burning impact plots indicate a monthly mean change greater than 3 ppb in southern Canada. The map for August 2012 in panel 5(e) reports hot spots in the Western Mountains and California, reproduced quite well by the model, with panel 5(f) indicating monthly mean forest fire impacts of 1–3 ppb in Idaho, California, Nevada, Montana, and Wyoming.

In the comparison for fall 2012 in Fig. 12.6, it is noticed that measured ozone concentrations in September were low on average (30–40 ppb) across the

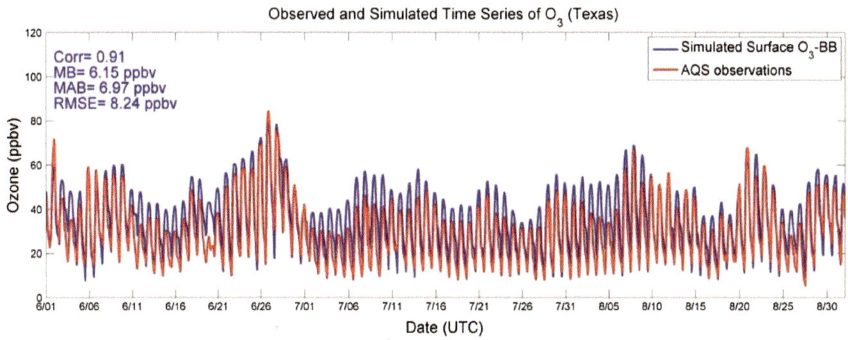

Fig. 12.8 Ozone time series for model-measurement comparison for Texas in summer 2012

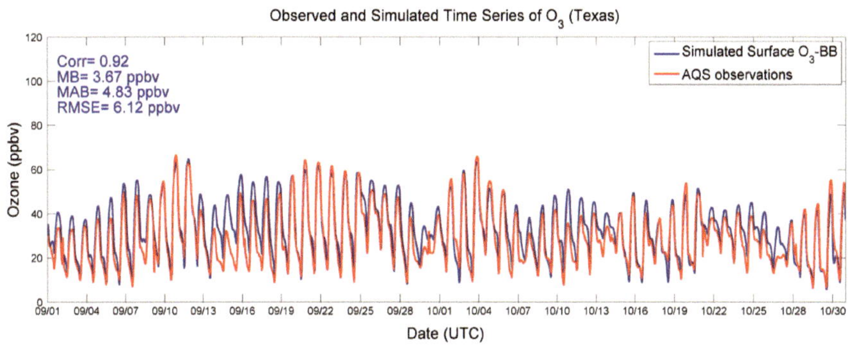

Fig. 12.9 Ozone time series for model-measurement comparison for Texas in fall 2012

continental USA, except for California (panel (a)). Surprisingly, the model overpredicts most of the low ozone across the country but underpredicts the high ozone in California. Panel (b) indicates an impact of 1–3 ppb over Idaho, Montana, and Wyoming. The October 2012 comparison in panel (c) indicates low ozone concentrations again across the country except for a few places in California and the Western Mountains regions. The model captures both spatial trends quite well and predicts a 1 ppb monthly mean impact from biomass burning in Mississippi, Alabama, and Louisiana.

Time series comparisons for Texas are plotted in Figs. 12.7, 12.8 and 12.9. The plots indicate that the comparisons for the state of Texas are very similar in spring and fall, with a bit more overprediction in summer. Overall, the model-measurement comparison demonstrates excellent correlation across all seasons, indicating that the model is able to capture the spatiotemporal trends of the in situ data. Similar behavior was observed for the ozone seasons in 2013 and 2014.

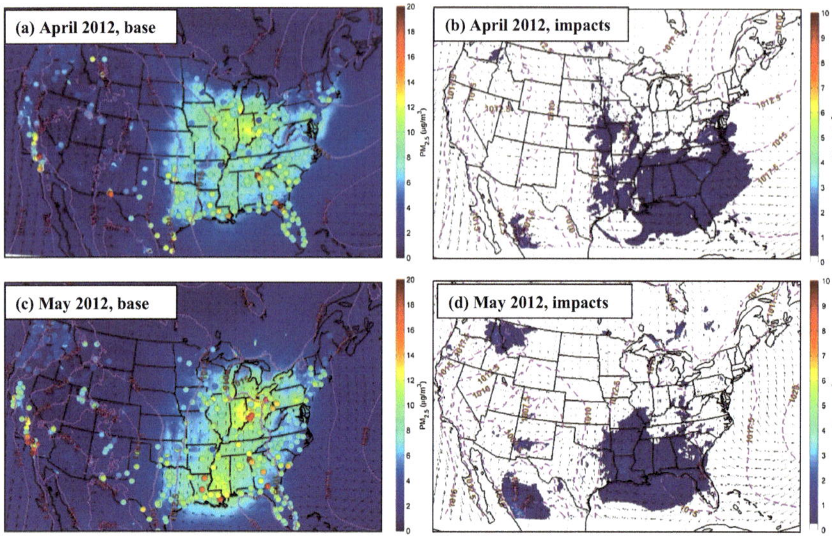

Fig. 12.10 $PM_{2.5}$ model-measurement comparison and biomass burning impacts in spring 2012

12.3.2 $PM_{2.5}$ Model-Measurement Comparisons and Biomass Burning Impacts in 2012

The $PM_{2.5}$ concentrations plotted in Figs. 12.10, 12.11 and 12.12 represent monthly means. As opposed to ozone, the spatial trends in $PM_{2.5}$ concentrations were reasonably consistent over the 2012 ozone season. For example, there were high in situ concentrations over the eastern USA and the west coast, the latter probably due to intense forest fires. The model broadly underpredicted over these regions but captured the spatial differences over the eastern USA.

The impact of biomass burning in the spring months of April and May 2012 was negligible, as shown in Fig. 12.10. Summertime biomass burning emissions impacted monthly mean $PM_{2.5}$ concentrations by 1 $\mu g\ m^{-3}$ across a significant portion of the southern USA in June, while for July, simulated $PM_{2.5}$ concentrations were higher by ~1 $\mu g\ m^{-3}$ across the eastern half of the USA due to biomass burning, with larger impacts of 5–10 $\mu g\ m^{-3}$ in southeastern Canada. Estimated impacts for August (Fig. 12.11) and September (Fig. 12.12) indicate a 1 $\mu g\ m^{-3}$ enhancement in monthly mean surface $PM_{2.5}$ across the western USA, with larger impacts up to 10 $\mu g\ m^{-3}$ predicted at the border of Idaho and Wyoming. September impacts in Fig. 12.12 also indicate 5–10 $\mu g\ m^{-3}$ hot spots in Montana and Idaho, while October predictions show monthly mean impacts from biomass burning of 1 $\mu g\ m^{-3}$ across most of the eastern USA and in Idaho and Wyoming, with larger impacts of ~4 $\mu g\ m^{-3}$ in Louisiana and Mississippi.

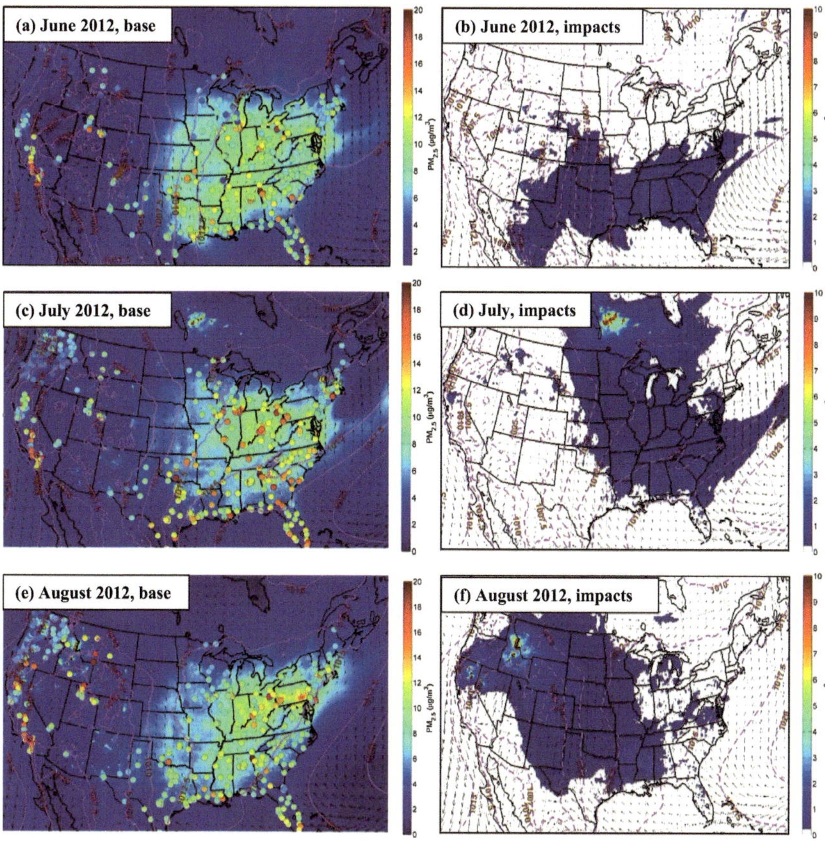

Fig. 12.11 $PM_{2.5}$ model-measurement comparison and biomass burning impacts in summer 2012

12.3.3 Ozone Model-Measurement Comparisons and Biomass Burning Impacts for 2013

Ozone season monthly model-measurement comparisons and impacts of biomass burning in 2013 are plotted in Figs. 12.13, 12.14 and 12.15. In situ data indicate 60–70 ppb hot spots over the Western Mountains region and California for almost the entire season, underpredicted by the model. In addition, there are several regions of the eastern USA with predicted monthly mean ozone concentrations ranging from 30 to 40 ppb. These concentrations stay at the upper limit ($\sim$40 ppb) during April but steadily decline to the lower ($\sim$30 ppb) as we move toward October. These are typically overpredicted by the model, although it captures the spatial trends nicely. Additionally, mean in situ ozone concentrations in October were the lowest in the season across the continental USA. Simulated biomass burning impacts on monthly mean ozone include the following:

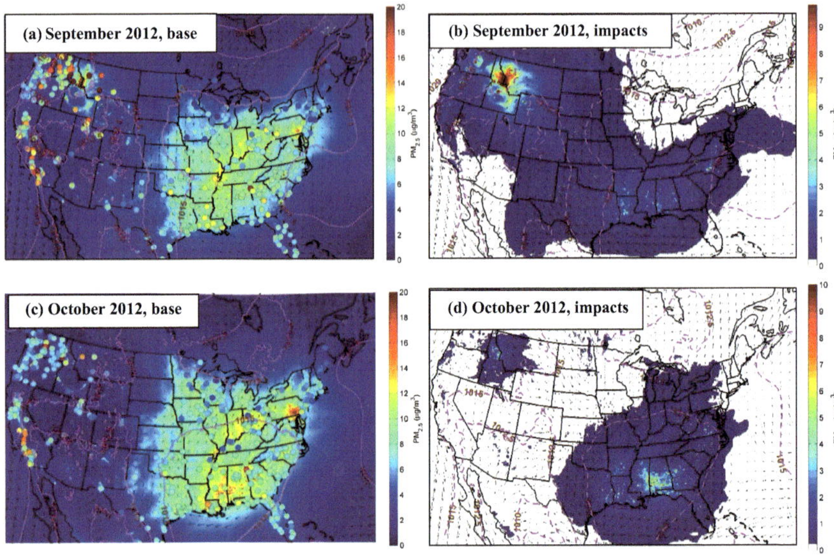

Fig. 12.12 $PM_{2.5}$ model-measurement comparison and biomass burning impacts in fall 2012

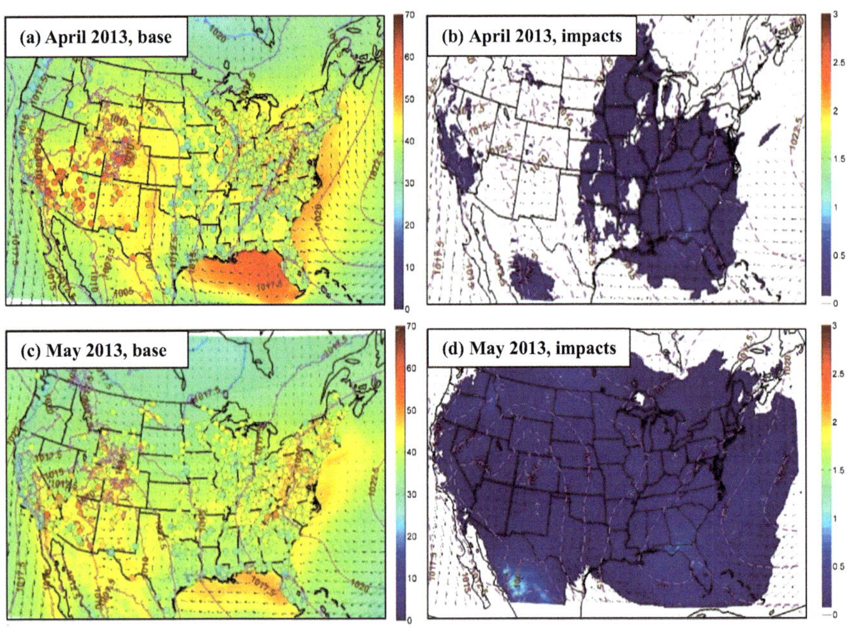

Fig. 12.13 Ozone model-measurement comparison and biomass burning impacts in spring 2013

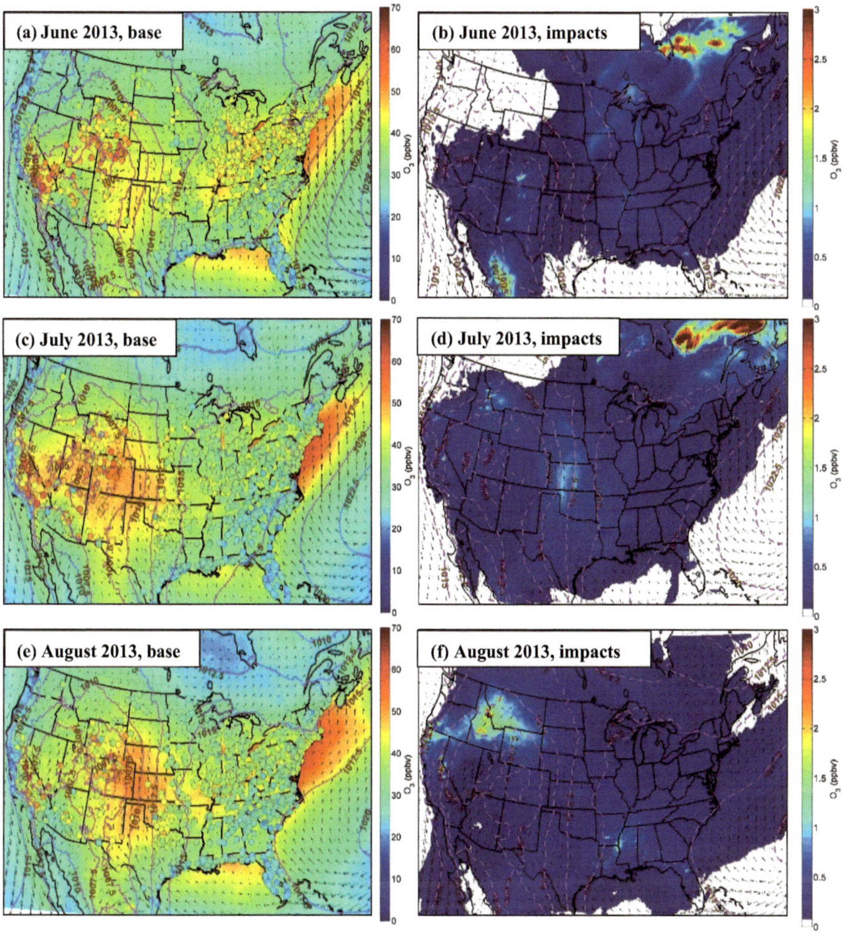

Fig. 12.14 Ozone model-measurement comparison and biomass burning impacts in summer 2013

(a) Spring 2013, Fig. 12.13: An impact of ~1 ppb at the border of Georgia and Florida in April; 1 ppb at the border of Georgia and Florida, and northern Mexico in May.
(b) Summer 2013, Fig. 12.14: 1–3 ppb in northern Mexico and southeastern Canada, and 1 ppb in Florida, New Mexico, and Colorado in June; 1–3 ppb impacts in southeastern Canada, 1 ppb in Oklahoma and Kansas in July; additions of ~2 ppb in Idaho and northern California, with 1 ppb in Louisiana, Arkansas, and Georgia in August.
(c) Fall 2013, Fig. 12.15: ~1 ppb in Alabama, Arkansas, Louisiana, and eastern Texas; and 2–3 ppb in northeastern California in September (panel (b)) and a 1 ppb impact over Alabama, Georgia, Idaho, Wyoming, and southern Canada in October (panel (d)).

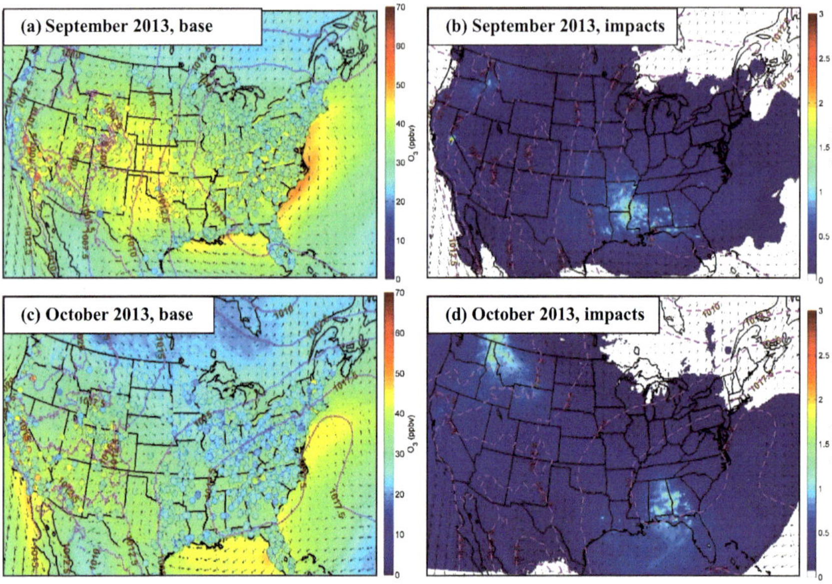

Fig. 12.15 Ozone model-measurement comparison and biomass burning impacts in fall 2013

12.3.4 $PM_{2.5}$ Model-Measurement Comparisons and Biomass Burning Impacts for 2013

Monthly model-measurement comparisons and impacts of biomass burning emissions in 2013 are plotted in Figs. 12.16, 12.17 and 12.18. Broadly, the in situ data show monthly mean concentrations of 8–10 μg m^{-3} over the eastern USA and Texas. While the model does a reasonably good job at capturing the trends over the eastern USA, it underpredicts in Texas. Additionally, it underpredicts at several places over the West Coast as we move toward the fall months. The biomass fires demonstrate varying spatial impacts.

(a) Spring 2013, Fig. 12.16: The fires contributed ~1 μg m^{-3} in Georgia, Alabama, eastern Texas, and southern Oklahoma in April. During May, their impacts were 1 μg m^{-3} in Georgia, South Carolina, Alabama, Louisiana, and most of northern and Central Florida plus a large portion of northern Mexico, western Idaho, and eastern Washington.
(b) Summer 2013, Fig. 12.17: Biomass fires elevated $PM_{2.5}$ monthly mean concentrations in southeastern Canada by 5–10 μg m^{-3} during June/July. In August, impacts of around 1 μg m^{-3} are predicted over Georgia, Alabama, and Louisiana, scattered locations in the Midwest, as well as Idaho, Wyoming, and California.
(c) Fall 2013, Fig. 12.18: Impacts of 1 μg m^{-3} over a significant part of the eastern USA, along with California, Idaho, Wyoming, and southwestern Canada during

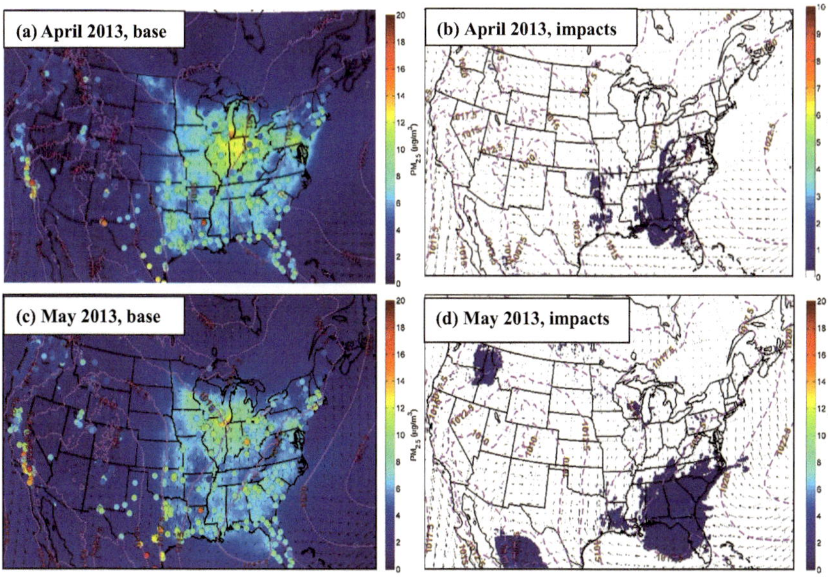

Fig. 12.16 $PM_{2.5}$ model-measurement comparison and biomass burning impacts in spring 2013

September 2013. Changes of 1 μg m^{-3} are predicted over most of the eastern USA and the Midwest, Idaho, Wyoming, Washington, California, and Oregon, with larger enhancements of 5–10 μg m^{-3} in eastern Texas, Louisiana, Alabama, Georgia, Idaho, and southwestern Canada.

12.3.5 Ozone Model-Measurement Comparisons and Biomass Burning Impacts for 2014

Contour maps highlighting the comparisons and impacts on monthly mean ozone due to biomass burning for 2014 are plotted in Figs. 12.19, 12.20 and 12.21. Broadly, several 60–70 ppb hot spots are observed over the Western Mountains region and California, the latter being consistently underpredicted by the model for all months. Following are the key points for the ozone modeling in 2014:

(a) Spring 2014, Fig. 12.19: In April, these hot spots are underpredicted by the model. In situ monthly mean concentrations over the eastern USA range between 35 and 40 ppb and are captured quite well by the model. The ozone impacts plot in the bottom panel shows that biomass events contributed 1–2 ppb enhancements to monthly mean ozone in Georgia, South Carolina, Florida, and Kansas. Plots for May 2014 indicate that the model can reasonably capture the Western Mountains trends. In situ monthly mean concentrations of

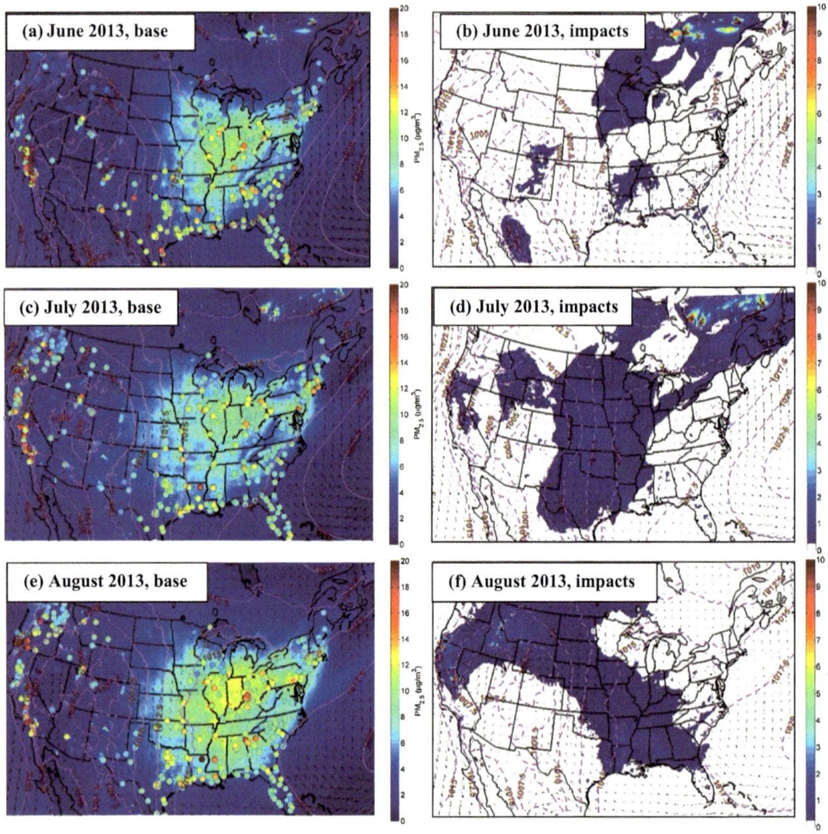

Fig. 12.17 $PM_{2.5}$ model-measurement comparison and biomass burning impacts for summer 2013

ozone over the East Coast range between 40 and 50 ppb; the model underpredicts these by 5–10 ppb. There were 1 ppb fire impacts in Georgia, Florida, Alabama, North and South Carolina, northern Mexico, and southwestern Canada.

(b) Summer 2014, Fig. 12.20: For June 2014, the model is able to reasonably capture the Western Mountains hot spots. Eastern USA in situ monthly mean ozone concentrations range from 30 to 40 ppb and are overpredicted quite substantially, by 10–20 ppb. Forest fires show negligible impacts for this month. In July 2014, low in situ monthly mean ozone concentrations were observed everywhere except California. The model reproduced the low concentrations reasonably well but underpredicted in California. The impacts of biomass burning on monthly mean ozone concentrations were 1–2 ppb in southeastern Oregon, northwestern Nevada, and southern Idaho. The evaluation impacts for August 2014 show the model reasonably reproduced the Western

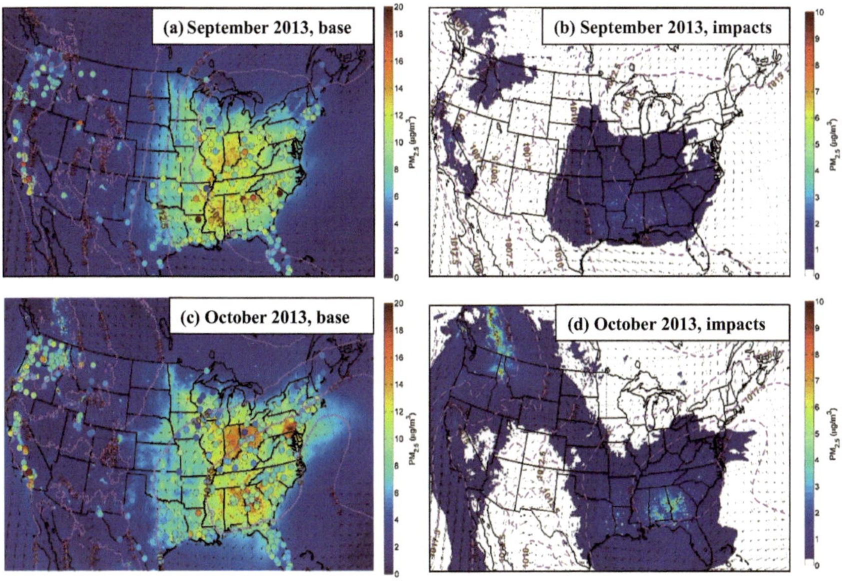

Fig. 12.18 $PM_{2.5}$ model-measurement comparison and biomass burning impacts in fall 2013

Mountains hot spots. Monthly mean ozone concentrations over the eastern USA ranged from 30 to 35 ppb, which the model significantly overpredicted by 10–20 ppb. The impacts plot in the bottom panel reveals 1–3 ppb additions occurring in western California, central Oregon, Idaho, southwestern Canada, and southern Georgia.

(c) Fall 2014, Fig. 12.21: In September 2014, low in situ monthly mean ozone concentrations were observed everywhere except for California. The model overpredicted most of the low in situ values. The impacts plot indicates negligible changes. In situ data for October 2014 showed low monthly mean ozone concentrations all across the USA. While these were overpredicted by the model over the western half, the model was able to capture the trends over the eastern USA reasonably well. A monthly mean impact of 1 ppb was predicted over Georgia, Alabama, and Mississippi.

12.3.6 $PM_{2.5}$ Model-Measurement Comparisons and Biomass Burning Impacts for 2014

The monthly mean $PM_{2.5}$ model-measurement comparisons for 2014 are plotted in Figs. 12.22, 12.23 and 12.24. Elevated concentrations of ~10–20 μg m^{-3} are observed in the in situ measurements over the eastern USA, Texas, and some

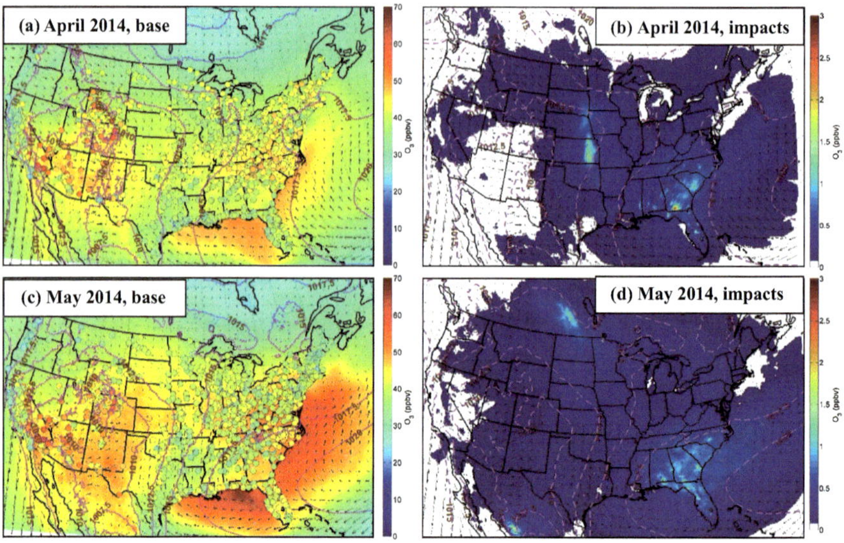

Fig. 12.19 Ozone model-measurement comparison and biomass burning impacts in spring 2014

stations in California. Higher in situ concentrations were reported over these sites during summertime, possibly attributed to biomass burning activities. Model underprediction is seen over a majority of these sites (especially the West Coast), although the model captured the spatial impacts over the eastern USA quite well. The salient features of the biomass burning impacts are summarized below.

(a) Spring 2014, Fig. 12.22 indicates the impacts of biomass burning in April to be $\sim 1\ \mu g\ m^{-3}$ on monthly mean $PM_{2.5}$ concentrations across the eastern USA, with higher impacts of 5–6 $\mu g\ m^{-3}$ predicted over the Midwest. For May 2014, the impacts plot predicts a 1 $\mu g\ m^{-3}$ enhancement in monthly mean $PM_{2.5}$ over the southeastern USA, certain parts of the Midwest, southern Canada, and northern Mexico.

(b) Summer 2014, Fig. 12.23: In June 2014, the impacts plot shows a 1 $\mu g\ m^{-3}$ enhancement of monthly mean $PM_{2.5}$ due to biomass burning was predicted in some parts of Georgia, South Carolina, Florida, and Mexico. For July 2014, the impacts of 1 $\mu g\ m^{-3}$ due to biomass burning were predicted for monthly mean $PM_{2.5}$ concentrations in Louisiana, northern Nevada, southern Idaho, northern Washington, and southwestern Canada. In August 2014, Fig. 12.23, panel (f) depicts a 1 $\mu g\ m^{-3}$ contribution of biomass burning emissions to monthly mean $PM_{2.5}$ concentrations over the southern states of Georgia, North and South Carolina, Alabama, Mississippi, Louisiana, northern Texas, Oklahoma, Missouri, Arkansas, and Tennessee; and also over northern California, Washington, Oregon, Idaho, Wyoming, and Nevada.

(c) Fall 2014, Fig. 12.24: A monthly mean $PM_{2.5}$ impact of 1 $\mu g\ m^{-3}$ due to biomass burning is predicted over almost the entire continental USA.

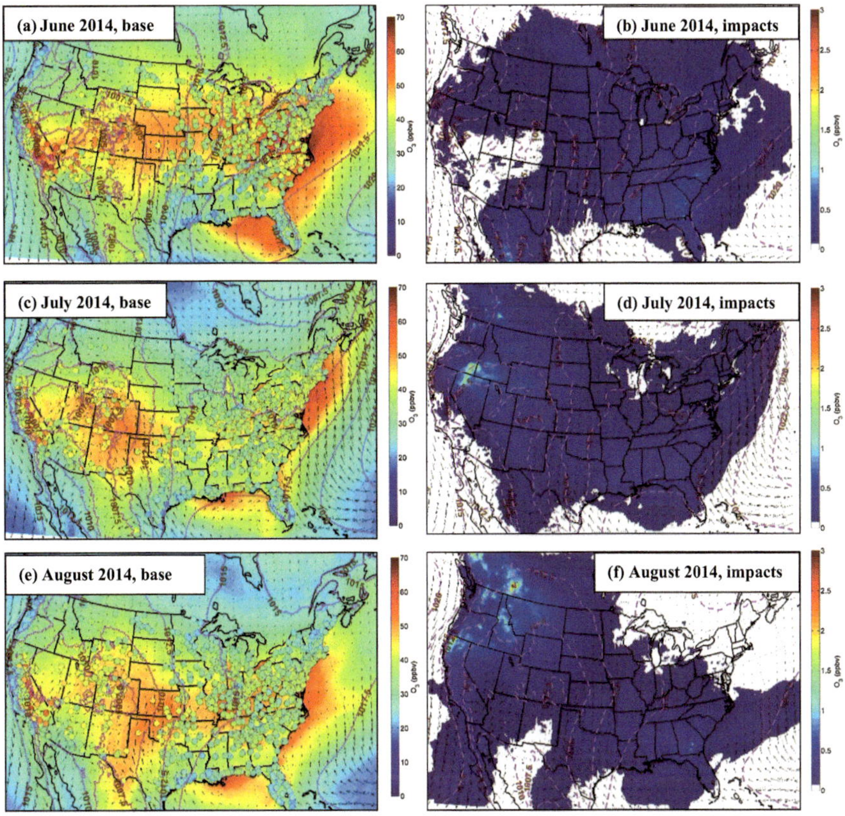

Fig. 12.20 Ozone model-measurement comparison and biomass burning impacts in summer 2014

(d) The time series for the fall season of 2014 for the state of Texas is plotted in Fig. 12.25. The correlation and mean bias are reasonable ($R = 0.65$ and $MB = 0.86\ \mu g\ m^{-3}$) compared to other years and seasons.

12.4 Summary and Conclusions

The NCAR's FINN biomass burning inventory (Wiedinmyer et al. 2011) was used in conjunction with the CMAQ chemical transport model and WRF meteorological model to assess the impact of biomass burning emissions on ozone and $PM_{2.5}$ concentrations over the continental USA, during the ozone seasons of 2012, 2013, and 2014. Anthropogenic and biogenic emissions were provided by the SMOKE and BEIS models, respectively. Observational data were taken from the TCEQ's

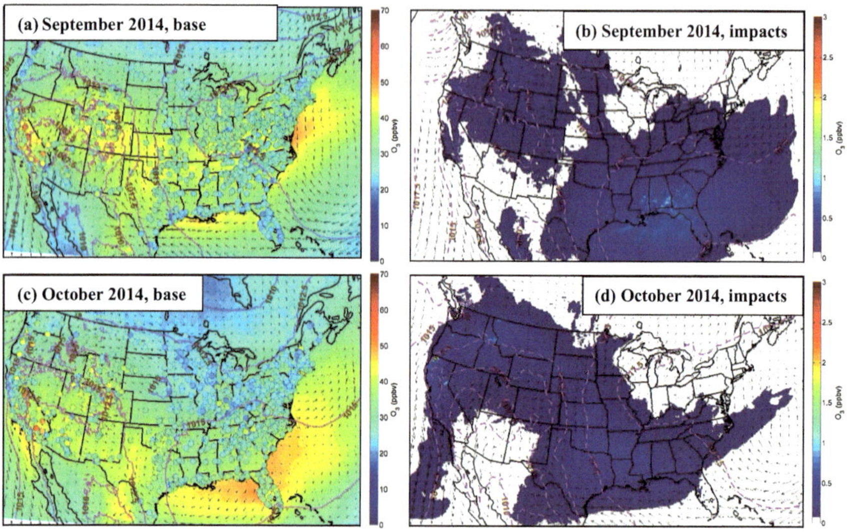

Fig. 12.21 Ozone model-measurement comparison and biomass burning impacts in fall 2014

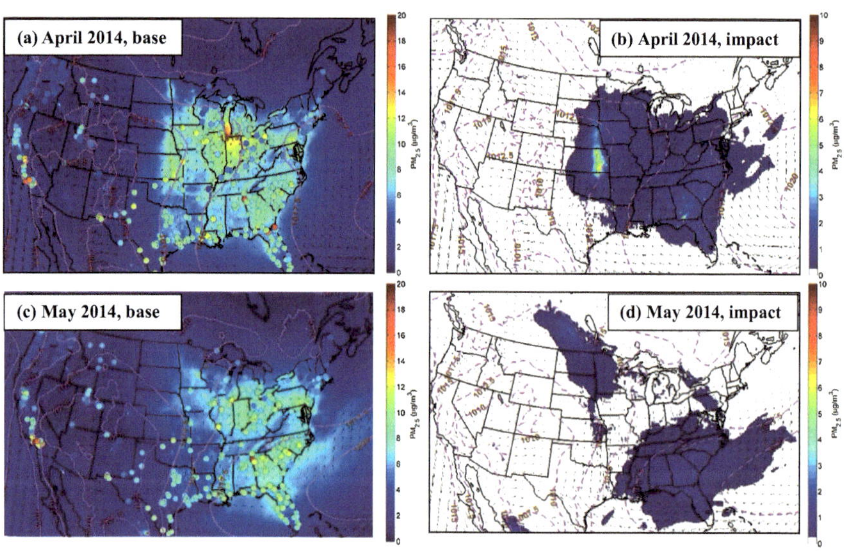

Fig. 12.22 $PM_{2.5}$ model-measurement comparison and biomass burning impacts in spring 2014

CAMS and USEPA's AQS networks. This study has marked improvements over previous biomass burning evaluations:

(a) A previous study (McDonald-Buller et al. 2015) which built the FINN inventory produced model runs for a 45-day episode (May 16–June 30, 2012).

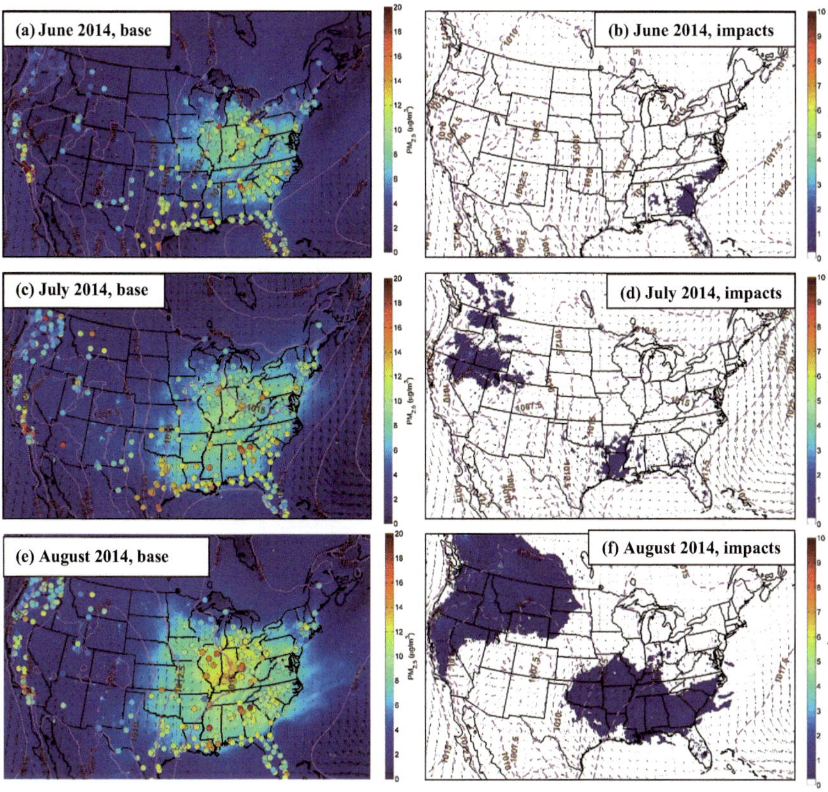

Fig. 12.23 $PM_{2.5}$ model-measurement comparison and biomass burning impacts in summer 2014

This study extended it by doing an evaluation for the ozone seasons of three years.

(b) The regional CTMs such as CMAQ model hitherto used static chemical boundary conditions with no diurnal variation. We generated more physically relevant dynamic boundary conditions using the global GEOS-Chem model.

(c) Previous studies (Li et al. 2015; Jeon et al. 2015) indicated that improving modeled meteorological fields using grid nudging and objective analysis (OA) could improve air quality modeling for ozone and $PM_{2.5}$. Hence, we used grid nudging for our current run.

(d) Previous biomass burning studies using CMAQ modeling used a parameterized version of plume rise height developed by WRAP. This approach did not exhibit physically relevant trends, predicting no residual nocturnal height. Hence, we used the WRF-Chem plume rise module which tracked nighttime plume rise height better.

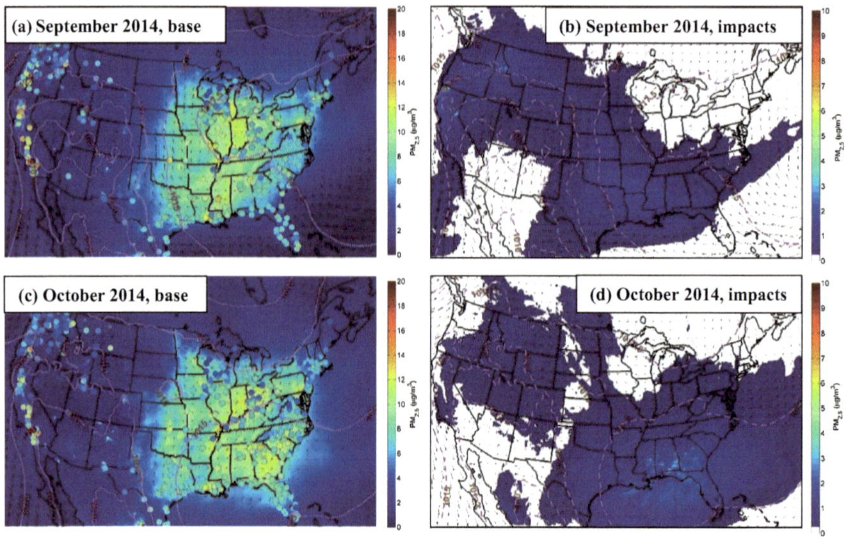

Fig. 12.24 $PM_{2.5}$ model-measurement comparison and biomass burning impacts in fall 2014

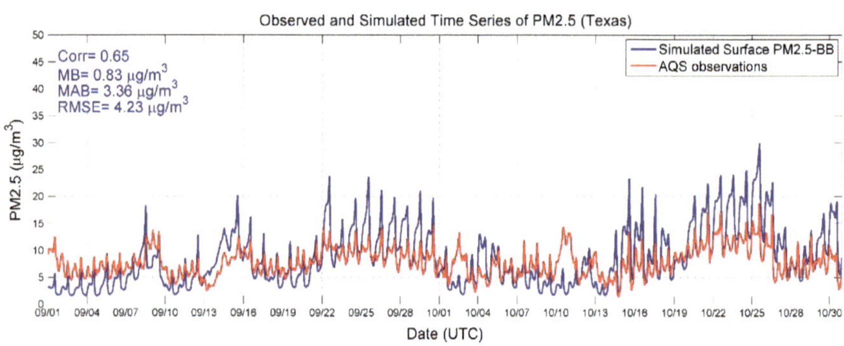

Fig. 12.25 $PM_{2.5}$ time series for model-measurement comparison during fall 2014 for the state of Texas

12.4.1 Model-Measurement Comparisons and Biomass Burning Impacts of Ozone

During 2012, broadly, in situ data showed monthly mean daylight (6:00–18:00 CST) ozone hot spots of 60–70 ppb across the Western Mountains region and California, which the model did a good job reproducing. In situ data for May 2012 showed that in addition to the Western Mountains region and California, there were a couple of hot spots over Ohio and Pennsylvania in the northeast with monthly mean (daylight) ozone concentrations around 60 ppb. The model did a good job of capturing these hot spots. Biomass burning emissions contributed 1–2 ppb to monthly mean ozone predictions across northern Mexico for this month.

Observations for June and July 2012 indicated the model captured the 60–70 ppb hot spots in the Western Mountains, Midwest, East North Central, and Northeastern regions, but was unable to capture hot spots higher than 70 ppb in Central California. Biomass burning during this period added ~1 ppb to monthly mean ozone in northern Mexico in June, and 1–3 ppb in southern Canada. August forest fires in Idaho and California added 1–3 ppb to monthly mean ozone in these areas, with a 1 ppb impact at the state boundary. In situ ozone concentrations in September on average were low (30–40 ppb) across the continental USA, except for California. Surprisingly, the model overpredicted most of the low ozone across the country but underpredicted high ozone in California. The reasons behind these discrepancies are unknown and being investigated.

Similar to 2012, in situ data for 2013 showed hot spots over the Western Mountains region and California. The model was able to capture the Western Mountains hot spots for July and August but consistently underpredicted in California. In situ data over the eastern USA showed quite low monthly mean concentrations, 30–40 ppb. The model significantly underpredicted these in May and overpredicted in September, capturing these trends reasonably well in other months. Biomass burning emissions contributed 1 ppb to monthly mean ozone at the border of Georgia and Florida in April and May. Additionally, a 1 ppb enhancement was predicted over northern Mexico in May.

Wildfire impacts in June were estimated to account for 1–3 ppb of monthly mean ozone in northern Mexico and southeastern Canada, along with 1 ppb in Florida, New Mexico, and Colorado. Biomass impacts in August indicated additions of ~2 ppb to monthly mean ozone in Idaho and northern California, with 1 ppb impacts predicted for Louisiana, Arkansas, and Georgia. Impacts of ~1 ppb in Alabama, Arkansas, Louisiana, and eastern Texas and 2–3 ppb impact in northeastern California were predicted for September of 2013. In October, a 1 ppb change was predicted over Alabama, Georgia, Idaho, Wyoming, and southern Canada. Time series comparisons for Texas show a good correlation in spring (0.81), but excellent correlation in summer and fall (0.93 and 0.92, respectively). However, the model did tend to overpredict more in 2013 than in 2012. Overprediction could potentially be due to combined problems in emissions and meteorological fields, as indicated by Pan et al. (2016).

Similar to 2013, in situ data for 2014 again showed ozone hot spots over the Western Mountains region and California. While the model was able to capture the Western Mountains hot spots, it once again underpredicted in California. Additionally, the eastern USA showed low monthly mean concentrations of ozone overall during the ozone season, 30–40 ppb. The model broadly overpredicted these concentrations but reproduced the spatial impacts.

For the month of April 2014, the model predicted 1–2 ppb impacts to monthly mean ozone in Georgia, South Carolina, Florida, and Kansas. Impacts of 1 ppb in Georgia, Florida, Alabama, North and South Carolina, northern Mexico, and southwestern Canada were predicted for May, with minimal impacts for June 2014.

The model predicted a 1–2 ppb impact on monthly mean ozone in southeastern Oregon, northwestern Nevada, and southern Idaho during July 2014, while in August, 1–3 ppb enhancements were predicted in western California, central Oregon, Idaho, southwestern Canada, and southern Georgia. Minimal impacts were predicted in September, but 1 ppb additions over Georgia, Alabama, and Mississippi were modeled for October.

12.4.2 Model-Measurement Comparisons and Biomass Burning Impacts of $PM_{2.5}$

Broadly in 2012, the model was unable to capture the high $PM_{2.5}$ concentrations across the continental USA. For example, in situ data for April 2012 indicate concentrations of 10–12 μg m^{-3} all across the eastern USA, including southeastern Texas. Additionally, there are a couple of places in Georgia, New Mexico, and California where the in situ $PM_{2.5}$ monthly mean concentration was greater than 20 μg m^{-3}. The model underpredicted in most of these places but was broadly able to capture the spatial variations across the eastern USA. However, it was unable to capture the extreme events resulting in $PM_{2.5}$ concentrations greater than 20 μg m^{-3}. It will be fruitful to more fully understand what is causing model-measurement error and poor correlation therein. Hence, as part of an ongoing study, we will perform model-measurement comparisons of the speciated components of $PM_{2.5}$, which include sulfate, nitrate, and primary and secondary organic aerosol, with the goal of identifying the source or sources of error (emissions, chemistry, meteorology, deposition) using a deep learning approach. Large monthly mean $PM_{2.5}$ impacts of up to 10 μg m^{-3} in southeastern Canada were predicted for July 2012. In 2013, in situ data showed monthly mean $PM_{2.5}$ concentrations of 8–10 μg m^{-3} over the eastern USA, reproduced well by the model. However, the model still underpredicted in Texas and at several places in California, and also in Washington during the fall months. Biomass fires in southeastern Canada elevated monthly mean $PM_{2.5}$ concentrations there by 5–10 μg m^{-3} during the summer months of June and July 2013. In October 2013, enhancements of 5–10 μg m^{-3} were predicted in eastern Texas, Louisiana, Alabama, Georgia, Idaho, and southwestern Canada.

In April 2014, elevated monthly mean $PM_{2.5}$ concentrations of 10–20 μg m^{-3} were observed in the in situ measurements over the eastern USA and some stations in California. The model did a good job of reproducing monthly mean $PM_{2.5}$ concentrations less than about 12 μg m^{-3} but was unable to capture higher concentrations. Impacts of 5–6 μg m^{-3} were predicted over the Midwest for this month. The model significantly underpredicted mean $PM_{2.5}$ in June when the biomass impact was minimal, indicating that error in biomass emissions was not the potential cause for model-measurement error.

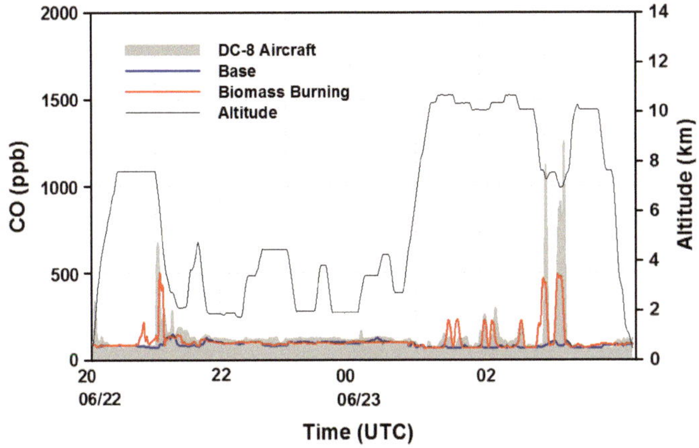

Fig. 12.26 Comparison of CO concentrations (left scale) using base model (blue) and STOPS (red; indicated by biomass burning). The STOPS version captures the peaks better

12.5 Future Work

12.5.1 Extending the Current Simulation Episode

We will simulate the 2011 and 2015 summer ozone seasons, which reported major fires in Texas. Additionally, we will conduct an evaluation of speciated $PM_{2.5}$ components as compared to aggregate $PM_{2.5}$ done in this study.

Acknowledgements The authors would like to thank Christine Wiedinmyer at NCAR for help with FINN, and Greg Yarwood and Zhen Liu at Ramboll Environ for their help on re-mapping the model species from MOZART to CB05.

References

Alexander B, Park R, Jacob D, Gong S (2009) Transition metal-catalyzed oxidation of atmospheric sulfur. Global implications for the sulfur budget. J Geophy Res 114(D2)

Bertschi IT, Jaffe DA (2005) Long-range transport of ozone, carbon monoxide, and aerosols to the NE Pacific troposphere during the summer of 2003: observations of smoke plumes from Asian boreal fires. J Geophys Res 110:D05303. https://doi.org/10.1029/2004JD005135

Bey I, Jacob DJ, Yantosca RM, Logan JA, Field B, Fiore AM, Li Q, Liu H, Mickley LJ, Schultz M (2001) Global modeling of tropospheric chemistry with assimilated meteorology: model description and evaluation. J Geophys Res 106:23,073–23,096

Byun D, Schere KL (2006) Review of the governing equations, computational algorithms, and other components of the Models-3 Community Multiscale Air Quality (CMAQ) modeling system. App Mech Rev 59(2):51–77

Choi Y, Jeon W, Roy A, Souri AH, Diao L, Pan S, Eslami E (2016) CMAQ Modeling Archive for Exceptional Events Analysis, Final report to the Texas Commission on Environmental Quality (TCEQ). https://www.tceq.texas.gov/assets/public/implementation/air/am/contracts/reports/pm/5821554181FY1609-20160829-uh-CMAQModelingArchiveForExceptionalEventsAnalyses.pdf

Czader BH, Choi Y, Li X, Alvarez S, Lefer B (2015a) Impact of updated traffic emissions on HONO mixing ratios simulated for urban site in Houston, Texas. Atmos Chem Phys 15:1253–1263. doi: https://doi.org/10.5194/acp-15-1253-2015

Czader BH, Percell P, Byun D, Kim S, Choi Y (2015b) Development and evaluation of the screening trajectory ozoneprediction system (STOPS, version 1.0). Geosci Model Dev 8: 1383–1394

Fiore AM, Pierce RB, Dickerson RR, Lin M (2014) Detecting and attributing episodic high background ozone events. Environ Mgt (EM; Special Issue for NASA AQAST) 22–28

Fountoukis C, Nenes A (2007) ISORROPIA II: a computationally efficient thermodynamic equilibrium model for K+–Ca_2+–Mg_2+–NH_4+–Na+–SO_{42}−–NO_3−–Cl−–H_2O aerosols. Atmos Chem Phy 7(17):4639–4659

Freitas SR, Longo KM, Chatfield R, Latham D, Silva Dias MAF, Andreae MO, Prins E, Santos JC, Gielow R, Carvalho JA Jr (2007) Including the sub-grid scale plume rise of vegetation fires in low resolution atmospheric transport models. Atmos Chem Phys 7 (13):3385–3398. https://doi.org/10.5194/acp-7-3385-2007

Hill GE (1974) Factors controlling the size and spacing of cumulus clouds as revealed by numerical experiments. J Atmos Sci 31:646–673

Houyoux MR, Vulkovich JM, Coats CJ, Wheeler NM, Kasibhatla PS (2000) Emission inventory development and processing for the seasonal model for regional air quality (SMRAQ) project. J Geophys Res 105:9079–9090

Jaffe D, Bertschi I, Jaegle L, Novelli P, Reid JS, Tanimoto H, Vingarzan R, Westphal DL (2004) Long-range transport of Siberian biomass burning emissions and impact on surface ozone in western North America. Geophys Res Lett 31:L16106. https://doi.org/10.1029/2004GL020093

Jeon W, Choi Y, Lee HW, Lee SH, Yoo JW, Park J, Lee HJ (2015) A quantitative analysis of grid nudging effect on each process of PM 2.5 production in the Korean Peninsula. Atmos Environ 122:763–774

Jiang X, Wiedinmyer C, Carlton AG (2012) Aerosols from fires: an examination of the effects on ozone photochemistry in the Western United States. Environ Sci Tech 46(21):442–460. https://doi.org/10.1021/es301541k

Kota SH, Schade G, Estes M, Boyer D, Ying Q (2015) Evaluation of MEGAN predicted biogenic isoprene emissions at urban locations in Southeast Texas. Atmos Environ 110:54–64

Lapina K, Honrath RE, Owen RC, Martin MV, Pfister G (2006) Evidence of significant large-scale impacts of boreal fires on ozone levels in the midlatitude Northern Hemisphere free troposphere. Geophys Res Lett 33:L10815

Latham D (1994) PLUMP: a one-dimensional plume predictor and cloud model for fire and smoke managers. General Technical Report INT-GTR-314, Intermountain Research Station, USDA Forest Service

Leung F-YT, Logan JA, Park R, Hyer E, Kasischke E, Streets D, Yurganov L (2007) Impacts of enhanced biomass burning in the boreal forests in 1998 on tropospheric chemistry and the sensitivity of model results to the injection height of emissions. J Geophys Res-Atmos 112 (D10):D10313. https://doi.org/10.1029/2006JD008132

Li X, Choi Y, Czader B, Roy A, Kim H, Lefer B, Pan S (2015) The impact of observation nudging on simulated meteorology and ozone concentrations during DISCOVER-AQ 2013 Texas campaign. Atmos Chem Phys 16:3127–3144; 2016. https://doi.org/10.5194/acp-16-3127-2016

Lilly DK (1962) On the numerical simulation of buoyant convection. Tellus XIV(2):148–172

McDonald-Buller E, Kimura Y, Wiedinmyer C, Emery C, Liu Z, Yarwood G (2015) Targeted improvements in the fire INventory from NCAR (FINN) model for Texas airquality planning. AQRP Project 14-011. http://aqrp.ceer.utexas.edu/projectinfoFY14_15%5C14-011%5C14-011%20Final%20Report.pdf

McKeen SA, Wotawa G, Parrish DD, Holloway JS, Buhr MP, Hübler G, Fehsenfeld FC, Meagher JF (2002) Ozone production from Canadian wildfires during June and July of 1995. J Geophys Res 107:4192

Morris GA, Hersey S, Thompson AM, Pawson S, Nielsen JE, Colarco PR, McMillan WW, Stohl A, Turquety S, Warner J Johnson BJ, Kucsera TL, Larko DE, Oltmans SJ, Witte JC (2006) Alaskan and Canadian forest fires exacerbate ozone pollution over Houston, Texas, on 19 and 20 July 2004. J Geophys Res 111:D24S03

Morton R, Taylor G, Turner J (1956) Turbulent gravitational convection from maintained and instantaneous sources. Proc Roy Soc A 234:1–23

Pan S, Choi Y, Jeon W, Roy A, Westenbarger DA, Kim HC (2016) Impact of high-resolution sea surface temperature, emission spikes and wind on simulated surface ozone in Houston, Texas during a high ozone episode. Atmos Environ 152(2017):362–376

Pfister GG, Wiedinmyer C, Emmons LK (2008a) Impacts of the fall 2007 California wildfires on surface ozone: integrating local observations with global model simulations. Geophys Res Lett 35:L19814

Pfister GG, Emmons LK, Hess PG, Lamarque J-F, Orlando J-J, Walters S, Guenther A, Palmer P, Lawrence PJ (2008b) Contribution of isoprene to chemical budgets: a model tracer study with the NCAR CTM MOZART-4. J Geophys Res 113:D05308

Pleim JE (2007) A combined local and nonlocal closure model for the atmospheric boundary layer. Part II: Application and evaluation in a mesoscale meteorological model. J App Meteor Climat 46(9):1396–1409

Pouliot G, Pierce T, Benjey W, O'Neill SM, Ferguson SA (2005) Wildfire emission modeling: integrating BlueSky and SMOKE. In: 14th international emission inventory conference "transforming emission inventories meeting future challenges today," Las Vegas

Raffuse SM, Craig KJ, Larkin NK, Strand TT, Coe Sullivan D, Wheeler NJM, Solomon R (2012) An evaluation of modeled plume injection height with satellite-derived observed plume height. Atmosphere 3(1):103

Simpson J, Wiggert S (1969) Models of precipitating cumulus towers. Mon Wea Rev 97:471–489

Singh HB, Herlth D, Kolyer R, Chatfield R, Viezee W, Salas LJ, Chen Y, Bradshaw JD, Sandholm ST, Talbot R, Gregory GL, Anderson B, Sachse GW, Browell E, Bachmeier AS, Blake DR, Heikes B, Jacob D, Fuelberg HE (1996) Impact of biomass burning emissions on the composition of the South Atlantic troposphere: reactive nitrogen and ozone. J Geophys Res 101(D19):24203–24219

Sinha P, Jaegle L, Hobbs P, Liang Q (2004) Transport of biomass burning emissions from southern Africa. J Geophys Res-Atmos 109:D20204

Smagorinsky J (1963) General circulation experiments with the primitive equations. Part I, the basic experiment. Mon Wea Rev 91:99–164

Tai E, Jimenez M, Nopmongcol O, Wilson G, Mansell G, Koo B, Yarwood G (2008) Boundary conditions and fire emissions modeling. ENVIRON International Corp, Novato, CA

Turner JS (1973) Buoyancy effects in fluids. Cambridge University Press, Cambridge, p 368

Val Martin M, Honrath RE, Owen RC, Pfister G, Fialho P, Barata F (2006) Significant enhancements of nitrogen oxides, black carbon, and ozone in the North Atlantic lower free troposphere resulting from North American boreal wildfires. J Geophys Res-Atmos 111: D23S60

Venkatram A, Pleim J (1999) The electrical analogy does not apply to modeling dry deposition of particles. Atmos Environ 33(18):3075–3076

Warneke C, de Gouw JA, Del Negro L, Brioude J, McKeen S, Stark H, Kuster WC, Goldan PD, Trainer M, Fehsenfeld FC, Wiedinmyer C, Guenther AB, Hansel A, Wisthaler A, Atlas E, Holloway JS, Ryerson TB, Peischl J, Huey LG, Hanks AT (2010) Biogenic emission measurement and inventories determination of biogenic emissions in the eastern United States and Texas and comparison with biogenic emission inventories. J Geophys Res 115:D00F18. doi:https://doi.org/10.1029/2009JD012445

Western Governors Association/Western Regional Air Partnership (2005) 2002 Fire emission inventory for the WRAP Region—Phase II. Air Sciences Inc., Portland, OR

Wiedinmyer C, Akagi SK, Yokelson RJ, Emmons LK, Al-Saadi A, Orlando JJ, Soja AJ (2011) The fire INventory from NCAR (FINN): a high resolution global model to estimate the emissions from open burning. Geosci Model Devel 4(3):625–641

Yarwood G, Rao S, Yocke M, Whitten GZ (2005) Final report: updates to the carbon bond mechanism: CB05. http://www.camx.com/files/cb05_final_report_120805.aspx

Zhao C, Wang Y, Choi Y, Zeng T (2009) summertime impact of convective transport and lightning NO_x production over North America: modeling dependence on meteorological simulations. Atmos Chem Phys 9:4315–4327

Part III
Control

Chapter 13
A Review on Membrane Fouling in Membrane Bioreactors: Control and Mitigation

Isha Burman and Alok Sinha

Abstract The growing interest toward the effective treatment of wastewater of different origin has increased worldwide, where membrane bioreactor (MBR) technology appears as the most appropriate treatment system. Membrane provides solid–liquid separation in MBRs and releases high-quality effluents which could be further reused for different purposes. The conventional biological treatment system often faces challenge of sludge disposal and requires larger footprint for their operation. These issues can be combated using MBR systems. However, MBR system associates with the membrane fouling problems, which limits its widespread application in wastewater treatment systems. Membrane fouling requires repeated cleaning that may reduce the membrane life span and frequent membrane replacement which incurs high cost. Membrane fouling leads to increase in transmembrane pressure (TMP), rapid declination of permeate flux, and increase in filtration resistance. The main factors that influence the membrane fouling are membrane characteristics, biomass characteristics, operating conditions, and membrane configuration. To address the recent advances made in MBR fouling, this paper reviews the mechanism, influencing factors, and methods for controlling and preventing membrane fouling. In addition, this chapter also deliberates the use of various adsorbents for controlling membrane fouling in MBRs.

Keywords Membrane bioreactor (MBR) · Wastewater · Fouling
Adsorbents

Abbreviations

AT	Ambient temperature
AIEX	Anion exchange
AnMBR	Anaerobic membrane bioreactor
CA	Cellulose acetate

I. Burman · A. Sinha (✉)
Department of Environmental Science and Engineering, Indian Institute of Technology (ISM), Dhanbad 826004, Jharkhand, India
e-mail: aloksinha11@yahoo.com

T. Gupta et al. (eds.), *Environmental Contaminants*, Energy, Environment, and Sustainability, https://doi.org/10.1007/978-981-10-7332-8_13

CFV	Cross-flow velocity
CST	Capillary suction time
DM	Dissolved matter
DO	Dissolved oxygen
DWW	Domestic wastewater
EDTA	Ethylene diamine tetra acetic acid
EPS	Extracellular polymeric substance
FS	Flat sheet
HCl	Hydrochloric acid
HRT	Hydraulic retention time
MAR	Mesoporous adsorbent resin
MBR	Membrane bioreactor
MIEXR	Magnetic ion-exchange resin
MLSS	Mixed liquor suspended solids
MMM	Mixed matrix membrane
MWCNT	Multiwalled carbon nanotube
MWW	Municipal wastewater
NaOCl	Sodium hypochlorite
NOM	Natural organic matter
PAC	Powdered activated carbon
PE	Polyethylene
PS	Polyether sulfone
PVDF	Polyvinylidene difluoride
SDS	Sodium dodecyl sulfate
SMP	Soluble microbial product
SRF	Sludge resistance to filtration
SRT	Sludge retention time
SS	Suspended solids
SWW	Synthetic wastewater
TMP	Transmembrane pressure
VFA	Volatile fatty acid

13.1 Introduction

Membrane bioreactors (MBRs) are the most promising technology (Deowan et al. 2016) for treatment and reuse of municipal, industrial, and commercial wastewater. MBRs are used as an alternative technology (Wong et al. 2016) to conventional treatment plants as they provide direct solid–liquid separation, smaller footprints, high-quality effluents, and comparatively low sludge generation (Lin 2012). The membrane separation provides a complete retention of biomass inside the bioreactor. As a result, the duration of sludge retention time (SRT) in the reactor is increased (Jeison and Van Lier 2007). The MBR technology is advantageous over the conventional activated sludge process (ASP) as it provides high volumetric loading, short hydraulic retention time (HRT), long SRT, less sludge production, simultaneous nitrification/denitrification in reactors, and high-quality effluent (Lin 2012). The key drawback of MBR technique is fouling of membrane (Le-Clech 2006; Meng 2009), which reduces the flux and increases the transmembrane pressure (TMP) (Trzcinski and Stuckey 2016). This results in the frequent cleaning and replacement of membrane and high pumping cost, thus escalating the overall operating cost of MBR. Membrane fouling is the main obstacle to the MBRs for its wider application in the different fields (Ishizaki et al. 2016).

Membrane fouling is the deposition and accumulation of flocs, colloids, solutes, microorganisms, and cell debris on the membrane surface and inside the membrane pores (Le-Clech 2010), which leads to increase in TMP, rapid declination of permeate flux, and increase in filtration resistance (Meng et al. 2009). Moreover, membrane fouling also aggravates feed pressure, reduces product quality, deteriorates membrane performance, and ultimately shortens membrane life. Fouling control approaches further lead to an increment in energy demand. This results in increment of overall operation and maintenance cost of the MBR. Membrane fouling requires repeated cleaning which may reduce the membrane life span leading to membrane replacement and thus the operation cost. So, it has become necessary to either reduce the price of the membrane (Meng et al. 2009) or look for such techniques which reduces or controls the fouling issues for effective use of MBR in future (Meng et al. 2010).

To address the recent advances made in membrane fouling control, this chapter reviews classification, mechanism, and factors responsible for membrane fouling and different methods for controlling and preventing membrane fouling in MBRs. In addition, this chapter deliberates the use of various adsorbents for controlling membrane fouling.

13.2 Classification of Membrane Fouling

13.2.1 Reversible and Irreversible Fouling

Based on cleaning techniques applied in membranes, fouling can be classified as reversible and irreversible types. Reversible fouling primarily occurs due to loosely bound external material that deposits on the membrane surface which causes cake layer formation. In contrast, the irreversible fouling may be caused due to strongly attached foulant components and pore blocking of the membrane during membrane filtration. Reversible fouling can be controlled by physical cleaning methods such as back flushing, while irreversible fouling (sometime called permanent fouling) (Tsuyuhara et al. 2010) is difficult to clean by any cleaning methods. However, reversible fouling may turn into an irreversible fouling layer by continuous filtration process.

13.2.2 Biological, Organic, and Inorganic Fouling

Based on the fouling components, membrane fouling can be classified as biofouling (Malaeb et al. 2013), inorganic (metal oxides, hydroxides, calcium salts) fouling, and organic (macromolecules, biopolymers) (Meng et al. 2009) fouling.

Biofouling is a consequence of biological growth, metabolism, and its secretion such as soluble microbial product (SMP) and extracellular polymeric substance (EPS) on the membrane surface (Meng et al. 2009). Bacterial growth and deposition on membrane and its pores lead to the formation of biocake. The metabolic products adsorb or accumulate on the surface and inside the pores of the membrane, leading to biofouling (Ramesh et al. 2007; Malaeb et al. 2013). For detail influence of bacteria and its metabolites to membrane fouling, organic fouling is studied separately. Some researchers, however, treat bacteria or flocs as biofoulants whose deposition, growth, and metabolism on the membrane results in fouling (Iorhemen et al. 2016). However, the organic substances produced by microorganisms are considered as organic foulants, and thus, organic fouling may be considered as a subset of biofouling (Malaeb et al. 2013).

Organic fouling implies fouling by organic matter present in the activated sludge and microbial metabolic by-products. Deposition of macromolecular species and biopolymers such as protein, polysaccharides, and humic substances on the membrane surface causes organic fouling. Colloidal particles and cell debris initiate pore clogging which ultimately leads to fouling of the membranes (Guo et al. 2012). Compared to the biological flocs, organic foulants are smaller, readily deposited on the membrane surface, and difficult to remove. MBR sludge may also contain large-sized free organic solutes called biopolymer clusters (BPCs). BPCs are composed of loose EPS and SMP, contain few microorganisms, and hence may not be classified as biofoulants. Due to their large size and accumulation on membrane surface, BPCs may cause severe fouling (Iorhemen et al. 2016).

Inorganic fouling is the chemical precipitation of inorganic flocs which includes cations and anions such as Ca^{2+}, Mg^{2+}, Al^{3+}, PO_4^{3-}, CO_3^{2-}, SO_4^{2-}, OH^-. Struvite ($MgNH_4PO_4 \cdot 6H_2O$) is the most dominated inorganic foulant in MBR systems. Struvite is formed due to the production of phosphate and ammonium ions during anaerobic decomposition of organic material present in wastewater (Lin et al. 2014). Other inorganic foulants such as $K_2NH_4PO_4$ and $CaCO_3$ also contribute as an important component in inorganic fouling. Fouling by $CaCO_3$ occurs when activated sludge has a highly alkaline condition. Calcium and phosphorus start precipitating on the membrane surface at higher pH conditions (Trzcinski and

Stuckey 2016). Inorganic fouling leads to an irremovable type of fouling as they show a cohesive property with the membrane and hence requires chemical cleaning. Ca^{2+} below 280 mg/L may reduce biofouling due to binding and bridging of EPS (Zhang et al. 2016); however, at higher concentrations, inorganic precipitation may increase. Two mechanisms for inorganic fouling have been proposed, viz. crystallization and particulate fouling. In crystallization, precipitation of ions takes place on membrane surface, whereas colloidal particulate deposition is the major pathway for particulate fouling (Iorhemen et al. 2016).

Among all types of fouling, biofouling and organic fouling are the principal reason for membrane fouling in MBRs system. However, some studies have also shown that metal substance may have a considerable effect on membrane fouling compared to biopolymers (Lyko et al. 2007). All the three types of fouling take place simultaneously and react with each other, which hamper filtration system. A study conducted by An et al. (2009) showed that the EPS extracted from fouling layer consisted of two components, i.e., protein and the fulvic acid, which forms the major component of the fouling layer (Guo et al. 2012).

13.3 Mechanism of Membrane Fouling

Fouling in membrane systems is initiated with pore narrowing, pore clogging/ blocking (Fig. 13.1a) followed by gel layer formation (Fig. 13.1b) and cake layer formation (Fig. 13.1c) (Meng et al. 2007). Cake layer is formed by deposition of large-sized particles such as sludge flocs with continuous MBR operation, while gel layers are formed by colloids, solutes, and SMP with long-term MBR operation (Wang et al. 2008; Lin et al. 2014). Mechanism of fouling (Fig. 13.2) involves adsorption and deposition of biomass fractions such as colloids or solute, sludge floc/biofloc, EPS, SMP on the membrane surface, interaction of which facilitates the growth of biofilm. This biofilm growth leads to pore clogging and cake/gel layer depositions (Hwang et al. 2008) which leads to decrease ineffective pore size and membrane surface area. Pore clogging study by Hong (2014) revealed that if sludge suspension contains more SMP, gel layer formed more easily than cake layer in submerged system. Another study revealed that cake/gel layer is formed on the membrane surface because of solute accumulation and adsorption of SMP (Wang et al. 2008). As size of SMP molecules is smaller as compared to membrane pores, it results in irreversible type of fouling (Costa et al. 2006). It has been identified by several researches that the dispersed supernatant SMP, together with the mixed liquor suspended solids (MLSS), strongly affects the fouling mechanism (Defrance et al. 2000; Wu et al. 2006). Pore narrowing, pore clogging/blocking, and cake layer formation reduce the pore size and hence the surface area of membrane.

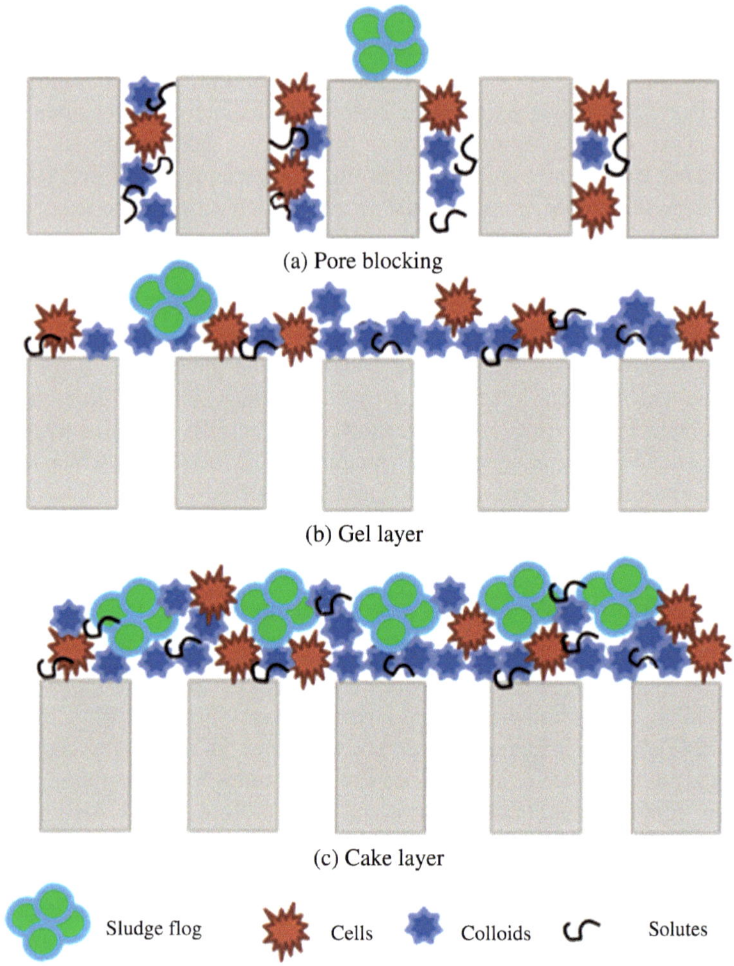

Fig. 13.1 Mechanism of **a** pore blocking, **b** gel layer, and **c** cake layer formation

13.4 Factors Influencing Membrane Fouling in MBRs System

Membrane fouling is the result of interaction between membrane material and sludge suspension. Recent literature related to AnMBR showed that parameters such as SMP, EPS, and hydrodynamic conditions directly affect the membrane fouling (Meng et al. 2009). Some operating parameters such as dissolved oxygen (DO), sludge retention time (SRT), and F/M ratio influence the sludge property and thus indirectly affect the fouling (Meng et al. 2009). A study conducted by Garcia

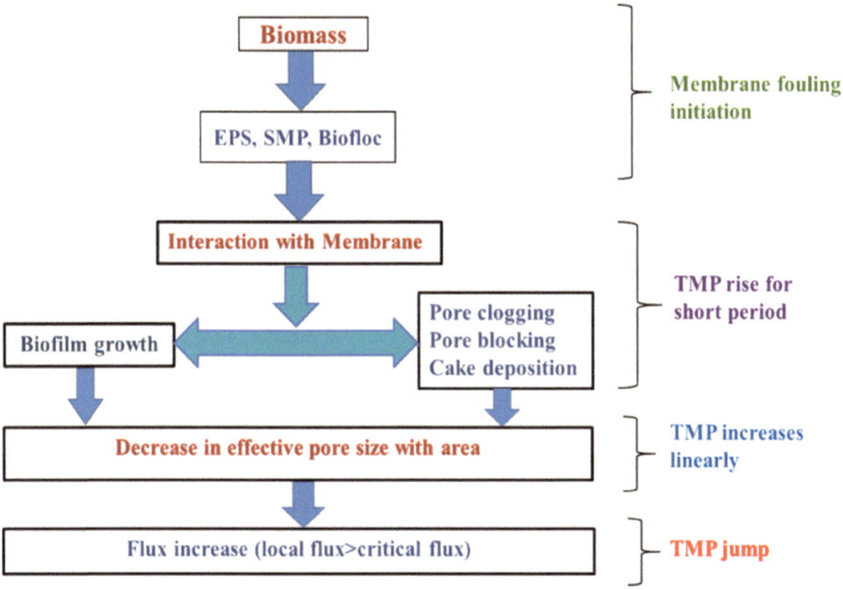

Fig. 13.2 Mechanism of membrane fouling

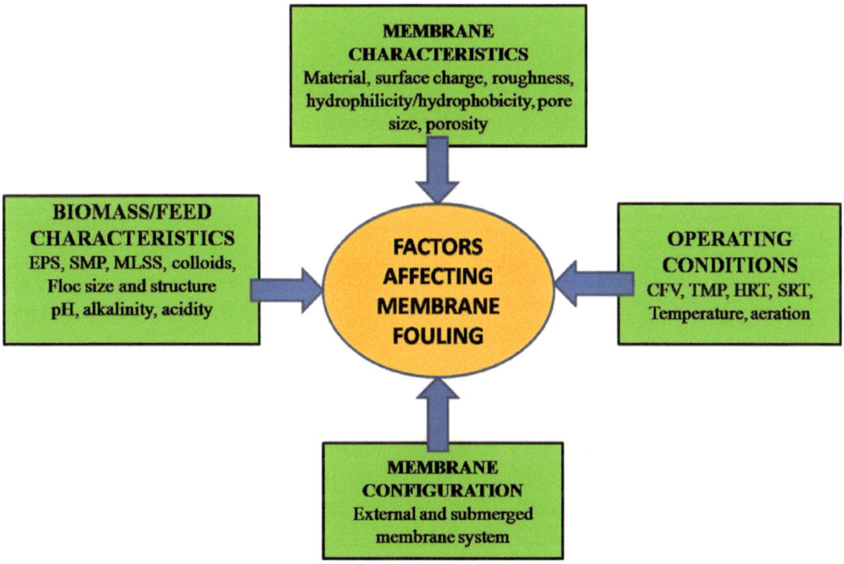

Fig. 13.3 Factors affecting membrane fouling in MBRs

(2011) found that under the same operating system, the supernatant SMP in anaerobic MBR was much higher than that of aerobic MBR and thus may cause higher fouling in AnMBR. The four membrane fouling factors (Fig. 13.3) that highly influence membrane fouling include membrane characteristics, feed/biomass characteristic, operational parameters, and membrane configurations.

13.4.1 Membrane Characteristics

Membrane characteristics, such as material and surface charge, membrane roughness and hydrophilicity/hydrophobicity, and pore size and porosity, have a direct impact on membrane fouling. MBRs require membranes to configure in such a way that there will be the availability of large membrane area-to-bulk volume ratio, energy requirements are low, and cleaning is simple (Judd 2011).

13.4.1.1 Membrane Material and Surface Charge

Membrane materials are categorized as organic (polymeric), inorganic (ceramic), and metallic types (Lin et al. 2012). Different membranes show different fouling characteristics. The accumulation of struvite ($MgNH_4\ PO_4–6H_2O$) mostly prevails in inorganic membranes. Organic membrane fouling mainly occurs due to the formation of a thick cake layer composed of biomass and inorganic compounds that cause flux decline. Stainless steel membranes come with the advantage of higher permeate flux but at the same time incur a higher cost (Meng et al. 2009). However, research is continuing in the development of new membrane material that fouls less and has good filtration capacity with ease of cleaning process (Santos 2011).

In micro-filter membranes, the charged particles form a filter cake layer which affects the filterability of the membranes. Increased particle surface charge increases compressibility and resistance of filter cake, while lower charge particles increase the cake growth and thickness (Lorenzen et al. 2016). Due to deposition of colloidal particles (Rana and Matsuura 2010), most membranes are negatively charged under normal conditions (Guglielmi and Andreottola 2010). This may attract positively charged cations such as Ca^{2+} and Al^{3+} in MLSS and enhance inorganic fouling.

13.4.1.2 Hydrophobicity and Membrane Roughness

It is believed by several researchers that hydrophobic membrane have more potential to cause fouling than hydrophilic one (Kim 2004; Santos and Judd 2010; Zhang et al. 2015). Hydrophobic character of membrane enhances the interfacial interaction between foulants and the membrane surface which results in severe fouling (Su et al. 2013; Zhang et al. 2015).

This is due to the fact that foulant molecules get attracted close to the membrane surface by hydrodynamic forces (a short-ranged interfacial force), which is responsible for adhesion of foulant to the membrane surface (Hong et al. 2013). The problem can be mitigated by the enhancement of hydrophilicity using surface modification of hydrophobic membranes (Liu et al. 2013). The hydrophilicity of the membrane surface could be achieved by means of grafting more hydrophilic polymers (Stuckey 2010), which enhances the antifouling property of the membrane. However, under harsh chemical and high thermal conditions, hydrophobic membrane shows high durability. Still, a hydrophilic membrane is normally preferred as they are less prone to organic fouling. Ceramic membranes (inorganic membranes) are highly hydrophilic in nature which imparts more fouling resistance. Polymeric membranes are mostly hydrophobic in nature which makes them susceptible to fouling (Iorhemen et al. 2016). It has been reported by several researchers that membrane roughness has more propensity to bind organic matters/sludge foulants (Jin et al. 2010; Myat et al. 2014) and stimulate more fouling as compared to a smoother surface.

13.4.1.3 Pore Size and Porosity

Pore size distribution plays a key role in determining the fouling property of the membrane. In MBR process, the pore size of the membrane generally ranges from 0.01 to 0.4 μm. Larger and wider distribution of pores allows the particle to entrap inside the voids resulting in pore blocking (Judd 2010) and thus causes irreversible/internal fouling (Shang et al. 2015). Larger pore size initially shows greater flux and subsequently subjected to internal fouling/pore blocking, and their permeability get decreased with time due to settling of foulants on the surface and pores of the membrane (Mafirad et al. 2011). Narrow pore size membranes are preferred in MBRs for fouling control to some extent (Meng et al. 2009).

13.4.2 Feed and Biomass Characteristics

Activated sludge is a heterogeneous suspension, which is composed of organic and inorganic compounds, suspended solids, various types of microorganisms, and their metabolites. These matters are either coming from raw wastewater or generated in consequence of biological growth and decay.

13.4.2.1 Role of Sludge Fractions

Different fractions of sludge (suspended solids, colloids, and dissolved matter/solutes) play an important role in membrane fouling. Wisniewski et al. (1996) studied the effect of MBR filtration on bacterial suspensions and considered three fractions: the suspended solids (SS), the colloidal fraction, and the dissolved molecules (DM). They concluded that the dissolved fraction accounted for 52% of the membrane resistance, the colloidal fraction 25%, and the particulate fraction

23%. Another study by Defrance et al. (2000) reported that the relative contributions of suspended solids (SS), colloidal solids, and dissolved matter to the filtration resistance caused by fouling were 65, 30, and 5%, respectively. However, they also revealed that the total resistance was less as compared to the sum of the resistance posed by individual component. The calculated sum of individual components was found to be 50% higher than the measured total resistance, which indicates that fouling resistances caused by individual constituents cannot be added. A similar study conducted by Bouhabila et al. (2001) showed the relative biomass contribution on fouling to be 24, 50, and 26% for SS, colloids, and solutes, respectively. Among these fractions, colloids are known to cause pore clogging and solutes cause cake layer formation (Bae and Tak 2005). They concluded that the major part of fouling was contributed by colloids and suspended solids (Sun and Liu 2013), i.e., 75% (Table 13.1). Another study by Lee et al. (2003) revealed that the contribution of supernatant (colloids and solutes) to membrane fouling depends on SRT; i.e., overall fouling propensity is relatively higher at shorter SRT.

13.4.2.2 Role of MLSS Concentration

MLSS of activated sludge mainly comprises of floc cell, EPS, and dissolved matter. As a general trend, it was shown and now accepted that the shorter the HRT and the longer the SRT, the higher the MLSS concentration. In the case of MBR, it can be operated at high concentration of MLSS as it utilizes membrane, which helps in solid–liquid separation (Trussel et al. 2005). Other researchers reported that higher MLSS concentration enhances fouling by increasing sludge viscosity (Delrue 2011). Membrane fouling increases with MLSS concentration as the viscosity of sludge increases and the filterability decreases. Filamentous bulking increases the production of SMPs which in turn enhances membrane fouling (Pan et al. 2010). However, there is no clear correlation between MLSS and membrane fouling as some researchers have studied that at higher MLSS concentration, i.e., up to 15 g/L, the fouling reduced and, however, above this concentration, the fouling rates increased (Rosenberger et al. 2005); and some reports that there is no or little effect of MLSS concentration on membrane fouling (Rosenberger et al. 2006; Iorhemen et al. 2016). Similarly as the MLSS concentration increases, the viscosity of the sludge increases which limits the oxygen transfer in MBRs. A critical MLSS concentration ranging from 10 to 17 g/L may result in exponential increase in sludge viscosity which results in higher membrane fouling (Le-Clech et al. 2006; Trusslee et al. 2007; Iorhemen et al. 2016).

13.4.2.3 Role of EPS and SMP

The flocs of activated sludge primarily comprise of microbial products such as EPS (extracellular polymeric substances), a kind of biopolymer attached to the bacterial surface, SMP (soluble microbial product) released by the bacteria (Hao 2016), and

Table 13.1 Relative resistance of different sludge fractions in membrane filtration

Type of membrane	Pore size (μm)	SRT (d)	TMP (kPa)	HRT (h)	MLSS (g/L)	Flux (L/h m^2)	% Resistance during filtration by various sludge fractions			References
							SS	Colloids	DM	
Ceramic	0.2	–	100	–	1–1.5	–	23	25	52	Wisniewski et al. (1998)
Ceramic	0.1	60	80	24	10	120	65	30	5	Defrance et al. (2000)
Hollow fibers	0.1	10–30	–	3.3	17.2–27	20	24	50	25	Bouhabila et al. (2001)
Hollow fibers	0.4	20–60	–	7.8	3	9	63–71	29–37[a]	29–37[a]	Lee et al. (2003)
CA (polymeric)	–	20	100		3.1–3.4	34.1–282.2	83–72	4–14	13–14	Bae and Tak (2005)
Polypropylene	0.2	300	50–200	24	0.32–1.2	5.78	66–80	12–19	23–4	Sun and Liu (2013)

[a]Contribution of supernatant to fouling

CA cellulose acetate; *SS* suspended solids; *DM* dissolved matter inert biomass. Microbial EPS and SMP secreted by microbial cells have high molecular weight and highly mucous in nature. EPS are waste products of bacteria resulting from microbial metabolites, cell lysis, or non-metabolized wastewater components, while SMPs are organic compounds released from substrate metabolism and biomass decay. The EPS and SMP are heterogeneous in nature which constitutes mainly polymeric materials such as polysaccharides, proteins, humic acids, nucleic acids, and lipids. Protein component in EPS is mainly hydrophobic in nature, whereas polysaccharides are hydrophilic in nature. Thus, polysaccharide fractions of EPS have more potential for membrane fouling if hydrophilic membranes are utilized. EPS can be subdivided into two types, i.e., bound and soluble EPS (also referred as SMPs). Bound EPS has a multilayer structure which contains loosely (LB-EPS) and tightly (TB-EPS) bound layers (Su et al. 2013). Loosely bound EPS are soluble in nature and move freely in the suspension (Rosenberger and Kraume 2002). However, it has been concluded by various researchers (Ramesh et al. 2007; Zhang et al. 2014a, b) that loosely bound EPS have strong potential of fouling than tightly bound EPS and are primarily responsible for irreversible fouling. The tightly bound EPS has a dominant role in the reduction of permeate flow as it forms a barrier layer having highly hydrated gel matrix containing microbes (Hwang et al. 2008). In recent years, numerous studies on MBR have acknowledged that the EPS is the most important biological factor that causes membrane fouling. EPS are amphoteric in nature as they contain both hydrophobic and hydrophilic groups. They contain charged groups, such as carboxyl, hydroxyl, phosphoric, sulfhydryl, and phenolic groups; hydrophobic regions in carbohydrates; and aliphatics in proteins and aromatics (Iorhemen et al. 2016). As the ratio of hydrophilic fraction plays an important role in fouling of hydrophilic membranes, the ratio of proteins (hydrophobic) to polysaccharides (hydrophilic) in EPS governs membrane fouling in MBRs, especially cake layer formation (Mukai et al. 2010).

SMPs produced due to substrate utilization are termed as substrate utilization-associated products (UAPs), and those produced from biomass decay are termed biomass-associated products (BAPs). SMPs contribute more to fouling as compared to colloids, MLSS, and bound EPS. SMP concentration increases with decreasing SRT and DO concentrations. As SMPs can easily penetrate the membrane pores in comparison with bound EPS, it may be a key factor in controlling membrane fouling (Iorhemen et al. 2016).

13.4.2.4 Floc Size

Smaller flocs tend to adhere more to the membrane surface (Shen et al. 2015), and thus for reducing the membrane fouling, larger flocs may be beneficial. Current studies are focusing on increasing the floc size by aerobic granulation (Li et al. 2012), activated carbon (Remy et al. 2009), and zeolite addition (Rezaei and Mehrnia 2014).

13.4.2.5 Alkalinity, pH, and Salinity

Low pH values increase the adsorption of EPS on membrane and flocculation of EPS and reduce the repulsion between foulant and membrane surface, thereby enhancing the fouling (Sweity et al. 2011; Zhang et al. 2014a, b); Sanguanpak et al. 2015). Hence, alkalinity may be added for preventing the lowering of pH in MBRs and thus combating the problem of membrane fouling (Iorhemen et al. 2016). Higher pH (8–9) may also result in chemical precipitation such as precipitation of $CaCO_3$ and enhance inorganic fouling (Meng et al. 2009). Higher salinity values (up to 5 g/L) may alter the biomass properties such as EPS, SMPs, floc formation, charge which resulted in higher membrane fouling (Elimelech et al. 1997; Reid et al. 2006; Iorhemen et al. 2016). High salt concentration in MLSS may cause chemical precipitation and enhanced electrostatic attraction toward membrane surface which increases fouling (Jang et al. 2013). EPS values may also be increased due to lower respiration rates of microorganism which in turn enhances fouling (Di Bella et al. 2013).

13.4.3 Operating Conditions

13.4.3.1 Operating Mode

MBR system operates in two modes (i) constant transmembrane pressure (TMP) with variable flux and (ii) constant permeate flux with variable TMP. Generally, the constant flux approach is followed as it can handle the variation in hydraulic loadings on membrane. The membrane in constant flux system must be operated at sustainable flux, i.e., below critical flux. Critical flux is the flux above which the cake or gel layer formation on membrane becomes evident. For sustainable flux, the rise in TMP is gradual and prevents excessive deposition of foulants on membrane surface. Thus, chemical cleaning may not be required under such operating conditions and backwashing may be sufficient (Iorhemen et al. 2016).

13.4.3.2 Cross-Flow Velocity

Cross-flow velocity (CFV) is an important hydrodynamic factor and has a significant impact on membrane fouling. Fluctuation in CFV affects particle settling (Lorenzen et al. 2016) and thus indirectly affects fouling phenomena. MBR with external membrane configuration usually shows high cross-flow velocity along the membrane surface. This reduces fouling (Lin et al. 2012) by inducing higher shear force on the membrane surface which reduces foulant deposition (She 2016). As shown in Table 13.2, Strohwald and Ross (1992) suggested that the rate of fouling could be reduced if cross-flow velocity is higher than 1.5 m/s. Another study by

Table 13.2 Effect of cross-flow velocity (CFV) on fouling

Cross-flow velocity (m/s)	Effect	References
4.5	Fouling decrease linearly	Choi et al. (2005)
1.5	Reduced fouling	Strohwald and Ross (1992)
2–3	Prevent formation of reversible fouling	Strohwald and Ross (1992)
0.1	Less particle deposition	Hwang et al. (2008)

Choi (2005) found that fouling decreased linearly with increasing CFV (up to 4.5 m/s). Detailed study of various calculated hydraulic resistances revealed that CFV values of 2–3 m/s were large enough to prevent the formation of reversible fouling without affecting the biological activities (Hwang et al. 2008). Finally, it could be concluded that high CFV is more effective in reducing fouling in the membranes rather than low cross-flow velocity.

13.4.3.3 Transmembrane Pressure (TMP)

Transmembrane pressure affects the membrane filtration, which has a direct impact on the fouling property of the membrane, especially in submerged MBR system (Hwang et al. 2008). In constant flux operating system, a sudden rise in TMP denotes severe membrane fouling and is referred to as "TMP jump." A recent trend for MBRs systems comprises three stages of TMP operation. TMP initially rises slowly due to initial pore blocking and fouling (stage 1); then, TMP rises gradually either in linear or in weakly exponential manner (stage 2); sudden rise in TMP is due to severe fouling (stage 3) (Zhang et al. 2006; Cho and Fane 2002). For sustainable operation in MBR, the cleaning of membrane must be done before stage 3 arises. Delay in cleaning the membrane during stage 3 may reduce the efficiency of MBR (Cho and Fane 2002). Thus, sludge characteristics' modification is necessary which helps to slow down the "TMP jump," which eventually leads to fouling control.

13.4.3.4 HRT, SRT, OLR, and F/M Ratio

As HRT influences the sludge characteristics like MLSS concentration, EPS production and sludge viscosity, which increases as the HRT is reduced, hence decreasing HRT may lead to increase in the fouling of membrane. Similarly, lowering of SRT results in higher EPS concentrations and thus enhances fouling. However, very high SRT may result in high MLSS concentrations and may increase fouling. SRT above 50 days tends to increase fouling (Iorhemen et al. 2016). High OLR rates result in production of hydrophilic substances which tend to increase the fouling (Johir et al. 2012). Similarly, variable OLR reactors showed higher fouling, in initial stages, in comparison with constant OLR reactors (Zhang et al. 2010). Low values of F/M are advisable to reduce the membrane fouling as high F/M ratios result in increased EPS concentration and protein-based foulants (Iorhemen et al. 2016).

13.4.3.5 Temperature

Temperature is considered as a key factor that plays an important role in influencing the efficiency of the MBR system. Variation in temperature affects the biological activity of the system which hampers the permeate quality and causes major fouling problems (Miyoshi et al. 2009). Aerobic MBRs usually operate at 20–30 °C, while anaerobic MBRs operate at 30–40 °C (Lin et al. 2012). An investigation was performed by Ma et al. (2013), and they reported that temperature fluctuation affects various parameters of the system. At low temperature (10 °C), fouling rate was higher as compared to normal temperature (20 °C) because the low temperature enhances the formation of organic materials such as protein, polysaccharide, and EPS in the mixed liquor that are responsible for severe fouling.

13.4.3.6 Configuration of MBR System: Fouling Impacts

MBR system may be classified based on the position of the membrane module as (i) external or side stream and (ii) internal/submerged. Figure 13.3 depicts the configuration of the membrane system in MBRs. In an external system (Fig. 13.4a), membranes are placed externally to the bioreactor and effluent is to be pumped from the bioreactor to the membrane unit. However, this type of configuration helps in reducing fouling by disrupting the cake layer formation due to cross-flow velocity of the effluent on the membrane surface. With the advantage of easier membrane replacement, the externally configured membrane also serves to provide more hydrodynamic fouling control (Khanal 2008). In the submerged system (Fig. 13.4b), the membrane is directly immersed in the bioreactor and permeate is pumped out or withdrawn through the membrane by using suction pumps. In such systems, the membranes are more prone to fouling and submerged conditions also result in less hydrodynamic fouling control. A comparative study between internal and external membrane system conducted by Andrade et al. (2014) revealed that MBR with external membrane system exhibited better performance than MBR with internal membrane system. The fact was related to high MLSS concentration in internal MBR system that aggravates production of EPS/SMP which led to cake formation and thus enhanced fouling.

The summary of various MBR systems treating different types of wastewater along with their operating conditions, membrane characteristics, and method adopted for cleaning of membrane is provided in Table 13.3.

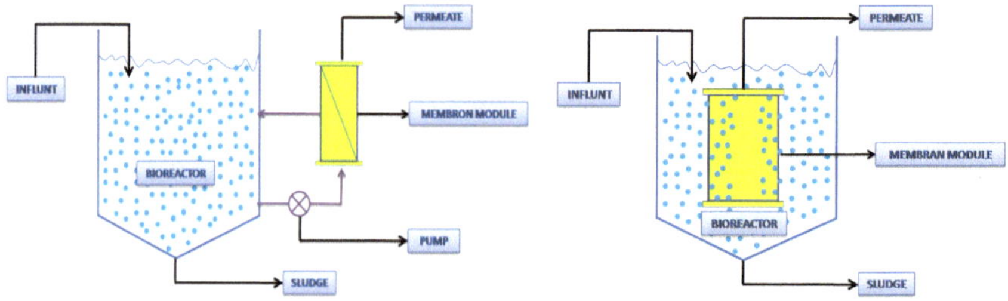

(a) External/ side stream MBR (b) Submerged MBR

Fig. 13.4 External/side-stream configuration and submerged configuration **a** external/side-stream MBR and **b** submerged MBR

13.5 Membrane Fouling Control—Prevention and Reduction

Fouling affects the TMP (transmembrane pressure) and permeates flux under varying operational conditions that pose a major obstacle, which could limit the widespread application of technique. Although the fouling mechanism is inevitable, it can be cured to some extent by investigating the factors that are liable to cause fouling. It is not possible to avoid the membrane fouling completely, but improvement in fouling control may enhance the life of the membrane along with a decrease in the overall cost. The control strategy for fouling has been summarized based on various fouling factors such as chemical control (e.g., chemically enhanced backwashing, use of coagulants/adsorbents), biological control (e.g., SRT, SMP, EPS, MLSS), and maintenance of hydraulics condition (e.g., HRT, backwashing, flux operations, aeration, permeability) of the MBR systems. Based on the parameters affecting membrane fouling, these strategies for MBR systems can be categorized into following groups: influent pretreatment, alteration in membrane properties, improvement of activated sludge, modification in operational conditions, membrane cleaning, and addition of adsorbents or additives.

13.5.1 Pretreatment of Influent Water

The quality of feed plays an important role in determining the impact of membrane fouling in any membrane system and improves final feed quality to a greater extent (Kabsch-Korbutowicz 2006). For example, the extreme pH condition of wastewater hampers the biological performance, enhances the precipitation of inorganic foulants, affects the permeability of the membrane, and reduces the life span. The wastewater typically contains diverse types of inorganic materials such as Al, Mg, Fe, Ca, and Si. However, these elements can be eliminated to some extent by pretreatment of wastewater and thus can reduce fouling. The pretreatment includes

Table 13.3 Summary of various MBR processes with their operating conditions

Source specification	Meng et al (2007)	Jaison et al. (2007)	Wo et al. (2008)	An et al. (2009)	Jin et al. (2010)	Tian et al. (2011)	Poostchi et al. (2012)	Duan et al. (2014)	Deng et al. (2015)	Miyoshia et al. (2015)	Deowan et al. (2016)	Hao et al. (2016)	Zhang et al. (2016)	Hong et al. (2016)
Configuration	SMBR	An-SMBR	SMBR	An-SMBR	SMBR	SMBR	SMBR	MBR	SMBR	SMBR	SMBR	SMBR	MBR	SMBR
Type of influent	MWW	SWW	SWW	MWW	DWW	SWW	SWW	–	SWW	SWW	SWW	SWW	SWW	MWW
Membrane module	HF	TB	HF	TB	FS	HF	FS	HF	HF	HF	FS	FS	HF	–
Material	PE	PS	PVDF	PET	Ceramic	PVDF	Polyster	PE	PVDF	PVDF	PVDF	PVDF	PVDF	PVDF
Area (m^2)	0.1	0.042	0.05	0.98	0.08	1	0.016	0.03	0.1	0.015	–	–	0.1	1.0
Pore size (μm)	0.1	0.2	0.2	0.64	0.08	0.2	30	0.4	0.2	–	0.3	0.3	0.2	0.1
T (°C)	25.0	30–55	18	15–20	25–30	12–24	23–25	–	–	–	35	35	AT	–
HRT (h)	10–12	–	8	2.6	6	7.6	4.7–12.5	6	6.67	35	1	1	8	5.5
SRT (d)	60	–	36	–	10	30	33.4	3-10	–	90	15	15	30	45
MLSS (g/L)	6	40	–	–	–	9.56	6.1–8.7	2	5	–	8.4–6.3	8.45	10–12	–
Mode of Cleaning	BW	CC/PC	BW	CC	CC	–	PC	–	BW/ CC	–	PC/CC	PC/CC	PC	–
TMP (kPa)	–	15–20	–	–	–	4-45	–	–	35	–	–	–	30	–
Flux (L/m^2h)	10	5–20	–	5	–	8	–	–	12	–	10	10	10	30

MBR membrane bioreactor; *T* temperature; *AT* ambient temperature; *TMP* transmembrane pressure; *SMBR* submerged MBR; *AnSMBR* anaerobic submerged MBR; *DWW* domestic wastewater; *MWW* synthetic municipal wastewater; *SWW* synthetic wastewater; *HF* hollow fiber; *TB* tubular; *FS* flat sheet; *PE* polyethylene; *PS* polyether sulfone; *PVDF* polyvinylidene difluoride; *BW* backwash; *CC* chemical cleaning; *PC* physical cleaning

Table 13.4 Different methods for pretreatment of feed water and their effect on membrane fouling

Method of pretreatment	Types	Effect on fouling	Limitations	References
Adsorption	PAC	Adsorb organic micro-pollutant and NOM	Difficulties in removing PAC	Matsui et al. (2003)
Chemical oxidation	Chlorine, permanganate, and ozone	Partially oxidize organic micro-pollutants and remove biofouling/NOM	May cause membrane damage	Crittenden et al. (2005)
Pre filtration	Packed bed filters or other membranes	Remove foulant particles and thus reduce biofouling and colloidal fouling	To enhance efficiency and may require pretreatment	Koh et al. (2006)
Anion exchange	[a]AIEX	Remove NOM, particles, anions	May lead to irreversible fouling	Humbert et al. (2007) and Hofs et al. (2012)
Coagulation	Al or Fe salts	Remove colloidal/pore blocking particle and NOM	Improper dosing may aggravate fouling	Huang et al. (2009) and Wong et al. (2016)

[a]*AIEX* Anion exchange

various methods such as coagulation, adsorption, ion exchange, filtration, chemical oxidation (Wong et al. 2016; Carroll et al. 2000; Grundestam and Hellstrom 2007). However, the addition of coagulants and adsorbents to feed water causes further chemical sludge production as a secondary problem that needs to be reduced or recovered (Lee et al. 2008; Prakash et al. 2004). In order to minimize the production of sludge, activated carbon adsorbents are being used. They have potential to remove micro-pollutants as well as NOM (natural organic matter) effectively. However, these coagulants when added to the influent water, prior to membrane filtration, show significant reduction in fouling (Holbrooket et al. 2004). Table 13.4 shows various methods employed for pretreatment of wastewater to reduce membrane fouling.

13.5.2 Alteration in Membrane Properties

The selection of membrane employed in MBR systems is done very precisely. Many researchers have investigated methods to improve membrane properties by modifying membrane surface to enhance hydrophilicity and antifouling property (Kumar et al. 2015) by bulk surface modification and blending (Li et al. 2015). Hydrophilicity can be achieved by grafting it with hydrophilic monomers. Kim and Bruggen (2010) have carried out an investigation to improve the membrane property using nanoparticle. They developed ceramic and polymeric membrane by adding nanoparticles having a size range from 4 to 100 nm during membrane synthesis. These membranes were found to be effective in flux enhancement and fouling control. Another study conducted by Bai et al. (2015) in surface modification by using MWCNT (multiwalled carbon nanotubes) to obtain hydrophilicity. They investigated that the UF membrane coated with MWCNT showed improvement in antifouling properties as coating layer diminished direct contact of foulant materials (NOM) to the membrane surface. Bilad et al. (2015) have developed mixed matrix membrane (MMM), a novel membrane made of polyvinyl chloride (PVC) and silica tube, for the treatment of wastewater. The PVC- and silica-based MMM is commonly used as a lead battery separator and has highly porous structure with high chemical and thermal stability. Recent study by Kim and Rhim (2016) showed that water flux could be enhanced by hydrophilization of hydrophobic membranes surface via coating by neutral-, cationic-, and anionic-based water-soluble polymers to modify the membrane property. Among them, cationic-based polymer was best suited in low concentration (150–300 mg/L) to enhance water flux. After compared with commercial membranes, MMM showed severe fouling in both short- and long-term filtration tests. The fouling was mainly irreversible type, dominated by pore blocking. However, other types of fouling were less in MMM which maintains its applicability with commercial membranes.

13.5.3 Modification of the Activated Sludge Property

Sludge property can be modified by the addition of certain types of additives including coagulants, flocculants, adsorbents, carriers, suspended particles, and other chemical agents (Lin et al. 2013). These additives act in different manners to control or mitigate fouling of the membrane to some extent. These mechanisms include adsorption, formation of cross-linking between flocs, coagulation, or a combination of these methods (Drews 2010).

13.5.4 Optimization of Operational Conditions

The chief operational condition mainly comprises temperature, pH, HRT, SRT, hydrodynamic conditions, flux, cross-flow velocity, flux/critical flux, and aeration (aerobic MBR). An important strategy for membrane fouling control is to operate the membrane at sustainable flux. Other mentioned parameters directly affect the sludge property except HRT and SRT that has been accepted that they have an indirect impact on fouling. It suggested that HRT and SRT cannot be considered as direct fouling causes, but rather parameters influencing factors such as MLSS, EPS, particle size distribution, which can then be directly related to fouling rates. These factors can be varied and controlled as discussed in the previous section to reduce the membrane fouling.

13.5.4.1 Optimization of pH

The pH is the primary operating factor that affects the membrane characteristics (Nanda et al. 2010), sludge properties (Gao et al. 2010), and microbial activities (Wu et al. 2010) in MBR systems. Several researchers have conducted experiments by varying the pH to show the effect of pH variation on fouling in MBR system. Brinck et al. (2000) investigated that no flux reduction was observed in high pH, whereas in low pH, flux reduced drastically. Another study by Sanguanpak et al. (2015) on varying pH of mixed liquor showed that in low pH condition, protein composition increased as dominating fouling component, while in higher pH condition, inorganic components (struvite, $CaCO_3$) dominated. The effect of variation of pH on MBR fouling is shown in Table 13.5.

13.5.4.2 Aeration

Aeration plays a major role in aerobic MBR. Aeration is specifically employed to deliver oxygen to the biomass in MBRs, which enhances the biodegradability of organic matter and cell synthesis. Increasing aeration rates may reduce cake layer

Table 13.5 Effect of varying pH in different property of MBR system on fouling

Parameters influence by pH	Low pH (0–6)	High pH (7–14)	References
Flux	Flux reduction	No effect on flux	Brinck et al. (2000)
EPS adsorption	Adsorption increase	Less adsorption observed	Sweaty et al. (2011)
Sludge	Enhance the sludge floc adherence to the membrane	Reduces the adherence of sludge floc to the membrane	Zhang et al. (2014)
Microbial activity	Decline in microbial activity (pH < 4)	Insignificant effect on microbial activity (pH 6–9)	Wu et al. (2010)
COD removal	No significant affect	Adversely affect the COD removal and membrane permeability (pH 9–10)	Gao et al. (2010)
Mixed liquor	Most severe membrane fouling (pH 5.5)	Lessen the fouling effect (pH 8.5)	Sanguanpak et al. (2015)
Foulant component	Removable fouling fraction dominated, composed of mainly protein	Irremovable fouling fraction dominated, composed of inorganic PPT, and humic-like substances	Sanguanpak et al. (2015)

formation on the membrane surface, which consequently alleviates fouling (Meng et al. 2007). In the case of aerobic MBR, aeration helps in controlling fouling of the membrane. However, in an anaerobic MBR, aeration hampers the biological activity of anaerobes, and consequently, an alternative method must be employed to reduce the fouling. However, aeration imparts additional cost in MBR as it consumes significant energy. A recent study conducted by Kurita et al. (2014) reported an alternative method to decrease the energy consumption due to aeration and to control membrane fouling in MBRs. They introduced granules in submerged MBR system which provided mechanical cleaning effect on the surface of the membrane. They showed that the granules drastically reduced the aeration rate and significantly increased the critical flux by 40%. However using granules, reversible fouling could be control, but at the same time, it promoted the irreversible fouling in long-term operation. It has been suggested that enhancement of aeration rates beyond the optimum range does not have any effect on fouling reduction (Ueda et al. 1997).

13.5.4.3 Sludge Retention Time (SRT)/Hydraulic Retention Time (HRT)

SRT plays an important role as longer SRT enhances the endogenous respiration in biomass, which helps in reducing sludge production. The longer SRT also favors the growth of some specialized bacteria, which mineralize complex organic matter effectively (Gander et al. 2000). Moreover, variation in SRT affects the MLSS concentration; i.e., with increasing SRT, MLSS concentration also increases which further enhances SMP concentration, thus enhancing the fouling (Nagaoka and Nemoto 2005; Han et al. 2005). Most of the investigators observed that too short or too long SRT either harms membrane property or enhances fouling action. Table 13.6 depicts the effect of SRT on membrane fouling. At shorter SRT, the concentration of bound EPS increases and vice versa (Ahmed et al. 2007). They revealed that shorter SRT is responsible for biofouling on membrane but at longer SRT biofouling rate declines gradually. Longer SRT often enhances fouling as it gradually leads to increase in foulant components as well as sludge viscosity, which

Table 13.6 Impact of SRT on fouling

SRT (days)	Effect on membrane fouling	References
10–20	Enhance membrane permeability within creasing SRT	Ng et al. (2006)
10–53	Bound EPS content decreased as SRT increased	Masse et al. (2006)
20–60	High concentration of bound EPS decreased gradually	Ahmed et al. (2007)
40–60	Fouling resistance increased within creasing SRT	Lee et al. (2003)
40–80	CST[a] and SRF[b] values decreased	Pollice et al. (2008)
30–100	Membrane fouling increased as SRT increased	Han et al. (2005)

[a]Capillary suction time (CST)
[b]Sludge resistance to filtration (SRF)

adversely affects membrane performance in MBRs (Han et al. 2005). However, Meng et al. (2009) have suggested that the SRT should be maintained at 20–50 days for MBRs depending on the value of HRT and feed water characteristics. High HRT in the reactor allows the accumulation of more foulants on the membrane surface, and low HRT helps in growth of filamentous bacteria, which leads to the production of large flocs, thus enhancing the fouling (Meng et al. 2007).

13.5.5 Membrane Resistance

The extent of fouling resistance can be measured by Darcy's law where flux through membrane can be expressed as follows:

$$J = \frac{\Delta P}{\mu R_t} \tag{13.1}$$

where J is the membrane permeate flux (m^3/m^2 s), ΔP is the transmembrane pressure difference (P_a), m is the permeate viscosity ($P_{a.s}$), and R_t is the total membrane fouling resistance (m^{-1}). The degree of membrane fouling is calculated using series resistance model where intrinsic resistance caused by membrane material, resistance due to formation of cake layer (reversible fouling), and resistance due to pore blocking and irreversible adsorption of foulants on membrane pores and surface (irreversible fouling) contribute to total resistance during filtration. Hence, the total filtration resistance (R_t) can be expressed as follows:

$$R_t = R_m + R_r + R_{ir} \tag{13.2}$$

where R_m is the intrinsic membrane resistance (m^{-1}), R_r is the reversible fouling resistance (m^{-1}), and R_{ir} is the irreversible fouling resistance (m^{-1}).

Field et al. (2011) proposed that the four most common fouling mechanisms can be described by solving Eq. (13.3) as follows:

$$-\frac{dJ}{dt} = K_n A^{2-n} J^{3-n} \tag{13.3}$$

where the values of n for different proposed mechanism are as follows:

(i) Complete pore blocking (n = 2),
(ii) Standard blocking (n = 1.5),
(iii) Intermediate fouling (n = 1),
(iv) Cake fouling (n = 0).

The measurement protocol of filtration resistances is as follows: (1) R_m was obtained by filtering distilled water with a cleaned membrane before MBR operation; (2) at the end of MBR operation, R_t was determined by filtering distilled water with a fouled membrane; (3) the filtration resistance obtained after removing cake layer on membrane surface yielded $R_m + R_{ir}$. Thus, R_r was calculated from Eq. (13.2) by subtracting $R_m + R_{ir}$ from R_t. R_{ir} was given by deducting R_m from $R_m + R_{ir}$ (Deng et al. 2015).

13.5.6 Membrane Cleaning

The membrane fouling problem may not be avoided completely, but the fouled membrane can be restored by different cleaning methods such as physical, chemical, or biological methods. The cleaning of membrane enhances membrane life to some extent that helps in reduction of overall cost of the system.

13.5.6.1 Physical Cleaning

To reduce membrane fouling, the basic and foremost technique applied is physical cleaning. Physical cleaning techniques that involve back flushing and relaxation or a combination of both may apply intermittently. Back flushing is applied by reversing the effluent back through the membrane. Back flushing reduces removable fouling as loosely bound material gets detached from the membrane (Wu et al. 2008). Periodic physical cleaning (backwashing) increases permeability and, in addition, reduces fouling that consequently leads to stable hydraulic operating conditions. However, the physical cleaning method does not produce any secondary contaminations, but energy consumptions are increased. If the physical method will not be satisfactory, it is essential to employ alternative methods, i.e., chemical cleaning for membrane cleaning.

13.5.6.2 Chemical Cleaning

To avoid fouling, various types of cleaning techniques are employed; among them, chemical cleaning is widely applied. Diverse types of cleaning agents are available that are actively employed for cleaning of fouled membrane in MBRs. These chemicals may include sodium hypochlorite (NaOCl), hydrochloric acid (HCl), nitric acid (HNO_3), citric acid, sodium hydroxide (NaOH), sodium dodecyl sulfate (SDS), and ethylene diamine tetra acetic acid (EDTA). Among them, NaOCl and citric acid are effective in removal of organic and inorganic foulant, respectively (Shi et al. 2014). While using these chemical agents as a cleaner, the consideration of some parameters such as pH, the concentration of chemicals, washing time, and some operational parameters such as temperature, pressure and cross-flow velocity are very important. However, chemical cleaning entirely recovers membrane flux, but frequent cleaning should be avoided because it reduces membrane life span. The chemical agent must have the ability to target principal compounds that are liable to cause fouling and do not hamper the membrane itself. Among such chemicals, NaOH is widely employed and known by many scientists as most effective cleaning agent (Urbanowska 2016). It has an ability to solubilize the protein (Mohammadi 2002) at high concentration and temperature by breaking the bonds between the membrane materials and fouling substance. Cleaning using chlorine is not recommended for polyamide surfaces or polymeric membranes (Cheryan 1998). Likewise, acid solutions are efficient in dissolving the fouling matrix by dissociating salts in organic matrix (Xing et al. 2003). On the other hand, EDTA is effective in solubilizing metal ions as it has strong binding ability and forms strong complexes with metal ions (such as Ca, Mg, Fe, Mn) and removal of these metal ions leads to breaking metal-associated structures including metal organic foulant complexation and inorganic scales (Lin et al. 2013). The combination of EDTA and NaOCl is reported to be very efficient in reduction of inorganic scaling and organic fouling (You et al. 2006; Wang et al. 2014). Enzymes can be used to solubilize the organic matrix of a fouling layer (Ross et al. 1992). In contrast, physical cleaning does not cause such changes as chemical cleaning does. Neither it produces filtrate nor does it take off line of the membrane. Therefore, it can be performed frequently at the interval of days, while chemical cleaning can be performed in the interval of days, month, or years as per requirement (Ueda and Hata et al. 1999). However, cleaning also affects the membrane as it reduces the permeability and enhances aging of the membrane (Drews 2010). Table 13.7 lists various types of chemical used for cleaning of fouled membrane.

Table 13.7 Different chemical agents and their action on fouling

Chemical agents	Action on fouling	Sources
EDTA	Increases the solubility of metal ions and breaks the interactions in metal–organic complexes and inorganic scales	Lin et al. (2013) and Shi et al. (2014)
Caustic solutions	Breaks bonds between the membrane surface and the fouling material and helps solubilize proteins	Cheryan (1998)
Chlorine	removes effectively pores clogging by organic matter	Cheryan (1998)
Acid solutions	Dissolves the organic fouling matrix	Xing et al. (2003)
Enzymes	Solubilizes the organic matrix of fouling layer	Ueda and Hata (1999)
Sodium hydroxide	Dissolves proteins and fats component	Mohammadi (2002)
Sodium bicarbonate	Decreases the EPS content in foulant components	Hu et al. (2015)

13.5.7 Role of Different Adsorbents and Their Combination in Fouling Control

There are varieties of approaches adopted to control membrane fouling as mentioned in the above section, but they may involve high operational and capital cost. Recently, adsorbents have been widely used to control and mitigate the fouling problem associated with the membrane systems. As these additives are chemically non-reactive and have an excellent capacity to sequester organic polymers, they enhance organic removal efficiency.

In MBR, adsorbents such as PAC (powdered activated carbon) are widely used in water treatment process. PAC commonly known as flux enhancer improves the efficiency of COD removal as well as adsorbs molecules having low molecular weight. The PAC also provides medium for bacterial growth, enhances bacterial population, adsorbs recalcitrant compounds (Skouteris 2015), and alters EPS concentration (Thuy and Visvanathan 2006). The PAC has a special property of adsorbing dissolved foulants directly, which decreases internal fouling/pore blocking. It helps in increasing the porosity of cake layers, enhancing the flux, and decreasing the compressibility of sludge flocs (Aquino et al. 2006; Lesage et al. 2007). Another advantage of the PAC application in MBR includes a low production of sludge, adsorption of toxic substance, and its cost effectiveness.

Mesoporous adsorbent resin (MAR) is a kind of adsorbent specifically developed to control the membrane fouling. It has been showed that MAR significantly reduces both the reversible fouling and irreversible fouling. Recently, Kai et al. (2014) have done a comparative study on MAR and PAC in the reduction of ultrafiltration membrane fouling with different fractions of NOM. They showed that after pretreatment, the adsorption capacity of MAR (61.5%) was significantly better than PAC (9.3%). However, Kang and Choo (2010) conducted a systematic study to compare the efficiency of PAC (powdered activated carbon) and FH (ferrihydrite) as adsorbent in controlling the membrane fouling. FH effectively adsorbed large NOM molecules (>100 kDa), while PAC was effective in removing smaller molecules (3–9 kDa). They analyzed hydraulic filtration resistance and revealed that the reversible fouling could be effectively removed by these adsorbents. However, adsorbent particles easily get attach to the membrane surface irreversibly so it is required to remove it periodically from the membrane system. Semblante et al. (2013) developed a very simple and low-cost technique of production of magnetite by co-precipitation which was capable of adsorbing protein and thus helps in reducing protein-based fouling. An advantage of using magnetite as foulant reducer is that it is recoverable and reusable through re-activation. Magnetic ion-exchange resin (MIEXR), a novel idea developed recently, was found to be effective in removal of negatively charged particles from water (Huang et al. 2009). Several researchers have reported that MIEXR removes NOM (natural organic matter) efficiently from wastewater and thus helps to reduce membrane fouling. An investigation carried out by Kabsch-Korbutowicz et al. (2006) to study the effect of MIEXR in pretreatment of water, prior to the membrane process, found MIEXR very effective in reducing fouling and enhancing the permeate quality. However, Humbert et al. (2007) found that MIEXR selectively adsorbed the low and medium molecular weight fractions of the NOM, while having relatively little effect on the high MW molecules that were primarily responsible for reversible membrane fouling. However, MIEXR was unable to remove turbidity and generated secondary pollutant. A recent study by Ding et al. (2014) revealed that the use of granular activated carbon (GAC) in membrane-coupled expanded granular sludge bed reactor reduces fouling effectively by the action of adsorption of various organics, such as SMP, protein, and polysaccharides. Table 13.8 shows various types of adsorbents used in MBR systems for reducing fouling of membranes.

A study was carried out by Zhang et al. (2004) to study the effect of membrane precoating by ferric hydroxide and sludge flocculation by ferric chloride. They revealed that this method showed excellent capability in fouling reduction but, at the same time, it also reduces the pH, thereby hampering the biological activity. Song et al. (2008) studied the effect of coagulants such as ferric chloride and alum in MBR on membrane fouling. They reported that both the coagulants improve filtration resistance and at same time remove phosphorus also. Further, the authors exclude the ferric chloride as it strongly decreases the pH. A study conducted by Kim et al. (2008) utilized three adsorbents, i.e., heated iron oxide particles (HIOPs), heated aluminum oxide particles (HAOPs), and powdered activated carbon (PAC) for removal of NOM. They reported that PAC selectively adsorbed non-foulant NOM molecules, whereas HAOPs and HIOPs selectively adsorbed

Table 13.8 Characteristic of the adsorbents and their combination in fouling effect

Additives	Surface area (m^2)	Particle size (μm)	Dosage (mg/l)	Effect on fouling	% of organic removal	Sources
FH	244	3.5	300	Reduces fouling by NOM	50	Kang and Choo (2010)
PAC		23.3	1			
MAR	1500	25.2	50	Mitigates NOM fouling	–	Kai et al. (2014)
Magnetite	108	0.00042	1	Reduces protein-induced fouling	60	Semblante et al. (2013)
		0.00058				
HIOPs	130–180	13.3	10	Selectively adsorbs foulant molecules	40	Song et al. (2008)
HAOPs		10.4	200			
PAC	–	33.9	300	Adsorbs non-foulant molecules	50	Jin et al. (2009)
	–		0–20			
MIEX resin	–	150–180	20	Reduces organic fouling	–	Jin et al. (2009)
GAC	867	500–1700	5000	Reduces SMP, polysaccharide and protein	24.7–27.8	Ding et al. (2014)

foulant molecules. Zhang et al. (2004) and Kim et al. (2008), who have worked on the conjunctive use of mineral oxide particles (e.g., HIOPs or heated aluminum oxide particles) with membranes, explained that mineral oxide particle layers on top of the membrane surface protected the membrane from fouling as they adsorbed NOM before it reaches the surface of the membrane. Periodic removals of saturated adsorbents by backwashing and reinjecting fresh particles may keep the membrane clean during continuous membrane system operation.

13.5.7.1 Limitation of Adsorbents

However, it is important to remove the adsorbent particles occasionally from the membrane, because they have potential to attach to the membrane surface irreversibly (Song et al. 2008). Once adsorbent particles attached to the membrane, it is sometimes difficult to dislodge all the particles away from the membrane. Although activated carbon adsorbents help to enhance the quality of treated water, they do not necessarily reduce membrane fouling. There are a few reports about the negative effects of activated carbon adsorbents on fouling control. Moreover, Lee et al. (2005) have reported that the adsorption capacity varies according to the material from which PAC is prepared. The favorable dose of an adsorbent also needs to be determined so that no further adverse effect could occur. Although not all adsorbents show the same removal efficiency in removal of foulant components, the selection of an adsorbent regarding organic removal efficiency must be taken into account for employing it to the membrane process.

13.6 Conclusion

The growing interest toward the effective treatment of wastewater of different origin has increased worldwide, where MBR technology appears as the most appropriate treatment system. However, MBR system associates with the membrane fouling problems which limit its widespread application in wastewater treatment systems. There have been extensive development and advancement in control of membrane fouling. This study presents the recent development, and the recent researches involving the type, mechanism, factors of fouling, and its control and mitigating measures are summarized below.

- Fouling caused by organic, inorganic, and biological origins such as SMPs and EPS results in cake layer formation and clogs the membrane which results in decrease in the surface area of the membrane and increase in the transmembrane pressure (TMP).
- Fouling in AnMBR is more intense than aerobic MBRs as aeration in aerobic MBR dislodges the cake layer formation and hence reduces fouling. Membrane

characteristics such as surface charge, roughness and hydrophilicity, pore size and porosity have a direct impact on the membrane fouling.

- Hydrophobicity of membrane may cause severe fouling which can be overcome by grafting hydrophilic polymers on membrane surface. Narrow pore sizes are preferred over wider ones as they cause less fouling.
- The major portion of biomass (75%) which causes fouling is colloids and solutes present in wastewater.
- MLSS concentration increases with shorter HRT and longer SRT and thus enhances fouling, while other studies concluded no or little effect on fouling.
- Low HRT enhances the growth of filamentous bacteria and produces large flocs, which enhances fouling, while extremely high HRT enhances accumulation of foulant components. EPS and SMP increases the fouling in MBRs, and high concentration may cause irreversible fouling.
- High cross-flow velocity reduces the fouling of membranes by inducing shear force.
- The preferred temperature to operate successfully for aerobic and anaerobic MBRs is 20–30 and 30–40 °C, respectively. Low temperature enhances the fouling rates.
- Provision of granules along with aeration can increase the critical flux by 40% due to mechanical cleaning. This may control reversible fouling but may cause irreversible fouling in long run.
- To control fouling, chemical cleaning of membranes can apply as they recover the membrane flux completely but reduce the membrane life span.
- Adsorbents can be used as flux enhancer by adsorbing molecules of low molecular weight, and they provide medium for bacterial growth, alter EPS concentrations, lower production of sludge, and adsorb toxic substances. However, attachment of adsorbents to membrane surface may cause difficulty in their dislodging and may enhance membrane fouling.

References

Ahmed Z, Cho J, Lim BR, Song KG, Ahn KH (2007) Effects of sludge retention time on membrane fouling and microbial community structure in a membrane bioreactor. J Membr Sci 287(2):211–218

An Y, Wang Z, Wu Z, Yang D, Zhou Q (2009) Characterization of membrane foulants in an anaerobic non-woven fabric membrane bioreactor for municipal wastewater treatment. Chem Eng J 155:709–715

Andrade LHD, Mendes FDDS, Espindola JC, Amaral MCS (2014) Internal versus external submerged membrane bioreactor configurations for dairy wastewater treatment. Desalination Water Treat 52(16–18):2920–2932

Aquino SF, Hu AY, Akram A, Stuckey DC (2006) Characterization of dissolved compounds in submerged anaerobic membrane bioreactors (SAMBRs). J Chem Technol Biotechnol 81:1894–1904

Bai L, Liang H, Crittenden J, Qu F, Ding A, Maa J, Du X, Guo S, Li G (2015) Surface modification of UF membranes with functionalized MWCNTs to control membrane fouling by NOM fractions. J Membr Sci 492:400–411
Bae TH, Tak TM (2005) Interpretation of fouling characteristic of ultrafiltration membranes during the filtration of membrane bioreactor mixed liquor. J Membr Sci 264(1–2):151–160
Bilad MR, Marbelia L, Laine C, Vankelecom IFJ (2015) A PVC–silica mixed-matrix membrane (MMM) as novel type of membrane bioreactor (MBR) membrane. J Membr Sci 493:19–27
Bouhabila H, Aim Ben BRH (2001) Fouling characterization in membrane bioreactors. Sep Purif Technol 22–23:123–132
Brinck J, Jonsson AS, Jonsson B, Lindau J (2000) Influence of pH on the adsorptive fouling of ultrafiltration membranes by fatty acid. J Membr Sci 164:187–194
Carroll T, King S, Gray SR, Boltom BA, Booker NA (2000) The fouling of microfiltration membranes by NOM after coagulation treatment. Water Res 34:2861–2868
Cheryan M (1998) Ultrafiltration and microfiltration handbook. Technomic Lancaster, PA
Cho BD, Fane AG (2002) Fouling transients in nominally sub-critical flux operation of a membrane bioreactor. J Membr Sci 209:391–403
Choi H, Zhang K, Dionysiou DD, Oerther DB, Sorial GA (2005) Influence of cross-flow velocity on membrane performance during filtration of biological suspension. J Membr Sci 248:189–199
Costa AR, de Pinho MN, Elimelech M (2006) Mechanisms of colloidal natural organic matter fouling in ultrafiltration. J Membr Sci 281:716–725
Crittenden JC, Trussell RR, Hand DW, Howe KJ, Tchobanoglous G (2005) Water treatment: principles and design, 2nd edn. Wiley, Hoboken, p 1948
Defrance L, Jaffrin MY, Gupta B, Paullier P, Paullier V (2000) Contribution of various constituents of activated sludge to membrane bioreactor fouling. Biores Technol 73 (2):105–112
Delrue F, Stricker AE, Mietton-Peuchot M, Racault Y (2011) Relationships between mixed liquor properties, operating conditions and fouling on two full-scale MBR plants. Desalination 272:9–19
Deng L, Guo W, Ngo HH, Zuthi MFR, Zhang J, Liang S, Li J, Wang J, Zhang X (2015) Membrane fouling reduction and improvement of sludge characteristics by bioflocculant addition in submerged membrane bioreactor. Sep Purif Technol 156:450–458
Deowan SA, Galiano F, Hoinkis J, Johnson D, Altinkaya SA, Gabriele B, Hilal N, Drioli E, Figoli A (2016) Novel low-fouling membrane bioreactor (MBR) for industrial wastewater treatment. J MembSci 510:524–532
Di Bella G, Di Trapani D, Torregrossa M, Viviani G (2013) Performance of a MBR pilot plant treating high strength wastewater subject to salinity increase: analysis of biomass activity and fouling behaviour. Biores Technol 147:614–618
Ding A, Liang H, Qu FS, Bai L, Li GB, Ngo HH, Guo W (2014) Effect of granular activated carbon addition on the effluent properties and fouling potentials of membrane-coupled expanded granular sludge bed process. Biores Technol 171:240–246
Drews, A. (2010) Membrane fouling in membrane bioreactors characterization, contradictions, cause, and cures. J Membr Sci 363:1–28
Duan L, Song Y, Yu H, Xia S, Hermanowicz SW (2014) The effect of solids retention times on the characterization of extracellular polymeric substances and soluble microbial products in a submerged membrane bioreactor. Biorestechnol 163:395–398
Elimelech M, Xiaohua Z, Childress AE, Seungkwan H (1997) Role of membrane surface morphology in colloidal fouling of cellulose acetate and composite aromatic polyamide reverse osmosis membranes. J Membr Sci 127:101–109
Field RW, Wu JJ (2011) Modelling permeability loss in membrane filtration: Re- examination of fundamental fouling equations and their link to critical flux. Desalination 283:68–74
Gander M, Jefferson B, Judd S (2000) Aerobic MBRs for domestic wastewater treatment: a review with cost considerations. Sep Purif Technol 18:119–130

Gao WJJ, Lin HJ, Leung KT, Liao BQ (2010) Influence of elevated pH shocks on the performance of a submerged anaerobic membrane bioreactor process. Biochem 45:1279–1287
Garcia M, Monsalvo V, Pidou M, Le-Clech P, Judd SJ, McAdam EJ, Jefferson B (2011) Impact of membrane configuration on fouling in anaerobic membrane bioreactors. J MembrSci 382:41–49
Grundestam J, Hellstrom D (2007) Wastewater treatment with anaerobic membrane bioreactor and reverse osmosis. Water Sci Technol 56:211–217
Guglielmi G, Andreottola G (2010) Selection and design of membrane bioreactors in environmental bioengineering. In: Environmental biotechnology. Humana Press, pp 439–516
Guo W, Ngo HH, Li J (2012) A mini-review on membrane fouling. BioresTechnol 122:27–34
Han SS, Bae TH, Jang GG, Tak TM (2005) Influence of sludge retention time on membrane fouling and bioactivities in membrane bioreactor system. Process Biochem 40:2393–2400
Hao L, Liss SN, Liao BQ (2016) Influence of COD: N ratio on sludge properties and their role in membrane fouling of a submerged membrane bioreactor. Water Res 89:132–141
Hofs B, Vries D, Siegers WG, Beerendonk EF, Cornelissen ER (2012) Influence of water type and pretreatment method on fouling and performance of an Al_2O_3 microfiltration membrane. Desalination 299:28–34
Holbrook RD, Higgins MJ, Murthy SN, Fonseca AD, Fleischer EJ, Daigger GT, Grizzard TJ, Love NG, Novak JT (2004) Effect of alum addition on the performance of submerged membranes for wastewater treatment. Water Environ Res 76:2699–2702
Hong H, Peng W, Zhang M, Chen J, He Y, Wang F, Weng X, Yu H, Lin H (2013) Thermodynamic analysis of membrane fouling in a submerged membrane bioreactor and its implications. Biores Technol 146:7–14
Hong H, Zhang M, He Y, Chen J, Lin H (2014) Fouling mechanisms of gel layer in a submerged membrane bioreactor. BioresTechnol 166:295–302
Hong H, Lin H, Mei R, Zhou X, Liao BQ, Zhao L (2016) Membrane fouling in a membrane bioreactor: a novel method for membrane surface morphology construction and its application in interaction energy assessment. J Membr Sci 516:135–143
Hu D, Zhou Z, Shen X, Wei H, Jiang LM, Lv Y (2015) Effects of alkalinity on membrane bioreactors for reject water treatment: performance improvement, fouling mitigation and microbial structures. BioresTechnol 197:217–226
Huang H, Schwab K, Jacangelo JG (2009) Pretreatment for low- pressure membranes in water treatment: a review. Environ Sci Technol 43:3011–3019
Humbert H, Gallard H, Jacquemet V, Croue JP (2007) Combination of coagulation and ion exchange for the reduction of UF fouling properties of a high DOC content surface water. Water Res 41:3803–3811
Hwang BK, Lee WN, Yeon KM, Park PK, Lee CH, Chang IS, Drews A, Kraume M (2008) Correlating TMP increases with microbial characteristics in the bio-cake on the membrane surface in a membrane bioreactor. Environ Sci Technol 42:3963–3968
Iorhemen OT, Hamza RA, Tay JH (2016) Membrane bioreactor (MBR) technology for wastewater treatment and reclamation: membrane fouling. Membranes 6(2):33
Ishizaki S, Fukushima T, Ishii S, Okabe S (2016) Membrane fouling potentials and cellular properties of bacteria isolated from fouled membranes in a MBR treating municipal waste water. Water Res 100:448–457
Jang D, Hwang Y, Shin H, Lee W (2013) Effects of salinity on the characteristics of biomass and membrane fouling in membrane bioreactors. Bioresour Technol 141:50–56
Jeison D, Van Lier JB (2007) Cake formation and consolidation: main factors governing the applicable flux in anaerobic submerged membrane bioreactors (AnSMBR) treating acidified waste waters. Sep Purif Technol 56:71–78
Jin L, Ong SL, Ng HW (2010) Comparison of fouling characteristics in different pore-sized submerged ceramic membrane bioreactors. Water Res 44:5907–5981
Johir MAH, Vigneswaran S, Sathasivan A, Kandasamy J, Chang CY (2012) Effect of organic loading rate on organic matter and foulant characteristics in membrane bio-reactor. Bioresour Technol 113:154–160

Judd S (2010) The MBR book: principles and applications of membrane bioreactors for water and wastewater treatment. Elsevier Oxford

Judd S (2011) The MBR book: principles and applications of membrane bioreactors for water and wastewater treatment. Second Edition Elsevier

Kabsch-Korbutowicz M, Bilyk A, Molczan M (2006) The effect of feed water pretreatment on ultra filtration membrane performance. Polish J Environ Studies 15:719–725

Kang SK, Choo KH (2010) Why does a mineral oxide adsorbent control fouling better than powdered activated carbon in hybrid ultrafiltration water treatment? J Membr Sci 355:69–77

Khanal SK (2008) Anaerobic biotechnology for bio-energy production: principles and applications. Wiley, Ames Iowa

Kim DS, Kang JS, Lee YM (2004) The influence of membrane surface properties on fouling in a membrane bioreactor for wastewater treatment. Sep Sci N Technol 39(4):833–854

Kim J, Cai Z, Benjamin MM (2008) Effects of adsorbents on membrane fouling by natural organic matter. J Membr Sci 310:356–364

Kim J, Bruggen BV (2010) The use of nano-particles in polymeric and ceramic membrane structures: Review of manufacturing procedures and performance improvement for water treatment. Environ Pollut 158:2335–2349

Kim KY, Rhim JW (2016) Hydrophilization of hydrophobic membrane surfaces for the enhancement of water flux via adsorption of water-soluble polymers. Membrane Water Treat 7 (2):101–113

Koh LC, Ahn WY, Clark MM (2006) Selective adsorption of natural organic foulants by polysulfone colloids: effect on ultrafiltration fouling. J Membr Sci 281(1–2):472–479

Kumar RS, Arthanareeswaran G, Paul D, Kweon JH (2015) Modification methods of polyethersulfone membranes for minimizing fouling—review. Membrane Water Treat 6 (4):323–337

Kurita T, Kimura K, Watanabe Y (2014) The influence of granular materials on the operation and membrane fouling characteristics of submerged MBRs. J Membr Sci 469:292–299

Le-Clech P, Chen V, Fane TAG (2006) Fouling in membrane bioreactors used in wastewater treatment. J Membr Sci 284:17–53

Le-Clech P (2010) Membrane bioreactors and their uses in wastewater treatments. Appl Microbiol Biotechnol 88(6):1253–1260

Lee W, Kang S, Shin H (2003) Sludge characteristics, and their contribution to microfiltration in submerged membrane bioreactors. J Membr Sci 216(1):217–227

Lee JW, Chun JI, Jung HJ, Jung DH, Ramesh T, Shim WG, Moon H (2005) Comparative studies on coagulation and adsorption as a pretreatment method for the performance improvement of submerged MF membrane for secondary domestic wastewater treatment. Sep Sci Technol 40:2613–2632

Lee SJ, Lee YJ, Nam SH (2008) Improvement in the coagulation performance by combining Al and Fe coagulants in water purification. Korean J Chem Eng 25:505–512

Lesage N, Sperandio M, Cabassud C (2007) Study of a hybrid process: adsorption on activated carbon/membrane bioreactor for the treatment of an industrial wastewater. Chem Eng Process 47(3):303–307

Li WW, Wang YK, Sheng GP, Gui YX, Yu L, Xie TQ, Yu HQ (2012) Integration of aerobic granular sludge and mesh filter MBR for cost-effective wastewater treatment. Bioresour Technol 122:22–26

Li Kai, Liang H, Qu F, Shao S, Yu H, Han ZH, Du X, Li G (2014) Control of natural organic matter fouling of ultra-filtration membrane by adsorption pretreatment: Comparison of mesoporous adsorbent resin and powdered activated carbon. J Membr Sci 471:94–102

Li H, Shi W, Zhang Y, Zhou R (2015) Comparison study of the effect of blending method on PVDF/PPTA blend membrane structure and performance. Membrane Water Treat 6(3):205–224

Lin H, Gao W, Meng F, Qiangliao B, Leung KT, Zhao L, Chen J, Hong H (2012) Membrane bioreactors for industrial wastewater treatment: a critical review. Environ Sci Technol 42:677–740

Lin H, Peng W, Zhang M, Chen J, Hong H, Zhang Y (2013) A review on anaerobic membrane bioreactors: applications, membrane fouling and future perspectives. Desalination 314:169–188

Lin H, Zhang M, Wang F, Meng F, Liao BQ, Hong H, Chen J, Gao W (2014) A critical review of extracellular polymeric substances (EPSs) in membrane bioreactors: characteristics, roles in membrane fouling and control strategies. J Memb Sci 460:110–125

Liu L, Shao B, Yang F (2013) Polydopamine coating – surface modification of polyester filter and fouling reduction. Sep Purif Technol 118:226–233

Lorenzen S, Yea Y, Chen V, Christensen ML (2016) Direct observation of fouling phenomena during cross-flow filtration: Influence of particle surface charge. J Membr Sci 510:546–558

Lyko SD, Halbouni Al, Wintgens T, Janot A, Hollender JW, Melin DT (2007) Polymeric compounds in activated sludge supernatant characterization and retention mechanisms at a full-scale municipal membrane bioreactor, Water Res 41:3894–3902

Ma C, Yu S, Shi W, Heijman SGJ, Rietveld LC (2013) Effect of different temperatures on performance and membrane fouling in high concentration PAC–MBR system treating micro-polluted surface—water. Biores Technol 141:19–24

Mafirad S, Mehrnia MR, Sarrafzadeh MH (2011) Effect of membrane characteristics on the performance of membrane bioreactors for oily waste water treatment. Water Sci Technol 64 (5):1154–1160

Masse A, Sperandio M, Cabassud C (2006) Comparison of sludge characteristics and performance of a submerged membrane bioreactor and an activated sludge process at high solids retention time. Water Res 40:2405–2415

Matsui Y, Fukuda Y, Inoue T, Matsushita T (2003) Effect of natural organic matter on powdered activated carbon adsorption of trace contaminants: characteristics and mechanism of competitive adsorption. Water Res 37:4413–4424

Malaeb L, Le-Clech P, Vrouwenvelder JS, Ayoub GM, Saikaly PE (2013) Do biological-based strategies hold promise to biofouling control in MBRs? Water Res 47(15):5447–5463

Meng F, Zhang H, Yang F, Liu L (2007) Characterization of Cake Layer in submerged membrane bioreactor. Environ Sci Technol 41:4065–4070

Meng F, Chae SR, Drews A, Kraume M, Shin HS, Yang F (2009) Recent advances in membrane bioreactors (MBRs): membrane fouling and membrane material. Water Res 43:1489–1512

Meng F, Liao B, Liang S, Yang F, Zhang H, Song L (2010) Morphological visualization, componential characterization and microbiological identification of membrane fouling in membrane bioreactors (MBRs). J Membr Sci 361:1–14

Miyoshi T, Tsuyuhara T, Ogyu R, Kimura K, Watanabe Y (2009) Seasonal variation in membrane fouling in membrane bioreactors (MBRs) treating municipal wastewater. Water Res 43:5109–5118

Miyoshia T, Yuasaa K, Ishigamia T, Rajabzadeha S, Kamioa E, Ohmukaia Y, Saekia D, Nib J, Matsuyamaa H (2015) Effect of membrane polymeric materials on relationship between surface pore size and membrane fouling in membrane bioreactors. Appl Surf Sci 330:351–357

Mohammadi T, Madaeni SS, Moghadam MK (2002) Investigation of membrane fouling. Desalination 153:155–160

Mukai T, Takimoto K, Kohno T, Okada M (2010) Ultrafiltration behavior of extracellular and metabolic products in activated sludge system with UF separation process. Water Res 34:902–908

Myat DT, Mergen M, Zhao O, Stewart MB, Orbell JD, Merle T, Croue JP, Gray SR (2014) Membrane fouling mechanism transition in relation to feed water composition. J Membr Sci 471:265–273

Nagaoka H, Nemoto H (2005) Influence of extracellular polymeric substances on nitrogen removal in an intermittently-aerated membrane bioreactor. Water Sci Technol 51(11):151–158

Nanda D, Tung KL, Li YL, Lin NJ, Chuang CJ (2010) Effect of pH on membrane morphology, fouling potential, and filtration performance of nano filtration membrane for water softening. J Membr Sci 349:411–420

Ng HY, Tan TW, Ong SL (2006) Membrane fouling of submerged membrane bioreactors: impact of mean cell residence time and the contributing factors. Environ Sci Technol 40(8):2706–2713
Pan JR, Su Y, Huang C (2010) Characteristics of soluble microbial products in membrane bioreactor and its effect on membrane fouling. Desalination 250:778–780
Pollice A, Laera G, Saturno D, Giordano C (2008) Effects of sludge retention time on the performance of a membrane bioreactor treating municipal sewage. J Membr Sci 317 (1–2):65–70
Poostchi AA, Mehrnia MR, Rezvani F, Sarrafzadeh MH (2012) Low-cost monofilament mesh filter used in membrane bioreactor process: Filtration characteristics and resistance analysis. Desalination 286:429–435
Prakash P, Hoskins D, Sengupta AK (2004) Application of homogeneous and heterogeneous cation-exchange membranes in coagulant recovery from water treatment plant residuals using Donnan membrane process. J Membr Sci 237:131–144
Ramesh A, Lee DJ, Lai JY (2007) Membrane biofouling by extracellular polymeric substances or soluble microbial products from membrane bioreactor sludge. Appl Microbiol Biotechnol 74:699–707
Rana D, Matsuura T (2010) Surface modifications for antifouling membranes. Chem Rev 110 (4):2448–2471
Reid E, Liu X, Judd SJ (2006) Effect of high salinity on activated sludge characteristics and membrane permeability in an immersed membrane bioreactor. J Membr Sci 283:164–171
Remy M, van der Marel P, Zwijnenburg A, Rulkens W, Temmink H (2009) Low dose powdered activated carbon addition at high sludge retention times to reduce fouling in membrane bioreactors. Water Res 43:345–350
Rezaei M, Mehrnia MR (2014) The influence of zeolite (clinoptilolite) on the performance of a hybrid membrane bioreactor. Bioresour Technol 158:25–31
Rosenberger S, Kraume M (2002) Filterability of activated sludge in membrane bioreactor. Desalination 151:195–200
Rosenberger S, Evenblij H, TePoele S, Wintgens T, Laabs C (2005) The importance of liquid phase analyses to understand fouling in membrane assisted activated sludge processes—Six case studies of different European research groups. J Membr Sci 263:113–126
Rosenberger S, Laabs C, Lesjean B, Gnirss R, Amy G, Jekel M, Schrotter JC (2006) Impact of colloidal and soluble organic material on membrane performance in membrane bioreactors for municipal wastewater treatment. Water Res 40:710–720
Ross WR, Barnard JP, Strohwald NKH, Grobler CJ, Sanetra J (1992) Practical application of the ADUF process to the full-scale treatment of a maize processing effluent. Water Sci Technol 25 (10):27–39
Sanguanpak S, Chiemchaisri C, Chiemchaisri W, Yamamoto K (2015) Influence of operating pH on biodegradation performance and fouling propensity in membrane bioreactors for landfill leachate treatment. Int Biodeterior Biodegradation 102:64–72
Santos A, Ma W, Judd SJ (2011) Membrane bioreactors: two decades of research and implementation. Desalination 273:148–154
Santos A, Judd S (2010) The commercial status of membrane bioreactors for municipal waste water. Sep Sci Technol 45(7):850–857
Semblante GU, Sanggam DRT, You SJ, Lin YF, Chang TC, Yen FC (2013) Fouling reduction in membrane reactor through magnetic particles. J Membr Sci 435:62–70
Shang R, Vuong F, Hu J, Li S, Kemperman AJB, Nijmeijer K, Cornelissen ER, Heijman SGJ, Rietveld LC (2015) hydraulically irreversible fouling on ceramic MF/UF membranes: comparison of fouling indices, foulant composition and irreversible pore Narrowing. Sep Purif Technol 147:303–310
She Q, Wang R, Fane AG, Tang CY (2016) Membrane fouling in osmotically driven membrane processes: a review. J Memb Sci 499:201–233
Shen LG, Lei Q, Chen JR, Hong HC, He YM, Lin HJ (2015) Membrane fouling in a submerged membrane bioreactor: impacts of floc size. Chem Eng J 269:328–334

Shi X, Tal G, Hankins NP, Gitis V (2014) Fouling and cleaning of ultrafiltration membranes: a review. J Water Process Eng 1:121–138

Skouteris G, Saroj D, Melidis P, Hai FI, Ouki S (2015) The effect of activated carbon addition on membrane bioreactor processes for wastewater treatment and reclamation–a critical review. Biores Technol 185:399–410

Song KG, Kim Y, Ahn KH (2008) Effect of coagulant addition on membrane fouling and nutrient removal in a submerged membrane bioreactor. Desalination 221:467–474

Strohwald NKH, Ross WR (1992) Application of the ADUF process to brewery effluent on a laboratory scale. Water Sci Technol 25(10):95–105

Stuckey DC (2010) Anaerobic membrane reactors. In: Fang HHP (ed) Environmental anaerobic technology: applications and new developments. Imperial College Press London, UK, pp 137–161

Su X, Tian Y, Li H, Wang C (2013) New insights into membrane fouling based on characterization of cake sludge and bulk sludge: an especial attention to sludge aggregation. BioresTechnol 128:586–592

Sun DD, Liu S (2013) Comparison study on membrane fouling by various sludge fractions with long solid retention time in membrane bioreactor. Membrane Water Treat 4(3):175–189

Sweity A, Ying W, Belfer S, Oron G, Herzberg M (2011) pH effects on the adherence and fouling propensity of extracellular polymeric substances in a membrane bioreactor. J Membr Sci 378:186–193

Thuy QTT, Visvanathan C (2006) Removal of inhibitory phenolic compounds by biological activated carbon coupled membrane bioreactor. Water Sci Technol 53(11):89–97

Tian Y, Chen L, Zhang S, Shuai CC (2011) Correlating membrane fouling with sludge characteristics in membrane bioreactors: An especial interest in EPS and sludge morphology analysis. Biores Technol 102:8820–8827

Trussell RS, Adham SR, Trussell RR (2005) Process limits of municipal wastewater treatment with the submerged membrane bioreactor. J Environ Eng 131(3):410–416

Trussell RS, Merlo RP, Hermanowicz SW, Jenkins D (2007) Influence of mixed liquor properties and aeration intensity on membrane fouling in a submerged MBR at high mixed liquor suspended solids concentrations. Water Res 41:947–958

Trzcinski AP, Stuckey DC (2016) Inorganic fouling of an anaerobic membrane bioreactor treating leachate from the organic fraction of municipal solid waste (OFMSW) and a polishing aerobic membrane bioreactor. Biores Technol 204:2–17

Tsuyuhara T, Hanamoto Y, Miyoshi T, Kimura K, Watanabe Y (2010) Influence of membrane properties on physically reversible and irreversible fouling in membrane bioreactors. Water Sci Technol 61(9):2235–2240

Ueda T, Hata K (1999) Domestic wastewater treatment by a submerged membrane bioreactor with gravitational filtration. Water Res 33(12):2888–2892

Ueda T, Hata K, Kikuoka Y, Seino O (1997) Effects of aeration on suction pressure in a submerged membrane bioreactor. Water Res 31:489–494

Urbanowska A, Kabsch-Korbutowicz M (2016) Cleaning agents efficiency in cleaning of polymeric and ceramic membranes fouled by natural organic matter. Membrane Water Treat 7 (1):1–10

Wang Z, Wu Z, Yin X, Tian L (2008) Membrane fouling in a submerged membrane bioreactor (MBR) under sub-critical flux operation: membrane foulant and gel layer characterization. J Memb Sci 325(1):238–244

Wang Z, Ma J, Tang CY, Kimura K, Wang Q, Han X (2014) Membrane cleaning in membrane bioreactors: a review. J MembSci 468:276–307

Wisniewski AG (1996) Membrane bioreactor for treatment of polluted effluents: Biological performance and advantages. Med Fac Landbouw Univ Gent 61, (4b)2017–2024,44

Wisniewski AG (1998) Floc size distribution in a membrane bioreactor and consequences for membrane fouling. Colloids Surf A 138:403–411

Wong LY, Nga CA, Bashir MJK, Koo CK, Humaire N (2016) Enhancement of membrane fouling control in hybrid aerobic membrane bioreactor system for domestic waste water application: effect of alum concentration. Proc Eng 148:726–734

Wu JL, Chen FT, Huang X, Geng WY, Wen XH (2006) Using inorganic coagulants to control membrane fouling in a submerged membrane bioreactor. Desalination 197(1–3):124–136

Wu J, Le-Clech P, Stuetz RM, Fane AG, Chen V (2008) Effects of relaxation and backwashing conditions on fouling in membrane bioreactor. J Membr Sci 324:26–32

Wu J, Zhuang Y, Li H, Huang X (2010) pH adjusting to reduce fouling propensity of activated sludge mixed liquor in membrane bioreactor. Sep Sci Technol 45:890–895

Xing C, Wen XH, Qian Y, Sun D, Klose PS, Zhang XQ (2003) Fouling and cleaning of microfiltration membrane in municipal wastewater reclamation. Water Sci Technol 47 (1):263–270

You HS, Huang CP, Pan JR, Chang SC (2006) Behavior of membrane scaling during cross-flow filtration in the anaerobic MBR system. Sep Sci Technol 41(7):1265–1278

Zhang Y, Bu D, Liu CG, Luo X, Gu P (2004) Study on retarding membrane fouling by ferric salts dosing in membrane bioreactors. Proceedings of the IWA specialty conference WEMT, Seoul, 7–10 June

Zhang J, Zhou J, Liu Y, Fane AG (2010) A comparison of membrane fouling under constant and variable organic loadings in submerge membrane bioreactors. Water Res 44:5407–5413

Zhang Y, Zhang M, Wang F, Hong H, Wang A, Wang J, Weng X, Lin H (2014a) Membrane fouling in a submerged membrane bioreactor: effect of pH and its implications. Biores Technol 152:7–14

Zhang A, Liu Z, Chen Y, Kuschk P, Liu Y (2014b) Effects of EPS on membrane fouling in a hybrid membrane bioreactor for municipal wastewater treatment. Membrane Water Treat 5 (1):1–14

Zhang M, Liao BQ, Zhou X, He Y, Hong H, Lin H, Chen J (2015) Effects of hydrophilicity/ hydrophobicity of membrane on membrane fouling in a submerged membrane bioreactor. Biores Technol 175:59–67

Zhang H, Fan X, Wang B, Song L (2016) Calcium ion on membrane fouling reduction and bioflocculation promotion in membrane bioreactor at high salt shock. Biores Technol 200:535–540

Chapter 14
Eco-Friendly Treatment Strategies for Wastewater Containing Dyes and Heavy Metals

M. P. Premkumar, K. V. Thiruvengadaravi, P. Senthil Kumar, J. Nandagopal and S. Sivanesan

Abstract In the present scenario, people are suffering due to the scarcity of freshwater and clean drinking water and it remains a worldwide problem. Fast growth of industries and urban localities and change in lifestyle lead to the increase in the necessity of different kinds of synthetic materials and have led to pollution of water. Water pollution is a major environmental issue faced by the modern world, which leads to ecological disequilibrium that can cause harmful effect on flora and fauna of the ecosystem. Heavy metals and synthetic dyes are the major pollutants to be prioritized in wastewater treatment because of their lethal toxicity. Heavy metal poisoning in human being could have a direct impact on the drinking water, on the very air for breathing, and on the food chain. Synthetic dyes which are in use are highly poisonous and difficult to degrade because of their complex form. At present, researchers are focused on the treatment of wastewater containing toxic and complex organic contaminants. Adsorption is a promising technology for treating wastewater with heavy metal contaminants. Recovery and reusability of the adsorbents make adsorption an eco-friendly and cost-effective technology. Photocatalysis is a highly proven treatment technique for dyes intoxicated wastewater. Conversion of non-biodegradable and complex organic dyes into simple biodegradable molecules by photocatalysis is a greater addition in wastewater treatment. Concentration of heavy metal ions could be measured throughout the treatment process using Atomic Absorption Spectrophotometer (AAS), and for dyes, UV–Visible Spectrophotometer could be employed. Kinetic modeling and adsorption isotherms would pave the way for the better understanding

M. P. Premkumar · K. V. Thiruvengadaravi · S. Sivanesan (✉)
Department of Applied Science and Technology,
Alagappa College of Technology, Anna University, Chennai 600025, India
e-mail: siva@annauniv.edu; sivanesh1963@gmail.com

P. Senthil Kumar
Department of Chemical Engineering, SSN College of Engineering,
Old Mahabalipuram Road, Kalavakkam 603110, India

J. Nandagopal
Department of Chemistry, Velammal Engineering College, Chennai,
Tamil Nadu 600066, India

T. Gupta et al. (eds.), *Environmental Contaminants*, Energy, Environment, and Sustainability, https://doi.org/10.1007/978-981-10-7332-8_14

of the rate and nature of the adsorption of heavy metals. Biochemical oxygen demand (BOD), chemical oxygen demand (COD), and eco-toxicity studies are used to monitor the treatment processes.

Keywords Heavy metals · Dyes · Adsorption · Photocatalysis
Toxicity

14.1 Introduction

Life on planet Earth depends on various natural resources, and one such major natural resource is water. Water is the most important resource for all living organisms on Earth. Water is needed for various purposes, which include household, agricultural, industrial, recreational, and environmental activities. Planet Earth is a place full of water, with a total volume of 1.386 million cubic kilometers, and approximately 71% of its surface is covered by it, being 97% in oceans and seas. Only 3% of the water on Earth is freshwater, and the clear majority is trapped underground or on the surface in the form of glaciers and polar ice caps, which results in freshwater scarcity.

On Earth, 20% of the current population of about 7 billion people live in countries that lack this vital liquid (Brennan et al. 2001). In general, the maximum dependency on freshwater is for domestic, agriculture, and industrial purposes followed by the conservation of aquatic ecosystem. Currently, it has been estimated by the World Health Organization (WHO) that about 2.3 million people are below the "Hardship Threshold" and it is expected to reach about 3.5 million people in the year 2025. The WHO considers that the appropriate amount of water for human consumption (drinking, cooking, personal hygiene, household cleaning) is about 50 L per capita per day. Considering all the above-mentioned parameters, it is considered that the minimum amount of water consumption could be about 100 L per capita per day for daily use by a human being in a society.

On the other hand, due to environmental pollution, a significant fraction of the available freshwater is getting contaminated beyond recovery. Natural water sources have self-purging processes, but when used in excess and under poor waste management, the waste ended up being dispersed in water resources. This kind of large-scale waste dumping breaks the balance of elimination of contaminants by natural agents, worsening the quality of the resources available for consumption. According to the WHO, more than 1.2 million people consume water without health guarantees, which causes about 30,000 daily deaths and of many widespread diseases.

Rapid industrialization, urbanization, and increase in the number of use of different kinds of synthetic materials have led to pollution of water. In the present scenario, millions of people worldwide are suffering due to the unavailability of freshwater and clean potable water. Water pollution is a major environmental issue faced by the modern world, which leads to ecological disequilibrium which can cause harmful effect on flora and fauna of the ecosystem (Hernandez-Ramirez et al. 2008). The main pollutants of water that arise due to industrial activities include dyes and phenolic compounds, nitrates, phosphates, pesticides, biodegradable organic matter, recalcitrant organic matter, sediments, heavy metal ions, and even micropollutants emerging from pharmaceutical preparations, personal care products, and others (Table 14.1).

Water consumption is increased in the recent years with the increase in the global population, and apart from that, the growth of the level of life requires a greater demand of water. If responsible measures were not taken in sustainable management of water and water policy, future generations will face the adverse effects of the lack of water resource.

Polluted water is unsuitable for drinking, agriculture, recreation, and industrial utilization. Heavy metal ions and dyes adversely affect the potable quality of rivers, lakes, and groundwater supplies. Water pollution adversely affects the aquatic life and reduces its life-giving ability and the economy depending on it (Aklil et al. 2004) which is depicted in Fig. 14.1.

14.1.1 Environmental Issue

Environmental pollution is one of the serious concerns to be taken care of the modern world (Karthik et al. 2016). Rapid growth in agriculture, medicine, energy sources, and all chemical industries is mandatory to fulfill the needs and demands of the enormously growing human population. Most of the human activities to produce goods and services result in the accumulation of environmental pollutants.

Table 14.1 Freshwater distribution on earth's surface

Water fountain	Volume of water in kilometers cubic	Percentage of freshwater	Percentage of total water
Oceans and seas	1,338,000,000	–	96.54
Polar ice caps, glaciers, and permanent snow	24,064,000	68.7	1.74
Groundwater	23,400,000	–	1.69
Freshwater	10,530,000	30.1	0.76
Saline	12,870,000	–	0.93
Soil moisture	16,500	0.05	0.001
Underground ice	300,000	0.86	0.022
Lakes	176,400	–	0.013
Atmosphere	12,900	0.04	0.001
Swamp water	11,470	0.03	0.0008
Rivers	2120	0.006	0.0002
Biological water	1120	0.003	0.0001

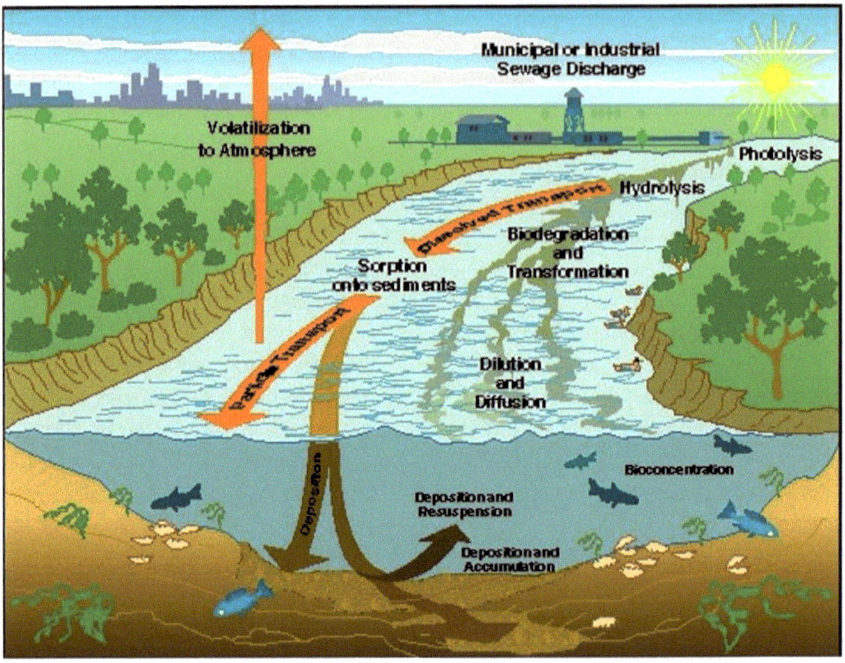

Fig. 14.1 Discharge of effluents from industries into waterbodies

These pollutants in the air, water, and soil have a direct impact on the health of humans, plants, animals, and microbes. Hence, green technologies and eco-friendly methodologies are in a desperate need to maintain a sustainable human society. Over the last two decades, there has been a worldwide awareness of the toxic and carcinogenic effects of many polluting chemicals, which were not considered hazardous in the past (Premkumar et al. 2013).

14.1.2 Industrial Organic Pollutants

A large part of the contamination of water resources and the consequent loss of water quality has its origin in the human activity through industrial activity. The industries with the greatest potential of contamination are those related to the chemical industry that processes a wide variety of organic and inorganic products, for example, the processing industry oil, paper industry, fertilizer industry, and the metallurgical industry, among others. The main pollutants of water due to human industrial activity include dyes and phenolic compounds, nutrient substances such as nitrates and phosphates, pesticides, biodegradable organic matter, recalcitrant organic matter, sediments, and even micropollutants emerging as pharmaceutical compounds, personal care products, pesticides, and others.

14.1.2.1 Synthetic Dyes and Its Effect on Ecosystem as Pollutants

In 1856, the beginning of synthetic dyes started with the unexpected discovery of the first aniline dye called Mauveine by the Englishman William Henry Perkin. A dye is an organic compound which produces color because of the conjugate chemical bonds. Synthetic dyes are widely used in textile dyeing, paper printing, color photography, pharmaceutical, food, cosmetic, and leather industries (Rafii et al. 1990; Kuhad et al. 2004). Since 1856, 7×10^5 metric tons of 105 different dyes have been produced worldwide annually (Chen et al. 2003). Generally, the materials to be colored include textile fibers, polymers, food products, leather products, and many other similar materials (Rangnekar and Singh 1980). Chromophores and auxochromes are major color-producing moieties of the dye molecule. Dyes possess an unsaturated group which absorbs and emits a wavelength known as chromophore ("*chroma*" means color and "*phore*" means bearer). Auxochromes ("*Auxo*" means augment) are the characteristic groups, which intensify color and improve the dye affinity to the substrate (Rangnekar et al. 1980). Most synthetic dyes are harmful and highly resistant to degradation because of their complicated chemical structures (Lu et al. 2009).

Industrial activities with potential for discharge of wastewater with high content of organic matter include livestock, wood-processing industries such as paper mills and industry of cork, wine industry, the food, etc. This type of wastewater treatment depends on their physicochemical and microbiological characteristics. In general, biological treatments are the most recommended for the treatment of biodegradable wastewater (Meena et al. 2005). However, many compounds generated by these industries are extremely toxic, recalcitrant, and non-biodegradable. Therefore, conventional biological treatment is not feasible for decontamination of this type of pollutants and a pretreatment or treatment with oxidizing chemicals is typically applied. Among the different treatments, the so-called advanced oxidation processes (AOPs) are the most promising technologies for decontamination of toxic, recalcitrant, and non-biodegradable pollutants.

One of the common features of non-biodegradable water is the presence of aromatic compounds. Among them, the phenolic compounds are the most characteristic and, therefore, are used frequently as compound model to evaluate the efficiency of these AOPs. While the toxicity of phenolic compounds is not as high as in the case of pesticides or heavy metals, its high concentration (up to several grams per liter) can cause inhibition or even eliminate populations of bacteria in wastewater treatment plants (Mohammed-Azizi et al. 2013).

14.1.2.2 Environmental Significance of Heavy Metals

"Heavy metals" are elements with a specific gravity of minimum 5 times the specific gravity of water. Heavy metals are the elements having a specific gravity above 5.0 and atomic weights between 63.5 and 200.6 (Srivastava and Majumder 2008). Certain heavy metals (e.g., copper, zinc, selenium) are necessary to maintain the natural metabolism of the human body, even though, above a certain concentration, the heavy metal ions will cause serious health problems in human being. Heavy metal toxicity originates from drinking water, air, and agricultural soil. Heavy metals such as nickel, iron, zinc, copper, mercury, cadmium, lead, thorium, and chromium focus on industrial wastewater treatment (Zaki et al. 2007; Demirbas 2008).

Heavy metals such as lead, copper, cadmium, zinc, and nickel are extremely harmful to human beings. These metals are widely found in the soil in trace amounts, which causes few problems. The toxic effects of these five heavy metals are as follows.

Lead

Lead is naturally found in its divalent state from several industrial and mining sources. Lead has its wide application in storage batteries, galvanization, petroleum refining, printing and pigment industries, paints, paper and pulp, and electrodes in electrochemistry and chemical industries. Lead is harmful to many organs and tissues including the heart, bones, intestines, kidneys, and the reproductive and nervous systems. It has a direct impact on the development of the nervous system and is therefore highly harmful to children which results in learning and behavioral disorders. Symptoms include abdominal pain, confusion, headache, anemia, irritability, and, in severe cases, seizures, coma, and death. The permissible limit of lead in drinking water is 0.01 mg/L (BIS 1994) according to the Bureau of Indian Standards (BIS).

Copper

Copper is one of the widely used heavy metals in many industries including metal cleaning and plating baths, paints and pigments, fertilizer, paperboard, wood pulp, printed circuit board production. The industrial effluents from the above-mentioned sectors normally contain considerable quantity of copper, which pollutes the environment and finally occupies the food chain causing health hazards. At higher dosages, copper will cause severe mucosal irritation, widespread capillary damage, hepatic and renal damage, and the central nervous system damage followed by depression. The maximum permissible concentration of copper in drinking water as regulated by the BIS is 0.05 mg/L (BIS 1992).

Cadmium

Cadmium is one of the heavy metals with high toxicity. Cadmium ions are non-biodegradable and easily get accumulated in living tissues and can be readily absorbed into the human body. An excessive level of cadmium ions in the potable water can pose a harmful threat to human beings. Cadmium toxicity causes adverse health effects such as bone lesions, cancer, lung insufficiency, and hypertension. The drinking water guideline designed by the BIS for this element is 0.003 mg/L (BIS 1992).

Zinc

Another important heavy metal which is released into the environment from various natural and anthropogenic activities, such as acid mine drainage, galvanizing plants, ores, and municipal wastewater treatments is zinc. Accumulation of zinc ions can cause hazardous effects in plants and animals. BIS has regulated the maximum permissible limit for zinc in drinking water as 5 mg/L (BIS 1994).

Nickel

Nickel gets dumped into water stream from electroplating, refining, and welding industries. Enormous amounts of nickel in water can affect the human beings. Nickel toxicity can lead to cancer, skin allergy, and lung fibrosis. The most widely seen health problems due to nickel and its compounds are allergic dermatitis (nickel itch) and increased incidence of cancers. The BIS regulates the nickel concentration not to exceed 0.02 mg/L in drinking water (BIS 2003).

Effects of Heavy Metals on Aquatic Organisms

Aquatic ecosystem is completely deteriorated by the adverse effect of heavy metals in the environment. The metals are entering the food chain through bioaccumulation by the microorganisms, which in turn is consumed by phytoplankton and further by aquatic organisms. The metals are biomagnified several times by the aquatic organisms that enter the food chain of human beings from the contaminated water.

- Histological or morphological changes in aquatic organisms happen by the increase in metal ion concentration.
- Physiological changes in the aquatic organisms such as suppression of growth and development, poor swimming performance, and changes in circulation are sub-lethal effects.
- Biochemical alterations, enzyme activity, and blood chemistry of the aquatic organisms will be seriously affected.
- Alterations in behavior and reproduction cycle.

Metal ions enter the biological systems of aquatic organisms via three main pathways:

(a) Absorption through the respiratory surface (e.g., gills) are readily diffused into the bloodstream.
(b) Adsorbed onto the body surface and passively diffused into the bloodstream.
(c) Metals ingested as free ions.

It is estimated that 70–80% of sickness in the developing countries is due to water contamination, especially for women and children, who are extremely prone to diseases caused by water pollution. Because of the development of advanced analytical techniques and better health monitoring technologies, the acceptable minimum concentration of these heavy metals is progressively decreasing. Stringent regulations have been implemented in many countries for the heavy metal contamination in water which forces the industries to treat their effluents properly before discharging them into the natural waterbodies.

14.2 Measurement of Pollutants

Mostly, all the wastewater treatment technologies use basic measurement techniques such as Atomic Absorption Spectroscopy, UV–Visible Spectroscopy, and High-Performance Liquid Chromatography. Some brief introduction about the instruments and equipment used in these measurement techniques is discussed under this section.

14.2.1 Atomic Absorption Spectrophotometer (AAS) Analysis

This is a quantitative analysis for measuring the amount of heavy metal ions present in the wastewater. Atomic Absorption Spectroscopy of atoms or ions transition involves excitation of electrons from the ground state to the excited state (Fig. 14.2).

The basic principle of the AAS is to pass the light through the atoms in its gaseous state, the wavelength of the light has energy corresponding to the energy difference between the energy levels in the atoms, and a portion of the light will be absorbed. The concentrations of metal ions can be calculated using Beer-Lambert law.

14.2.2 Ultraviolet–Visible Spectroscopy (UV) Analysis

Ultraviolet–Visible spectroscopy is based on the absorption and emission of certain wavelengths due to the excitation of electrons from the ground state to the higher energy state. It is used as a qualitative as well as quantitative analysis to measure the

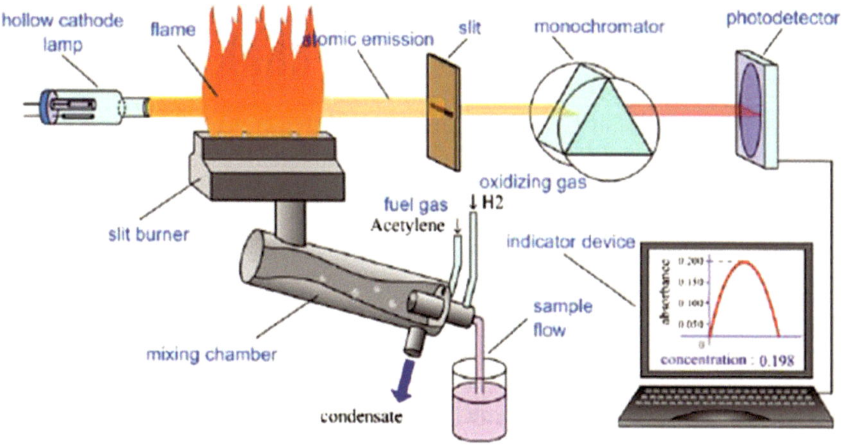

Fig. 14.2 Atomic absorption spectrophotometer

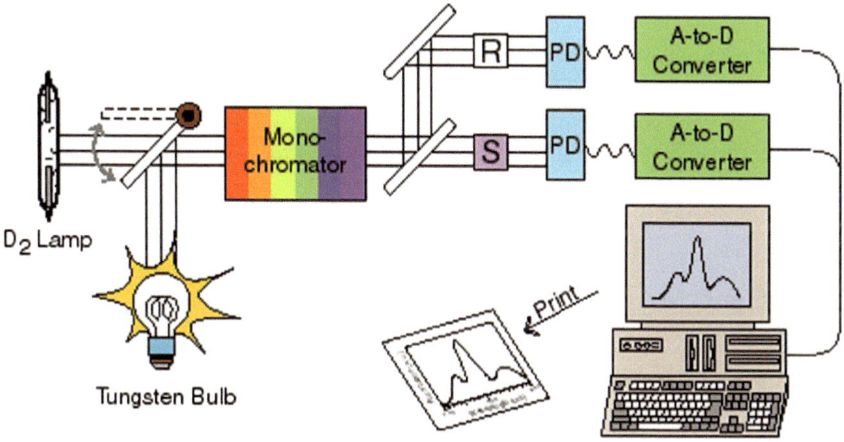

Fig. 14.3 Ultraviolet–Vis spectrophotometer

number of conjugated double bonds and aromatic conjugation within various pollutants (Fig. 14.3).

The efficiency of the adsorbent can be determined from the adsorption studies by calculating the percentage removal of pollutants using the following relationship:

$$\%\,\text{removal} = \frac{C_o - C_e}{C_o} \times 100 \quad (14.2.1)$$

where C_o and C_e are the concentrations (mg/L) of pollutants before and after adsorption.

14.2.3 High-Performance Liquid Chromatography (HPLC) Analysis

High-performance liquid chromatography is an analytical technique used to separate, identify, and quantify each component in a mixture of compounds. Pumps are used to pass a pressurized liquid solvent containing the sample mixture through a column filled with a solid adsorbent material. Each component in the sample interacts slightly differently with the adsorbent material, causing different flow rates for the different components and leading to the separation of the components as they flow out the column (Fig. 14.4).

The schematic of an HPLC instrument typically shows a sampler, pumps, and a detector. The sample mixture is passed into the mobile-phase stream by means of a sampler which carries it into the column. The flow and composition of the mobile phase through the column are delivered by the pump. The detector generates a signal proportionate to the concentration of sample elutes from the column, hence allowing for quantitative and qualitative analysis of the sample components. The HPLC instrument is controlled by a digital microprocessor and user software to provide data analysis. Mechanical pumps in a HPLC instrument can generate a composition gradient in the mobile phase by mixing multiple solvents together in ratios changing in time. UV/Vis photodiode array (PDA) based on mass spectrometry are the various detectors used generally. Column oven is used for adjusting the temperature in most of the HPLC instruments.

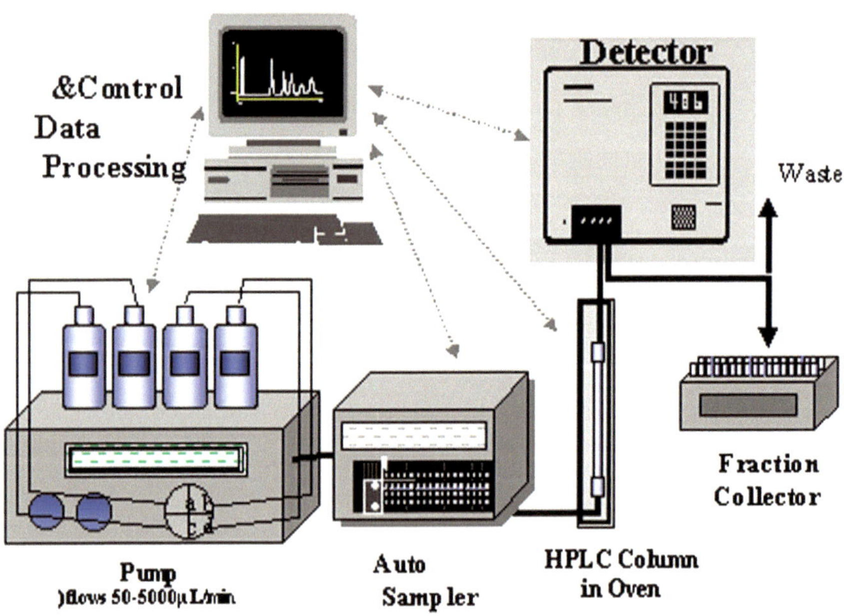

Fig. 14.4 HPLC

14.3 Modeling and Control of Pollutants

14.3.1 Wastewater Treatment Methods

The main objective of all the wastewater treatment methods is to treat the wastewater according to the permissible quality standards. The quality standards of treated water are usually designed according to the reuse of the treated water or discharging into water stream (Al-Shammiri et al. 2005). Wastewater treatment process is affected by many factors such as the reliability of the process equipment, quality of treated water, the construction space for treatment facilities, waste disposal standards, implementation, and maintenance. Physical, chemical, and biological processes are widely used for the wastewater treatment process (Bhatnagar et al. 2010).

14.3.1.1 Biological Treatment Process

Fungal decolorization, microbial degradation, adsorption by microbial biomass, and bioremediation systems are commonly applied to the treatment of industrial effluents (Sarioglu et al. 2010; Fu et al. 2001; Kikuchi et al. 2012). Bacteria, algae, yeast, and fungi are widely studied for their ability to degrade pollutants such as metal ions and dyes. The colloidal and dissolved carbonaceous organic matter has been converted into various gases and into biomass by the microorganisms. The biomass can be easily removed from the wastewater by the density gradient through gravity settling process. Biomass of the microorganisms will contribute to the biological oxygen demand (BOD), and hence, it is necessary to remove the cell tissue from the effluent after treatment.

Anaerobic–Aerobic-Activated Sludge Process

The anaerobic–aerobic-activated sludge process is a suitable method for removing heavy metal ions from the wastewater. Chromium (VI) can be biologically removed by fresh activated sludge acclimation (Chen et al. 2005). Two different kinds of air drying of aerobic and anaerobic-activated sludge were used for the removal of Cu (II), Zn (II), Mn (II), and Fe (III) from aqueous solution. The activated sludge which is obtained from the oxidation tank of Jiangning Economic Development Zone Sewage Treatment Plant was washed, air dried under room temperature and used as a biosorbent material (Wu et al. 2012). Municipal Solid Waste Incineration (MSWI) process generates bottom ash and shredded low-organic residues from which many heavy metals such as copper (Cu), lead (Pb), zinc (Zn), manganese (Mn), iron (Fe), boron (B), arsenic (As), molybdenum (Mo), and vanadium (V) were leached in co-disposal. Continuous-flow As-biofilm pilot plant was used to remove the cadmium, zinc, and nickel via conducting a series of biosorption batch tests, and biosorption capability of biomass was evaluated (Chang et al. 2006).

14.3.1.2 Chemical Treatment Processes

Chemical treatment processes consist of using one or more chemical reactions to improve the water quality. The most commonly used chemical processes are ozonation, precipitation, ion exchange, and electro-coagulation, among others.

Ozonation

Ozone is an extremely powerful oxidant and found to be efficient disinfectant (Monier et al. 2010). Ozone gets consumed in the oxidation of various organic and inorganic compounds, either by a direct oxidation of molecular ozone or through formation of secondary hydroxyl radical produced by the ozone decomposition in water (Baig et al. 2001).

Precipitation

Precipitation is one of the common chemical processes employed for the removal of metal ions or dyes from water (Ratna Kumar et al. 2004). It is widely applied to treat the hardness in potable waters caused by enormous presence of calcium and magnesium salts. Heavy metals can be removed by the addition of alkaline reagents such as limestone, lime, ferrous salts, or other compounds (e.g., $Mg(OH)_2$, $MgCO_3$, $BaCl_2$, $CaCl_2$), by means of increasing the pH of acid water to alkaline values (Morrison et al. 1998).

Ion Exchange

Ion exchange has been used extensively to remove hardness and iron and manganese salts from drinking water supplies (Rudzinski et al. 1988). It has also been used to selectively remove specific impurities and to recover valuable trace metals (chromium, copper, lead, cadmium, nickel, etc.,) and dye molecules from industrial waste discharges. Exchange of cations or anions from the surrounding materials is happening in an ion exchanger. Ion exchange resins are available selectively for certain metal ions. The sulfonic acid group (SO_3H) and other possible groups such as carboxylic acid group (–COOH) are the major active ion groups present in the cation exchange resins. Natural materials such as zeolites can be used as ion exchange resin in modified zeolites such as zeocarb and chalcarb show a greater affinity for metals such Ni(II) and Pb(II) ions (Van et al. 1977; Groffman et al. 1992). Ahmed et al. (1998) reported removal of cadmium and lead from an aqueous solution by the batch ion exchange with a solid Na–Y zeolite. Inglezakis et al. (2005) investigated the effects of competitive cations NH_4(I), K(I), Ca(II), Na(I), Mg(II), and Li(I) on the ion exchange of heavy metals Pb(II), Fe(III), Cr(III), and Cu(II) on clinoptilolite.

Electro-Coagulation

Electro-coagulation is an electrochemical process, which uses electrical current and is frequently employed to remove the heavy metals from a solution (Sakaguchi et al. 1986). This treatment methodology is widely used in the treatment of wastewater containing suspended solids, dissolved metals, tannins, and dyes. Most of the pollutants in wastewater are sustained in the solution by electrical charges. Neutralization of these ions and other charged particles by the counterions provided by the electro-coagulation system results in the destabilized state of ions and get precipitated (Escobar et al. 2006).

The removal of Zn(II) and Cd(II) ions coexisting with Fe^{2+}/Fe^{3+} in HNO_3 by electro-coagulation process was investigated (Youn et al. 2008). Arsenic was removed from the aqueous solution using mild steel electrode by electro-coagulation method. The As(III) was found to be oxidized to As(V) during the process of electro-coagulation. Effect of electrolytes such as NaCl, $NaNO_3$, and Na_2SO_4 on anodic dissolution of iron and, in turn, on the arsenic removal was deliberated.

14.3.1.3 Physical Treatment Process

Physical treatment processes are based on the physical properties of the contaminants and are generally simplest forms of treatment. Physical treatment processes include flotation, reverse osmosis, and adsorption and can remove heavy metal ions from aqueous solution.

Flotation

Suspended particles can be removed by allowing them to float on the surface and then removed by skimming because of the specific gravity gradient. Gravity flotation can be done through the following methods using air and dissolved air flotation (DAF).

Chabazite is rich in Na(I), K(I), Ca(II) which are usually exchangeable with metal ions. By dissolved air flotation process, heavy metal ions (Zn(II), Ni(II), and Cu(II)) were removed after being adsorbed in natural zeolite (chabazite) (Rubio et al. 1997).

Reverse Osmosis

Salts, metal ions, and dye molecules can be effectively removed by reverse osmosis (RO). RO is one of the physical treatment processes that use a semipermeable membrane for the removal of ions by diffusion controlled transport, a mechanism where one molecule at a time diffuses through pores in the membrane material. Migration of water (or any solvent) from a diluted solution to a concentrated solution through a semipermeable membrane is defined as osmosis. When the freshwater and saltwater are separated by a membrane, the solvent (i.e., pure water) will pass through the membrane until an equilibrium is attained in the salt concentration. The migration of water across the membrane is caused by a pressure gradient and continues until the pressure on both sides attained equilibrium. The pressure gradient with which the osmosis happens is called as the osmotic pressure. The solvent has been made to migrate from the concentrated solution to the diluted by exerting pressure higher than the osmotic pressure on the concentrated solution by reversing the direction of flow across the membrane in RO (McCray et al. 1987; Peters et al. 1991).

In reverse osmosis process, a thin-film composite spiral wound-type membrane was used for the removal of Cu(II) and Ni(II) from single-salt and mixed-salt systems and its removal efficiency was increased by adding a suitable chelating agent Na_2EDTA (Mohsen-Nia et al. 2007) (Table 14.2).

The various physical, chemical, and biological methods for treating the wastewater containing metal ions and dyes discussed above have more limitations like being economically unachievable and incapable of complete removal of the targeted metal ions, dyes, and and/or their organic pollutants. The color refuge, stability and the resistance of dyes to degradation; production of substantial amount of sludge that may cause secondary pollution; the cost of these treatment methods; are the complicated procedures (Sud et al. 2008; Febrianto et al. 2009; Wang et al. 2006).

14.3.1.4 Adsorption

An alternative method for the removal of heavy metal ions is adsorption, which is an effective and competitive process. Adsorption is the exchange of materials at the interface between two immiscible phases in contact with one another. Adsorbent is a porous material which possesses high surface area and adsorbs organic and inorganic matter by means of intermolecular forces to its surface. The substance adsorbed or attached is called the adsorbate (e.g., metal ions, dye molecules). Although there are certain phenomena associated with adsorption process known in ancient times, the uptake of glass by charcoal and clay was the first set of experiments carried out. In this case, the interaction of H^+ and metal ions between the solid and the surrounding aqueous solution is the focus. Mechanism of adsorption is given in Fig. 14.3. Adsorption by porous adsorbents proceeds through the three basic steps (McKay et al. 1980) given below.

Table 14.2 Merits and demerits of the wastewater treatment methods

Methods	Merits	Demerits
Physical		
Adsorption on activated carbon	High removal efficiency	Regeneration of support, cost-effective
Adsorption on bagasse	Utilization of waste to treat another waste	Post-treatment disposal
Adsorption on peat	Effective adsorbent, no activation required	Lower surface area than activated carbon
Adsorption on fly ash	Simple and effective adsorption, no activation required, inexpensive	Regeneration of support is difficult, highly toxic
Membrane filtration	Effective removal of wide range of colorants at low volumes	Required high dissolved oxygen, ineffective for light-resistant colorants
Chemical		
Ozonation	Effective for azo dye removal	Unsuitable for disperse dyes, releases aromatic amines
Sodium hypochlorite	Low-temperature requirement	Cost-intensive process
Coagulation and precipitation	Short detention time, low capital costs, good removal efficiencies	High cost of chemicals for pH adjustment, problems of dewatering and sludge handling
Biological		
Aerobic process	Color removal along with biological oxygen demand (BOD) and chemical oxygen demand (COD) removal	Long retention time,, low efficiency for removal of recalcitrant pollutants
Anaerobic process	Resistant to wide variety of complex colorants, produced biogas used for steam generation	Long acclimatization phase
Emerging technologies		
Advanced oxidation processes	High mineralization, effective pretreatment methodology in integrated systems, enhancement of biodegradability, removal of toxicity	High consumption of oxidants
Electrochemical oxidation	Decolorization of wide range wastes, no alternation in volume	Sludge disposal problem, requirement of low pH values
Sonication	Highly effective for integrated systems	Relatively new method, awaiting full-scale application
Enzymatic treatment	Effective for specific compounds, unaffected by shock loadings, required shorter contact times	Cost of enzymes, enzyme stability, product inhibition
Redox mediators	Easily available, enhancement of the process by increasing electron transfer efficiency	Mediator may give antagonistic effect and depend on biological activity of the system

(continued)

Table 14.2 (continued)

Methods	Merits	Demerits
Engineered wetland systems	Cost-effective technology, applicable for huge volumes of wastewater	High installation cost, requires expertise, and difficult to manage during monsoon

- Transport of the adsorbate molecule from the bulk to the exterior of the adsorbent and adsorption at the exterior surface (external diffusion).
- Migration of the adsorbate into the pores of the adsorbent (gradual adsorption stage).
- Interaction of the adsorbate with the available sites on the interior surfaces, bounding with the functional group, the pore and capillary spaces of the adsorbent (intraparticle diffusion).

Adsorption is widely used for industrial applications and in water purification. The exact nature of the bonding between the adsorbent and the adsorbate depends on the species involved in the adsorption process. The rate of adsorption is defined as the rate at which substances are transferred from the liquid phase to the solid phase (Fig. 14.5).

From the available literature, the advantages of adsorption over several other conventional treatment methods are as follows:

- Less land area (half to quarter of what is required in a biological system).
- Lower sensitivity to diurnal variation.
- Greater flexibility in the design and operation, and superior removal of organic contaminants.
- Fully controlled operation.
- Scope for expansion and modular design.
- Economical and efficient method of TDS removal.

Types of Adsorption

Adsorption can be categorized based on the following:

(i) **Concentration of adsorbate**:

The concentration of adsorbate is different at the surface of the adsorbent as compared to connecting substance or interior phase. It is further subdivided into:

- positive adsorption and
- negative adsorption.

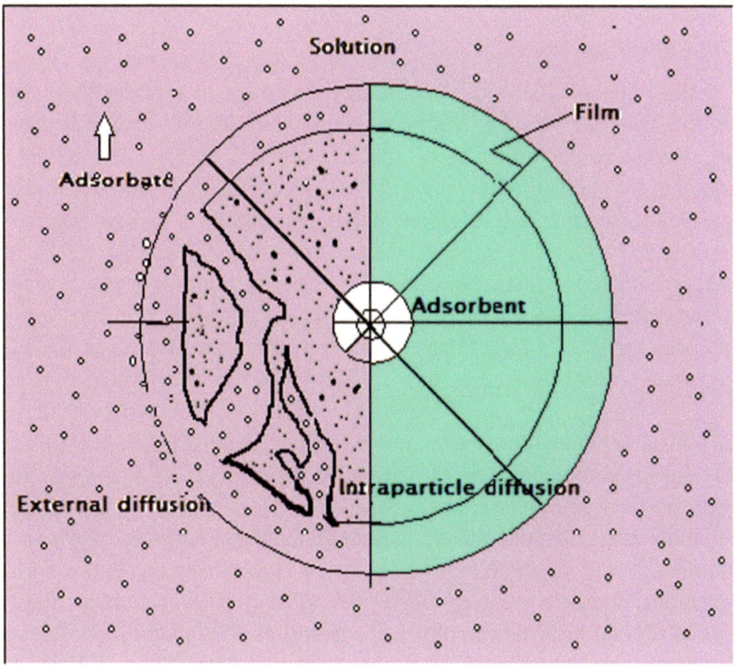

Fig. 14.5 Adsorption mechanism

(ii) **Nature of the force existing between the adsorbate and adsorbent molecule**:

It is further classified into:

- physical adsorption and
- chemical adsorption.

Physical and Chemical Adsorption

Aqueous suspensions of finely divided solids exhibit charged solid surfaces. The origin of this charge may be either due to the interaction of the solid surface functional groups with the aqueous medium (e.g., ion exchange) or due to the broken edges of microcrystallites that may constitute the sorbent surface. Most sorbent surfaces are heterogeneous, and it is difficult to identify the individual types of sites and sorbate interactions (Suzuki 1990).

The nature and strength of these interactions form the basis for a broad categorization of the adsorption process, namely:

i. physical adsorption and
ii. chemical adsorption.

When the molecules are adsorbed onto the surface of the adsorbent by weak physical or van der Waal's forces, it is termed as physical adsorption (or physisorption) or van der Waal's adsorption. Such adsorption is characterized by a comparatively low heat of adsorption, viz. 20–40 kcal/mol. This kind of adsorption is reversible, since the molecules are not tightly retained by the adsorbent and desorption (removal of adsorbed substance from the surface) processes that are closely approximated to the adsorption curves as their mirror images, though the driving force may be less. Forces responsible for physisorption are very weak. When the van der Waal's forces can extend from one layer to another, the adsorbed layer becomes several molecules thick, which is termed as multilayer adsorption (Ibach 2006) (Fig. 14.6).

Chemical adsorption (chemisorption) or activated adsorption, the force which holds the adsorbed molecules, is of valence type. This type of adsorption involves interaction of the free valences at the surface molecules of the adsorbent and the adsorbed molecules. Complete interaction of all available valences on the surface of the adsorbent would produce a monomolecular layer over the entire surface. In chemisorptions, the heat evolved (40–400 kcal/mol) is considerably higher than that evolved (20–40 kcal/mol) in physical adsorption. Chemisorption is irreversible, since molecules are tightly retained by adsorbent. The transition of physical sorption to chemisorptions may be possible, particularly at elevated temperatures, when

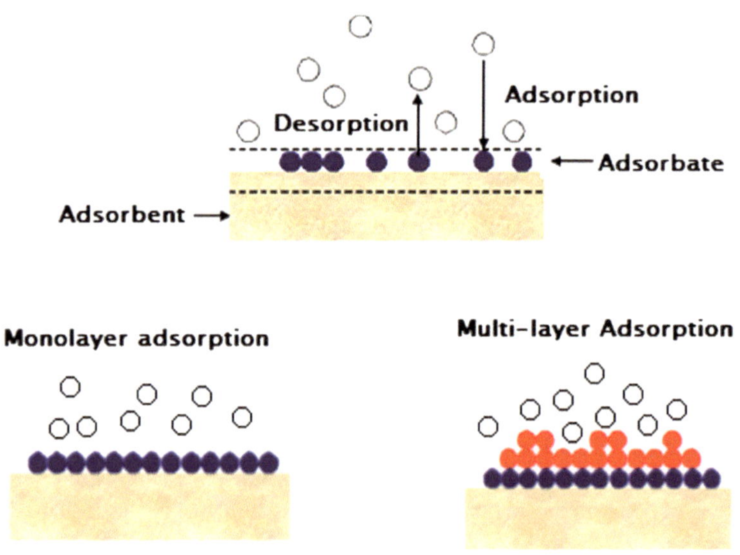

Fig. 14.6 Monolayer and multilayer adsorption

the adsorbed molecules acquire sufficient energy to involve in chemical interaction with the sorbent surface (Ibach 2006) (Fig. 14.7).

Commercial Adsorbents

Several materials have been extensively examined as adsorbents in mitigation of water pollution. Few commercial adsorbents in use are silica gel, activated alumina, zeolites, activated carbon, etc.

Silica Gel

Silica gels are majorly divided into three classes based on their density: regular-, intermediate-, and low-density gels. Regular density silica gel is synthesized in an acid medium with a high surface area of 750 m^2/g. Intermediate- and low-density silica gels possess a low surface area 300–350 and 100–200 m^2/g, correspondingly. Silica gel is a widely used adsorbent in most of the industries (Ahmed and Ram 1992; Backhaus et al. 2001). Improved forms of silica have also been widely investigated for the treatment of different pollutants (Moriguchi et al. 2005; Saad et al. 2008; Wang et al. 2009).

Activated Alumina

Activated alumina contains several non-equilibrium forms of moderately hydroxylated alumina oxide, Al_2O_3. In general, activated alumina is prepared by heating the precursor hydrous alumina to remove the hydroxyl groups, leaving a porous solid structure. Activated alumina possesses a surface area ranging from 200 to 300 m^2/g. It is used in the treatment of wastewater containing organic pollutants, gasoline, kerosene, oils, aromatic hydrocarbons, and many chlorinated hydrocarbons (Ku and Chiou 2002; Kasprzyk-Hordern 2004; Singh and Pant 2004; Naiya et al. 2009).

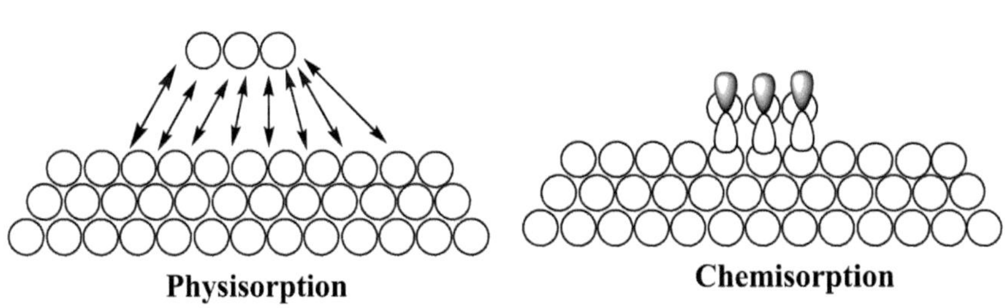

Fig. 14.7 Physisorption and chemisorption

Zeolites

There are about 40 natural and 100 synthetic zeolites with Si/Al ratios between 1 and infinity which are selectively used as adsorbents. Zeolite-based materials are widely used in detergent manufacture, water softening resins (i.e., water softeners), petroleum industry, separation processes (i.e., molecular sieves), and as adsorbents for carbon dioxide and hydrogen sulfide containing wastewater (Ellis and Korth 1993; Okolo et al. 2000; Metes et al. 2004; Motsi et al. 2009).

Activated Carbon

Activated carbon has undeniably been used as the most popular and widely used adsorbent in wastewater treatment plants. Charcoal is the widely used activated carbon which is identified as the oldest adsorbent known in wastewater treatment. Activated carbon is synthesized by dehydration of raw material followed by activation through carbonization. Activated carbon has a highly porous structure with a large surface area, ranging from 600 to 2000 m^2/g. Activated carbon has been used as an adsorbent for the treatment of varied types of pollutants such as metal ions (Perez-Candela et al. 1995; Gabaldon et al. 1996, 2000; Sanchez-Polo and Rivera-Utrilla 2002), anions (Bao et al. 1999; Huang and Cheng 2008; Mahmudov and Huang 2010), dyes (Pelekani and Snoeyink 2000; Al-Degs et al. 2001; Pereira et al. 2003; Gomez et al. 2007), phenols (Chern and Chien 2002; Haydar et al. 2003; Dabrowski et al. 2005), detergents (Bele et al. 1998; Malhas et al. 2002), pesticides (Foo and Hameed 2010), humic substances (McCreary and Snoeyink 1980; Lee et al. 1983), chlorinated hydrocarbons (Sotelo et al. 2002), and many other chemicals and organisms (Saito 1984; McKay et al. 1985; Najm et al. 1993; Rivera-Utrilla et al. 2003).

Despite of the abundant uses of activated carbon, its applications are sometimes constrained because of their high cost. Hence, researchers are on the lookout for low-cost materials for water pollution control predominantly, where the cost factor plays a major role. For quite some time, efforts have been directed toward the development of alternative adsorbents which are efficient and cheap. They can be prepared from a wide variety of raw materials, which are abundant and possess high organic (carbon) content and low inorganic content and can be easily activated (Moreno-Castilla and Rivera-Utrilla 2001).

Several advantages have been found in preparation of low-cost adsorbents from waste materials, mainly of environmental and economic importance. A wide variety of low-cost adsorbents have been prepared from different waste materials including agricultural, industrial, and municipal wastes.

Adsorbents from Agricultural Wastes

Agricultural materials exhibit potential sorption capacity toward various pollutants. The basic components of the agricultural waste materials are cellulose, hemicellulose, lignin, lipids, proteins, simple sugars, water, hydrocarbons, and starch. Agricultural waste materials are cost-effective and eco-friendly due to their unique chemical composition and renewable nature and are a sustainable resource for wastewater remediation (Ahmedna et al. 2000). As a result, the conversion of agricultural wastes into low-cost adsorbents is a promising alternative to solve environmental problems and to minimize the complications included in the preparation of the adsorbents. In the last several decades, several agricultural wastes have been explored as low-cost adsorbents. Some of them include the shells and/or stones of fruits like nuts (Nguyen et al. 1995; Ahmadpour and Do 1997; Toles et al. 1998), peanuts (Wafwoyo et al. 1999), olive wastes (Nyazi et al. 2005), almonds (Christopher and Wayne 2002), apricots stones (Soleimani and Kaghazchi 2008), and cherries (Lessier et al. 1994); wastes resulting from the production of cereals such as rice (Khalil 1996), maize (Elizalde-Gonzalez et al. 2008), and corn (Tsai et al. 2001) as well as sugarcane bagasse (Girgis 1994) and coir pith (Namasivayam and Sangeetha 2006). These agricultural waste materials have been used either in their natural form or after certain physical or chemical modifications.

Adsorbents from Rice and Wheat Waste

Rice is the seed of a monocot plant, *Oryza sativa*. It is one of the major crops grown throughout the world. As a cereal grain, it is the most important staple food for a large part of the world's human population. Thus, the rice industry produces several by-products, such as rice husk, rice hull, rice husk/hull ash, and rice bran. Wheat (Triticum spp.) is also an important staple food, generally used to make flour. Similar to the rice industry, the wheat industry also produces certain by-products, such as wheat bran and wheat husk, whose use as adsorbents has been explored.

Rice husk has been tested for arsenic removal from water by Nasir et al. (1998). The adsorption behavior of rice husk for the removal of antimony ions from aqueous solutions has been investigated by Khalid et al. (2000). The same group of researchers have also studied the removal of Pb(II) and Hg(II) ions using rice husk as an adsorbent (Khalid et al. 1998, 1999). The sorption of Cd(II) ion from an aqueous solution by rice husk has been studied by Kumar and Bandyopadhyay (2006). Rice hull ash has also been explored as an adsorbent for the removal of Pb (II) ion from an aqueous solution by Wang and Lin (2008). Acid-treated (H_3PO_4) rice bran was used as a low-cost sorbent for the removal of nickel by Zafar et al. (2007). Wheat bran has been utilized for the removal of Cd(II) ion from wastewater by Singh et al. (2006). The adsorption of Pb(II) ion from an aqueous solution on wheat bran has also been reported by Bulut and Baysal (2006).

Adsorbents from Tea and Coffee Waste

Tea is the product of the leaves, leaf buds, and internodes of the *Camellia sinensis* plant, which is prepared and cured by various methods, and it is the most popular drink in the world in terms of its consumption, and the world tea production was 4.73 million tonnes annually in 2008. Waste/spent tea leaves have been investigated as adsorbents by various researchers for the removal of different types of pollutants from water.

Waste tea leaves have been used for the removal of zinc, iron, lead, and nickel ions from water by Ahluwalia and Goyal (2005). The adsorption of copper and lead ions onto tea waste from an aqueous solution has also been studied by Amarasinghe and Williams (2007). Mozumder et al. (2008) have explored the potential of tea leaves waste for Cr(VI) ion removal from an aqueous solution. The adsorption ability of Turkish tea waste (fibrous) was investigated for the removal of Cu(II) and Cd(II) ions from single (non-competitive) and binary (competitive) aqueous systems by Cay et al. (2004). Batch adsorption studies have been performed for evaluating the suitability of tea factory waste as a low-cost adsorbent for the removal of zinc (Wasewar et al. 2009).

Globally, coffee is another most popular beverage and 6.7 million metric tonnes of coffee was produced annually in 1998–2000, and the forecast is a rise to 7 million metric tonnes annually in future. Wastes from the coffee industry have been explored as adsorbents for water treatment. The adsorption behavior of heavy metals on arabica and robusta roasted coffee beans was investigated by Minamisawa et al. (2005). Coffee residue has been used as the raw material in the preparation of powder-activated carbon by chemical activation with zinc chloride for the sorption of Pb(II) ion from dilute aqueous solutions (Boudrahem et al. 2009).

Adsorbents from Coconut Waste

Coconut (*Cocos nucifera*) is highly nutritious and rich in fiber, vitamins, and minerals. Coconut palms are grown in more than 80 countries around the world. Coconut wastes such as coir pith, coconut bunch waste, coconut husk, copra meal, and male flowers of the coconut tree have been widely explored as adsorbents for the removal of various pollutants from water. Coir pith waste has been used for the adsorptive removal of Pb(II) ion from an aqueous solution by Kadirvelu and Namasivayam (2000).

Coir pith has also been investigated as an adsorbent for Co(II), Cr(III), and Ni(II) ions adsorption in both single- and multi-component systems by Parab et al. (2006). Namasivayam et al. (2001) have studied the use of coir pith as an adsorbent for the adsorption of rhodamine B and acid violet dyes. Coir pith has also been examined for Congo red dye removal (Namasivayam and Kavitha 2002). Coconut copra meal, a waste product of coconut oil production, was investigated as a sorbent for cadmium removal from an aqueous solution by Ho and Ofomaja (2006).

Adsorbents from Peanut or Groundnut Waste

Peanut, or groundnut (Arachis hypogaea), is a species in the legume "bean" family, and its world production of 34.43 million metric tonnes of peanuts has been reported in 2008–2009. India and China are the world's largest producers of peanuts. Peanut/groundnut's husk/hull has been largely used as an adsorbent for the removal of various pollutants from water.

Okieimen et al. (1991) studied the removal of cadmium and lead ions from an aqueous solution by groundnut husk modified with EDTA. Peanut hull has been converted into an adsorbent by treatment with concentrated sulfuric acid by Namasivayam and Periasamy (1993). Chamarthy et al. (2001) also prepared adsorbents from peanut shell by thermal treatment in the presence of phosphoric acid or citric acid and used it for the adsorption of Cd(II), Cu(II), Ni(II), Pb(II), and Zn(II) ions. Untreated and silver impregnated groundnut husk carbon were tested for the removal of Cr(VI) ion by Dubey and Gopal (2007). Activated carbon prepared from peanut shell was used for the removal of Pb^{2+} ion from an aqueous solution by Xu and Liu (2008). The use of peanut hulls for copper and lead removal was studied by Oliveira et al. (2009). Zhu et al. (2009) have explored the use of peanut hull as an adsorbent for the removal of Cu(II) ion from an aqueous solution.

Adsorbents from Peels of Different Agricultural Products

Peel, also known as skin, is the outer protective layer of a fruit or vegetable, currently gaining wide attention as an adsorbent in water treatment. Peels of different fruits, such as orange, banana, watermelon, cassava, and mango, have been used as adsorbents for the removal of different pollutants from water. Orange peel was studied for Ni(II) ion removal from electroplating wastewater by Ajmal et al. (2000). The preparation of chemically modified orange peel adsorbents and their sorption behavior for Co(II), Ni(II), Zn(II), and Cd(II) ions have also been studied by Li et al. (2008).

Banana peel, a commonly produced fruit waste, was examined as adsorbent for the removal of Cd(II) ion from environmental and industrial wastewater by Memon et al. (2008). Activated carbons prepared from waste cassava peel (an agricultural waste from the food processing industry) using physical and chemical methods have been investigated for their efficiency in the removal of dyes and metal ions from aqueous solutions (Rajeshwarisivaraj et al. 2001). The ability of pomelo peel to remove Cd(II) ion from aqueous solutions was investigated by Saikaew et al. (2009).

Adsorbents from Shells of Different Agricultural Products

Shells of different agricultural products have also been explored as adsorbents by different researchers for the removal of toxic pollutants from water. Bael fruit shell was used to prepare activated carbon and used as an efficient low-cost adsorbent to remove Cr(VI) ion from an aqueous phase (Anandkumar and Mandal 2009). Acid formaldehyde-pretreated chestnut shell was used as an adsorbent, and the effect of initial cation concentration, temperature, and pH was studied toward the optimization of Pb^{2+}, Cu^{2+}, and Zn^{2+} ions removal from aqueous solutions (Vazquez et al. 2009). The sorption of Cu(II) ion onto the chestnut shell using a batch adsorber has been studied (Yao et al. 2010). The potential of shells of hazelnut

(*Corylus avellana*) and almonds (*Prunus dulcis*) to remove Pb^{2+} ion from aqueous solutions was investigated by batch experiments (Pehlivan et al. 2009).

Adsorbents from Miscellaneous Agricultural Waste

Besides the various agricultural by-products mentioned above, different miscellaneous agricultural wastes have also been investigated as low-cost adsorbents. The utilization of grape bagasse as an adsorbent to remove Cd(II) and Pb(II) ions has been investigated by Farinella et al. (2004). The adsorption of iron and manganese to remove either each metal ion as single or both metals as a multi-component system was studied using maize cob and palm fruit bunch (Nassar et al. 2004). The adsorption potential of black gram husk was tested for the removal of Pb, Cd, Zn, Cu, and Ni ions from water (Saeed et al. 2005). Dried sugar beet pulp was used as a sorbent for the removal of Cu(II) ion from an aqueous solution (Aksu and Isoglu 2005). Sugarcane bagasse was converted into a carbonaceous adsorbent and used for the removal of cadmium and zinc from wastewater (Mohan and Singh 2002).

Adsorbents from Industrial and Municipal Wastes

Common industrial activities produce huge amount of solid waste materials as by-products. Some of these materials are being used, while others find no proper utilization and are dumped elsewhere. Industrial waste materials are available almost free of cost. If these solid wastes could be used as adsorbents, it will provide a twofold advantage to environmental pollution. Firstly, the volume of waste materials could be partly reduced, and secondly, adsorbent, if developed from such wastes, can reduce the wastewater treatment cost. In view of the low cost of such adsorbents, it may not be essential to regenerate the spent materials. Thus, a few industrial wastes have been investigated with or without treatment as adsorbents for the removal of pollutants from wastewater. Some of them are listed and discussed below.

Adsorbents from Fly Ash

The major solid by-product of thermal power plants based on coal burning is fly ash. The main uses of fly ash include construction of roads and manufacture of bricks and cement. The high percentage of silica and alumina in fly ash makes it a good option for utilization as an inexpensive adsorbent for bulk use. Panday et al. (1985) used fly ash without any pretreatment for the removal of Cu^{2+} ions. Sen and Arnab (1987) investigated the potential of fly ash for Hg(II) ion removal. The removal of lead and copper ions from aqueous solutions by fly ash was investigated by Alinnor (2007).

Adsorbents from Steel Industry Waste

Steel industries also produce several wastes in large quantities, such as blast furnace slag, dust and sludge, and they have been investigated as adsorbents. Dimitrova (1996) investigated ungranulated blast furnace slag for the removal of Cu^{2+}, Ni^{2+}, and Zn^{2+} ions from water. Slag columns were utilized by the same worker (Dimitrova 2002) for lead ion removal. Kanel et al. (2006) used blast furnace slag for As(III) ion remediation.

Adsorbents from Miscellaneous Industrial Wastes

Namasivayam and Yamuna (1999) have explored the use of waste biogas residual slurry for Cr(III) ion removal from an aqueous solution. Bhatnagar et al. (2007) explored battery industry waste used as an adsorbent for the removal of Pb, Cu, Cr, and Zn from aqueous solution.

Complicate regeneration process is not needed because of this low rate and high availability materials. The adsorption capacities of many adsorbents are not too long so the research of many newer adsorbents from waste is under process. The present research is to develop inexpensive and effective adsorbents from cashew nut shell, a common agricultural waste, as an alternative to the existing commercial adsorbents.

Modeling Strategies Used in Adsorption Studies

Adsorption Isotherms

Adsorption isotherm models play an important role in the determination of the maximum capacity of adsorption and clearly depict the relationship of the amount of metal ion adsorbed by a unit weight of the adsorbent at equilibrium. The equilibrium adsorption isotherm describes how the adsorbate interacts with the adsorbent and how the adsorbate segregates between the solution and the solid phase when the adsorption process reaches an equilibrium state. Adsorption isotherm models such as the Langmuir (1918), Freundlich (1906), Sips (1948), and Temkin (1940) can be fit for the experimental data. It is essential to find out the best-fitting isotherm model to calculate the efficiency of the prepared adsorbent and to improve the adsorber design.

The Langmuir Adsorption Isotherm

Assumption for the Langmuir adsorption isotherm is mainly based on the decrease in the intermolecular forces with distance, and accordingly, the monolayer of the adsorbate is existed at the outer surface of the adsorbent. Likewise, the main assumption of Langmuir equation for a structurally homogenous adsorbent is, all adsorption sites are energetically equivalent and identical. This indicates that the completion of adsorption occurs till the monolayer of adsorption and there will not be any interaction between the adsorbent and adsorbate molecules.

The nonlinear equation of the Langmuir isotherm model is expressed as follows:

$$q_e = \frac{q_m k_L C_e}{1 + k_L C_e} \tag{14.3.1}$$

where q_m is the maximum amount of adsorbent to give a complete monolayer on the surface (mg/g) and K_L is the energy constant representing maximum capacity of the adsorbent.

The dimensionless constant R_L can be expressed as follows:

$$R_L = \frac{1}{1 + k_L C_o} \tag{14.3.2}$$

The values of R_L implies that adsorption is favorable, unfavorable, linear, and irreversible when $0 < R_L < 1$, $R_L > 1$, $R_L = 1$ and $R_L = 0$, respectively.

The Freundlich Isotherm

The important assumption of Freundlich isotherm is the adsorption followed on a heterogeneous adsorption surface with varying energies of adsorption and is given as follows:

$$q_e = K_F \cdot C_e^{1/n} \tag{14.3.3}$$

where Freundlich constants k_f connected to adsorption capacity and n related to adsorption intensity, respectively. The value of $1/n$ that varies between 0.1 and 1.0 for the favorable adsorption of the metal ion and n is measure of the deviation of the adsorption isotherm linearity. If $n = 1$, the adsorption is linear, i.e., homogenous adsorption sites, and adsorbed species are not interacted; if $n < 1$, the adsorption is favored by the chemisorption process; if $n > 1$, then adsorption is preferred by the physisorption process.

The Sips Model

The combination of Langmuir and Freundlich isotherm models is Sips model. At low metal ion concentrations, Sips model reduces to a Freundlich isotherm, and at high metal ion concentrations, it forecasts a monolayer adsorption characteristic of Langmuir isotherm. The expression for Sips model is as follows:

$$q_e = \frac{Q_{\max} K_s C_e^{1/n}}{1 + K_s C_e^{1/n}} \tag{14.3.4}$$

where K_S is the Sips constant connected to the affinity constant $(\text{mg/L})^{-1/n}$, n is the index of the heterogeneity, and $Q_{\max}$ is the Sips maximum adsorption capacity (mg/g)

The Temkin Adsorption Isotherm Model

The Temkin adsorption isotherm model assumes that the linear decrease in the heat of adsorption of all the molecules with the interactions covered by adsorbate/adsorbate. Temkin isotherm is expressed as follows:

$$q_e = \text{B}\ln(AC_e) \tag{14.3.5}$$

where A and B are the Temkin constants.

Dubinin-Radushkevich Adsorption Isotherm

Based on the potential theory, which assumes heterogeneous surface, Dubinin-Radushkevich isotherm is applied for adsorption mechanism. The expression for Dubinin-Radushkevich isotherm is:

$$q_e = q_{m,D}\,\text{e}^{-\beta_D\varepsilon^2} \tag{14.3.6}$$

where $q_{m,D}$ is monolayer capacity of Dubinin-Radushkevich (mg/g), β_D is the constant associated with the adsorption energy, and ε is the Polanyi potential linked to equilibrium concentration as follows:

$$\varepsilon = \text{RT}\ln\left(1 + \frac{1}{C_e}\right) \tag{14.3.7}$$

where gas constant R = 8.314 J/mol K and T is the absolute temperature in Kelvin.

The adsorption per molecule of the adsorbate, when it is transferred from infinity to the surface of the solid in the solution, the mean free energy (E) can be calculated with constant β_D using the relation,

$$E = \frac{1}{\sqrt{2\beta_D}} \tag{14.3.8}$$

The final form of the Dubinin-Radushkevich isotherm is expressed as follows:

$$q_e = q_{m,D}\exp\left(-\beta_D(\text{RT}\ \ln(1 + 1/C_e))^2\right) \tag{14.3.9}$$

The Dubinin-Radushkevich isotherm constants, $q_{m,D,}$ which is the monolayer capacity and adsorption energy (β_D) which is calculated from the plot q_e versus C_e. When the E value is less than 8 kJ/mol, one mole of an ion is transferred to the adsorbent surface and it specifies physical adsorption (Rieman and Walton 1970). When the value of E is between 8 and 16 kJ/mol, it denotes ion exchange process (Helfferich 1962), and when E value lies between 20 and 40 kJ/mol, it indicates chemisorption process (Rieman and Walton 1970).

Adsorption Kinetics

Kinetic study in the adsorption process plays an important role, because it shows the uptake rate of the adsorbate and controls the residential time of the whole process. For validating the experimental data, pseudo-first order, pseudo-second order, and intraparticle kinetic models can be recycled.

The Lagergren Pseudo-first Order Kinetic Model

The Lagergren pseudo-first order (Lagergren 1898) rate expression is based on the adsorption of the solid/liquid system, and its formula is given as follows:

$$\log(q_e - q_t) = \log q_e - \frac{k_{\text{ad}}}{2.303}t \qquad (14.3.10)$$

where q_e and q_t are the quantities of metal ion adsorbed on the adsorbent (mg/g) at equilibrium and at time t, respectively, and k_{ad} and q_e are calculated from the slope and intercept of the plot of $\log(q_e - q_t)$ versus t.

Ho's Pseudo-second Order Kinetic Model

The experimental data are also analyzed using Ho's pseudo-second order kinetic model (Ho and McKay 1999). It can be written as follows:

$$\frac{t}{q_t} = \frac{1}{h} + \frac{1}{q_e}t \qquad (14.3.11)$$

where h can be observed as the initial adsorption rate ($h = kq_e^2$ (mg g^{-1}min^{-1}) as $t \rightarrow 0$ and k is the rate constant (k = g mg^{-1}min^{-1}) of the pseudo-second order model. The plot of t/q_t versus "t" should give a straight line, and q_e and h are determined from the slope and intercept of the plot, respectively.

The Weber and Morris Intraparticle Diffusion Model

The adsorption kinetics of the adsorbate migration from bulk solution to the adsorbent surface and further into the pores of the adsorbent could be explained by intra particle diffusion. This model is explained by three different steps of the solid–liquid interaction. The first step is film diffusion, where the adsorbate molecules diffuse through the solution to the external surface of the adsorbent. The second step is the gradual adsorption stage, in which the adsorbate molecules are transported to the pores of the adsorbent from the exterior surface of the adsorbent. In the third step, the intraparticle diffusion slows down, due to the decrease in the adsorption sites and the decrease in the concentration of the adsorbate in the solution (Weber and Morris 1963).

The intraparticle diffusion during the adsorption process is defined by the equation:

$$q_t = k_{\mathrm{ad}} t^{1/2} + C \tag{14.3.12a}$$

where k_{ad} is the intraparticle diffusion rate constant (mg g^{-1} $min^{-1/2}$) and intercept "C" is the boundary layer effect. The larger the intercept is, the greater the boundary layer effect will be.

Boyd Kinetic Model

The third step in the adsorption dynamics is assumed to be very rapid and can be considered negligible. For design purposes, it is required to distinguish between the film diffusion and particle diffusion of adsorbate molecules. In order to identify the slowest step in the adsorption process, the Boyd kinetic equation could be applied, which is expressed as follows:

$$F = 1 - \frac{6}{\pi^2}\exp(-\mathrm{Bt}) \tag{14.3.12b}$$

$$F = \frac{q_t}{q_e} \tag{14.3.13}$$

where q_e is the amount of metal ions adsorbed at equilibrium (mg/g), q_t represents the amount of metal ions adsorbed at any time t, F represents the fraction of metal ions adsorbed at any time t, and Bt is a mathematical function of F. Equation (4.10) can be rearranged by taking the natural logarithm to obtain the equation:

$$\mathrm{Bt} = -0.4977 - \ln(1 - F) \tag{14.3.14}$$

The plot of Bt against time t can be employed to test the linearity of the experimental values. If the plots are linear and pass through the origin, then the slowest step in the adsorption process is the internal diffusion. The calculated B values were used to calculate the effective diffusion coefficient, D_i (m^2/s), using the following relationship:

$$B = \frac{\pi^2 D_i}{r^2} \tag{14.3.15}$$

where D_i is the effective diffusion coefficient of metal ions in the adsorbents surface and r is the radius of the adsorbent particle calculated by the sieve analysis and by assuming spherical particles.

Elovich Kinetic Model

The Elovich kinetic model can be stated as:

$$q_t = \frac{1}{\beta}\ln(\alpha\beta) + \frac{1}{\beta}\ln t \tag{14.3.16}$$

where α denotes the initial adsorption rate in mg/(g min) and $\beta\,(\mathrm{g/mg})$ is the desorption constant correlated to the extent of the surface coverage and activation energy for chemisorption.

These are the various adsorption isotherm and the kinetic models currently in use to study the adsorption behavior of various adsorbents on various pollutants. By using these models, selection and synthesis process of absorbents can be optimized in the future to improve the efficiency of the overall process.

Boyd Kinetic Model

The adsorption dynamics of the third step is assumed to be very fast and considered to negligible. It is required to distinguish between the film diffusion and particle diffusion of adsorbate molecules for the design purposes. In order to identify the slowest step in the adsorption process, the Boyd kinetic equation could be applied, which is written as:

$$F = 1 - \frac{6}{\pi^2}\exp\,(-\mathrm{Bt}) \tag{14.3.12b}$$

$$F = \frac{q_t}{q_e} \tag{14.3.13}$$

where q_e (mg/g) is the quantities of metal ions adsorbed at equilibrium, q_t represents the total amounts of metal ions adsorbed at any time t, F represents the fraction of metal ions adsorbed at any time t, and Bt is a mathematical function of F. Equation (4.10) can be rearranged by taking the natural logarithm to obtain the equation:

$$\mathrm{Bt} = -0.4977 - \ln\,(1 - F) \tag{14.3.14}$$

The plot of Bt against time t can be employed to test the linearity of the experimental values. The slowest step in the adsorption process is the internal diffusion, if the plots are linear and lead through the origin. To calculate the effective diffusion coefficient, D_i (m^2/s), the following relationship is used:

$$B = \frac{\pi^2 D_i}{r^2} \tag{14.3.15}$$

where D_i is the effective diffusion coefficient of metal ions in the surface of the adsorbent and r is the radius of the adsorbent particle calculated by the sieve analysis and by assuming spherical particles.

14.4 Photocatalysis for Wastewater Treatment

14.4.1 General Information

As previously commented, the main causes of the pollution of surface water and groundwater reside in industrial discharges, excessive use of pesticides and fertilizers and the dumping of household waste. In general, the wastewater treatments are based on various mechanical, biological, physical, and chemical processes. They consist of a combination of many operations such as filtration, sedimentation, flocculation, sterilization, or chemical oxidation processes. After pretreatment and primary treatment, aerobic biological treatment is the favorite process. However, as we have commented above, non-biodegradable and recalcitrant organic pollutants are not eliminated with these conventional treatments. For this reason, water pretreatments with processes generate very reactive chemical species, which degrade more recalcitrant molecules in biodegradable compounds. These processes are known as advanced oxidation processes (Malato et al. 2002).

AOPs are among the most useful technologies for the remediation of wastewaters containing recalcitrant contaminants or resisting pathogens (Pera-Titus et al. 2004). AOPs are based on the generation of reactive oxygen species, such as highly aggressive hydroxyl radicals, superoxide or others, able to oxidize organic contaminants to less-toxic biodegradable intermediates or even to produce their partial or complete mineralization. The following table shows a list of redox potentials of some oxidizing chemicals in acidic medium (Litter et al. 2005). For example, AOPs have been extensively applied to the treatment of textile wastewaters characterized by the production of large volumes of wastewater containing colored, toxic, and/or mutagenic dyes. It has been estimated that about of 15% of the produced dyes are released in textile wastewaters (Langhals et al. 2004). Color and chemical oxygen demand (COD) removal represents one major environmental concern in textile wastewaters due to the inefficacy of conventional biological treatments (Robinson et al. 2001) (Table 14.3).

Hydroxyl radicals ($^{\bullet}OH$) radicals can be generated by photochemical means (such as sunlight) or by other forms of energy and have a high effectiveness for the oxidation of organic matter (Domenech et al. 2001; Litter et al. 2005). Table 14.4 presents some common AOPs.

Table 14.3 Standard oxidation–reduction potentials in an acid medium of some oxidizing agents

Species	E^0 (V, 25 °C)
Fluorine	3.03
Hydroxyl radical	2.80
Atomic oxygen	2.42
Ozone	2.07
Hydrogen peroxide	1.78
Radical perhidroxilo	1.70
Permanganate	1.68
Chlorine dioxide	1.57
Chlorine	1.36
Bromine	1.09
Iodine	0.54

The AOP is employed in the degradation of pollutants in wastewater due to the following characteristics (Malato et al. 2011):

- They are able to treat contaminants in the ranks of micrograms to grams per liter.
- They have a non-selective character allowing to act on complex matrices of different pollutants.
- They are responsible for the elimination of toxicity of the effluents due to organic pollution.
- They are able to transform non-biodegradable contaminants into smaller molecules that are biodegradable.
- They are a viable complement in case of non-biodegradable wastewater with high toxicity.

The main disadvantages include the energy costs of operation, and it is necessary to optimize them according to the needs of the resulting effluent and thus make them economically viable (Sanz et al. 2012).

The following analysis (Malato et al. 2011) should be done to study the parameters of the AOP-treated water:

- COD: The amount of oxygen necessary to oxidize the organic and inorganic matter subjected to oxidation contained in a sample.
- BOD: The amount of oxygen used for the biodegradation of organic and inorganic matter contained in a sample. This measure allows to follow the evolution of biodegradable compounds.
- Total organic carbon (TOC): The amount of carbon dioxide produced in the total mineralization of a sample. Frequently, TOC measurements are employed to determine the mineralization degree of a sample containing organic matter that has been submitted to an AOP.
- Toxicity: The determination of toxicity is key in a photocatalytic degradation process efficiency. One commonly performed analysis to the effluent of

Table 14.4 Advanced oxidation technologies

Non-photochemical processes	Photochemical processes
Ozonation in alkaline medium (O_3/OH^-)	Direct photolysis
Ozone with hydrogen peroxide (O_3/H_2O_2)	Photolysis of water in the ultraviolet vacuum (UVV)
Fenton reaction (Fe^{2+}/H_2O_2)	UV/H_2O_2
Electrochemical oxidation	UV/$Ø_3$
Treatment with electron beams	Photo-Fenton reaction
Non-thermal plasma	Heterogeneous photocatalysis
Download electro-hydraulic–ultrasound	Radiolysis γ
Oxidation of water sub- and supercritical	

determine toxicity is based on toxicity to the bacterium *Vibrio Fischeri* bioluminescent method.

In present scenario, photochemical technologies using natural sunlight are one of the most promising strategies for the reduction of operating costs of the AOPs (Domenech et al. 2001; Litter et al. 2005). Among the different photochemical technologies, particular importance due to their effectiveness and easy implementation at industrial scale could be achieved with the heterogeneous photo-Fenton process assisted by natural sunlight. Therefore, many research efforts are currently aimed at the development of photocatalysts that are capable of harnessing the irradiation with visible light from the Sun. It is further discussed in a detailed way the fundamentals and most important features of these processes.

14.4.1.1 Heterogeneous Photocatalysis

Heterogeneous photocatalysis is one of the AOP processes. This process is based on the absorption of light (UV or Visible) by a solid (heterogeneous photocatalyst, which tends to be a broadband semiconductor) (Domenech et al. 2001; Litter et al. 2005; Malate et al. 2011). The photocatalysis is attracting great attention from science to various practical applications. In a pioneering work, Fujishima and Honda (Fujishima et al. 2008) succeeded in the dissociation of water induced by ultraviolet light using titanium dioxide (TiO_2) as photoanode in combination with a counterelectrode platinum dipped in an aqueous electrolyte solution. This fact opened the possibility of conversion of solar energy into chemical energy using semiconductor. In recent years, semiconductor-based photocatalysts have been commercialized for water or atmospheric decontamination processes. In this way, photocatalytic processes have been designed that are capable of generating hydrogen from water and photoreductive CO_2 in the so-called solar fuels or carry out decontamination or disinfection process in contaminated water.

14.4.1.2 Heterogeneous Photocatalysts

The power of an ideal crystal spectrum consists of groups of states of energy filled (valence band) and empty energy states (conduction band). Between these two states is an area in which electrons cannot remain due to lack of energy states, and it is a forbidden band or *band gap*. Depending on the of the band gap value, materials are classified as dielectrics. Substances with large band gaps are generally insulators, those with smaller band gaps are semiconductors, while conductors either have very small band gaps or none, because the valence and conduction bands overlap (Candal et al. 2005). Semiconductor materials can also be classified into three groups:

- Reductive: It may result in the reduction of water and generate H_2, but not oxidize it to molecular O_2.
- Redox: It has the ability to oxidize and reduce the water.
- Oxidative: It can cause oxidation of water and generation of O_2, but not to reduce it to hydrogen.

The following are the criteria for selecting a good photocatalyst (Candal et al. 2005):

- Adequate redox potentials of electrons and holes to make feasible the mineralization of organic matter.
- Photoactivation in the range of UV–Visible light close (less than 3.1 eV to take advantage of the sunlight).
- Be resistant to the photocorrosion, low toxicity, and high active area.

Titanium oxide (TiO_2) is the benchmark semiconductor material in photocatalysis. TiO_2 has been widely studied and used in many applications due to various reasons including their ability to degrade to a wide range of contaminants, organic and inorganic, chemical stability, their photostability, durability, non-toxicity, low cost, abundance, and transparency to visible light (Nakataa et al. 2012; Schneider et al. 2014).

Although the mechanism of photocatalysis can vary from one pollutant to another depending on their physical and chemical properties, primary reactions responsible for the photocatalytic effect are redox interfacial reactions of electrons and holes that are generated when semiconductor catalyst is exposed to the light of sufficient energy (Schneider et al. 2014).

The main steps of the mechanism of photocatalysis are as follows:

1. Direct or indirect absorption of photons of UV or visible light, with equal energy or greater than the power of the forbidden band gap semiconductor.
2. Separation of charges due to the promotion of an electron to the conduction band results in formation of a hole in the valence band.
3. Recombination of charge carriers.
4. Oxidation reaction through a hole in the valence band.
5. Reduction reaction by an electron in the conduction band.
6. Thermal and photocatalytic reactions to produce products mineralization.

Despite its numerous advantages, the use of the TiO_2 as photocatalyst presents a series of disadvantages, which are:

- The photoactivity under solar radiation due to the minimum absorption of visible light.
- Fast recombination of the electron/hole pairs.

14.4.1.3 (Photo-)Fenton Reaction

Fenton (described by Fenton at the end of the nineteenth century) reagent is an aqueous solution of hydrogen peroxide (oxidizing agent) and ferrous ion (catalyst) that generates hydroxyl radicals. The Fenton reaction is one of the most important processes of decontamination of wastewater with high contaminant loads due to its easy implementation to industrial level (Stasinakis et al. 2008). The Fenton reaction is mainly used in the decontamination of sewage.

Fenton reaction is an autocatalytic process at pH 3, since the Fe (III) breaks down the hydrogen peroxide into O_2 and H_2O through a process chain. For this purpose, it is necessary that hydrogen peroxide is in excess with respect to the amount of iron added (Domenech et al. 2001). However, it is worth to note the big difference between the constants of speed of reactions shown in Eqs. (14.4.1) and (14.4.2) with values of 70 and 0.01 $M^{-1}s^{-1}$, respectively.

The Fenton process has several advantages, among which are:

- The easy handling of the reagents involved.
- Iron is an abundant element and non-toxic.
- The design of reactors for the technological application is simple.
- External source of energy is not required.

However, the existing disadvantages include:

- The use of large quantities of chemical agents to acidify the medium at pH around 3.
- The generation of iron sludge resulting from removal by precipitation of iron used as a reagent.
- Incomplete mineralization may generate toxic reaction intermediates such as quinones, and therefore, they must be carefully controlled.

The Fenton reaction generates $^{\bullet}OH$ radicals that can lead to the complete mineralization of organic substrates, i.e., degradation compounds giving CO_2 and H_2O according to 14 equivalents of hydrogen peroxide that are required for the complete mineralization of phenol.

When the Fenton reaction is carried out under UV–Vis irradiation, the degradation of contaminants present in the water is accelerated considerably. This process is called photo-Fenton. As already commented, under acidic conditions, around pH 3, Fenton reagent is a powerful oxidizer of organic compounds, and when the process is supplemented with UV–Vis radiation ($\lambda < 380$ nm), degradation rate increases significantly. The photo-Fenton process involves the formation of hydroxyl radicals $^{\bullet}OH$ by direct reduction of H_2O_2 by Fe^{2+} salts. In the presence of UV radiation, aqueous complexes of $Fe(OH)^{2+}$ decompose, leading to the formation of Fe^{2+} ions and hydroxyl radicals. The process can be summarized according to Domenech and Litter (2001), and Stasinakis (2008) represented the following Eqs. (14.4.1) and (14.4.2):

$$Fe^{2+} + H_2O_2 \xrightarrow[pH3(H_2SO_4)]{} Fe^{3+} + {}^{\bullet}OH + OH^- \tag{14.4.1}$$

$$Fe(OH)^{2+} \xrightarrow[pH3(H_2SO_4)]{UV-Vis} Fe^{2+} + {}^{\bullet}OH \tag{14.4.2}$$

The main advantage of photo-Fenton process is the ferrous ions photocatalytic regeneration. However, this process is limited to the following criteria (Espinosa et al. 2015):

- It needs a pH value of around 3 that allows obtaining of iron species required for the efficiency of the photo-Fenton reaction.
- The high amount of hydrogen peroxide is consumed due to the intermediate reactions.
- The large amount of sludge is generated by the precipitation of iron salts.
- The need for UV–Visible light is to promote the photolysis of the absorbing species $Fe(OH)^{2+}$ ($\lambda \sim 300$ nm).

In order to overcome some of the drawbacks of the homogeneous (photo-) Fenton process, many efforts have been done to develop the heterogeneous (photo-) Fenton reaction under visible light irradiation.

14.4.1.4 Heterogeneous Catalysts for the (Photo-)Fenton Reaction

Various solid materials such as zeolites, silica, resins, or activated carbon have been used for preparing heterogeneous catalysts (photo-)Fenton. In heterogeneous phase, both physical and chemical catalysis stages take place in active sites that are found on the surface of the catalyst. At the end of the reaction, the product molecules are desorbed and leave active centers available for a new set of reagents. The use of iron in the heterogeneous phase facilitates the separation of the catalyst and reusability, and the overall cost of the process might be reduced. The limitation of a large part of this type of (photo) catalysts is their low efficiency to selectively decompose hydrogen peroxide into hydroxyl radicals.

Bearing in mind that hydrogen peroxide is a relatively expensive reagent, it is necessary to optimize the amount of its use during the process of (photo-)Fenton. In this context, (photo)catalysts based on metallic nanoparticles such as Au or Ag that present a surface plasmon band under visible light irradiation or nanoparticles based

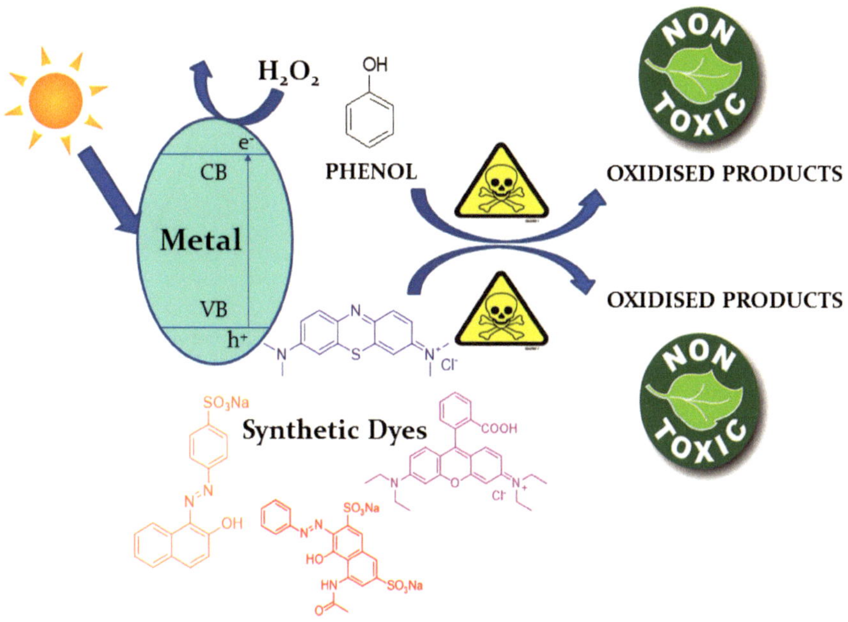

Fig. 14.8 Schematic representation of photocatalysis of various pollutants

on iron oxides, e.g., Ag–Fe_2O_3, have proven to be efficient for the process of photo-Fenton heterogeneous using visible light irradiation (Fig. 14.8).

14.4.1.5 Metallic Nanoparticles as Photocatalysts

In recent years, the use of supported metal nanoparticles has attracted a special interest in the area of photocatalysis (Uçurum et al. 2009). Firstly, the importance of nanoparticles is its inherent properties which differ from the macroscopic forms of the same material. The activity and selectivity of nanoparticles as catalysts rely heavily on their size, shape, and structure of the surface, as well as its composition. They have a relationship between surface areas and very interesting catalytic activity for the development of highly active heterogeneous catalysts. Hence, there are numerous heterogeneous catalysts based on the deposition of these nanoparticles as active on a stand with a large surface (Dhakshinamoorthy et al. 2012).

In the case of gold, silver, and copper nanoparticles, they have an absorption band of energy known as resonance of surface plasmon in the visible region due to the vibration of free electrons. The photocatalytic activity of catalysts that use this type of nanoparticles for heterogeneous photo-Fenton reaction is attributed to the photoexcitation of electrons after irradiation at wavelengths corresponding to the surface plasmon band. These photoejected electrons are trapped by the H_2O_2 with the consequent generation of hydroxyl radicals. Figure 14.2 (Scaiano et al. 2012)

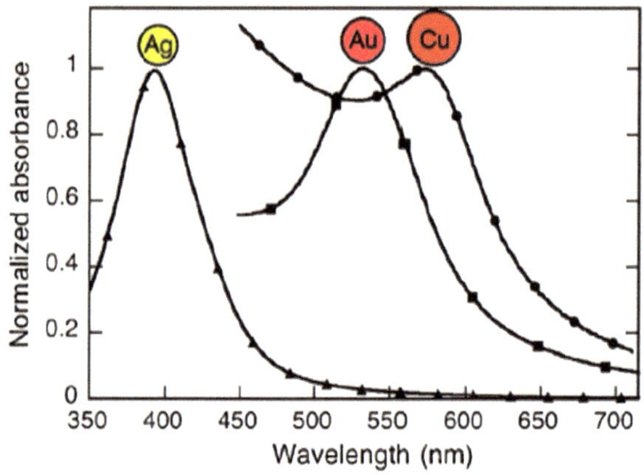

Fig. 14.9 Plasmon absorption of Au, Ag, and Cu nanoparticles

shows the wavelengths of absorption of gold, silver, and copper are around 530, 400, and 560 nm, respectively. The presence of these bands of surface plasmon allows the use of visible light to promote the reaction (Scaiano et al. 2012; Espinosa et al. 2015) (Fig. 14.9).

In this context, it has been reported that Au or Ag NPs supported on hydroxylated diamond nanoparticles (D NPs) is an efficient photocatalyst for degradation of phenol as well as its reaction intermediates (hydroquinone, catechol and *p*-benzoquinone) using less than 5.5 equivalents of H_2O_2 under natural sunlight irradiation (Espinosa et al. 2015). Therefore, it would be of interest to evaluate the scope of these Fenton catalysts by testing their efficiency in the degradation of other recalcitrant contaminants, particularly decolorization of synthetic dyes that are frequently employed in the textile industry.

Similarly, Cu NPs supported on hydroxylated D NPs have been reported as an efficient catalyst to promote the (photo-)Fenton reaction under visible light irradiation. Unfortunately, the catalyst deactivates upon reuse due to the oxidation of the reduced copper species to Cu(II).

In terms of the nature of the metals used, the iron and manganese are also good choices of metallic nanoparticles as heterogeneous catalysts for radical generation $^{\bullet}OH$ from H_2O_2 due to its availability and lack of toxicity. However, it is worth commenting that the mechanism of action of nanoparticles based on iron oxides to promote the heterogeneous photo-Fenton reaction is related to the semiconductor character of them. Indeed, Fe_2O_3 is a semiconductor n-type whose band gap is 2.3 eV, so it can be photoexcited by UV-Vis light for pollutant decomposition organic in wastewater according to the following mechanism (Wang et al. 2012).

14.4.1.6 Carbonaceous Supporting Materials

On the one hand, the carbon is most commonly used to remove organic compounds in treatments of wastewater purification. On the other hand, hydrogen peroxide is a disinfectant which, when combined with a catalyst, can react to provide hydroxyl radicals that affect degradation and mineralization of pollutants. It is interesting to develop heterogeneous catalysts that generate $^{\bullet}OH$ by reducing H_2O_2 by the action of the catalytic sites and on the other hand to restore such active centers through the oxidation of hydrogen peroxide (Navalón et al. 2011).

Activated carbon is widely used as support in heterogeneous catalysis due to its large surface area and the good adsorption capacity. In this context, iron nanoparticles supported on activated carbon have been reported has heterogeneous (photo-)Fenton catalysts, but, however, metal leaching from the solid to the solution occurs in a large extent.

Diamond nanoparticles are constituted by a network of covalent bonded sp^3 carbons. Diamond NPs is an inert material, affordable, and cheap and is available commercially obtained by the detonation of explosive (Krueger et al. 2007). Usually, these nanoparticles agglomerate because of the covalent bonding between the nanoparticles and at the same time are embedded in amorphous carbon. In order to avoid agglomeration of diamond nanoparticles and remove the amorphous carbon, diamond nanoparticles can be subjected to Fenton treatment under aggressive conditions. In addition, this chemical treatment also leads to the generation of a large surface population of hydroxyl groups useful for subsequent deposition of metallic nanoparticles. The use of diamond as support for metal nanoparticles has given good results for (photo-)Fenton reaction since the lining of the surface of the support and the inert nature of this make that hydroxyl radicals remain in solution due to the low affinity of these by the catalyst (Espinosa et al. 2015).

References

Ahmed MN, Ram RN (1992) Removal of basic dye from waste-water using silica as adsorbent. Environ Pollut 77:79–86

Ahmed S, Chughtai S, Keane MA (1998) The removal of cadmium and lead from aqueous solution by ion exchange with NaY zeolite. Sep Purif Technol 13(1):57–64

Aklil A, Mouflih M, Sebti S (2004) Removal of heavy metal ions from water by using calcined phosphate as a new adsorbent. J Hazard Mater 112(3):183–190

Al-Degs Y, Khraisheh MAM, Allen SJ, Ahmad MNA (2001) Sorption behavior of cationic and anionic dyes from aqueous solution on different types of activated carbons. Sep Sci Technol 36:91–102

Al-Shammiri M, Al-Saffar A, Bohamad S, Ahmed M (2005) Waste water quality and reuse in irrigation in Kuwait using microfiltration technology in treatment. Desalination 185(1–3):213–225

Backhaus WK, Klumpp E, Narres HD, Schwuger MJ (2001) Adsorption of 2,4-dichlorophenol on montmorillonite and silica: influence of nonionic surfactants. J Colloid Interface Sci 242:6–13

Baig S, Liechti PA (2001) Ozone treatment for biorefractory COD removal. Water Sci Technol J Int Assoc Water Pollut Res 43(2):197–204

Bao ML, Griffini O, Santianni D, Barbieri K, Burrini D, Pantani F (1999) Removal of bromate ion from water using granular activated carbon. Water Res 33:2959–2970

Bele M, Kodre A, Arc I, Grdadolnik J, Pejovnik S, Besenhard JO (1998) Adsorption of cetyltrimethylammonium bromide on carbon black from aqueous solution. Carbon 36:1207–1212

Bhatnagar A, Sillanpää M (2010) Utilization of agro-industrial and municipal waste materials as potential adsorbents for water treatment—a review. Chem Eng J 157(2–3):277–296

Brennan JK, Bandosz TJ, Thomson KT, Gubbins KE (2001) Water in porous carbons. Colloids Surf A 187–188:539–568

Candal R, Estrada W (2005) El fotocatalizador: síntesis, propiedades y limitaciones. Tecnologíassolares para la desinfección y descontaminación del agua, pp 129–145

Chang WC, Hsu GS, Chiang SM, Su MC (2006) Heavy metal removal from aqueous solution by wasted biomass from a combined AS-biofilm process. Biores Technol 97(13):1503–1508

Chen JP, Wu S, Chong K-H (2003) Surface modification of a granular activated carbon by citric acid for enhancement of copper adsorption. Carbon 41(10):1979–1986

Chen Y, Liu Y, Zhou Q, Gu G (2005) Enhanced phosphorus biological removal from wastewater effect of microorganism acclimatization with different ratios of short-chain fatty acids mixture. Biochem Eng J 27(1):24–32

Chern JM, Chien YW (2002) Adsorption of nitrophenol onto activated carbon: isotherms and breakthrough curves. Water Res 36:647–655

Dabrowski A, Podkoscielny P, Hubicki Z, Barczak M (2005) Adsorption of phenolic compounds by activated carbon-a critical review. Chemosphere 58:1049–1070

Demirbas A (2008) Heavy metal adsorption onto agro-based waste materials: a review. J Hazard Mater 157(2–3):220–229

Dhakshinamoorthy A, Navalon S, Alvaro M, Garcia H (2012) Metal nanoparticles as heterogeneous fenton catalyst. Chem Sus Chem 5:46–64

Ellis J, Korth W (1993) Removal of geosmin and methylisoborneol from drinking water by adsorption on ultrastable zeolite-Y. Water Res 27:535–539

Escobar C, Soto-Salazar C, Inés Toral M (2006) Optimization of the electrocoagulation process for the removal of copper, lead and cadmium in natural waters and simulated wastewater. J Environ Manage 81(4):384–391

Espinosa JC, Navalón S, Primo A, Moral M, Fernández Sanz J, Álvaro M, García H (2015) Graphenes as efficient metal-free fenton catalysts. Chem Eur J 21:11966–11971

Febrianto J, Kosasih AN, Sunarso J, Ju Y-H, Indraswati N, Ismadji S (2009) Equilibrium and kinetic studies in adsorption of heavy metals using biosorbent: a summary of recent studies. J Hazard Mater 162(2–3):616–645

Foo KY, Hameed BH (2010) Detoxification of pesticide waste via activated carbon adsorption process. J Hazard Mater 175:1–11

Freundlich HMF (1906) Over the adsorption in solution. J Phys Chem 57:385–470

Fu Y, Viraraghavan T (2001) Fungal decolorization of dye wastewaters: a review. Biores Technol 79(3):251–262

Fujishima A, Zhang X, Donald B, Trykc A (2008) TiO2 photocatalysis and related surface phenomena. Surf Sci Rep 63:515–582

Gabaldon G, Marzal P, Ferrer J, Seco A (1996) Single and competitive adsorption of Cd^{2+} and Zn^{2+} onto a granular activated carbon. Water Res 30:3050–3060

Gabaldon G, Marzal P, Seco A, Gonzalez JA (2000) Cadmium and copper removal by a granular activated carbon in laboratory column systems. Sep Sci Technol 35:1039–1053
Gomez V, Larrechi MS, Callao MP (2007) Kinetic and adsorption study of acid dye removal using activated carbon. Chemosphere 69:1151–1158
Groffman A, Peterson S, Brookins D (1992) Removing lead from wastewater using zeolite. Water Environ Technol 4:4–59
Haydar S, Ferro-Garcia MA, Rievera-Utrilla J, Joly JP (2003) Adsorption of p-nitrophenol on an activated carbon with different oxidations. Carbon 41:387–395
Helfferich F (1962) Ion Exchange. McGraw-Hill Book Co., New York
Hernandez-Ramirez O, Holmes SM (2008) Novel and modified materials for wastewater treatment applications. J Mater Chem 18(24):2751–2761
Ho YS, McKay G (1999) Pseudo-second order model for sorption processes. Process Biochem 34:451–465
Huang WJ, Cheng YL (2008) Effect of characteristics of activated carbon on removal of bromated. Sep Purif Technol 59:101–107
Ibach H (2006) Physics of surfaces and interfaces. Springer
Inglezakis VJ, Zorpas AA, Loizidou MD, Grigoropoulou HP (2005) The effect of competitive cations and anions on ion exchange of heavy metals. Sep Purif Technol 46(3):202–207
Kasprzyk-Hordern B (2004) Chemistry of alumina, reactions in aqueous solution and its application in water treatment. Adv Colloid Interface Sci 110:19–48
Khalil LB (1996) Adsorption characteristics of activated carbon obtained from rice husks by treatment with phosphoric acid. Adsorp Sci Technol 13:317–325
Kikuchi T, Tanaka S (2012) Biological removal and recovery of toxic heavy metals in water environment. Crit Rev Environ Sci Technol 42(10):1007–1057
Krueger A (2007) The structure and reactivity of nanoscale diamond. J Mater Chem 1485–1492
Ku Y, Chiou HM (2002) The adsorption of fluoride ion from aqueous solution by activated alumina. Water Air Soil Pollut 133:349–361
Kuhad RC, Sood N, Tripathi KK, Singh A, Ward OP (2004) Developments in microbial methods for the treatment of dye effluents. Adv Appl Microbiol 56:185–213
Lagergren S (1898) About the theory of so-called adsorption of soluble substances. Kungliga Svenska Vetensk Handl 24:1–39
Langhals H (2004) Color chemistry synthesis, properties and applications of organic dyes and pigments, 3rd revised edn by Heinrich Zollinger
Langmuir I (1918) The adsorption of gases on plane surfaces of glass, mica and platinum. J Am Chem Soc 40:1361–1403
Litter M (2005) Tecnologíasavanzadas de oxidación: tecnologíassolares. Safe Water, Photocatalytic Tecnology, pp 73–90
Lu H, Zhu L, Zhu N (2009) Polycyclic aromatic hydrocarbon emission from straw burning and the influence of combustion parameters. Atmos Environ 43(4):978–983
Mahmudov R, Huang CP (2010) Perchlorate removal by activated carbon adsorption. Sep Purif Technol 70:329–337
Malato S, Blanco J, Vidal A, Ritcher C (2002) Photocatalysis with solar energy at a pilot-plant scale: an overview. Appl Catal B Environ 1–15
Malato S, Blanco J, Estrada C, Bandala E (2011) Purificación de aguasporfotocatálisisheterogénea: estado del arte
Malhas AN, Abuknesha RA, Price RH (2002) Removal of detergents from protein extracts using activated charcoal prior to immunological analysis. J Immunol Methods 264:37–43

McCray SB, Ray RJ (1987) Concentration of synfuel process condensates by reverse osmosis. Sep Sci Technol 22(2–3):745–762
McCreary JJ, Snoeyink VL (1980) Characterization and activated carbon adsorption of several humic substances. Water Res 14:151–160
McKay G, Otterburn MS, Sweeney AG (1980) The removal of colour from effluent using various adsorbents—III. Silica: rate processes. Water Res 14(1):15–20
McKay G, Bino MJ, Altameni AR (1985) The adsorption of various pollutants from aqueous solutions onto activated carbon. Water Res 19:491–495
Meena AK, Mishra GK, Rai PK, Rajagopal C, Nagar PN (2005) Removal of heavy metal ions from aqueous solutions using carbon aerogel as an adsorbent. J Hazard Mater 122(1–2):161–170
Metes A, Kovacevic D, Vujevic D, Papic S (2004) The role of zeolites in wastewater treatment of printing inks. Water Res 38:3373–3381
Mohammed-Azizi F, Dib S, Boufatit M (2013) Removal of heavy metals from aqueous solutions by Algerian bentonite. Desalin Water Treat 51(22–24):4447–4458
Monier M, Nawar N, Abdel-Latif DA (2010) Preparation and characterization of chelating fibers based on natural wool for removal of Hg(II), Cu(II) and Co(II) metal ions from aqueous solutions. J Hazard Mater 184(1–3):118–125
Moreno-Castilla C, Rivera-Utrilla J (2001) Carbon materials as adsorbents for the removal of pollutants from the aqueous phase. Mater. Res. Soc. Bull. 26:890–894
Moriguchi T, Yano K, Tahara M, Yaguchi K (2005) Metal-modified silica adsorbents for removal of humic substances in water. J Colloid Interface Sci 283:300–310
Morrison S (1998) Research and application of permeable reactive barriers, U.S. Department of Energy
Motsi T, Rowson NA, Simmons MJH (2009) Adsorption of heavy metals from acid mine drainage by natural zeolite. Int J Miner Process 92:42–48
Naiya TK, Bhattacharya AK, Das SK (2009) Adsorption of Cd(II) and Pb(II) from aqueous solutions on activated alumina. J Colloid Interface Sci 333:14–26
Najm IN, Snoeyink VL, Richard Y (1993) Removal of 2,4,6-trichlorophenol and natural organic matters from water supplies using PAC in floc-blanket reactors. Water Res 27:551–560
Okolo B, Park C, Keane MA (2000) Interaction of phenol and chlorophenols with activated carbon and synthetic zeolites in aqueous media. J Colloid Interface Sci 226:308–317
Pelekani C, Snoeyink VL (2000) Competitive adsorption between atrazine and methylene blue on activated carbon: the importance of pore size distribution. Carbon 38:1423–1436
Pera-Titus M, García-Molina V, Baños MA, Giménez J, Esplugas S (2004) Degradation of chlorophenols by means of advanced oxidation processes: a general review. Appl Catal B-Environ 47:219–256
Pereira MFR, Soares SF, Orfao JJM, Figuerredo JL (2003) Adsorption of dyes on activated carbon: influence of surface chemical groups. Carbon 41:811–821
Perez-Candela M, Martin-Martinez JM, Torregrosa-Macia R (1995) Chromium(VI) removal with activated carbons. Water Res 29:2174–2180
Periyasamy Karthik, Laishram Santhalembi, Mortha Gérard, Aurousseau Marc, Subramanian Sivanesan (2016b) Carrier-free co-immobilization of xylanase, cellulase and β-1,3-glucanase as combined cross-linked enzyme aggregates (combi-CLEAs) for one-pot saccharification of sugarcane bagasse; m RSC. Advances 6(39):32849–32857
Peters TA (1991) Desalination and industrial waste water treatment with the ROCHEM disc module DT. Desalination 83(1–3):159–172
Premkumar MP, Vinoth Kumar V, Senthil Kumar P, Baskaralingam P, Sathyaselvabala V, Vidhyadevi T, Sivanesan S (2013) Kinetic and equilibrium studies on the biosorption of textile dyes onto *Plantago ovata* seeds. Korean J Chem Eng 30(6):1248–1256
Rafii F, Fraeankalin W, Cerniglia CE (1990) Azo reductase activity of anaerobic bacteria isolated from human intestinal microflora. Appl Environ Microbiol 56(7):2146–2151
Rangnekar DW, Singh PP (1980) Introduction to synthetic dyes. Himalaya Pub, Ho, India

Ratna Kumar P, Chaudhari S, Khilar KC, Mahajan SP (2004) Removal of arsenic from water by electrocoagulation. Chemosphere 55(9):1245–1252

Rieman W, Walton H (1970) Ion exchange in analytical chemistry, international series of monographs in analytical chemistry, vol 38. Pergamon Press, Oxford

Rivera-Utrilla J, Sanchez-Polo M, Carrasco-Marin F (2003) Adsorption of 1,3,6-naphthalenetrisulfonic acid on activated carbon in the presence of Cd(II), Cr(III), and Hg(II): importance of electrostatic interactions. Langmuir 19:10857–10861

Robinson T, McMullan G, Marchant R, Nigam P (2001) Remediation of dyes in textile effluent: a critical review on current treatment technologies with a proposed alternative. Biores Technol 77:247–255

Rubio J, Tessele F (1997) Removal of heavy metal ions by adsorptive particulate flotation. Miner Eng 10(7):671–679

Rudzinski WE, Guthrie SR, Cassidy PE (1988) Poly (Schiff base) polymers based on substituted biphenyl. J Polym Sci Part A Polym Chem 26(6):1677–1680

Saad R, Hamoudi S, Belkacemi K (2008) Adsorption of phosphate and nitrate anions on ammonium-functionalized mesoporous silicas. J Porous Mater 15:315–323

Saito I (1984) The removal of hexacyanoferrate (II) and (III) ions in dilute aqueous solution by activated carbon. Water Res 18:319–323

Sakaguchi T, Nakajima A (1986) Recovery of uranium by immobilized polyhydroxyanthraquinone. Sep Sci Technol 21(5):519–534

Sanchez-Polo M, Rivera-Utrilla J (2002) Adsorbent–adsorbate interactions in the adsorption of Cd (II) and Hg(II) on ozonized activated carbons. Environ Sci Technol 36:3850–3854

Sarioglu M, Akkoyun S, Bisgin T (2010) Inhibition effects of heavy metals (copper, nickel, zinc, lead) on anaerobic sludge. Desalin Water Treat 23(1–3):55–60

Scaiano J, Stamplecoskie K, Hallett-Tapley G (2012) Photochemical Norrish type I reaction as a tool for metal nanoparticle synthesis: importance of proton coupled electron transfer. Chem Comm 4798–4808

Schneider Jenny, Matsuoka Masaya, Takeuchi Masato, Jinlong Zhang Yu, Horiuchi Masakazu Anpo, Bahnemann Detlef W (2014) Understanding TiO_2 photocatalysis: mechanisms and materials. Chem Rev 114(19):9919–9986

Singh TS, Pant KK (2004) Equilibrium, kinetics and thermodynamic studies for adsorption of As (III) on activated alumina. Sep Purif Technol 36:139–147

Sips R (1948) Combined form of Langmuir and Freundlich equations. J Chem Phys 16:490–495

Sotelo JL, Ovejero G, Delgado JA, Martinez I (2002) Comparison of adsorption equilibrium and kinetics of four chlorinated organics from water onto GAC. Water Res 36:599–608

Srivastava NK, Majumder CB (2008) Novel biofiltration methods for the treatment of heavy metals from industrial wastewater. J Hazard Mater 151(1):1–8

Stasinakis AS (2008) Use of selected advanced oxidation processes (AOPS) for waste water treatment—a mini review. Global NEST J 10(3):376–385

Sud D, Mahajan G, Kaur MP (2008) Agricultural waste material as potential adsorbent for sequestering heavy metal ions from aqueous solutions—a review. Biores Technol 99 (14):6017–6027

Suzuki M (1990) Adsorption engineering. Kodansha

Temkin MJ, Pyzhev V (1940) Recent modifications to Langmuir isotherms. Acta Physicochim. URSS. 12:217–225

Tsai WT, Chang CY, Wang SY, Chang CF, Chien SF, Sun HF (2001) Utilization of agricultural waste corn cob for the preparation of carbon adsorbent. J Environ Sci Health B 36:677–686

Uçurum M (2009) A study of removal of Pb heavy metal ions from aqueous solution using lignite and a new cheap adsorbent (lignite washing plant tailings). Fuel 88(8):1460–1465

Van der Heen P (1977) The removal of traces of heavy metals from drinking water and industrial effluent with ion exchanger, The Regional Chemical Society Meeting, 1977

Wang J, Chen C (2006) Biosorption of heavy metals by *Saccharomyces cerevisiae*: a review. Biotechnol Adv 24(5):427–451

Wang H, Kang J, Liu H, Qu J (2009) Preparation of organically functionalized silica gel as adsorbent for copper ion adsorption. J. Environ. Sci. 21:1473–1479

Wang C, Liu H, Sun Z (2012) Heterogeneous photo-Fenton reaction catalyzed by nanosized iron oxides for water treatment. J. Photoenergy, Int, p 801694

Weber WJ, Morris JC (1963) Kinetics of adsorption on carbon from solution. J Sanit Eng Div Am Soc Civ Eng 89:31–60

Wu Y, Zhou J, Wen Y, Jiang L, Wu Y (2012) Biosorption of heavy metal ions (Cu^{2+}, Mn^{2+}, Zn^{2+}, and Fe^{3+}) from aqueous solutions using activated sludge: comparison of aerobic activated sludge with anaerobic activated sludge. Appl Biochem Biotechnol 168(8):2079–2093

Youn SJ, Myoungho P (2008) Removal of heavy metal ions by electrocoagulation for continuous use of Fe^{2+}/Fe^{3+}—mediated electrochemical oxidation solutions. Bull Korean Chem Soc 29 (5):974–978

Zaki NG, Khattab IA, Abd El-Monem NM (2007) Removal of some heavy metals by CKD leachate. J Hazard Mater 147(1–2):21–27

Chapter 15
Common Mistakes in Planning of Sewer Networks and STPs

Sudhir Kumar Arora and Colonel Naresh Sharma

Abstract Planning and design of sewer networks and STPs are outsourced by works department, Military Engineering Services (MES), due to shortage of designers. The peculiarities of defence establishments, which are small townships with sparse population density, i.e. low initial discharges, large distances, wide fluctuations in discharge, high ingress of rainwater and extreme climatic conditions, etc., are generally ignored by consultants, and DPRs are prepared based on average parameters normally prevalent with civil population. Further, consultants do not carry out cost–benefit analysis in totality including cost of conveyance, treatment and reuse. All these issues lead to delays in clearance of DPRs/Estimates when they are sent for scrutiny. Attempt has been made to list out such mistakes, analyse and suggest methods to mitigate these. Suggestions regarding higher peak flow factor (suggested 4–6 instead of 3), restrictive use of flushing tanks, reuse of treated wastewater for flushing, use of energy efficient DWCPE sewers and restrictions on depth of sewers for all stations. For extreme cold conditions, insulation of sewers and manholes has been reiterated, and for hot climates, higher self-cleansing velocities >0.8 m/s have been suggested. Cost analysis of STPs in defence establishments has also been done to facilitate designers for selecting appropriate capacity of STP. These suggestions if made part of special conditions of consultancy contract will ensure efficient designs by consultants and faster approvals of DPRs/Estimates.

S. K. Arora (✉)
Indian Defence Service of Engineers, Military Engineer Services (MES), Kashmir House, Rajaji Marg, 110011 New Delhi, India
e-mail: arora.mes@gmail.com

Colonel N. Sharma
Corps of Engineer, Indian Army, E-in-C's Branch, Kashmir House, Rajaji Marg, 110011 New Delhi, India
e-mail: narry29@gmail.com

T. Gupta et al. (eds.), *Environmental Contaminants*, Energy, Environment, and Sustainability, https://doi.org/10.1007/978-981-10-7332-8_15

Keywords Sewage treatment plant (STP) · Detailed project report (DPR) Self-cleansing velocity (SCV) · Automatic flushing tank (AFT) Double-walled corrugated polyethylene sewer (DWCPE sewer) Manning's constant (n) · Peak factor (PF) · Zero liquid discharge (ZLD) Estimating data (ED-2016) · Stoneware glazed (SWG)

15.1 Introduction

In this chapter, existing method of planning of network of sewers has been critically examined. Aim of sewage disposal is to treat wastewater so that pollutants are reduced to a desired degree without causing adverse impact on health/environment. It has three components, i.e. collection, transportation and treatment. Gravity sewers, lift wells and pressure sewers are three important sections for transport of sewage. Efficient planning and functioning of network of gravity sewers are vital for success of entire sewage disposal system.

Low population densities, fluctuating populations, high sanitation standards, remote locations, large unpaved areas and extreme climatic conditions make planning of network of gravity sewers a challenge for PHE engineers in defence establishments. In normal city conditions, planners rarely encounter a problem of low discharge, while in defence establishments all initial stretches of networks in plains have this problem. The consultants hired by chief engineers are mostly having experience of working in normal city like conditions where density is high, and they tend to ignore the peculiarities in defence establishments.

Traditional solutions, of providing automatic flushing tanks and accepting lower range of self-cleansing velocity only once in a day, have been found to be insufficient. Blockages, premature failures of sewers and septic conditions are not only creating unhygienic conditions but also taking large share of maintenance funds.

Consultants not only ignore vital slope requirements in small diameter sewers at initial stretches, but they select traditional material of sewers which have high roughness coefficient. Often they suggest automatic flushing tanks (AFTs) to overcome the requirement of SCV without realizing the additional requirement of freshwater for flushing and maintenance issue of AFTs. Sometimes steep slopes of sewers are proposed; thus, large number of lift wells are required making the entire proposal very costly. Further, deep depths of sewers with small diameters are very difficult to maintain.

Planning of lift wells by consultants without considering realistic lean and peak flow discharges is another grey area. This leads to stagnation of sewage in lift well, and hence problem of septic conditions and sludge deposit occur.

Mandatory reuse of treated effluent and concept of ZLD have given a new dimension to planning of sewer network. The concept of decentralized STPs may prove to be more appropriate as it saves cost of laying sewers and also minimizes cost of distribution network of treated effluent for its reuse. However, in defence establishments where small STPs are required, cost–benefit study in totality, including cost of laying sewers, cost of STPs and that of pipe network to be laid for reuse must be done. This exercise is necessary because per litre cost of very small STPs or package type STPs is 3–6 higher than mid-size or large capacity plants.

15.2 Contour Superimposed on Layout Plans

Consultants tend to ignore importance of contours and layout plans for planning of network of sewers as most of them blindly bank upon the general sewer design software available in market rather than visualizing and appreciating ground issues involved in each branch of sewer. Cases have been reported where sewers have been proposed from lower invert levels to higher invert level without any engineering application. Such proposals should get rejected at initial stages itself.

In some cases, it may be better to take a longer route to collect more discharge downstream and thus obtain milder slopes, rather than planning sewers via shortest routes, but with deeper excavations. Thus, holistic understanding of topography is vital for designer for planning efficient network of gravity sewers.

Preferably, layout plans should be prepared on such a scale that entire station is easily covered in one A_0 drawing sheets. In addition, zoomed up detailed plans with scales 1:1000 or higher on A_3 sheets may be prepared along with key plans for ease of understanding and readability. Contours, existing networks, water supply lines, underground cables and other services should be marked. Details of existing sewers including size of sewers, load and invert level are essentially required for synchronization with new schemes. Colour schemes in line plans should facilitate easy readability and comprehension. Alternately SP-46 of Bureau of Indian Standards for preparation of engineering drawings may be followed.

15.3 Diameter of Sewer Influences Self-cleansing Velocity

As per new CPHEEO manual (2013) and MES Technical Instructions (2012), gravity sewer should be laid at a slope so that self-cleansing velocity (SCV) of 0.8 m/s velocity is obtained at least once in a day at ultimate population discharges. However, for initial periods, when sewage load is not enough but likely to increase in phases, reduction in SCV to 0.6 m/s can be accepted for the sake of economy. The velocities in gravity sewers are calculated using Manning's equation which is reproduced below.

$$V = (1/n) \times R^{(2/3)} \times S^{(1/2)}$$

where,

n Manning's coefficient of roughness and varies with smoothness of internal surface of sewer as listed in Table 3.11 of CPHEEO manual.
R Hydraulic mean radius, 'm'.
S Slope of hydraulic gradient.

In defence establishments, obtaining SCV of 0.6 m/s is a challenge due to sparse population density. Population density as low as 5–10 persons/hectare need to be effectively connected to central sewage system. For calculation of SCV, in the previous manual by CPHEEO (1993) para 3.4.3.1, it is mentioned that for small discharges up to 30 l/s diameter of sewer does not have much effect on velocity which is incorrect.

The new CPHEEO manual has also provided approximate values only for discharges and velocities using modified and simplified nomograms based on Manning's equation. However, since these values are approximate, fine tuning in selection of slope of sewers remains a challenge for designers. The National Building Code (NBC 2016)[1] has also given approximate values of discharges and velocities for certain slope conditions in Table-22 but for only one type of material of sewer (SWG with $n = 0.015$). Thus, exact calculations are missing in both the important documents. Exact calculations show that there is influence of diameter due to change in wetted perimeter of sewer. Using Manning's equation and a software developed in-house in MS Excel, graph has been plotted for very low discharges to show the effect of diameter on velocity. It can be seen in the graph at Fig. 15.1 that for same slope and same Manning's constant, higher discharge is required to obtain SCV if diameter is increased. It also shows that oversizing of sewers should be strictly avoided for low discharges. In fact, ideally in defence establishments or similar townships, egg shaped small diameter sewers should be used.

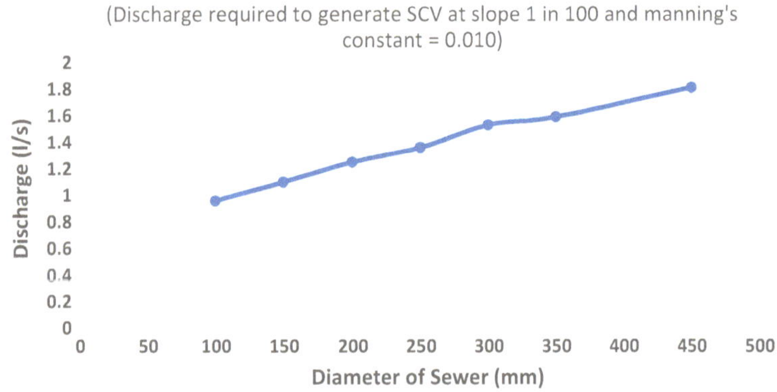

Fig. 15.1 Influence of sewer diameter on self-cleansing velocity

[1]SP-46 of Bureau of Indian Standards.

15.4 High Peak Factor to be Adopted for Low Population

CPHEEO (see Footnote 1) manual Table 3.2 recommends peak factor, ratio of peak hourly flow to average daily flow, to be adopted based on population being served. For populations less than 20,000 factor of 3 has been recommended and is being used by all consultants. Hitherto, in defence establishments a peak factor of 4 was recommended for smaller populations less than 5000 as suggested by Metcalf and Eddy (2003). However, for very small populations, peak factor of 4 is also unrealistic.

Literature survey on Internet was done, and one reference by IIT Kharagpur[2] was found where IIT in their Web courses has recommended peak factor of 6 for lateral sewers as laterals serve small populations. It appears to be realistic keeping in view the fact that large variation in discharges is observed for small population in defence establishments also. Therefore, for very small populations, peak factor of 6 is recommended in defence establishments. Thus, Table 15.1 may be referred for peak factors to be used in defence establishments or small townships.

15.5 Selection of Material of Sewer

The importance of material of sewer is sometimes ignored by consultants in defence establishments, and they tend to follow the traditional choice of stoneware glazed (SWG) sewers for small diameters and concrete (type NP_2) sewers for 200 mm and above. SWG sewers due to their inherent limitations of short length, high roughness

Table 15.1 Peak factor as recommended in defense establishments

Population	Peak factor
Below 500	6.00
Above 501–5000	4.00
Above 5001–20,000	3.00
Above 20,001–50,000	2.50
Above 50,001–750,000	2.25
Above 750,001	2.00

[2]IIT Kharagpur, Web Courses, Module 4: Quantity Estimation of Sewage Lecture 4: Quantity Estimation of Sewage.

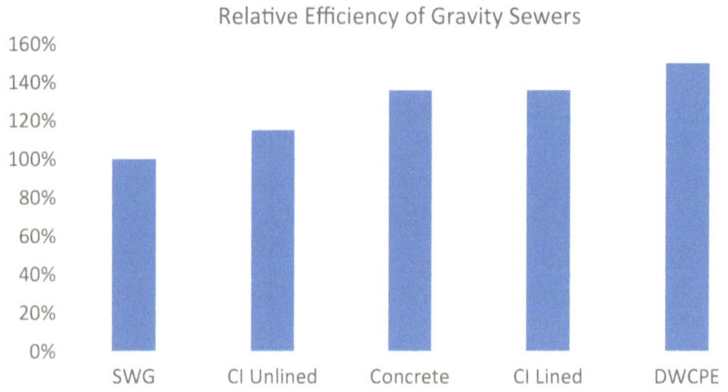

Fig. 15.2 Relative efficiency of sewers of different materials

coefficient and cumbersome jointing procedure are now banned in defence establishments except for acidic flows. NP_2 pipes have also been replaced by NP_3 due to requirement of stronger sewers to minimize in situ concreting for bedding or cradle.

Double-walled corrugated polyethylene (DWCPE) sewers have now been introduced which have least roughness coefficient (Manning's constant) amongst all type of sewers available in market. These are being manufactured in India as per IS Code (2013). A comparison of carrying capacity of various types of sewers has been made and reproduced above in Fig. 15.2. It can be seen that DWCPE sewers are 50% more efficient than SWG sewers. Other advantages of DWCPE sewers are lightweight, longer length, better strength, high water tightness, etc. However, being light in weight, they sometimes require anchoring against uplift due to ground water. Under normal conditions, no concrete bedding is required for these sewers. Since DWCPE pipes are available in 6.0 m length, number of joints are much less than SWG or concrete pipes. Jointing is fast and easy too, thereby need of expert jointers is minimized.

15.6 Use of Automatic Flushing Tanks (AFTs) for Very Small Populations

Designers of MES and consultants find it difficult to decide when AFTs are to be provided. There is a tendency amongst consultants to suggest provision for AFTs for all locations where SCV is not achieved. In certain cases, more than 100 AFTs have been suggested in small stations by consultants. Such high numbers of AFTs are undesirable as they entail high demand of additional water, apart from maintenance issues.

It is proposed that gravity sewers be designed optimally taking full advantage of gravity flows. Once inescapable requirement of AFTs is felt, one of the following approaches can be adopted.

Table 15.2 Automatic flushing tank at various slopes and populations

Population	Slope of 160 mm OD, DWCPE (SN8) sewer				
	1 in 50	1 in 80	1 in 120	1 in 150	1 in 170
>50 and up to 80	Yes	Yes	–	–	–
>80 and up to 100	No	Yes	Yes	Yes	–
>100 and up to 120	No	No	Yes	Yes	Yes
>120 and up to 150	No	No	No	Yes	Yes
>150 and up to 170	No	No	No	No	Yes
>170 and up to 200	No	No	No	No	No

(a) **AFTs for very small population**. For very small population up to 200, Table 15.2 above can be used to decide if AFTs are to be provided.
(b) **Treated wastewater used for flushing the sewers**. It is a new concept proposed for the first time in defence establishments (may be first time in the country) to use treated sewage for flushing of sewers to replace AFTs. It is being provided at three stations. To ensure that accidental cross-connections with drinking water pipes do not take place, violet or red coloured HDPE pipes are used and detachable spindles are provided for plug and use. In order to minimize human efforts in the system, the spindles can be strategically located at one or two points close to lift wells. Automation of the valve operations is also possible.

15.7 Restriction on Maximum Depth of Gravity Sewers

Since in defence establishments, the diameters of sub-main gravity sewers are small, it is difficult to provide maintenance to such sewers at deeper depths; hence, depth of gravity sewers has been restricted to 6.0 m.

15.8 Climate Specific Recommendations

(a) **Minimum Velocity for Preventing Hydrogen Sulphide in Sewers.**

In order to avoid formation of foul gas hydrogen sulphide (H_2S), the velocity shall not only be self-cleansing but also be sufficient to keep the submerged surfaces of the sewer free from slime and prevent the generation of hydrogen sulphide gas which can attack the cement concrete sewers. Stations which are located in warm areas should preferably be designed to have velocity of 0.8 m/s at least once in a day. However, this should not be the criteria for providing AFTs.

(b) **Extreme Climatic conditions.**

(i) *Extreme Cold Climates*. It has been seen that consultants ignore extreme cold climatic conditions when proposals are given for such areas although IS code (2001) exists since 1986. For extreme cold climates, sewers and manholes should always be insulated. The sewers should run below frost line, and manholes should have double opening for insulation. Building sewers which run from toilet to underground sewer should also be specially insulated. ED-2016 of MES published in 2016 caters for additional cost of these measures. For this, the basic cost is increased at the rate of 8% per 50°C fall in average ambient temperature below 200°C.

(ii) *Hot Climates*. In case of hot climates, the self-cleansing velocity should be taken as 0.8 m/s instead of 0.6 m/s to minimize formation of slime and H_2S gas in sewers. The lift wells in hot climates should be adequately ventilated. For this purpose, a station having average daily temperature more than 30 °C for at least three months may be defined as hot climate.

15.9 Lift Wells and Pumps for All Flow Conditions

When the invert level of gravity sewers reaches 6.0 m, it is better to provide lift well to pump the sewage. The sewage should not stagnate in lift well for more than 15–30 min to avoid development of septic conditions. Consultants do provide small lift wells, but they ignore the fact that pumps should also be provided with the same logic. Capacity of pumps should be such that for all three conditions, i.e. peak, average and lean flow conditions there is no stagnation for more than 15 min and pumps are running at least for 2 min at a stretch. For this, number of pumps may be 3–5 and level sensors and timer switches may be provided. Alternately, variable current pumps may be used.

In one case, consultant proposed connecting 2–3 lift wells with single delivery pressure main without non-return valves and without detailed analysis of system behaviour during parallel pumping. Such systems do not work for wastewater transportation. Direct pumping of sewage in gravity sewer downstream or at STP is reliable and a simple solution. The carrying capacity of gravity sewer downstream must be checked for lean as well peak discharges.

15.10 Decentralized STPs

These days, decentralized sewage systems are being preferred as cost of transport of sewage and that of treated effluent is reduced. In defence establishments since the size of STPs is generally much small in comparison to civil setups, therefore it may not be economical to use decentralized plants in every case. Therefore, designer

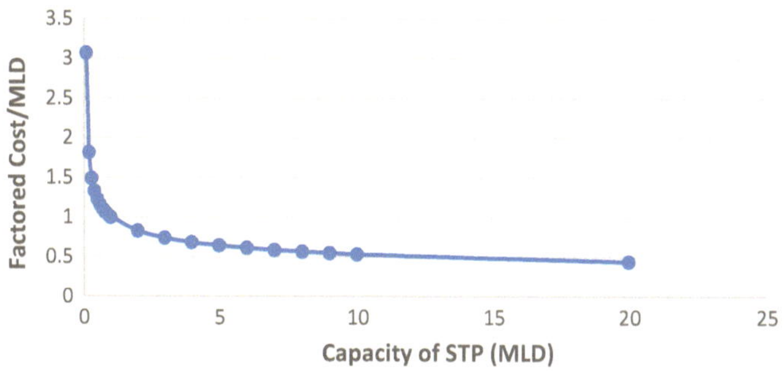

Fig. 15.3 Factored cost of STP/MLD

should ask consultants for planning different options in totality including cost of transportation of sewage and that of treated effluent for reuse. Cost–benefit analysis for all options should be done by consultant to facilitate selection of most efficient system. It may be noted that cost per million litre/day (MLD) increases inversely with reduction of size of STP as analysed based on cost of several departmental STPs and summarized in Fig. 15.3 above.

15.11 Conclusion

Common mistakes which have been discussed above are tabulated as follows:

S No.	Common mistakes
(a)	Inadequate attention to contours and layout plans
(b)	Selection of correct diameter of sewer which influences SCV is ignored
(c)	Application of wrong peak factor for discharges
(d)	Incorrect material of sewer
(e)	Non-judicious use of automatic flushing tanks (AFTs)
(f)	Deep excavations with small sewers
(g)	Ignoring extreme climatic conditions
(h)	Inappropriate design of lift wells
(i)	Improper selection of site and capacity of STP

It can be seen that attempt has been made in this chapter to analyse common mistakes made by consultants, and suggestions have been made to minimize these mistakes. Suggestions regarding higher peak flow factor, restrictive use of flushing tanks, reuse of treated wastewater for flushing, use of energy efficient DWCPE sewers and restrictions on depth of sewers for all stations. For extreme cold conditions, insulation of sewers and manholes has been reiterated, and for hot climates, higher velocities greater than 0.8 m/s have been suggested. Conditions for efficient lift wells have also been reiterated. Cost–benefit analysis for selecting best sewage disposal system is also recommended. These suggestions if made part of special conditions of consultancy contract will ensure efficient designs by consultants and faster clearances of DPRs/Estimates.

References

Ministry of Urban Development (2013) Manual on sewerage and sewage treatment system
Technical instructions on sewage management part I by E-in-C's Br, New Delhi (2012)
Ministry of Urban Development (1993) Manual on sewerage and sewage treatment system
National Building Code (2016) Bureau of Indian Standard, Group 5
Metcalf, Eddy (2003) Wastewater engineering treatment and reuse, 4th edn. Tata McGraw Hill
IS Code 16098 (part 2): 2013 Indian standard structured-wall plastics piping systems for non-pressure drainage and sewerage—specifications
IS Code 6295-1986 (Reaffirmed 2001) Indian Standard Code of practice for water supply and drainage in high altitudes and/or sub-zero temperature regions
Estimating Data (ED) for E/M and PHE schemes of Military Engineer Services (2016)

Chapter 16
Waste Containment System to Limit Landfill Gas Emission—Mechanism, Measurement, and Performance Assessment

S. Rajesh

Abstract The biodegradation of municipal solid wastes (MSW) within waste containment system (engineered landfill) generates various gases, primarily, methane and carbon dioxide, commonly known as landfill gases or biogas. Methane generation in landfills and the resulting emissions to the atmosphere have become most anthropogenic methane source. The generation of landfill gas could create a pressure gradient within the waste containment system that forces the gas to migrate. As reported by few case studies, the migration of landfill gas through landfill covers is unavoidable even with the provision of gas drainage layer in the landfill capping system. Hence, the performance of waste containment system, in specific, cap barriers under gas flow condition needs to be evaluated, monitored, and understood. This chapter essentially deals with the nature and movement of gases in various cap barriers due to transport mechanisms like advection and diffusion. The mechanism involved in the generation of landfill gas due to biodegradation of MSW, gas transport through cap barrier, and development of test apparatus for the determination of transport parameters have been explained. This chapter highlights the need for evaluation of gas permeability and the diffusion coefficient of the cap barriers experimentally. The factors governing the performance of various cap barriers under gas flow conditions have been explained.

Keywords Engineered landfill · Cap barriers · Landfill gas · Advection
Diffusion · Distortion · Gas permeability · Diffusion coefficient

---

S. Rajesh (✉)
Department of Civil Engineering, Indian Institute of Technology Kanpur,
Kanpur 208016, India
e-mail: hsrajesh@iitk.ac.in

T. Gupta et al. (eds.), *Environmental Contaminants*, Energy, Environment, and Sustainability, https://doi.org/10.1007/978-981-10-7332-8_16

16.1 Introduction

The municipal solid waste (MSW) generation rate, particularly in megacities of India, is showing an increasing trend with an increase in population and migration of people to the cities. The per capita waste generation in major cities ranges from 0.2 to 0.6 kg/day depending on the size of the population. It is estimated that about 150,000 metric tonnes per day of MSW are generated annually in megacities of India (CPCB 2012). In addition, it was estimated that the urban governing bodies spend about Rs. 500–1500 per tonne on collection, segregation, transportation, treatment, and disposal of MSW. The amount spent on waste collection, transportation, and final disposal was estimated to be 60–70%, 20–30%, and less than 5%, respectively. This fact clearly states the least importance given to scientific and safe disposal of waste (CPHEEO 2013). MSW is a complex material that comprises of various materials having different properties. Some of the components might degrade as a result of physical, chemical, and biological process. Developing countries often have MSW that contain more biodegradable materials and inconsistent grading whereas developed countries, adopting segregation of waste, have more uniform and consistent grading (Dixon and Jones 2005). The composition and hazardous nature of MSW generated in India cities differs from MSW generated in the Western countries. Typical composition of MSW generated in Indian cities comprises of organic fraction in the range of 40–60%, ash and fine earth in the range of 30–40%, paper and cardboard in the range of 3–6% and plastic, glass, and metals (each less than 1%). Organic matter content present in the MSW tends to decompose leading to the production of harmful gases and leachate. The landfill gas, primarily, methane, and carbon dioxide are sources of fire, explosion, and global warming to some extent if they are not controlled. However, if an attempt is made to collect the landfill gases securely, it can be used for energy generation.

The quantity and the composition of generated MSW form the basis on which the solid waste management system may be planned, designed, and operated. Moreover, geotechnical properties of MSW are of prime importance for the design and maintenance of waste containment system, such as engineered MSW landfill. Heterogeneity of waste adds the complexity in evaluating the engineering properties and hence to understand deformation behavior of MSW landfills (Rajesh et al. 2015). MSW landfill may be designed as a passive system or active system based on the extent of biodegradation. In the passive or conventional system, a cover system is composed of an impermeable material, and landfill needs to be closed as soon as it has ceased operation, to minimize the generation of leachate. In the active system/bioreactor landfills, leachate is reinjected into the waste to initiate the natural biodegradation process and to accelerate waste decomposition Viswanadham and Rajesh (2007). This process might bring the bioreactor landfill to field capacity many times quicker than conventional landfill. This emphasizes the importance of settlement rates while designing landfill lining systems. The schematic representation of conventional and bioreactor landfills is shown in Fig. 16.1. Base lining system and side lining system are provided to avoid migration of leachate into groundwater,

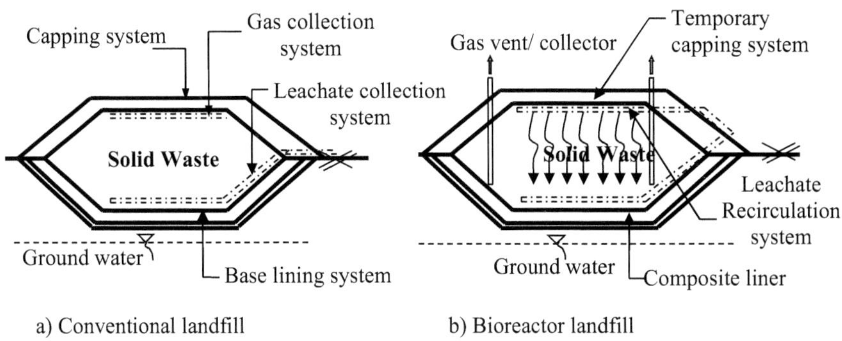

Fig. 16.1 Schematic representation of municipal solid water containment systems

whereas, the cover system is provided to avoid infiltration of rainwater into the landfill and mitigation of landfill gases to the atmosphere. A typical base lining system normally consists of a soil barrier, geomembrane, and drainage blanket along with a leachate collection system. A cover or capping system consists of a cap barrier, gas-ventilating layer, water drainage layer, cover soil, and vegetation. In all lining systems, an impermeable barrier in the form of a compacted clay liner, amended soil liner, geomembranes, geosynthetic clay liner, sand-bentonite barrier, polymer amended sand-bentonite barrier, or composite liner (combination of these) is commonly adopted. Leachate collection system is used to collect the leachate produced in a landfill and to drain the leachate to wastewater treatment plant. The gas collection system is used to collect the landfill gases generated during the decomposition of organic components of solid waste. The provision of gas drainage layer in engineered landfills can significantly reduce the gas migration through soil barrier, but eventually the migration of landfill gas through soil barriers is unavoidable. Hence, landfill capping system has to be designed to prevent the migration of landfill gas to the environment, in addition to limiting infiltration of rainwater.

16.2 Mechanism Involved in the Generation of Landfill Gas

Landfill gas (LFG) is generated in the MSW landfill due to the presence of organic fraction and bacterial activity occurring over a period of time. The LFG generation rate is governed by various factors like waste temperature, waste composition and density, pH within landfill, concentration of substrate, moisture content, and toxins. The components of landfill gas are primarily CH_4 (45–65%), CO_2 (35–55%), N_2 (2.4%), O_2 (0.16%), H_2 (0.05%), CO (0.001%), and some others in trace amount (Maciel and Jucá 2000). Methane generation in landfills and the resulting emissions to the atmosphere have become most anthropogenic methane source (EPA 2013).

Table 16.1 Properties of main constituents of landfill gas (At 1 atm and 35 °C).

Parameter	Methane	Carbon dioxide
Molecular weight (g/mol)	16.04	44.01
Volumetric weight (kg/m^3)	0.634	1.74
Relative vapor density	0.55	1.53
Viscosity (Pa.s)	11.439×10^{-6}	15.394×10^{-6}
Boiling point (^{o}C)	−161.0	−78.5
Vapor pressure (kPa)	–	5720
Water solubility (g/L)	0.0251–0.064	1.69

Adopted from Durmusoglu (2002)

The properties of main constituents of LFG (i.e., methane and carbon dioxide) are provided in Table 16.1.

The generation of gas within a landfill occurs due to both aerobic and anaerobic mechanisms. The aerobic reactions start just after the filling of waste in the landfill. These processes consume atmospheric or trapped oxygen and generate CO_2 and water vapor. The amount of water vapor generated by aerobic processes is usually considered to be negligible and is not included in the leachate content of the landfill. When a next MSW layer is added, it insulates the waste, thus cutting the oxygen penetration (EPA 2013). After a phase lag which lasts for a time equal to the acclimation period (rate of bacterial activity increases over a period of time and reaches to a maximum) of the waste, aerobic activity diminishes as trapped oxygen gradually depletes and anaerobic activity attains its peak. CH_4 and CO_2 are consistently generated at a gradually decreasing rate as a result of anaerobic activity for a very long period (decades). Durmusoglu (2002) stated that anaerobic activity occurs in four steps. The first step involves solubilization of cellulose to glucose as a result of hydrolysis. The glucose, so produced, is then passed through the action of acidogenic bacteria which produce alcohol, carbon dioxide, hydrogen gas, and fatty acids. The above products are converted into acetic acid, hydrogen, and carbon dioxide (known as acetogenesis phase). The generated acetic acid is converted into methane and carbon dioxide (referred as methanogenesis phase).

Several numerical and mathematical tools have been developed to estimate LFG generation due to difficulties and uncertainties in monitoring methane emissions in an engineered landfill. Extraction rates may typically range from 25 to 100 m^3/h for small landfill sites having MSW capacity of 100,000 m^3. For large sites with capacities of 1–10 million m^3, extraction rates may be within the range of 250–10,000 m^3/h (EPA 2009). The generated LFG can be collected by vertical and horizontal wells and gas drainage layer, which is commonly placed beneath the cap cover of the landfill. The collected LFG could be either flared or used to generate electricity and/or heat. Flares may reduce potential methane emissions drastically simply by converting methane to carbon dioxide. The uncollected methane may be partially oxidized in the cover soil. The quality of the landfill cap barrier material could dictate the rate at which LFG escapes through the cover soil of the engineered landfill to the atmosphere.

16.3 Landfill Cap Barriers

Municipal solid waste disposal by engineered landfills using liner and capping systems is the most popular method that is currently in practice in many parts of the world. Several types of landfill barrier material can be used to limit the contaminant migration in such a way that they will result in a negligible impact on the environment. The use of baseliner is to limit the release of leachate to the subsurface whereas the cap barrier is used to minimize the infiltration of water into the waste and to control the fugitive emission of harmful gases. Natural soils are often used as cap barriers because of its low hydraulic conductivity property. The thickness of compacted soil barriers may range from 1.2 to 3 m in the case of base lining system, whereas in the cover system, they may range between 0.6 and 1.2 m (EPA 2004; MoEF 2000). Benson et al. (1999) presented a database comprising of field properties of compacted soil barrier materials collected from different landfills. In order to attain hydraulic conductivity of soil barrier material criteria ($<1 \times 10^{-9}$ m/s), soil barrier material should possess percentage fines $\geq 30\%$, percentage clay fraction $\geq 15\%$, liquid limit $\geq 20\%$, and plasticity index $\geq 7\%$. The compacted soil barriers are constructed by compacting at wet side of the optimum, which often lies in unsaturated state. When impervious natural soil is not readily or economically available, then the mixture of sand or locally available soil (recently fly ash) and bentonite can be used as a soil barrier material. The amount of bentonite generally varies between 3 and 15% of the total dry weight of the mixture, depending upon the type of soil used, gradation of soil, and quality of bentonite (Chapuis et al. 1992). The integrity of compacted soil barriers may depend on: (i) the characteristics of clayey soil used; (ii) the method of compaction and in particular, the moisture content; (iii) the protection against desiccation after construction, and (iv) the level of distortion.

The second-preferred cap barrier, next to compacted soil barrier, is the geomembrane. However, while installation of geomembrane, material expansion caused due to temperature and placement of overlying materials could lead to wrinkles in the geomembrane. The tension in the geomembrane may be possible during construction of the landfill covers, wind uplift on uncovered areas, and movement of heavy vehicles employed for compaction. The frictional forces from the cover soil may result in tension in the geomembrane. Tension and wrinkles of geomembrane could affect the performance of geomembrane as a cap barrier material. Due to above-mentioned limitations, it is advised to avoid using geomembrane as a single barrier but may be laid above the soil barrier (composite barrier).

Difficulty in compacting soil on compressible waste, the high cost of clay at some locations and the limitations of using geomembrane as a single barrier material have resulted in increasing use of geosynthetic clay liners (GCLs). GCLs are thin prefabricated clay liners consisting of sodium bentonite encased between two geotextiles. The main advantages of this cap barrier are the limited thickness, easy installation, and low cost. However, limited thickness of GCL may lead to puncture or damage to the GCL, possible penetration of plant roots into the GCL and lesser sorption capacity (Benson et al. 2007). Moreover, when hydrated with some types of leachate instead of pure water, bentonite will show a minor swelling that will result in an inefficient hydraulic barrier.

Effectiveness of above-mentioned cap barriers as hydraulic barriers is well-researched and well-proven (Rajesh and Viswanadham 2009, 2011, 2012; Viswanadham et al. 2012). However, the effectiveness of the cap barrier as a gas barrier is yet to be understood completely. This made several researchers to study the gas transfer mechanism and method to assess the gas permeability of the cap barrier. Few researchers have recommended that gas permeability can be considered as an important design parameter, in addition to hydraulic conductivity for selecting suitable cap barriers (Vangpaisal and Bouazza 2004; Moon et al. 2008; Rajesh et al. 2014, 2016).

16.4 Gas Transport Mechanism

Gas transport in cap barriers is a complex phenomenon and usually described by advection and hydrodynamic dispersion. Hydrodynamic dispersion comprises of mechanical dispersion and molecular diffusion. The gas flow through unsaturated cap barrier can be due to concentration, pressure and temperature gradient, and various other factors. Advection is the movement of gases from a region of higher pressure to a region of lower pressure until the pressures are equalized. Diffusion phenomenon happens when diffusing species are transported as an effect of concentration gradient across a barrier. Mechanical dispersion is the product of dispersivity and advective velocity. Dispersion is generally neglected in the gas phase of unsaturated soil, as gas velocities are low. As the biodegradation of waste could generate LFG within the landfill, gas pressure will try to build up over time. Eventually, the gas pressure would cause higher pressure within the landfill than atmospheric pressure. From this behavior, it can be inferred that the primary driving force for gas migration through cover systems could be advective flow. Stark and Choi (2005) presented field evidence of the gas blow up (building up of high gas pressure) that develop in PVC geomembranes in landfill capping system (refer Fig. 16.2). Any defects/holes/seams in the geomembrane could result in fugitive

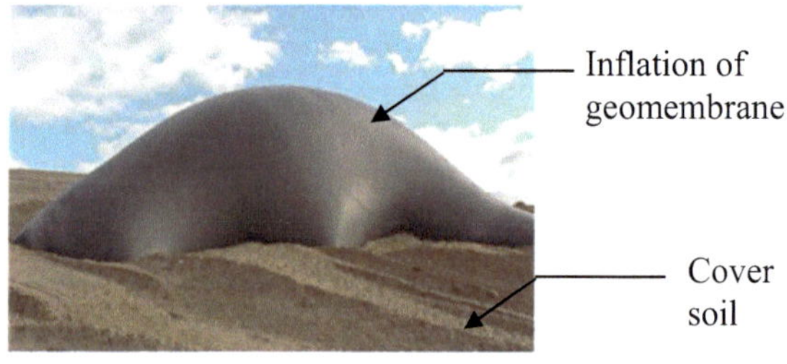

Fig. 16.2 Inflation of PVC geomembrane by LFG in landfill capping system. Modified after Stark and Choi (2005)

gas emission, so careful quality control procedure should be followed for landfill capping system to ensure containment of the landfill gas.

The gas flow in unsaturated cap barriers can be due to continuous air phase or occluded bubbles. According to Corey (1957), the air phase is discontinuous when the degree of saturation of soil is 85% or above. Advective gas transport readily occurs when the gas phase becomes continuous, which happens at a degree of saturation less than 85%. Diffusion may happen generally in nearly saturated GCLs or geomembrane, where the gas phase is discontinuous (Aubertin et al. 2000). Darcy's law is used to model advective gas transport, and Fick's law is used to model the diffusion gas transport. The parameter that represents such gas transport is intrinsic permeability (for advective flow condition) and diffusion coefficient (for diffusive flow condition), which can be determined experimentally or analytically. Few researchers have developed apparatus to measure gas permeability of soil/synthetic cap barrier considering advective and diffusive flow through the medium.

16.5 Test Apparatus and Governing Mechanism

The two prime parameters governing gas transport are intrinsic gas permeability (for advective flow condition) and diffusion coefficient (for diffusive flow condition). Gas permeability can be described as the ease/ability with which the medium conducts gas through it. The gas permeability of the cap barrier under advective transport mechanism may be determined considering steady or transient-state flow conditions. In steady state analysis, the hydraulic head and the coefficient of permeability at any point in the cap barrier material remain constant with respect to time. In unsteady-state flow analyses, the hydraulic head changes with respect to time. Few researchers like Shan and Yao (2000), Bouazza and Vangpaisal (2003), and Stolz et al. (2010) estimated the gas permeability of the cap barrier using steady-state flow condition. One of the main difficulties that arise by employing steady-state techniques is the extremely slow permeant flow rate due to the very low permeability of the unsaturated cap barrier. Hence, a longer time is required to achieve steady state. The separation of contact between the cap barrier specimen and the permeameter due to shrinkage is very difficult to handle. These limitations made researchers to use transient-state flow condition for determination of gas permeability (Li et al. 2004; Rajesh et al. 2014, 2016).

16.5.1 Steady-State Advective Flow

Darcy's law could be used for characterizing advective gas flow in a low permeability cap barrier material (Bouazza and Vangpaisal 2003). Hence, models developed for water flow can also be used for gas flow. Based on Darcy's law, the 1D volumetric flow rate Q (m^3/s) of gas in cap barrier material is given as:

$$Q = \frac{k}{\mu} A \frac{dP}{dx} \qquad (16.1)$$

where k is the intrinsic permeability of the cap barrier material (m^2); A is the cross section of the cap barrier (m^2); dP/dx is the pressure gradient; and μ is the dynamic viscosity of gas ($(N\ s)/m^2$). Assuming that LFG behaves as an ideal gas and that the continuity equation for a gas applies, then Eq. 16.1 becomes

$$PQdx = \frac{k}{\mu} APdP \qquad (16.2)$$

For a cap barrier of thickness L (Fig. 16.3a), the solution to the Eq. 16.2 is subjected to the boundary conditions like at $x = 0$, $P = P_1$ and at $x = L$, $P = P_2$: The volumetric flow rate Q_2 (m^3/s) at point 2 can be obtained by integrating Eq. 16.2

$$Q_2 = \frac{k}{\mu} A \frac{P_1^2 - P_2^2}{2P_2 L} \qquad (16.3)$$

Knowing the volumetric flow rate at the outlet, area of the specimen, inlet and the outlet gas pressure, thickness of the sample, and the dynamic viscosity of the applied gas, the intrinsic permeability of the cap barrier can be determined.

Bouazza and Vangpaisal (2003) developed a test setup adhering to the boundary conditions and principle outlined above to determine the gas permeability of the GCL. Test apparatus consists of base cylinder, upper cylinder with a piston, nitrogen container, pressure regulators, gas flow meter, and pressure gauge as shown in Fig. 16.3b. The piston is used to apply required overburden pressure to

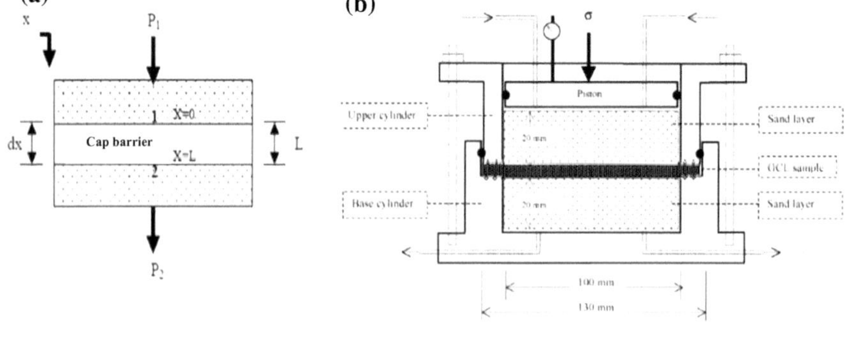

Fig. 16.3 **a** Boundary condition and **b** schematic view of gas permeability test apparatus developed by Bouazza and Vangpaisal (2003)

the cap barrier specimen. O-rings are used to provide an airtight seal at the several connections. The inlet and outlet gas ports are provided as shown in Fig. 16.3b for gas inflow and outflow. Nitrogen gas was used as the permeating gas, in place of landfill gas, because of its inert and hazardless nature and low water solubility. The gas permeability is determined by sending nitrogen gas to the top of the chamber, which permeated through the GCL specimen and flowed out from the base of the cell by way of a gas flow meter. The outflow gas port was opened to atmospheric pressure. The differential gas pressure was calculated from the pressure difference between the inflow and outflow. Typical variation of flow rate (Q_2) with $(P_1^2 - P_2^2)$ for a GCL with water content 121%, as reported by Bouazza and Vangpaisal (2003), is reproduced in Fig. 16.4. The intrinsic permeability of the GCL can be determined from Eq. 16.3 and was found to be 5×10^{-17} m^2 for the variation shown in Fig. 16.4. More details of the test setup, sample preparation, and testing procedure are given in Bouazza and Vangpaisal (2003).

16.5.2 Transient-State Advective Flow

The gas permeability of the cap barrier material can be determined by considering transient-state advective flow/falling gas pressure conditions (Barrel et al. 2010; Pitanga et al. 2011; Rajesh et al. 2014). Rajesh et al. (2014) developed a test setup to determine the gas permeability of the cap barrier material considering falling head technique under various distortion levels. Test apparatus comprises a motorized actuator of the triaxial frame, loading and support platform, gas chamber, nitrogen gas cylinder, and few sensors like pressure transducer, proving ring, and potentiometers, as shown in Fig. 16.5a. The moist-compacted cap barrier beam specimen was placed on the support platform, and then a sealing agent was applied between the beam and gas chamber interface to prevent gas leakage. The loading platform was placed on the cap barrier beam specimen along with the proving ring

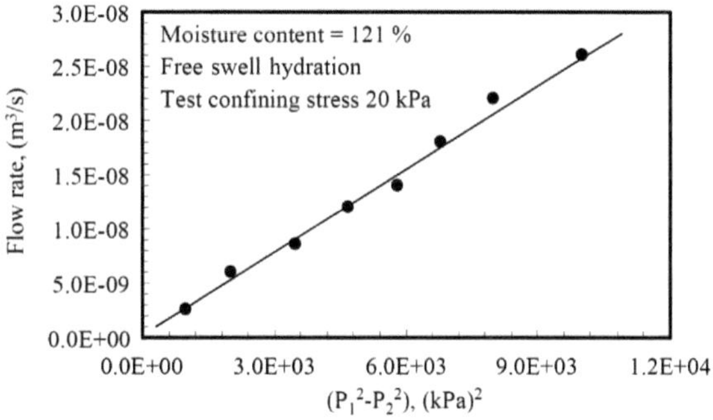

Fig. 16.4 Typical variation of flow rate with $(P_1^2 - P_2^2)$ for GCL cap barrier. Adopted from Bouazza and Vangpaisal (2003)

and potentiometer. The gas chamber was filled with nitrogen gas at a relative pressure of 3 kPa. After attaining equilibrium, the relevant valve was closed. A reduction in gas pressure within the gas chamber could be noticed primarily due to gas flow transfer through the soil beam. By measuring the drop in the gas pressure over a definite period of time, gas permeability of the soil beam can be determined. The gas permeability of the soil beam at various displacements is determined by loading the soil beam continuously at a displacement rate of 0.2 mm/min till the required displacement is reached. At any required displacement, gas permeability is measured as outlined above.

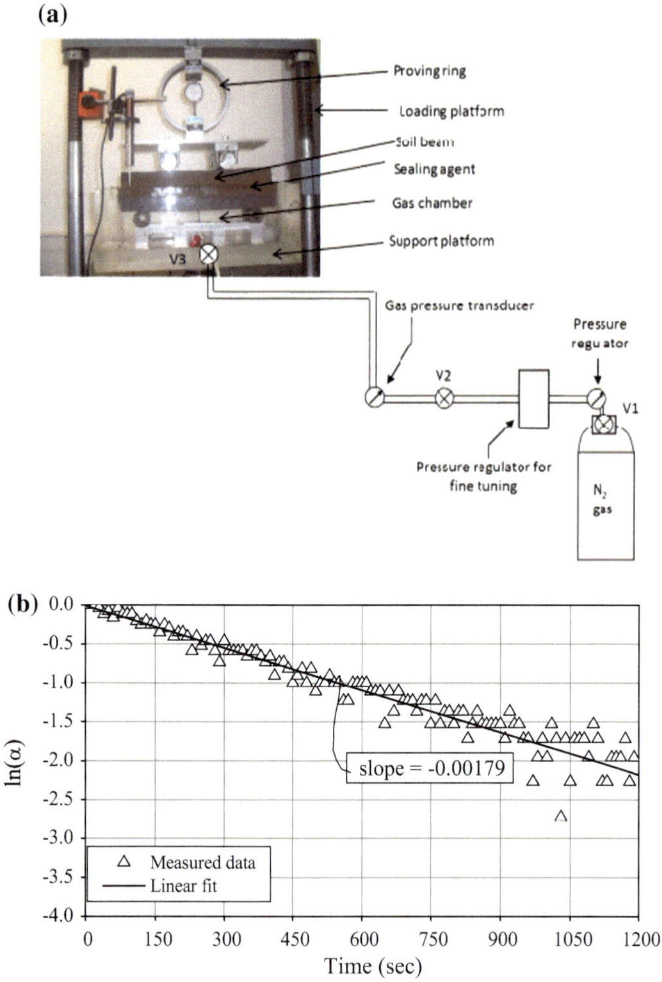

Fig. 16.5 **a** Gas permeability test apparatus developed by Rajesh et al. (2014); **b** typical variation of ln(α) with time for a compacted soil cap barrier

The analytical solution that describes the gas pressure variations over time was proposed by Li et al. (2004). The gas flow across the specimen tested is the mass flow rate of gas $\mathrm{d}m(t)/\mathrm{d}t$, which is induced from the pressurized container:

$$\frac{\mathrm{d}m(t)}{\mathrm{d}t} = v\frac{\mathrm{d}\rho_c(t)}{\mathrm{d}t} \tag{16.4}$$

where v is the volume of the intermediate container and $\rho_c(t)$ is the gas density in the container at any time t.

The isothermal condition is assumed thus,

$$\frac{p_c(t)}{\rho_c(t)} = \frac{p_c(0)}{\rho_c(0)} \tag{16.5}$$

where $p_c\ (t)$ is the differential gas pressure at any time t.

Combining Eqs. (16.4) and (16.5):

$$\frac{\mathrm{d}m(t)}{\mathrm{d}t} = v\frac{\rho_c(0)}{p_c(0)}\frac{\mathrm{d}\rho_c(t)}{\mathrm{d}t} \tag{16.6}$$

The differential equation governing the transient variation of the gas pressure can be obtained from the continuity equation, ideal gas law, and extended Darcy's law, which is given by

$$\frac{\mathrm{d}p_c(t)}{\mathrm{d}t} = \frac{Ak_g}{2v\mu}\frac{\mathrm{d}p^2(z,t)}{\mathrm{d}t} \quad \text{at z} = 0 \tag{16.7}$$

where k_g = gas permeability of the soil barrier material (m^2); μ = dynamic viscosity of nitrogen gas (Pa.s); A = cross section of the soil beam (m^2); v = net volume of gas entrapped in the container (m^3); and $p_c(t) = p(0,\ t)$. If the function p^2 (z) is approximated by a linear function, which is acceptable for a small specimen thickness d and a great chamber volume v, then Eq. (16.7) reduces to:

$$\frac{\mathrm{d}p_c(t)}{\mathrm{d}t} = \frac{Ak_g}{2v\mu d}\left(p_{\mathrm{atm}}^2 - p_c^2(t)\right) \tag{16.8}$$

If during the test, $p_c(t)$ is close to p_{atm} the approximation of Eq. (16.8) can be considered:

$$p_{\mathrm{atm}}^2 - p_c^2(t) \sim 2p_{\mathrm{atm}}\left(p_{\mathrm{atm}} - p_c(t)\right). \tag{16.9}$$

It leads to the simple analytical solutions

$$p_c(t) = p_{atm} + (p_c(0) - p_{atm})\exp\left(\frac{-t}{\tau}\right) \quad (16.10)$$

$$\text{With} \quad \tau = \frac{v\mu d}{k_g A p_{atm}} \quad (16.10a)$$

$$\text{If} \quad \alpha = \left[\frac{p_c(0) + p_{atm}}{p_c(0) - p_{atm}}\right]\left[\frac{p_c(t) - p_{atm}}{p_c(t) + p_{atm}}\right] \quad (16.11)$$

Then from Eqs. (16.10) to (16.11)

$$\ln(\alpha) = \frac{-t}{\tau} \quad (16.12)$$

The slope of the linear part of the plot ln α versus time t is denoted as s. Thus,

$$k_g = \frac{v\mu ds}{A p_{atm}} \quad (16.13)$$

A typical variation of ln(α) with time for a natural soil-based cap barrier compacted at 2% wet of optimum, as reported by Rajesh et al. (2014) is reproduced in Fig. 16.5b. From the slope of the linear part of the curve (s), gas permeability of the soil barrier material can be determined and was found to be 2.9×10^{-15} m^2.

16.5.3 Diffusion Gas Flow

Fick's laws can be used to model gas diffusion through cap barrier. The 1D transient diffusion equation is given by Fick's second law:

$$\varepsilon \frac{\partial C_g}{\partial t} = D_p \frac{\partial^2 C_g}{\partial z^2} \quad (16.14)$$

where ε is the air-filled porosity (m^3/m^3), D_p is the diffusion coefficient of the gas (m^2/s), C_g is the concentration difference in the gaseous phase (g/m^3), z is a thickness of the cap barrier (m), and t is the time (s). Considering uniform sample with respect to diffusion coefficient and constant air-filled porosity, Eq. (16.14) can be solved using the boundary conditions of the diffusion chamber. The solution to the Eq. (16.14) for the relative concentration (C_r) in the diffusion chamber is given by Carslaw and Jaeger (1959) can be given as:

$$C_r = \frac{C(L,t) - C_0}{C_i - C_0} = \sum_{n=1}^{\alpha} \frac{2h\ \exp(-D_p \alpha_n^2 t/\varepsilon)}{L(\alpha_n^2 + h^2) + h} \quad (16.15)$$

Considering $n \geq 2$ is negligible in the first term, the equation can be rewritten as:

$$\mathrm{Ln}(C_\mathrm{r}) = \mathrm{Ln}\left(\frac{C(L,t) - C_0}{C_i - C_0}\right) = -\frac{D_\mathrm{p}\alpha_1^2}{\varepsilon}t + \mathrm{Ln}\left[\frac{2h}{L(\alpha_1^2 + h^2) + h}\right] \quad (16.16)$$

where C_0 is the gas concentration of the source (atmosphere) (g/m^3), at $z < 0$; $C(L, t)$ is the gas concentration in the diffusion chamber at time $t > 0$ (g/m^3), at $z > L$; L is the thickness of the sample (m); C_i is the gas concentration in the diffusion chamber at $t = 0$ (g/m^3), at $z > L$; $h = \varepsilon/L_a$ (m^{-1}), where L_a is the sum of the height of diffusion chamber and length of the free space below the cap barrier specimen (m); α_1 is the first positive root of $hL = \alpha_n L_n \tan(\alpha_n L_n)$.

The slope of the linear part of the plot Ln (C_r) versus time t is denoted as S_D. Thus, diffusion coefficient can be obtained from Eq. (16.18)

$$D_\mathrm{p} = -\frac{S_\mathrm{D}\varepsilon}{\alpha_1^2} \quad (16.17)$$

Rouf et al. (2016) developed a test setup adhering to the boundary conditions and principle outlined above for determining diffusion coefficient. The developed test apparatus as shown in Fig. 16.6a can be used to estimate gas permeability and diffusion coefficient of the cap barriers sequentially. The test apparatus comprises bottom cylindrical diffusion chamber, top cylindrical upper chamber, and a middle cylinder to accommodate a cap barrier specimen. The three parts are held together with four threaded retaining rods. The diffusion chamber was equipped oxygen electrode and a pressure transducer. The oxygen electrode was used to measure the variation of oxygen concentration within the diffusion chamber. The pressure transducer monitors the variation in the gas pressure inside the diffusion chamber.

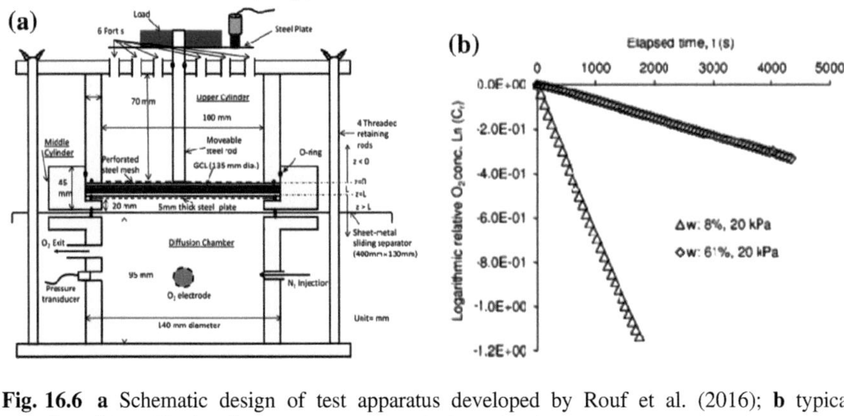

Fig. 16.6 **a** Schematic design of test apparatus developed by Rouf et al. (2016); **b** typical variation of logarithmic relative oxygen concentration with time

Gas flow meters were attached to the chamber to record gas outflow. Nitrogen gas was used as permeating fluid for the reason mentioned earlier.

After placing the cap barrier at the required location, target vertical stress was applied using the movable plate placed above the specimen. The nitrogen gas is introduced into the diffusion chamber to remove the oxygen present in the chamber. The chamber was sealed so that nitrogen gas filled in the diffusion chamber was not allowed to escape. The atmospheric air is allowed to pass through the specimen thereby the increase in the oxygen concentration in the diffusion chamber can be anticipated. The oxygen concentration in the diffusion chamber was recorded at required intervals using oxygen electrode. The gas permeability of the cap barrier (GCL) can be determined from the slope parameter $S_{D.}$ A typical variation of the logarithmic relative oxygen concentration of GCL at two different water content with time, as reported by Rouf et al. (2016) is shown in Fig. 16.6b. The slope of the linear part yields slope parameter S_D. Diffusion coefficient at a gravimetric water content of 8 and 12% was found to be 5.5×10^{-7} to 7.0×10^{-8} m^2/s, respectively.

16.6 Relative Permeability with Respect to Gas and Water Saturation

Hydraulic conductivity of the soil is used to denote the behavior of the water flow within the soil. The ability of the soil to retain the water under a specified pressure and hydraulic conditions can be determined from the soil–water characteristics curve (SWCC). The intrinsic permeability of the soil depends only on properties of the solid matrix, i.e., pore size, pore geometry, and pore size distribution not on the properties of the fluid flowing through it. Gas permeability (m/s) and intrinsic gas permeability (m^2) are related as follows:

$$K_g = k_g \frac{\rho_g g}{\mu_g} \quad (16.18)$$

where K_g = gas permeability (m/s), k_g = intrinsic gas permeability (m^2), ρ_g = density of the gas (kg/m^3), g = acceleration due to gravity (m/s^2), and μ_g = gas viscosity (kPa.s).

Relative permeability of gas (k_{rg}) is a function of the gas saturation (S_G) and is defined as the ratio of permeability of the unsaturated medium at particular gas saturation ($K_g S_g$) to its permeability at 100% saturation (K_g).

$$k_{rg} = \frac{k_g(S_g)}{k_g} \quad (16.19)$$

Relative permeability of cap barrier depends on the degree of saturation, nature of fluid (wetting or non-wetting), and hysteresis behavior during wetting and drying

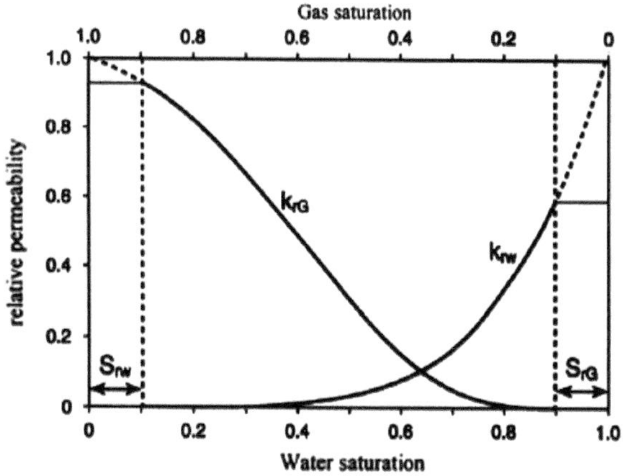

Fig. 16.7 Typical variation of relative permeabilities with water and gas saturation. Adopted from Scanlon et al. (2002)

cycle. It is common to develop expressions to estimate relative permeability considering water as a fluid medium. The expression can be modified to corresponding gas relative permeability by replacing the water saturation (S_e) by ($1-S_e$). Figure 16.7 shows a typical variation of relative permeabilities with water and gas saturation (Scanlon et al. 2002). The gas flow occurs only when the system attains minimum gas saturation (SrG). The region where the gas relative permeability is zero correspond the phase wherein random distribution of trapped gas is present thereby disrupting the gas connectivity due to water blockages. Similarly, zero water relative permeability region is observed, where the water does not begin to flow until minimum saturation is reached. These variations infer that the performance of cap barriers under various gas transport mechanism depends on the water content of the cap barrier, which indirectly suggests that the unsaturated soil properties of the cap barrier materials need to be evaluated, in addition to conventional geotechnical testing. The relative permeabilities with gas and water saturation may be used to effectively design the cap barriers as a gas barrier, in addition to a hydraulic barrier.

16.7 Performance of Cap Barriers Under Gas Flow Condition

The performance of cap barriers under gas flow condition depends on the type of cap barrier, relative gas pressure, placement water content and density, confining pressure, and the distortion levels. Rajesh et al. (2016) noticed that the time taken for the dissipation of the relative gas pressure through the compacted soil cap barrier material was found to increase considerably with an increase in the initial relative gas pressure. The rate of dissipation of the relative gas pressure at any time of the initial portion of the curve was found to vary considerably with an increase in the initial relative gas pressure; however, with increased elapsed time, there is no significant variation with an increase in the initial relative gas pressure. This indirectly suggests that gas permeability of the compacted soil cap barrier may not have a greater influence on the relative gas pressure. Similar observations were noticed for other cap barriers like GCL and geomembrane (Bouazza and Vangpaisal 2003; Stark and Choi 2005).

The gas flow characteristics of cap barrier material primarily depend on the initial moisture of the cap barrier material. Gas permeability was found to decrease with an increase in the degree of water saturation. The gas permeability of the cap barrier was found to attain maximum value in its dry state and least at its saturated state (Rajesh et al. 2014). Bouazza and Vangpaisal (2003) reported that higher intrinsic permeability of GCL was obtained when the GCL had a non-uniform gravimetric moisture content distribution within the GCL. For the same moisture content, with an increase in the confinement, gas permeability of GCL was found to decrease. Rouf et al. (2016) experimental observed that the gas permeability and gas diffusion remained constant up to 20–25% apparent degree of saturation, beyond which, both the gas permeability and gas diffusion coefficient reduced significantly with an increase in the moisture content. It was suggested that soil layer having sufficiently high gravimetric water content may be placed adjacent to GCL for hydration (at least the limiting levels) before coming in contact with the gas. These studies imply that it is important to ensure that the partially saturated cap barrier should be either sufficiently saturated or sufficiently confinement pressure need to be applied on cap barrier to qualify as efficient hydraulic cum gas cap barrier.

Stark and Choi (2005) found that landfill gas migration through geomembrane is negligibly small if a gas drainage layer is placed below the geomembrane. However, any defects in the geomembrane could result in gas migration; hence, careful quality control is essential for landfill covers to guarantee containment of the LFG.

Rajesh et al. (2014) noticed that performance of cap barrier as an effective hydraulic cum gas cap barrier depends on distortion level generated due to the biodegradation of the MSW. The heterogeneous nature of solid waste materials and varying rate of biodegradation does not typically provide a stable support structure for the overlying cover system; hence, possibility of occurrence of distortion at the base of cap barrier is inevitable. The gas permeability of the compacted soil cap barrier material was found to be the constant up to certain distortion levels followed by a steep increase in the gas permeability. A steep increase in the gas permeability implies gas breakthrough of the soil barrier material. This indirectly suggests that the cap barrier has failed to retain its gas intactness characteristics. It was also noticed that with an increase in the molding moisture content of the cap barrier material, a considerable delay in crack initiation and gas breakthrough has occurred. The performance of the compacted soil cap barrier under gas transport and distortion may be considerably increased with the inclusion of geo-fibers.

16.8 Summary and Conclusions

Waste disposal by engineered landfills is the most popular method that is currently in practice in many parts of the world. This involves the complete encapsulation of MSW with a low permeability material to prevent the infiltration of water into the waste body and the leakage of leachate into the surrounding ground. In recent times, it was noticed that even with the provision of gas drainage layer in the landfill capping system, the migration of landfill gas through cap barriers is unavoidable. As per EPA (2013) report, methane production in landfills and the resulting emissions to the atmosphere have become second-largest anthropogenic methane source. Hence, landfill capping system needs to be designed to prevent the migration of landfill gas to the environment, in addition to limiting infiltration of rainwater. Proper selection of cap barrier material, design guidelines, and construction of a lining system, in specific, for cap barrier play a key role in overseeing proper functioning of landfills, in terms of retarding gas migration to the environment. The effectiveness of various cap barriers as hydraulic barriers has been well researched and well proven; however, the effectiveness of the cap barriers as hydraulic cum gas barrier is still not yet understood completely. In this chapter, the mechanism involved in the estimation of the gas flow characteristics of the cap barriers has been addressed. The importance and necessity of estimating gas permeability and diffusion coefficient have been emphasized. The mechanism involved in the generation of landfill gas due to biodegradation of MSW is explained. Landfill gas may flow through cap barriers either by advection or diffusion, depending on the degree of saturation of cap barriers. Advective gas transport occurs when the gas phase becomes continuous, which happens at a degree of saturation less than 85%. Diffusion may happen generally in nearly saturated GCLs or geomembrane, where the gas phase is discontinuous. The parameter that represents advective gas flow transport and diffusive gas flow transport is intrinsic permeability and diffusion coefficient, respectively. Detailed description of the governing mechanism in the development of test apparatus and the methodology for measuring the intrinsic permeability and diffusion coefficient has been discussed. It was found that the performance of cap barriers under gas flow condition depends on the type of cap barrier, relative gas pressure, placement water content and density, confining pressure, and the distortion levels. Landfill gas migration through compacted soil barrier, and GCL was found to depend on the degree of saturation, whereas for geomembrane, it depends on defects in the geomembrane. The performance of the compacted soil cap barrier may be considerably increased with the inclusion of geo-fibers. This chapter provides the basic understanding of landfill gas generation mechanism, gas transport mechanism, test apparatus, and its governing mechanism for determining gas transport parameters and can be a reference for the researchers and policymakers.

References

Aubertin M, Aachib M, Authier K (2000) Evaluation of diffusive gas flux through covers with a GCL. Geotext Geomembr 18(2):215–233

Barral C, Oxarango L, Pierson P (2010) Characterizing the gas permeability of natural and synthetic materials. J Transp Porous Media 81(2):277–293

Benson CH, Daniel DE, Boutwell GP (1999) Field performance of compacted clay liners. J Geotech Geoenviron Eng ASCE 126(5):390–403

Benson CH, Thorstad PA, Jo H-Y, Rock SA (2007) Hydraulic performance of geosynthetics clay liners in a landfill final cover. J Geotech Geoenviron Eng ASCE 133(7):814–827

Bouazza A, Vangpaisal T (2003) An apparatus to measure gas permeability of geosynthetic clay liners. Geotext Geomembr 21(2):85–101

Carslaw HS, Jaeger JC (1959) Conduction of heat in solids, 2nd edn. Clarendon Press, Oxford

Chapuis RP, Lavoie J, Girard D (1992) Design, construction, performance, and repair of the soil-bentonite liners of two lagoons. Can Geotech J 29(4):638–649

Corey AT (1957) Measurement of water and air permeability in unsaturated soil. Proc Soil Sci Soc Am 21(1):7–10

CPCB (2012) Status report on municipal solid waste management. Central Pollution Control Board. New Delhi

CPHEEO (2013) Advisory on improving municipal solid waste management services. Central Public Health and Environmental Engineering Organization

Dixon N, Jones DRV (2005) Engineering properties of municipal solid waste. Geotext Geomembr 23(3):205–233

Durmusoglu E (2002) Municipal landfill settlement with refuse decomposition and gas generation. Ph.D. Dissertation, Texas A&M University, College Station, Texas

EPA (2004) Technical guidelines for RCRA/CERCLE final covers. United States Environmental Protection Agency, Washington DC

EPA (2009) Landfill gas energy basics. LFG energy project development handbook. United States Environmental Protection Agency, Washington DC

EPA (2013) Inventory of U.S. greenhouse gas emissions and sinks: 1990–2011. United States Environmental Protection Agency, Washington DC

Li H, Jiao JJ, Luk M (2004) A falling-pressure method for measuring air permeability of asphalt in laboratory. J Hydrol 286:69–77

Maciel FJ, Jucá JFT (2000) Laboratory and field tests for studying gas flow through MSW landfill cover soil (Geo-Denver 2000). In: Advances in unsaturated geotechnics, pp 569–585

MoEF (2000) Municipal solid waste management and handling rules. Ministry of Environment and Forests, Government of India, New Delhi

Moon S, Nam K, Kim JY, Hwan SK, Chung M (2008) Effectiveness of compacted soil liner as a gas barrier layer in the landfill final cover system. Waste Manag 28(10):1909–1914

Pitanga HN, Pierson P, Vilar OM (2011) Measurement of gas permeability in geosynthetic clay liners in transient flow mode. Geotech Test J ASTM 34(1):1–7

Rajesh S, Viswanadham BVS (2009) Evaluation of geogrid as a reinforcement layer in clay based engineered barriers. Appl Clay Sci 46(2):153–165 (Elsevier)

Rajesh S, Viswanadham BVS (2011) Hydro-mechanical behaviour of geogrid-reinforced soil barriers for landfill covers system. Geotext Geomembr 29(1):51–64 (Elsevier)

Rajesh S, Viswanadham BVS (2012) Modelling and instrumentation of geogrid reinforced soil barriers of landfill covers. J Geotech Geoenviron Eng ASCE 138(1):26–37

Rajesh S, Gourc JP, Viswanadham BVS (2014) Evaluation of gas permeability and mechanical behaviour of soil barriers of landfill cap covers through laboratory tests. Appl Clay Sci 97–98:200–214

Rajesh S, Rao BH, Sreedeep S, Arnepalli DN (2015) Environmental geotechnology: an indian perspective. Environ Geotech 2(6):336–348 (ICE Publishers)

Rajesh S, Naik AA, Khan V (2016) Performance of compacted soil barriers under advective gas flow condition. ASCE Geotechnical Sp. Pub. No. 271, Geo-Chicago 2016, pp 212–221
Rouf MA, Bouazza A, Rao MS, Gates WP, Rowe RK (2016) Gas flow unified measurement system for sequential measurement of gas diffusion and gas permeability of partially hydrated geosynthetic clay liners. Can Geotech J 53:1000–1012
Scanlon BR, Nicot JP, Massman JW (2002) Chapter 8: Soil gas movement in unsaturated systems. In: Warrick AW (ed) Boca raton, CRC Press, pp 297–341
Shan HY, Yao JT (2000) Measurement of air permeability of geosynthetic clay liners. Geotext Geomembr 18(2):251–261
Stark TD, Choi H (2005) Methane gas migration through geomembranes. Geosynth Int 12(2): 120–126
Stoltz G, Gourc JP, Oxarango L (2010) Liquid and gas permeabilities of unsaturated municipal solid waste under compression. J Contam Hydrol 118(1):27–42
Vangpaisal T, Bouazza A (2004) Gas permeability of partially hydrated geosynthetic clay liners. J Geotech Geoenviron Eng ASCE 130(5):93–102
Viswanadham BVS, Rajesh S (2007) Bioreactor landfills: future for environmental infrastructure in India. The Economic Times, Polymers, June–July, pp 16–21
Viswanadham BVS, Rajesh S, Bouazza A (2012) Effect of differential settlements on the sealing efficiency of GCL compared to CCLs: centrifuge study. Geotech Eng J SEAGS & AGSSEA 43 (3):55–61

Chapter 17
Effect of Emerging Contaminants from Paper Mill Industry into the Environment and Their Control

Farha Deeba, Vikas Pruthi and Yuvraj S. Negi

Abstract There has been growing concern regarding environment and human health owing to rise in environmental pollution. The paper industry released huge quantity of toxic effluent into the surrounding environment which inhibits the photosynthetic activity of aquatic biota, crop growth, etc., and thus has adverse effects on health, soil properties, and relative mobility in the soil profile. Hence, several aspects for management of pulp and paper effluent such as chemical, biological and mechanical treatments, recycling, monitoring, toxicology, environmental impact, and sustainable practice have been carried out. Further, various strategies need to be developed to determine the nature and extent of contaminants to natural resources. Presently, bioremediation methods using terrestrial and aquatic plants as well as microorganisms are considered more favorable for treating various industrial effluents. Although all bioremediation processes using microorganisms are appropriate for wastewater treatment, the integrated approach of yeast treatment process showed higher contaminant removal efficiencies with reduced energy costs and sustainable production of biodiesel. Thus, the utmost need of the hour is to synchronize environmental friendliness by using these methods over conventional methods owing to their reduced chemical and biological sludge production, more efficiency, and low cost.

Keywords Paper mill effluent · Pollutants · Environmental impact
Bioremediation · Removal efficiency

F. Deeba · Y. S. Negi (✉)
Department of Polymer and Process Engineering, IIT Roorkee Saharanpur Campus, 247001 Saharanpur, India
e-mail: yuvrajnegi@gmail.com

V. Pruthi
Department of Biotechnology, IIT Roorkee, 247667 Roorkee, India

T. Gupta et al. (eds.), *Environmental Contaminants*, Energy, Environment, and Sustainability, https://doi.org/10.1007/978-981-10-7332-8_17

17.1 Introduction

In recent years, nature is suffering from adverse effect of environmental pollution. Increased industrialization has resulted in indiscriminate release of highly concentrated toxic chemical wastes into the surrounding environment. Soil, ground water, and surface water are contaminated due to various industrial effluents when they are not discharged properly. Worldwide, the paper industry ranks fifth as prime consumer of energy and is the sixth major producer of ecological polluters following textile, steel, leather, oil, and cement industries (Ashrafi et al. 2015; Patel et al. 2016). There are totally 7745 pulp and paper mills, and demand is 402 million per annum around the world (Singh et al. 2016). However, in India, 759 pulp and paper mills produce about 10.11 million tons of paper per year, which is about 2.52% of the total world production. The present consumption of pulp and paperboard is about 11.15 million tons per year (Singh et al. 2016). Paper mill industry utilizes large amount of freshwater for paper manufacturing and produces equal quantity of highly toxic effluent more than any other industry (Pokhrel and Viraraghavan 2004). About 1 ton of paper production generates 150 m^3 of effluent heavily loaded with toxic contaminants (Singhal and Thakur 2009). The paper industry releases about 100 million kg of harmful contaminants every year which requires application of sustainable solutions for the effluent treatment solutions (Environmental Protection Agency, EPA, USA 2000). Recovery of paper leads to 15–25% reduction in harmful products responsible for global warming (NCASI et al. 2010).

For meeting the demand, the paper industry utilizes various raw materials like fiber, wood, husk, cellulose of agricultural products and more than 200 chemicals at different stages of paper manufacturing to produce paper (Sivakumar and Nouri 2015). Based on factors such as production process, amount of water consumed, raw materials used, and additive chemicals, the generated effluents commonly have a high concentrations of total solids (TS), chemical oxygen demand (COD), biological oxygen demand (BOD), total suspended solids (TSS), total dissolved solids (TDS), organic matter, sulfates, sodium, magnesium, calcium, phenols, adsorbable organic halogens (AOX), color, etc., and low biodegradability (Buyukkamaci and Koken 2010). COD and BOD of the waste stream are in the ranges of 20–200 kg/t and 10–40 kg/t of air dried pulp, respectively. Many of these chemicals as well as other additional pulp and paper industry by-products when released into air and water bodies poses serious risk to human health and the environment. Such chemicals are usually lethal or cancer causing, and some are even DNA destroying or genotoxic (Pokhrel and Viraraghavan 2004). Also the possible impact of these toxic chemicals is that they destroy or completely reshape the environments surrounding the paper mills.

Therefore, the effluent generated from paper mills may not be disposed as such to any medium without appropriate treatment. Effluents generated must be treated properly, and contaminants must be limited before being discharged into the environment below their acceptable limit. Effective wastewater management practices need to be undertaken to remove contaminants from paper mill effluents (PME) before they are discharged or disposed (Ince et al. 2011). In reality, most of the paper mills around the world do not follow suitable treatment techniques which lead to contamination in that disposal place. Soil medium is also affected along with water medium which in turn affects the growth of plants. Paper industry is under intense pressure by environmental groups, regulators, and others to reduce the destructive impacts caused by their solid wastes and effluents. However, paper mills usually search for the most efficient, fastest, and cheapest treatment processes to reduce costs, streamlining their production process and raising their profits (Kamali and Khodaparast 2015). Some paper mills have started using plants in the surrounding bodies of wastewater and effluents to soak up the chemicals and decrease their harmful impacts on the environment. But the chemicals when left unchecked enter into the ecosystem, move into the food chain, and have severe impacts on humans in the downstream and surrounding communities near the paper mills (Thompson et al. 2001; Raj et al. 2014).

In earlier stages, highly reliable and efficient method for treating the wastewater from paper mills such as physical, chemical, and biological process has been applied to reduce the contaminants from PME. The several treatment techniques for the amputation of pollutants from different industries are electrodialysis and electrocoagulation, reverse osmosis, ion exchange, membrane separation, chemical precipitation, ultrafiltration, adsorption, etc., but these methods generate more biological and chemical sludge, which further increase the cost of their treatment and disposal (Kamali and Khodaparast 2015). Presently, bioremediation methods using aquatic and terrestrial plants as well as various microorganisms are considered more positive for treating different industrial effluents. Bioremediation methods have many advantages compared to conventional methods in terms of less cost; more efficiency; reduced biological and chemical sludge production (Sivakumar and Nouri 2015). There are several studies conducted on germination of plants stating that using raw PME diminished the plant growth rate, whereas using diluted PME improved the growth (Kumar et al. 2015). Many studies on aquatic plant and microorganisms for removing contaminants from PME also have been carried out. In this chapter, the role of yeast microorganism in an integrated process for the effective reduction of extremely toxic chemicals from PME and simultaneous production of energy by-product has been discussed. In contrast to other methods, this would provide viable approach for lawful and proper discard of PME in an eco-friendly manner along with a sustainable production of biodiesel. The effect of harmful contaminant generated from PME and their impacts on ecosystem as well as the different treatment processes dealing with pulp and paper mill wastewater also has been discussed in this chapter.

17.2 Key Pollutants and Their Impact

Paper mills commence with the process of conversion of raw materials such as canes, wood, reeds, grasses, bamboos, and straw into paper with the addition of enormous amounts of chemicals at several points. During this process, chlorinated resin acids, juvaniones, resin acids, diterpene alcohols, unsaturated fatty acids, and other chemicals are taken away from the transitional and end products with the wastewater that becomes the effluent of the plant. In all the steps of the paper-making process, pulping generates the huge quantity of pollutants present in wastewater (Thompson et al. 2001). The bleaching stage of pulp and paper production also leads to addition of huge quantity of contaminants to the effluent, including acetone, chlorophenols, chlorate ions, furans, dioxins, and several others that are found to be more hazardous chemicals added to the surrounding environment (Fig. 17.1). After paper production process, the effluent generated is disposed into the environment i.e. into oceans, lakes, and rivers. These chemicals if left untreated in the effluent will harm the organisms and environments surrounding the paper mills.

17.2.1 *Sulfur, Sulfur Dioxide, and Hydrogen Sulfide*

Sulfur-related chemicals are used in sulfite as well as in kraft process of paper-making process. The water soluble sulfur dioxide (SO_2) is released during combustion of raw materials in papermaking process and is the main cause of acid rain. In Canada (2006), the paper industry liberated 60,000 tons of SO_2 into the atmosphere which is about 4% of the total sulfur oxides (SO_x) released from all Canadian industries (Air Pollutant Emissions for Canada 2006). Air emissions including dimethyl sulfide, hydrogen sulfide, methyl mercaptan, and various sulfur-related volatile compounds are responsible for odor coming from paper mills.

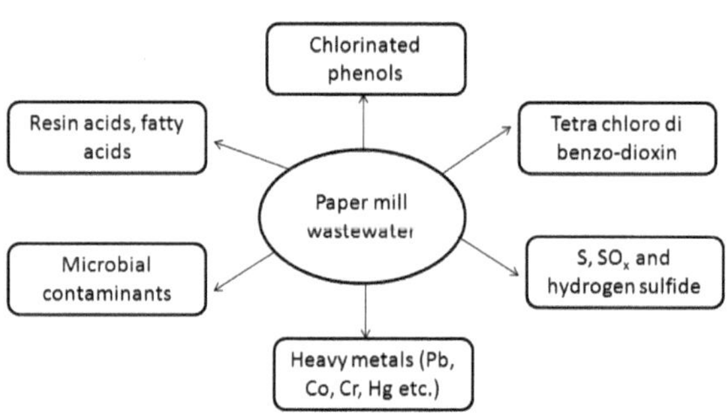

Fig. 17.1 Highly toxic compounds present in paper mill effluent

17.2.2 Tannins

Tannins ranging in weight from 500 to 3000 g/mol are polar phenolic polymeric compounds that are very reactive to proteins. Tannins are released along with wastewater of paper mills during debarking process and contribute about 50% of the COD. The tannins add color to processed water, absorb more heat and light, and maintain less oxygen compared to freshwater, thus hazardous to the aquatic flora and fauna. Tannins show methanogenic toxicity to microorganisms as well as to aquatic organisms like fish (Ali and Sreekrishnan 2001).

17.2.3 Resin Acids

They are weak hydrophobic acids present in the resin of tree bark and wood naturally with aqueous solubility of 3–6 mg L^{-1} and passed to wastewater during pulping process. They are tricyclic diterpenes and show toxicity to fish at 200–800 µg/L concentrations in wastewaters (McFarlane and Clark 1988). The commonly recorded resin acids found in pulp and paper mill disposes are isopimaric acid, palustric acid, neoabietic acid, sandaracopimaric acid, pimaric acid, levopimaric acid, dehydroabietic acid, and abietic acid. Isopimaric acid is one of the highly toxic and least soluble compared to other resin acids (Ali and Sreekrishnan 2001). They are highly toxic to aquatic organisms.

17.2.4 Fatty Acids (FA)

Unsaturated FA like linolenic FA, linoleic FA, and oleic FA are also monitored in PME which are toxic to fish, especially salmonoids. Long chain fatty acids also inhibit methanogenic bacteria (acetoclastic bacteria) that play an essential function in anaerobic wastewater treatment (Ali and Sreekrishnan 2001).

17.2.5 Tetra Chloro Di Benzo-Dioxin (TCDD)

TCDD is one of the most toxic pollutants present in the effluent and is generally consumed by drinking contaminated water or eating contaminated fish. Dioxins are most toxic chemicals known to man and are supposed to be limited to small amounts in PME. TCDD is very aggressive carcinogen, and the tolerable daily intake of dioxins is 1–4 pg/kg body mass (Van Leeuwen et al. 2000). It causes cancer, impaired reproductive health, and sterility. It requires multiple treatments every month for many years when get affected.

17.2.6 Chlorinated Phenols

The chlorinated phenols are the main contaminant affecting the flora and fauna of aquatic environment. They are generated during bleaching of papers in the effluent and are very harmful even at small concentration (5–25 mg L^{-1}). It has been studied that phenol concentration of 1 g/mL is fatal to humans and inhibits photosynthesis in blue algae and diatoms. Phenol long-term intake causes necrosis, dermal inflammation, enlarged liver, and gastrointestinal irritation (Patel et al. 2016). Numeric limitations for 12 adsorbable organic halides (AOX) and chlorinated phenolic pollutants present in effluent are given by EPA.

17.2.7 Heavy Metals

Mercury is very harmful element present at high concentration in PME. When exposed it causes birth defects, tumors, and even death in severe cases. Outcome of contact with mercury leads to mercury poisoning, i.e., hydrargyriasis which is a type of heavy metal poisoning. Lung disease, kidney failure, internal organs damage, and sensory impairment are the common symptoms due to mercury poisoning. Lead, cadmium, cobalt, nickel, chromium, etc., are also present in effluents which are highly toxic to humans and aquatic system. Chromium is carcinogenic and nephrotoxic in nature. These toxic metals possess potential hazards to life forms and bioaccumulate in the food chain causing life-threatening illness as well as damage to vital body system (Ahluwalia et al. 2007).

17.2.8 Microbial Contamination

PME have severe impact on the receiving aquatic ecosystem by microorganisms through production of slime such as *Sphaerotilus* sp. (Pellegrin et al. 1999) as well as scum formation causing toxicity to the communities which lead to aesthetical issues if treatment is incomplete (Pokhrel and Viraraghavan 2004). The pathogenic coliform bacteria found in PME cause hazardous effect such as fever, vomiting, nausea, abdominal pain, and diarrhea after consuming contaminated water (Singh et al. 2016).

17.2.9 Air Emissions

Carbon dioxide (CO_2), SO_2, and nitrogen oxide (NO) are major greenhouse gases released by paper industry that cause acid rain and lead to climate change

(Gavrilescu et al. 2012). Around 1% of global CO_2 emissions come from the print and paper industry (Ecofys 2013).

17.2.10 Others

Many other contaminants that are present in PME and have been found to affect humans are chelating agents, nitrite nitrogen, alcohols, chlorides, transition metals, solids and dissolved organic matter like lignin. Nutrients such as phosphorus and nitrogen can cause eutrophication of freshwater bodies such as rivers and lakes.

The reason these toxins are so harmful is because they are being pumped into the rivers and are in regular use. Direct contact with the infected water while swimming causes rashes and infections of the skin, eyes, and ears which can cause serious medical issues. The highest risk of becoming ill from these toxins is by eating food that came into contact or by drinking the contaminated water. Drinking contaminated water causes ulceration of internal organ linings, severe diarrhea, and if left untreated can lead to death. The larger fish slowly collect the toxins from all the organisms they eat through biomagnification and become even more dangerous to consume (Fig. 17.2).

17.3 Paper Mill Effluent Treatment Methods

Many studies have stated that effluents generated from paper mill can foster toxicity in aquatic system, mainly at their productive level (Kamali and Khodaparast 2015; Waye et al. 2014). Paper mill treatment methods have been developed, and toxicity of the ultimate disposed wastewater can be greatly reduced by applying the specific technique. Still various pollutants are continued to be monitored in the final effluents after treatment. This is mainly due to the economic drawbacks of some

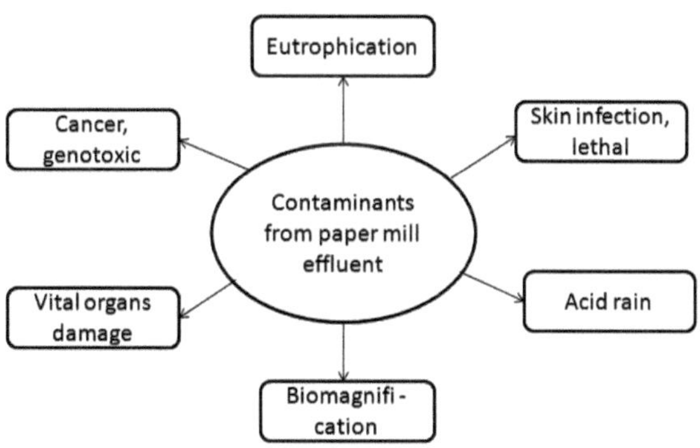

Fig. 17.2 Effect of contaminants released from paper mill effluent on environment

efficient effluent treatment techniques as well as some technical issues which can cause incomplete degradation. Recent studies, discussed in this chapter, have aimed to conquer drawbacks of such used paper mill treatments methods to fulfill the environmental safety needs and provide cost-effective method. Different categories of treatment processes are available for paper mills to treat their effluents before releasing the wastewater into the environment: physiochemical, biological, and integrated (Kamali and Khodaparast 2015). Each category has several process types; each specialized at removing certain chemicals from the effluent with varying rates of success (Table 17.1).

17.3.1 Physicochemical Methods

These processes are applied to eliminate colors, toxic compounds, colloidal particles, floating material, and suspended solids from effluent. It includes various

Table 17.1 Removal efficiency of PME parameters after different treatments

S. No:	Treatment	Removal efficiency (%)
1	Sedimentation and floatation	80% of suspended solids
2	Coagulation and precipitation	Color 80%, COD 20–96%, lignin 83.4%, turbidity 95.7%
3	Electrocoagulation (Al)	COD 75%, phenol 70%, lignin 80%
4	Nanofiltration	DOC 91%
5	Laccase-polymerized membrane filtration	COD $\geq$ 60%
6	Ozonation	Color 50%, COD 20%, DOC > 15%
7	*Bacillus* sp. and *Serratia marcescens*	Color 80%, PCP 94%, BOD 98%, TSS 99%
8	*Paenibacillus* sp.	Color 68%, COD 78%, BOD 83%, phenol 86%, lignin 54%
9	*Pencillium* sp.	Color 50%, AOX 50%
10	*Merulius aureus* + unidentified genus + *Fusarium sambucinum*	Color 78.6%, lignin 79.0%, COD 89.4%
11	*Aspergillus niger*	Color 43%, MTBE extracts 97%, COD 60%, turbidity 77%
12	*Azolla caroliniana*	TDS 82.3%, BOD 88.6%, COD 79.1%
13	*Chlamydomonas debaryana* IITRIND3	Total nitrogen 72.69%, Total phosphorous 65.64%, COD 62.5%, TOC 92.82%, Fl^- 86.20%, Cl^- 50.18%, SO_4^{-2} 53.34%
14	*Rhodosporidium kratochvilovae* HIMPA1	Color 89%, BOD 77.36%, TDS 84.59%, COD 94.22%, phenol 99.60%, lignin 94.27%, Total nitrogen 97.93%, Fl^- 78.19%, Cl^- 66.54%, SO_4^{-2} 87.20%

techniques such as reverse osmosis, filtration, adsorption, ozonation, wet oxidation, coagulation and precipitation, sedimentation and floatation, etc., for effluent treatment.

17.3.1.1 Sedimentation and Floatation

The wastewater contains high concentrations of floating matters and suspended solids. Hence, the usage of a primary treatment, generally sedimentation, is necessary for the treatment procedure. Sedimentation method removes about 80% of suspended solids from effluent (Thompson et al. 2001). The selection and effectiveness of these methods depends upon the employed paper production procedure and on the secondary treatment methods applied.

17.3.1.2 Coagulation and Precipitation

Coagulation and precipitation are applied in the paper industry for further polishing of the wastewater and for removing suspended solids as well as COD from the wastewater. The removal percentage of COD varied from 20–96% depending on the treatment process conditions and effluent characteristics while the removal percentage of color was about 80% after which the treated effluent can be reused. Though this system led to removal of suspended solids and color, the obtained removal efficiencies of BOD and COD were not acceptable under all examined states. Also because of their high cost, only some kinds of tertiary treatment processes are applied in paper industry.

Commonly, such methods involve addition of metal salts to the wastewater to produce bigger flocs from smaller particles. For example, aluminum chloride when used as coagulant along with a modified natural polymer as flocculant for wastewaters treatment, the removal efficiency of lignin and turbidity were 83.4 and 95.7%, and water recovery is 72.7%, respectively (Wang et al. 2011). Electrochemical methods are considered more economically and technically feasible in large-scale operation as compared to other physicochemical treatment techniques.

17.3.1.3 Membrane Technologies

The utilization of membrane technologies can be an efficient method for effluent treatment. Reverse osmosis technique is commonly used for destruction of lethal microorganisms (Asano and Cotruvo 2004). Membrane electrochemical reactor facilitates the reduction of TDS, COD, BOD, TSS, color and produces small amount of the sludge (Chanworrawoot and Hunsom 2012). However by using nanofiltration treatment technique, about 91% removal efficiency of the DOC (dissolved organic carbon) from biologically treated newsprint mill wastewater has been monitored (Ciputra et al. 2010). Also, when ultrafiltration membranes are applied for the treatment of effluent about 50, 95, 97, 89, and 83%, removals of conductivity, spectral absorption coefficient, sulfate, COD, and total hardness, respectively, have been reported (Gonder et al. 2012). The raw wastewater as well as secondary effluents pre-treatment by laccase polymerization before filtration by different molecular weights membranes leads to 60% reduction in COD (Ko and Fan 2010). Membrane bioreactors (MBRs) are generally used for reducing the amount of the produced sludge but membrane fouling leads to rise in operational costs and maintenance, limiting the utilization of MBRs. Though these techniques show improved removal efficiency due to its high cost, these methods are not common in paper industry.

17.3.1.4 Adsorption

A variety of adsorbents such as ash, fuller's earth, silica, coal, activated carbon, etc., have shown satisfactory performances for refractory pollutants removals and decolorization of wastewater. Using granular-activated carbon and ion exchange resin adsorption mechanisms, about 76 and 72% reductions in DOC have been reported (Ciputra et al. 2010).

17.3.1.5 Oxidation

The processes such as ozonation, hydrogen peroxide oxidation, combined ozonation, and Fenton's oxidation (FO) are used for removing color and COD from the effluent. Among these methods, FO process was more effective in reduction of both color and COD (Sevimli 2005). The chemical precipitation proceeded by ozonation for the effluent treatment depicts 96% of color and 60–70% of BOD removal efficiency, respectively (De los Santos Ramos et al. 2009). Ozone-treated nanofiltered wastewater after activated sludge method leads to more than 50% decline in lignin, turbidity, and color (Manttari et al. 2008). Chemical treatment by advanced oxidation processes is the most widely used method for the treatment of effluent containing large number of refractory organic pollutants with low biodegradability index (Babuponnusami and Muthukumar 2012).

Physicochemical techniques for effluent treatment have many limitations such as harmful by-products formation, high cost, large quantity of sludge production, and high chemical consumption (Boopathy and Senthilkumar 2014). Hence, biological treatment techniques can be more attractive in terms of small reactor volume requirement, low cost, and faster performance.

17.3.2 Biological Techniques

Biological treatment techniques can be applied for treatment of wastewater in an independent treatment step or in combination with other physicochemical methods. Biological techniques are measured to be eco-friendly, cost-effective, and appropriate for removal of the COD and BOD from the wastewater as compared to physicochemical methods. However, the conventional biological methods such as stabilization pond, anaerobic lagoon, activated sludge process, etc., do not perform efficiently for reduction of recalcitrant compounds and color from PME. Among all biological process for effluent treatment, bioremediation using microorganisms has fascinated the interest of researchers around the world (Bezuneh 2016). Microbial decolorization and degradation of PME seem to be environmental friendly and cost-effective substitute to physicochemical methods. Different kinds of microorganisms such as yeast, fungi, bacteria, and algae are used for potential decolorization, degradation, and removal of persistent contaminants from PME.

17.3.3 Plant Treatment

Bioremediation methods using aquatic and terrestrial plants are found more suitable for treating PME compared to physicochemical methods. The success of phytoremediation depends upon the growth rate and photosynthetic activity of plants. For example, water hyacinth, due to large biogas production and fast growth, has potential to remove phosphorus, heavy metals, ammonium, nitrate, and phenols from wastewaters (Xia and Ma 2006). Also, an aquatic plant, *Azolla caroliniana,* is used for removing various contaminants from PME. It can be used as sorbents for removing COD (79.1%), BOD (88.6%), and TDS (82.3%) along with other parameters in PME (Sivakumar and Nouri 2015). The process is cost-effective but contaminants bioaccumulate in plants and then pass into the food chain. Also, by applying plant treatment, the complete removal of pollutants from contaminated ground is not possible and still the problem does not resolve.

17.3.4 Bacterial Treatment

A broad variety of bacterial species have been utilized for bioremediation of PME. Bacteria show higher biodegradation capability due to environmental adaptability, biochemical versatility, and wide pH range tolerability (Chandra and Singh 2012). Hence, different biological treatment methods, like activated sludge and aerated lagoons, are used for treatment of PME. But many of the conventional processes are not efficient in biodegradation of the polymers like lignin, due to their complex structure and size. These methods cause incomplete degradation of recalcitrant pollutants which remain suspended in treated wastewater. Thus, the modified and extended techniques are developed which generate less sludge. Aerated lagoons applied for treatment of effluent were efficient in reduction of chlorinated phenols (85%), AOX (50%), COD (60–70%), and BOD (95%) (Pokhrel and Viraraghavan 2004). *Bacillus cereus* GN1 can degrade 2,4-dichlorophenol (2,4-DCP) found in PME up to 400 mM concentrations (Matafonova et al. 2006). The activated sludge bioreactors involved aerobic heterotrophic communities which effectively reduce nutrients, toxic compounds, organic substances as well as pathogens from the effluent (Wells et al. 2011). For instance, degradation of pentachlorophenol (PCP) found in PME with *Bacillus* sp. and *Serratia marcescens* mixed culture also results in considerable decrease of salts, total phenols, phosphate, nitrogen, sulfate, metals, TSS, TDS, TS, COD, and BOD (Menezes and Moo-Young 2009). The treatments of paper mill wastewaters by a *Paenibacillus* sp. producing laccase causes efficient reduction in several pollution parameters (COD 78%, BOD 83%, phenol 86%, lignin 54%, and color 68%) (Raj et al. 2014).

The anaerobic and aerobic methods have many limitations such as large production of sludge during aerobic method and anaerobic bacteria sensitivity to harmful compounds. The increased sulfur content in wastewater produced during chemical pulping had a harmful impact on the pollutant reducing ability of anaerobic methods even at low pH. The combination of different physicochemical and biological processes for the wastewater treatment is advantageous as compared to physicochemical or biological treatment methods independently.

17.3.5 Fungal Treatment

Fungi are commonly found in PME and secrete extracellular enzymes. In contrast to bacteria, fungi can survive at higher effluent load. Many white-rot fungi like *Phanerochaete chrysosporium* and *Trametes pubescens* can degrade phenolic or lignin compounds by their enzymes secretion (manganese peroxidases dependent peroxidases, laccases, and lignin peroxidases) (Chandra and Singh 2012; Kamali and Khodaparast 2015). Bioremediation of PME with deuteromycetous fungus (*Fusarium sambucinum*) and two basidiomycetous fungi (*Merulius aureus* and an unidentified genus) leads to reduction of color (78.6%), lignin (79.0%), and COD (89.4%), respectively (Malaviya and Rathore 2007). White-rot fungi along with soft-rot fungi lead to reduction in color and COD (74–81%) of effluents (Freitas et al. 2009). The *Aspergillus niger* treatment of mechanical pulping (poplar alkaline peroxide) wastewater causes reduction in 43% color, 77% turbidity, 60% COD, and 97% methyl tertiary butyl ether (MTBE) extracts, respectively (Liu et al. 2011).

Furthermore, fungal treatment can also be applied with other physicochemical methods for degradation of the persistent contaminants. For example, *T. pubescens* (white-rot fungus) biodegrades chlorophenols proceeded by TiO_2/UV application and allowed up to 100% removal of chlorophenol (Gonzalez et al. 2010). Though fungal treatment is effective in removing contaminants from effluent but a few drawbacks of fungal species such as oxygen limitations, high pH, high carbon requirement, and slow metabolism may restrict their utilization for the treatment of effluent under harsh environmental conditions (Tarlan et al. 2002a; Balcioglu et al. 2007; Chandra and Singh 2012). Few studies have investigated that algae can substitute fungi by removing AOX and color more effectively.

17.3.6 Algae Treatment

Microalgae can effectively uptake both inorganic and organic forms of nitrogen, phosphorous, and carbon accompanied by agglomeration of trace elements from effluent for their growth. Microalgae are unicellular microorganisms that have the ability to biodegrade and bioadsorb toxic contaminants such as xenobiotics, melanoidins, polycyclic aromatic hydrocarbons, heavy metals, phenols, pesticides from effluent (Bezuneh 2016). Bioremediation by using microalgae has several advantages in contrast to that of fungal and bacterial species in numerous ways.

(1) Microalgae have potential to utilize contaminants like phosphate, nitrate, and ammonium as a nutrient for their growth without any external nutrient supplementation compared to bacteria and fungi.
(2) Microalgae can rapidly grow and adapt harsh conditions for bioremediation activity compared to bacteria and fungi which require optimum condition to grow.
(3) Microalgae produce valuable by-products such as biodiesel, methane ethanol and because of its increased N:P ratio it can also be utilized as organic fertilizer.

Algae use a natural mechanism for the reduction of recalcitrant pollutants and color from PME. The key mechanism of organic and color reduction by algal strains is partially transformation and partially metabolism of chlorinated organic molecules (COM) and color to non-chlorinated organic molecules (NCOM) and non-colored ones (Tarlan et al. 2002a). For instance, high reduction efficiencies of AOX (82–93%), color (42–75%) and COD (60–85%) in PME is mainly because the algae removed both COM and NCOM from wastewater by metabolism. The biodegradation of COM was much faster than reduction of NCOM and colored ones (Tarlan et al. 2002b).

The potential of algae *Chlamydomonas debaryana* IITRIND3 for integrated bioremediation of wastewater as well as production of biodiesel has been investigated which showed the considerable lipid productivity (44 mg L^{-1} day^{-1}) along with effective phosphorous (65.64%), nitrogen (72.69%), and COD (62.5%) removal. This sustainable strategy is cost-benefit and potentially eliminates the use of expensive nutrients and freshwater required for biodiesel production accompanied by efficient phycoremediation by algae (Arora et al. 2016).

17.3.7 Yeast Treatment

To combat treatment challenges, yeast treatment method also has been developed as preferred process for treatment and reduction of the toxic pollutants present in generated effluents. Previous studies have demonstrated that various oleaginous yeasts such as *Yarrowia lipolytica, Candida tropicalis, Candida cylindracea, Candida rugosa, Trichosporon cutaneum, Saccharomyces* sp. have ability to remove several toxic components and phenolic compounds from the final processed wastewater (Deeba et al. 2016; Patel et al. 2016). For example, *Rhodotorula glutinis* shows about 89% reduction in concentration of phenol found in raw olive mill wastewater (Karakaya et al. 2012). Similarly, *Debaryomyces etchellsii* strain BM1, oleaginous yeast cultivated on the expired soft drinks and olive mill wastewater mixture shows significant reduction in the COD (58%) with simultaneous lipid yield production of 1.2 g/L (Arous et al. 2016). In other study, an oleaginous yeast *Rhodosporidium kratochvilovae* HIMPA1 is used in an integrated method to treat PME for the reduction of harmful pollutants as well as simultaneous production of energy by-product from their intracellular lipid compartment. The lipid harvested from yeast cells cultivated on PME can be transesterified to produce biodiesel. The cultivation of *R. kratochvilovae* HIMPA1 on PME leads to significant removal of TDS 84.59%, BOD 77.36%, COD 94.22%, lignin 94.27%, color 89%, and phenol 99.60% with 8.56 g/L of lipid yield (Patel et al., 2016). Oleaginous yeast shows the remarkable removal efficiency of phenols and lignin from PME with high decline in TDS, BOD, and COD compared to algae. The integrated method exhibits effective pollutant removal from PME along with sustainable production of biodiesel for transportation fuels (Fig. 17.3).

Some of the technical problems faced by microalgae can be overcome by utilizing oleaginous yeast for wastewater treatment and production of lipid. The utilization of oleaginous yeast for bioremediation has several advantages over microalgae in many ways.

(1) Oleaginous yeasts have potential to grow on a broad spectrum of carbon sources which make them economically interesting.
(2) Oleaginous yeasts have good growth characteristics with a broad temperature range and ability to utilize various contaminants from wastewaters.

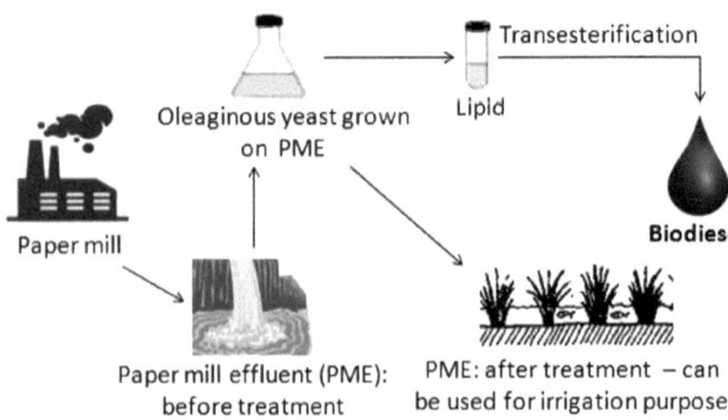

Fig. 17.3 Integrated method of PME treatment and biodiesel production by oleaginous yeast

(3) Oleaginous yeasts have short life cycle, hence required less time for effluent treatment and lipid production in contrast to microalgae.

Hence, the oleaginous yeast treatment can be applied as cost-benefit method with effective removal of contaminants from PME compared to microalgae and other conventional methods with simultaneous production of biodiesel.

17.4 Conclusion

In recent years, paper industry is facing reforms with the energy efficiency mechanisms, production processes and management of the contaminants released, environmental assessment consideration, and profitability of the competitive markets. The treatment of effluents generated from paper industry emerges as most serious issue in environmental protection. The wastewater disposed into aquatic system before its proper treatment causes health hazards in animal and human, eutrophication in aquatic system, and change in climate. To overwhelm the incomplete treatment of PME and produce cost-effective treatment techniques, several physicochemical and biological methods have been utilized. However, many of these processes are costly, and none are measured to be commercially feasible, and still the issue remains unsolved. Recently, bioremediation is found to be an alternative option for removing the contaminants from wastewater because of its economical impact and high efficiency compared to chemical remediation. Microorganisms such as bacteria, fungi, yeast, and algae play a main role in bioremediation of toxic pollutants from paper mill wastewater for safe disposal. The treatment of wastewater by adopting the integrated method, i.e., cultivation of oleaginous microorganisms on PME for biodiesel production not only effectively reduce contaminants from it but also leads to sustainable production of biofuel from wasteful resource. Such treatment methods seem to be environmentally and economically favorable to reduce environmental pollutants as well as for energy recycling. Furthermore, isolation of novel microorganisms from various sources that show more potential to overcome the toxic effects of PME with enhanced yield of lipid can be proved important for bioremediation. Also, the microorganisms can be genetically modified to get improved yield of bioremediation.

References

Ahluwalia SS, Goya D (2007) Microbial and plant derived biomass for removal of heavy metals from wastewater. Bioresour Technol 98:2243–2257

Air Pollutant Emissions for Canada (Tonnes) (2006) Environment Canada

Ali M, Sreekrishnan TR (2001) Aquatic toxicity from pulp and paper mill effluents: A review. Adv in Environ Res 5:175–196

Arora N, Patel A, Sartaj K, Pruthi PA, Pruthi V (2016) Bioremediation of domestic and industrial wastewaters integrated with enhanced biodiesel production using novel oleaginous microalgae. Environ Sci Pollut Res 23:20997–21007. https://doi.org/10.1007/s11356-016-7320-y

Arous F, Frikha F, Triantaphyllidou I-E, Aggelis G, Nasri M, Mechichi T (2016) Potential utilization of agro-industrial wastewaters for lipid production by the oleaginous yeast *Debaryomyces etchellsii*. J Clean Prod 133:899–909. https://doi.org/10.1016/j.jclepro.2016.06.040

Asano T, Cotruvo JA (2004) Ground water recharge with reclaimed municipal wastewater: Health and regulatory considerations. Water Res 38:1941–1951

Ashrafi O, Yerushalmi L, Haghighat F (2015) Wastewater treatment in the pulp-and-paper industry: A review of treatment processes and the associated greenhouse gas emission. J of Environ Manag 158:146–157. https://doi.org/10.1016/j.jenvman.2015.05.010

Babuponnusami A, Muthukumar K (2012) Removal of phenol by heterogenous photoelectro Fenton-like process using nano-zerovalentiron. Sep Purif Technol 98:130–135

Balcioglu IA, Tarlan E, Kivilcimdan C, Sacan MT (2007) Merits of ozonation and catalytic ozonation pre-treatment in the algal treatment of pulp and paper mill effluents. J Environ Manage 85:918–926

Bezuneh TT (2016) The Role of Microorganisms in Distillery Wastewater Treatment: A Review. J Bioremediat Biodegrad 7:1–6. https://doi.org/10.4172/2155-6199.1000375

Boopathy MA, Senthilkumar SNS (2014) Media optimization for the decolorization of distillery spent wash by biological treatment using *Pseudomonas fluorescence*. Int J Innov Eng Technol 4:1–8

Buyukkamaci N, Koken E (2010) Economic evaluation of alternative wastewater treatment plant options for pulp and paper industry. Sci Total Environ 408:6070–6078

Chandra R, Singh R (2012) Decolourisation and detoxification of rayon grade pulp paper mill effluent by mixed bacterial culture isolated from pulp paper mill effluent polluted site. Biochem Eng J 61:49–58

Chanworrawoot K, Hunsom M (2012) Treatment of wastewater from pulp and paper mill industry by electrochemical methods in membrane reactor. J Environ Manage 113:399–406

Ciputra S, Antony A, Phillips R, Richardson D, Leslie G (2010) Comparison of treatment options for removal of recalcitrant dissolved organic matter from paper mill effluent. Chemosphere 81:86–91

Deeba F, Pruthi V, Negi YS (2016) Converting paper mill sludge into neutral lipids by oleaginous yeast *Cryptococcus vishniaccii* for biodiesel production. Bioresour Technol 213:96–102. https://doi.org/10.1016/j.biortech.2016.02.105

De Los Santos Ramos W, Poznyak T, Chairez I, Cordova RI (2009) Remediation of lignin and its derivatives from pulp and paper industry wastewater by the combination of chemical precipitation and ozonation. J Hazard Mater 169:428–434

Ecofys (2013) World GHG Emissions Flow Chart 2010

Freitas AC, Ferreira F, Costa AM, Pereira R, Antunes SC, Gonçalves F, Rocha- santos TAP, Diniz MS, Castro L, Peres I, Duarte AC (2009) Biological treatment of the effluent from a bleached kraft pulp mill using basidiomycete and zygomycete fungi. Sci. Total Environ 407:3282–3289

Gavrilescu D, Puitel AC, Dutuc G, Craciun G (2012) Environmental impact of pulp and paper mills. Environ Eng and Manage J 11:81–85

Gonder ZB, Arayici S, Barlas H (2012) Treatment of pulp and paper mill wastewater using ultra filtration process: Optimization of the fouling and optimization of the fouling and rejections. Ind Eng Chem Res 51:6184–6195

Gonzalez LF, Sarria V, Sanchez OF (2010) Degradation of chlorophenols by sequential biological-advanced oxidative process using *Trametes pubescens* and TiO2/UV. Bioresour Technol 101:3493–3499

Ince BK, Cetecioglu Z and Ince O (2011) Pollution prevention in the pulp and paper industries. Environmental management in practice. In Tech 223–246

Kamali M, Khodaparast Z (2015) Review on recent developments on pulp and paper mill waste water treatment. Ecotoxicol Environ Saf 114:326–342. https://doi.org/10.1016/j.ecoenv.2014.05.005

Karakaya A, Laleli Y, Takaç S (2012) Development of process conditions for degradation of raw olive mill wastewater by *Rhodotorula glutinis*. Int Biodeterior Biodegr 75:75–82. https://doi.org/10.1016/j.ibiod.2012.09.005

Kumar V, Chopra AK, Kumar S, Singh J, Thakur RK (2015) Effects of pulp and paper mill effluent disposal on soil characteristics in the vicinity of Uttaranchal Pulp and Paper Mill, Haridwar (Uttarakhand), India. Int J Agric Sci Res 4:117–125

Ko C, Fan C (2010) Enhanced chemical oxygen demand removal and flux reduction in pulp and paper wastewater treatment using laccase-polymerized membrane filtration. J Hazard Mater 181:763–770

Liu T, Hu H, He Z, Ni Y (2011) Treatment of poplar alkaline peroxide mechanical pulping (APMP) effluent with *Aspergillus niger*. Bioresour Technol 102:7361–7365

Malaviya P, Rathore VS (2007) Bioremediation of pulp and paper mill effluent by a novel fungal consortium isolated from polluted soil. Bioresour Technol 98:3647–3651

Manttari M, Kuosa M, Kallas J, Nystrom M (2008) Membrane filtration and ozone treatment of biologically treated effluents from the pulp and paper industry. J Membr Sci 309:112–119

Matafonova G, Shirapova G, Zimmer C, Giffhorn F, Batoev V, Kohring G (2006) Degradation of 2,4-dichlorophenol by *Bacillus* sp. isolated from an aeration pond in the Baikalsk pulp and paper mill(Russia). Int Biodeterior Biodegrad 58:209–212

Menezes GB, Moo-Young HK (2009) Pulp and paper mill effluents management. Water Environ Res 81:1687–1695. https://doi.org/10.2175/106143009X12445568400250

McFarlane PN, Clark TA (1988) Metabolism of resin acids in anaerobic systems. Water Sci Technol 20:273–276

National Council for Air and Stream Improvement (NCASI), Life cycle assessment of North American printing and writing paper products. Final report prepared for American Forest and Paper Association (AF&PA), Forest Products Association of Canada (FPAC) (2010)

Patel A, Arora N, Pruthi V, Pruthi PA (2016) Biological treatment of pulp and paper industry effluent by oleaginous yeast integrated with production of biodiesel as sustainable transportation fuel. J of Clean Prod 142:2858–2864. https://doi.org/10.1016/j.jclepro.2016.10.184

Pellegrin V, Juretschko S, Wagner M, Cottenceau G (1999) Morphological and biochemical properties of a *Sphaerotilus* sp. isolated from paper mills limes. Appl Environ Microbiol 65:156–162

Pokhrel D, Viraraghavan T (2004) Treatment of pulp and paper mill wastewater—A review. Sci Total Environ 333:37–58

Raj A, Kumar S, Haq I, Singh SK (2014) Bioremediation and toxicity reduction in pulp and paper mill effluent by newly isolated ligninolytic *Paenibacillus* sp. Ecol Eng 71:355–362

Sevimli MF (2005) Post-treatment of pulp and paper industry wastewater by advanced oxidation processes. Ozone: Sci Eng 27:37–43

Singh C, Chowdhary P, Singh JS, Chandra R (2016) Pulp and paper mill wastewater and coliform as health hazards: A review. Microbiol Res Int 4:28–39

Singhal A, Thakur IS (2009) Decolourization and detoxification of pulp and paper mill effluent by *Cryptococcus* sp. Biochem Eng J 46:21–27

Sivakumar D, Nouri J (2015) Removal of contaminants in a paper mill effluent by *Azolla caroliniana*. Global J Environ Sci Manage 4:297–304. https://doi.org/10.7508/gjesm.2015.04.004

Tarlan E, Dilek FB, Yetis U (2002a) Effectiveness of algae in the treatment of a wood-based pulp and paper industry wastewater. Bioresour Technol 84:1–5

Tarlan E, Yetis U, Dilek FB (2002b) Algal treatment of pulp and paper industry wastewaters in SBR systems. Water Sci Technol 45:151–158

Thompson G, Swain J, Kay M, Forster CF (2001) The treatment of pulp and paper mill effluent: A review. Bioresour Technol 77:275–286

Van Leeuwen FR, Feeley M, Schrenk D, Larsen JC, Farland W, Younes M (2000) Dioxins: WHO's tolerable daily intake (TDI) revisited. Chemosphere 40:1095–1101

Wang J, Chen Y, Wang Y, Yuan S, Yu H (2011) Optimization of the coagulation–flocculation process for pulp mill wastewater treatment using a combination of uniform design and response surface methodology. Water Res 45:5633–5640

Waye A, Annal M, Tang A, Picard G, Harnois F, Guerrero-analco JA, Saleem A, Hewitt LM, Milestone CB, Maclatchy DL, Trudeau VL, Arnason JT (2014) Canadian boreal pulp and paper feedstocks contain neuroactive substances that interact invitro with GABA and dopaminergic systems in the brain. Sci Total Environ 468:315–325

Wells GF, Park H, Eggleston B, Francis CA, Criddle CS (2011) Fine-scale bacterial community dynamics and the taxa-time relationship within a full-scale activated sludge bioreactor. Water Res 45:5476–5488

Xia H, Ma X (2006) Phytoremediation of ethion by water hyacinth (*Eichhornia crassipes*) from water. Bioresour Technol 97:1050–1054

Chapter 18
Thermal Pollution: Mathematical Modelling and Analysis

Maneesh Punetha

Abstract Waste heat dispersion from power plants and heavy industries is always a major concern. Every industry needs a cheap source for cooling its necessary components, and water serves this purpose. This is due to ease of availability and high specific heat capacity of water. But after industrial use, the heated effluent is again discharged in the same water body from where it is taken. This not only disturbs the aquatic life but also affects the balance of the ecosystem. We sometimes forget that affecting one essential component of ecosystem will completely disturb the environment, since water is a necessary component out of three basic components of life (air, water and soil). This chapter presents the background of the thermal pollution, modelling approach and analysis methods. For primary analysis of thermal pollution, an analytical solution of two-dimensional thermal dispersion is discussed. Dispersion is considered over a surface with velocity in only one direction i.e. in the direction of the wind. A parabolic partial differential equation is solved analytically to predict temperature contours over a surface. Due to lack of adequate boundary condition, this solution is only capable of predicting far-field temperatures. For prediction of near-field temperatures, the same parabolic equation or a full three-dimensional energy and momentum equations can be solved numerically. A numerical problem formulation methodology is discussed for accurate prediction of thermal pollution. Finally, a scaling analysis is shown to develop an experimental model for proper validation of the numerical code and laboratory-scale experimental study.

Keywords Thermal effluent · Waste heat dispersion · Aquatic life
Modelling · Scaling

M. Punetha (✉)
Department of Mechanical Engineering, Indian Institute of Technology
Kanpur, Kanpur 208016, India
e-mail: maneeshp@iitk.ac.in

T. Gupta et al. (eds.), *Environmental Contaminants*, Energy, Environment, and Sustainability, https://doi.org/10.1007/978-981-10-7332-8_18

Nomenclature

A Constant of linearization of heat interaction with the atmosphere in (W)
B Constant of linearization of heat interaction with the atmosphere in (W/K)
C_{p} Specific heat capacity of fluid (kJ/kgK)
D_x Thermal diffusivity in X-direction (m^2/s)
D_y Thermal diffusivity in Y-direction (m^2/s)
D_i Diffusion coefficient for heat (m^2/s)
D Characteristic discharge dimension (m)
H_w Height from the point of emission of effluent to the free surface (m)
$\tilde{k}$ Turbulent kinetic energy (m^2/s^2)
q_{o} Heat added by effluent (W)
T Temperature (K)
T_0 Initial temperature of effluent discharge (K)
U Velocity of the fluid (m/s)

Greek Letters

ρ Density of water (kg/m^3)
ρ_o Overall density of the top layer of water (kg/m^3)
Δ Variation
η Y-axis distance
σ X-axis distance
μ_t Eddy viscosity
ϵ Turbulent dissipation rate (m^2/s^3)

Other symbols

∇ Gradient

Subscripts

w Fluid as water
i Gridcounts in the X-direction
j Gridcounts in the Y-direction
e Effluent flow property
a Ambient flow property
m Model
p Prototype

18.1 Introduction

All power plants and heavy industries need water for cooling. This cooling water takes heat from various components of the industry and finally releases to the nearby water body (lake, river or estuaries). This coolant water carried with excess heat is known as thermal effluent. Power plant companies and industries throw this effluent to the same water body from where it is taken, since these water bodies are considered as an infinite heat sink. Adding heated effluent may not significantly increase the overall temperature of water bodies, but it will increase the local temperature of that place where it is released. The rise in temperature of the aquatic environment has made to think of its impact on aquatic life. Each species becomes adapted to its seasonal changes, but they cannot adjust the sudden temperature change. This affects in respiration, reproduction, growth and wiping out of important food organism (Krishnakumar et al. 1991; Laws 2000; Punetha et al. 2013). Thus, it is necessary to develop a mathematical model that can correctly predict the effect of heated effluent in aquatic flora and fauna in terms of developing a thermal map.

18.1.1 Thermal Dispersion

In this energy constrained world, power is needed for progress. Power generation and consumption can be directly associated with a region's development. Thermal and nuclear power plants are the best feasible option. The inevitable that comes from these power plants is effluent (shown in Fig. 18.1), which drastically disturbs the local inhabitants. This issue of thermal effluent is very significant and is analysed in the present chapter.

As we know for a thermodynamic engine to work, there has to be a heat source and a heat sink. Efficiency of a typical coal-fired plant is around 38–44%, and in case of nuclear power plant it is 30–34%. Lower efficiency is owing to the technical limitation of the heat source temperature. This means more than 60% of heat is rejected, and this results in larger coolant (generally water) requirement. Because of this reason, the thermal discharge in power plant is substantial. Thus, most of the power plants are located near abundant water sources such as sea, lake and river.

18.1.2 Environmental Impact

The rise in temperature of the sea/river/lake environment has huge impact on aquatic life. Laws (2000) suggested that every 10 °C rise in temperature of water reduces the solubility of oxygen by 20%. Thus, without proper mixing of effluent with ambient water, there forms a stratification of heated water at the surface. This reduces the effective exchange of oxygen between the atmosphere and the water

Fig. 18.1 Vermont Yankee thermal discharge into Connecticut River, USA (Stilts 2012)

surface. Ineffective exchange of oxygen increases the rate of photosynthesis which increases the amount of plant growth, developing eutrophic conditions; increases metabolic rate of fish which increases their need for oxygen. These effects in marine life for some extreme cases may lead to complete annihilation of sensitive species. Figure 18.2 depicts one of such conditions where the local flora and fauna are destroyed due to heated effluent discharge. These conditions may support anaerobic environments. This in turn affects the ecosystem of that area.

An example of this impact was being found in one past study on thermal pollution: the thermal outlet of NPP-2 Northern Taiwan in 1993, where a number of thorn fishes were deformed due to increase in ambient water temperature. Researches show that a higher temperature than normal of the ocean is the main cause of species deformation (such as *Terapon jarbua* and *Liza macrolepis*) (Hung et al. 1998; Fang et al. 2004).

18.1.3 Nuclear Power Plants and Issue of Thermal Effluent

World is progressing towards the nuclear power due to increasing energy demands; depleting fossil fuel; and concern over increasing greenhouse gasses. There are 435 operable reactors throughout the world, 71 are under construction and 170 are planned (*Nuclear power* Available from: http://www.szwgroup.com/ind_nuc.asp). Such a huge power generation and lower efficiency of nuclear power plant requires

Fig. 18.2 Implication of thermal discharge: land pollution near the discharge of heated effluent in river coast (Poppendieck 2008)

more coolant (water) and increases the thermal effluent discharge. Though it is always necessary to prepare environmental impact report for the consideration in licensing actions of power plant facilities, more detailed analysis of thermal effluent is required. For proposed facility of a power plant, thermal effects from release of condenser cooling water or a closed-cycle blow-down to a natural water body can have a significant impact due to thermal discharges in natural water body. These thermal effects are not possible to be judged during the pre-construction stages of proposed plants, thus require more advanced analysis methods. Consecutively, a reasonable approximation of the interaction of thermal discharges with environment must be adopted to provide a basis for impact assessment.

18.1.4 Analysis of Thermal Dispersion

This chapter suggests the details of mathematical modelling that describes a procedure for completing such analysis. Thermal discharge mathematical model attempts to accurately simulate the dispersion of heated effluent within the receiving water body. Because of unique properties of each plant site, it is difficult to develop a model which can be applied universally to all site conditions. Also, the characterization of the region affected from effluent is a challenging problem, and one needs to consider multiple potential physical factors. For background purposes, a qualitative account of the basic physical principles and critical site factors are described in this section. Fundamental differential equations, mathematical approximations and solution techniques for simulating turbulent transport processes are discussed in consequent sections. For mathematical modelling, the principal types of characteristics pertinent to thermal discharge must be known. These characteristic can be subdivided into:

- Discharge characteristics
- Receiving water characteristics
- Discharge/receiving water interactions
- Model characteristics

18.1.4.1 Discharge Characteristic

Most descriptive properties of discharge that must be identified for correct mathematical modelling are the following: *(a) type of discharge, (b) shape, (c) location, (d) position, (e) direction, (f) volume flow rate, (g) discharge velocity and (h) effluent excess temperature.*

18.1.4.2 Receiving Water Characteristic

The affected water body from the heated effluent discharge must be properly known to define the domain characteristic and the boundary conditions. The properties of receiving water depend on: *(a) type, (b) depth of outfall, (c) bottom slope, (d) natural stratification and (e) natural current direction.*

18.1.4.3 Discharge/Receiving Water Interactions

The mixing of the heated effluent in the receiving water is determined mainly by: *(a) jet entrainment, (b) cross-flow, (c) natural turbulence, (d) buoyancy, (e) recirculation and (f) surface heat transfer.*

18.1.4.4 Model Characteristic

In order to expand, the basis for differentiation among models and their solution approach are as follows.

a. *Field*
   The dispersion field to which the model applies can be the following: (1) near field, (2) intermediate field, (3) far field and (4) complete field

b. *Mathematical approach*
   The solution technique utilized by each model can be the following: (1) phenomenological, (2) analytical and (3) numerical
c. *Approximation*
   Simplifying approximations to the mathematical formulation of the model should be indicated by one or more of the following: (1) steady state, (2) Boussinesq, (3) hydrostatic pressure, (4) buoyancy decoupling and (5) other
d. *Model verification*
   Verify the model by field and/or laboratory measurements.

18.1.5 Mathematical Formulation of Thermal Dispersion

All thermal dispersion problems are governed by the basic laws of mass, momentum and energy conservation and an equation of state. Individual models differ in formulation, to the extent that approximations and simplifying assumption are applied to the set of equations expressing these laws. If one considers a fluid which has velocity components $u_j\,(j = 1, 2, 3)$ with density (ρ) as a function of position $x_j\,(j = 1, 2, 3)$, the basic hydrodynamic and thermodynamic equations governing thermal dispersion may be written as follows:

a. *Continuity equation*

$$\vec{\nabla} \cdot (\vec{u}) = 0 \tag{1}$$

b. *Momentum Equations*

$$\rho\left(\frac{\partial \mathbf{u}}{\partial t} + \mathbf{u}\cdot\vec{\nabla}\mathbf{u}\right) = \mu\nabla^2\mathbf{u} - \vec{\nabla}p + \boldsymbol{f} \tag{2}$$

This is a vector equation and can be expressed in three directions such as X-momentum, Y-momentum and Z-momentum, and $\boldsymbol{f}$ is the buoyant force. This term comes with Boussinesq's approximation,

$$\boldsymbol{f} = (\rho - \rho_o)\,g = \rho_o g\,\beta(T - T_o) \tag{3}$$

c. *Energy Equation*

$$\rho\left(\frac{\partial T}{\partial t} + \mathbf{u}\cdot\vec{\nabla}T\right) = k\nabla^2 T + S \tag{4}$$

18.1.6 Modelling Overview

Next sections of this chapter contain an extensive study on the dispersion of heat in the water body. Analysis of thermal dispersion is mainly divided into three parts i.e. analytical study, numerical study and experimental study. The techniques described here have a general application. The analytical solution is developed for dispersion of heat from a thermal discharge in a water body. The problem of thermal dispersion is simplified into a two-dimensional surface dispersion by eliminating unnecessary complication. Thus, a mathematical model based on conservation of energy is formulated and an analytical solution is obtained. Since the analytical solution is two-dimensional and only gives surface heat dispersion, a full-scale mathematical model is described which includes incompressible Navier–Stokes equations, energy equation and two separate turbulence equations for its $\tilde{k}$ and ϵ. This model can be solved by CFD. In a parallel approach, a laboratory-scale experimental set-up is also discussed to study the effect of thermal pollution.

18.2 Mathematical Modelling and Analytical Solution

There are a number of ways to deal with the problem of thermal dispersion, and a detailed three-dimensional analysis of effluent gives a perfect understanding of the problem, thereby providing a very clear prediction. But the analytical solution for a fully three-dimensional mathematical model is very difficult, and its numerical solution for the large domain is computationally very expensive. For the first part of the analysis, a simplified approach is followed here, which contains a formulation of a two-dimensional mathematical model of the thermal dispersion problem. This mathematical problem is solved analytically by the method of separation of the variable. This approach takes few seconds in the computer but satisfies to a limited extent and gives only far-field impact. The area of influence by thermal effluent is well provided by this analysis.

18.2.1 Mathematical Approach

Heated effluent is being released from power plants through a discharge outlet in a water body either from the top or at certain depth inside water (Wada 1966). In both cases, the thermal plume is assumed as a line source (Zeller et al. 1971; Baldwin 1972). It is because of less density of thermal plume, a layer of hot water over the surface of the ambient water is created. This hot water layer will now advect in the direction of flow of water as well as diffuse over a surface. Here, the surface of the water is defined in the X–Y plane, where X is the direction of flow or longitudinal direction, and Y represents transverse direction. The domain of study is represented

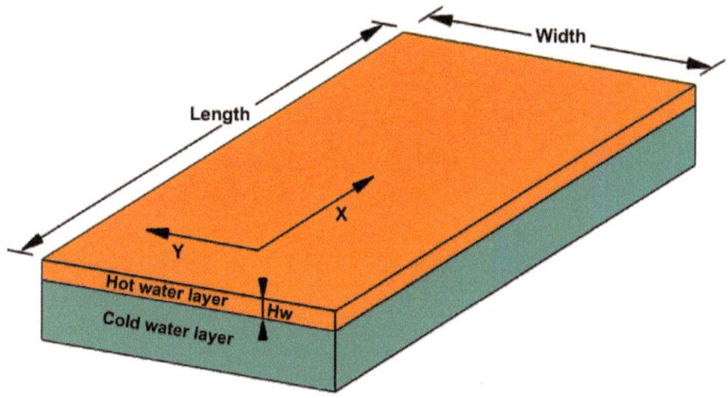

Fig. 18.3 Diagrammatic representation of domain for analytical study

in Fig. 18.3. For simplicity, it is assumed that a dispersed layer of hot water at the top is formed and separates from bottom fluid because thermal diffusivity in the *Z*-direction is very less as compared to thermal diffusivity in *X*-direction and *Y*-direction.

18.2.1.1 Introduction to Problem

A problem of thermal dispersion consists of steady-state energy equation. Velocity is assumed as laminar. Thus, the governing equation for an advective and diffusive system in three dimensions is as follows (Punetha et al. 2013; Muralidhar and Biswas 2005),

$$\vec{U}\cdot\vec{\nabla}T - D_i\nabla^2 T = \frac{Q(T)}{\rho_o C_p H_w} \tag{5}$$

where $Q(T)$ is defined as heat gain or heat loss by the system and function of surface temperature. It includes incident radiant flux, reflected radiant flux, emitted radiant flux, evaporative cooling and heat added by effluent (Baldwin 1972; James 1966) such that,

$$Q = Q\text{s} - Q\text{b} + Q\text{h} + Q\text{c} + Q\text{e} \tag{6}$$

where $Q\text{s}$ = Heating due to absorption of heat from sun and sky

$Q\text{b}$ = Cooling due to back radiation from water surface

$Q\text{h}$ = Convection of sensible heat to atmosphere

$Q\text{c}$ = Heat interaction due to condensation and evaporation of water vapour

$Q\text{e}$ = Heat added by effluent.

The above interaction process with water surface leads to heating and cooling of the top layer. Since due to heated effluent discharge, it can be considered that temperature of the top layer of water has a higher temperature as compared to the atmosphere. Hence, net heat transfer is from the water surface to atmosphere.

18.2.1.2 Assumptions in Mathematical Modelling and Analytical Solution

Certain parameters need modification in the mathematical model. The steady-state model which is given in Eq. (5) is further simplified, and differential equation is solved for temperature, *T*. The assumptions considered are the following:

1. Heat interaction with the ambient adds source term in the equation which is a negative gradient of temperature (heat follows from high temperature to low temperature). Thus, a linearized source term, $Q = A - BT$.
2. In water body, the direction of wind velocity decides the direction of water waves. Thus, advection occurs only in one direction. Thermal diffusion is considered in both directions where *X* is always in the direction of the velocity of water current. Thermal diffusion in the *Z*-direction is generally neglected because thermal diffusivity in *X*-direction and *Y*-direction is generally much higher than that of *Z*-direction. Hence, a reference depth is considered and temperature variation is found along the line within this reference depth.
3. Thermal plume is always considered as a line source, and this is realistic in a large domain. Due to density difference, plume rises upward. The additional velocity of volume discharge is neglected, and only a uniform mean current velocity in the *X*-direction is considered.

These assumptions lead to modification of governing differential equation as follows

$$U\frac{\partial T}{\partial x} - D_x\frac{\partial^2 T}{\partial x^2} - D_y\frac{\partial^2 T}{\partial y^2} = \frac{Q}{\rho_o C_p H_w} \tag{7}$$

The first term on the left-hand side of Eq. (7) represents advection of temperature in the X-direction with a velocity, U. The second and third terms govern the diffusion of temperature in *X*-direction and *Y*-direction with thermal diffusivity D_x and D_y. The right-hand side of the Eq. (7) is the net heat interaction of domain (lake or sea or river basin) with the atmosphere. Equation (7) is transformed from *X* and *Y* coordinate system to σ and η coordinate system, respectively, and $Q/(\rho C_p H_w)$ is linearized as a function of water surface temperature,

$$U'\frac{\partial T}{\partial \sigma} - \frac{\partial^2 T}{\partial \sigma^2} - \frac{\partial^2 T}{\partial \eta^2} = A' - B'T. \tag{8}$$

where the independent variables are transformed as $\sigma = x/D_x^{1/2}$, $\eta = y/D_y^{1/2}$, $A' = A/\rho_o C_p H_w$, $B' = B/\rho_o C_p H_w$ and $U' = U/D_x^{1/2}$.

To eradicate A', a new temperature variable, T', is defined, which is given as $T' = \frac{T - T_\infty}{T_0 - T_\infty}$. After simplification of this equation in terms of T', Eq. (8) can be written in the form,

$$U' \frac{\partial T'}{\partial \sigma} - \left(\frac{\partial^2 T'}{\partial \sigma^2} + \frac{\partial^2 T'}{\partial \eta^2} \right) = \frac{A' - B' T_\infty}{T_0 - T_\infty} - B' \cdot T' \tag{9}$$

In the far field, the net heat interaction is zero. This is due to the fact that at far fields, the temperature of water surface will be same as the ambient temperature and there will be no heat transfer in between. Therefore, $A - BT_\infty = 0$, and this leads to $A' - B'T_\infty = 0$. Now for removing the advection term and to make the Eq. (9) more recognizable, one can assume the solution as

$$T' = \Phi(\sigma, \eta) e^{U'\sigma/2} \tag{10}$$

Thus, by differentiating Eq. (10) twice with respect to and and incorporating these terms in differential Eq. (9), it becomes,

$$\left(\frac{\partial^2 \Phi}{\partial \sigma^2} + \frac{\partial^2 \Phi}{\partial \eta^2} \right) - k^2 \Phi = 0 \tag{11}$$

where $k^2 = \left[\frac{(U')^2}{4} + B' \right]$. This result leads to an elimination of the first term of Eq. (9). The new temperature field is now subjected to a variable Φ. Equation (11) is now free from the advection term. Also, the above equation is easily convertible into the cylindrical coordinate system, $\Phi(r, \theta)$ such that, $\sigma^2 + \eta^2 = r^2$ and $\theta = \tan^{-1}(\sigma/\eta)$. As temperature field does not vary with θ, therefore $\frac{\partial^2 \Phi}{\partial \theta^2} = 0$ and Eq. (11) becomes

$$\frac{\partial^2 \Phi}{\partial r^2} + \frac{1}{r} \frac{\partial \Phi}{\partial r} - k^2 \Phi = 0 \tag{12}$$

The above-mentioned second-order differential equation has a variable coefficient in a cylindrical coordinate system which is nothing but the Bessel differential equation. Thus, the required solution in terms of modified Bessel function becomes $\Phi = C_1 I_0(k \cdot r) + C_2 K_0(k \cdot r)$. In terms of T',

$$T' = e^{U'\sigma/2} [C_1 I_0(k \cdot r) + C_2 K_0(k \cdot r)] \tag{13}$$

18.2.1.3 Boundary Conditions

Applying far-field boundary condition as $T \to T_\infty$ at $r \to \infty$. $K_0(\infty) = 0$, Eq. (13) can be expressed as,

$$T' = e^{U'\sigma/2}[C_2 K_0(k \cdot r)] \tag{14}$$

For near-field boundary condition as $r \to 0$, $K_0(0) = \infty$. This leads to the temperature near heat source unrealistically tends to infinity. Hence, the solution is not valid for near field. To solve for C_2, Baldwin (1972) gave a simplified assumption as

$$-D_x(2\pi r)\frac{dT}{dr} = \frac{q_o}{\rho_o C_p H_w} \tag{15}$$

This equation is only adapted for near field where diffusion is dominant. Thus, the heat added by the effluent is equal to heat transfer due to conduction. In non-dimensional form, Eq. (15) can be written as

$$-D_x(2\pi r)(T_o - T_\infty)\frac{dT'}{dr} = \frac{q_o}{\rho_o C_p H_w} \tag{16}$$

Integrating Eq. (16) and applying far-field boundary condition, the solution becomes

$$T' = -\frac{q_o}{2\pi D_x \rho C_p H_w (T_o - T_\infty)} \ln r \tag{17}$$

Now by applying near-field boundary condition to Eqs. (14) and (17), thereby comparing these two equations, we get $C_2 = \frac{q_0}{2\pi H_w D_x \cdot (T_0 - T_\infty)}$. After putting the value of σ, k, r and C_2 in the Eq. (13), the final solution becomes

$$T = e^{Ux/2D_x} \frac{q_0}{2\pi H_w Dx} K_0 \left[\left(\frac{U^2}{4Dx} + \frac{B}{\rho_o C_p H_w} \right)^{1/2} \left(\frac{x^2}{D_x} + \frac{y^2}{D_y} \right)^{1/2} \right] + T_\infty \tag{18}$$

18.2.2 Discussion on the Analytical Solution

The above analytical expression corresponds to a steady-state solution for a 2-D surface dispersion model of thermal discharge in a water body. Effects of various parameter involved in the expression can be analysed to get a better interpretation of affecting variables. The exponential term increases as we go away from thermal discharge location in the downstream direction and represents the convection

Table 18.1 Affecting parameters of thermal effluent

Source, S	D_i (m^2/s)	U (m/s)	B (W/m^2)	q_o (kW)
1	0.87	0.0683	45.59	930.444
2	0.87	0.0683	45.59	1860.888
3	0.87	0.0683	45.59	465.222
4	0.87	0.0683	91.18	930.444
5	0.87	0.0683	22.80	930.444
6	0.87	0.1366	45.59	930.444
7	0.87	0.0342	45.59	930.444

part. The Bessel function term embodies the radial diffusion of heat and environmental interaction.

Temperature fields are calculated for the cases given in Table 18.1. Ambient temperature is 303 K. The density of water is taken as $\rho = 1000\ \mathrm{kg/m^3}$ and specific heat capacity, $C_p = 4187\ \mathrm{J/kg\,^\circ C}$. Other parameters are tabulated in Table 18.1.

For a fluid velocity of 0.0683 m/s, the surface temperature field is calculated. The results are obtained analytically. Figure 18.4a depicts that the far-field surface temperature is elliptical and well predicted by energy model. The near-field results of analytical solution do not suffice, as the near-field temperature unrealistically tends to infinity.

18.2.2.1 Results and Discussion

Calculation for temperature field is plotted for various cases mentioned in Table 18.1. Results are also discussed for each case. Figure 18.4b and c shows the temperature field for the first case.

According to the load in the thermal and nuclear power plant, cooling requirements in the condenser will also increase or decrease. This leads to variations in effluent heat disposed in a water body. As the heat released by the effluent increases, the temperature field will also increase. Figure 18.4b shows surface temperature field after increasing the heat released through the effluent by twice more than the heat released in case of Fig. 18.4a. One can see an increase in the area of isotherms in Fig. 18.4b as compared to Fig. 18.4a. Figure 18.4c shows a decrease in added heat through effluent by 50%, the area of isotherms contracts as compared to Fig. 18.4a.

During daytime, the ambient temperature is high, whereas during night-time it is low. This changes the heat interaction of water with the atmosphere. At night, temperature of the atmosphere drops down due to which heat interaction increases. Figure 18.4d shows an increase in heat interaction with the atmosphere by twice that of Fig. 18.4a. The isotherms in Fig. 18.4d are densely spaced as compared to Fig. 18.4a. During daytime, as the interaction between the water and atmosphere decreases, the heat loss to the surrounding will also decrease. Hence, more heat will

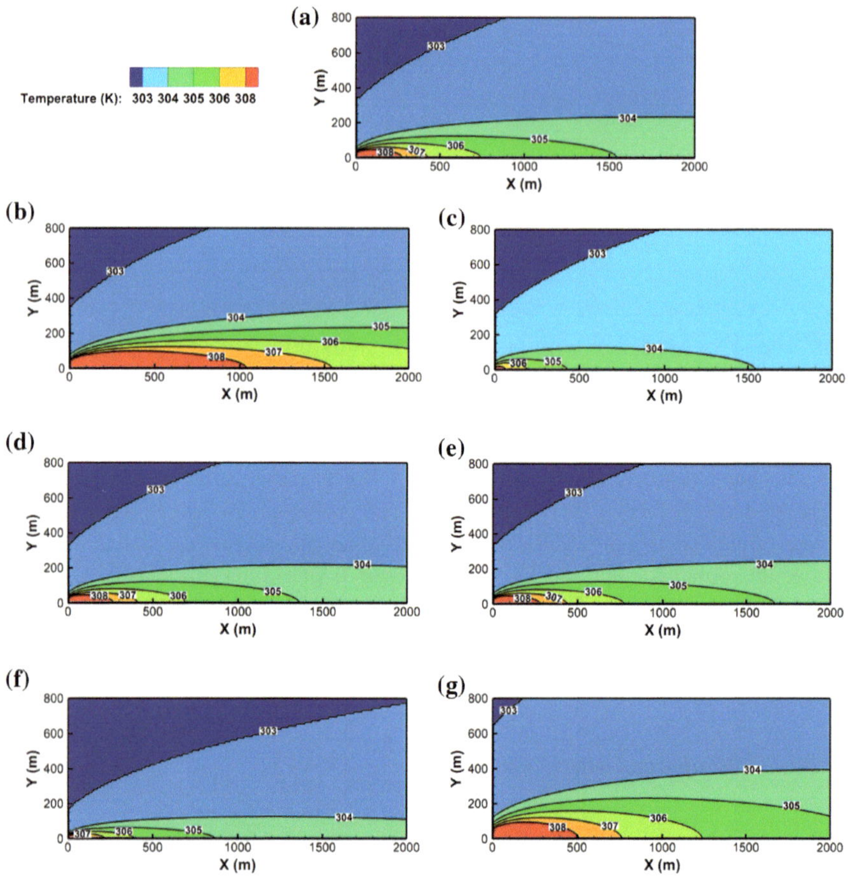

Fig. 18.4 **a** Surface temperature for case 1, **b** heat added through effluent is two times, **c** heat added through effluent is reduced by 50%, **d** doubling the heat interaction with atmosphere, **e** after decreasing the heat interaction with atmosphere by 50%, **f** for two times *U*, **g** for 0.5 times *U*

disperse in water. Figure 18.4e shows when heat interaction with atmosphere decreases by 50%, area of isotherms increases.

The velocity of the wind changes frequently and depends on the difference in barometric pressure between adjacent atmospheric zones. As the velocity of the wind varies, the water current velocity will also vary. Thus, it is important to study this parameter. Figure 18.4f shows an increase in the current velocity by twice, leads to lesser dispersion of temperature in far and near field as compared to Fig. 18.4a. Similarly, Fig. 18.4g shows that decrease in velocity of the current by 50%, the surface temperature field more dispersed as compared to Fig. 18.4a.

18.2.3 Need of Numerical and Experimental Studies

The far-field temperature is well described by this solution, but the near-field temperatures reach to infinity as shown in Fig. 18.5, which is unrealistic. This pertains to the fact that the temperature field near the effluent discharge cannot be true. This leads to the necessity of numerical and experimental analysis.

18.3 Numerical Approach

A full-scale mathematical model is required to eliminate the deficiencies of the analytical solution. This must include incompressible Navier–Stokes equations, energy equation and two separate turbulence equations for $\tilde{k}$ and ϵ (Yu and Zhu 1993; Sinha et al. 1998). Various modelling approaches are developed over the years for the different boundary and initial conditions. In these approaches, different factors are included to correctly predict the behaviour of thermal plume after discharge. The basic governing equations are modified, and the discretized equations are solved numerically. Once the numerical model is validated against known experiments, it can be further extended to observe the thermal characteristic of heated effluent in the existing prototype. The 3-D numerical solution of thermal dispersion problem will also give us the surface temperature for near field where the analytical solution does not satisfy. This also helps to understand the mixing process of heated discharge with the ambient flow in near field as well as in the far field.

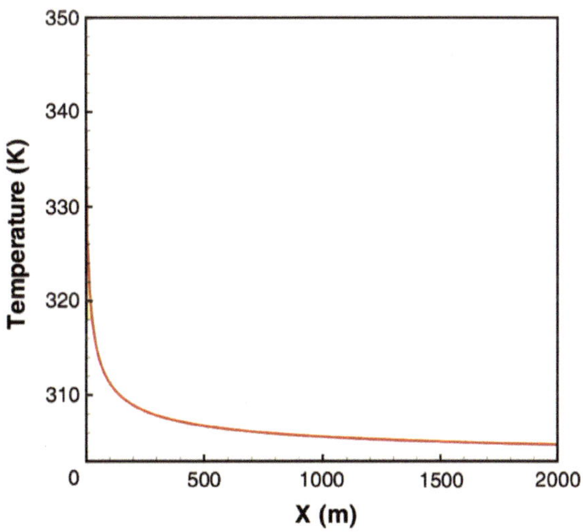

Fig. 18.5 Analytical solution of temperature at centreline with increasing distance

18.3.1 Basic Three-Dimensional Model

To study a complete thermal dispersion problem, a full three-dimensional mathematical model can be discretized for sample domain as shown in Fig. 18.6.

Furthermore, the model can be solved numerically in order to determine the flow field and temperature field. Equations (1–4) account for flow and energy, whereas turbulence pertains to (Zeng et al. 2002; Davidson et al. 2002; Punetha et al. 2014),

Turbulent kinetic energy

$$\rho\left(\frac{\partial \tilde{k}}{\partial t} + \mathbf{u}\cdot\vec{\nabla}\tilde{k}\right) = \vec{\nabla}\left[\frac{\mu_t}{\sigma_{\tilde{k}}}\vec{\nabla}\tilde{k}\right] + 2\mu_t G - \rho\epsilon \tag{19}$$

Turbulent dissipation rate

$$\rho\left(\frac{\partial \epsilon}{\partial t} + \mathbf{u}\cdot\vec{\nabla}\epsilon\right) = \vec{\nabla}\left[\frac{\mu_t}{\sigma_\epsilon}\vec{\nabla}\epsilon\right] + C_{1\epsilon}\frac{\epsilon}{\tilde{k}}2\mu_t G - C_{2\epsilon}\rho\frac{\epsilon^2}{\tilde{k}} \tag{20}$$

where $\mu_t = \rho C_\mu \frac{\tilde{k}^2}{\epsilon}$.

Constants in $\tilde{k}$–ϵ equations are $C_\mu = 0.09$, $\sigma_{\tilde{k}} = 1.00$, $\sigma_\epsilon = 1.30$, $C_{1\epsilon} = 1.44$ and $C_{2\epsilon} = 1.92$ which have been obtained by a number of experiments available in the literature (Wiles et al. 2006).

18.3.2 Other Numerical Models

Over the years, numerical models are reformed for better handling of complex geometry, the effect of salinity, geophysical flow dynamics and reduction of

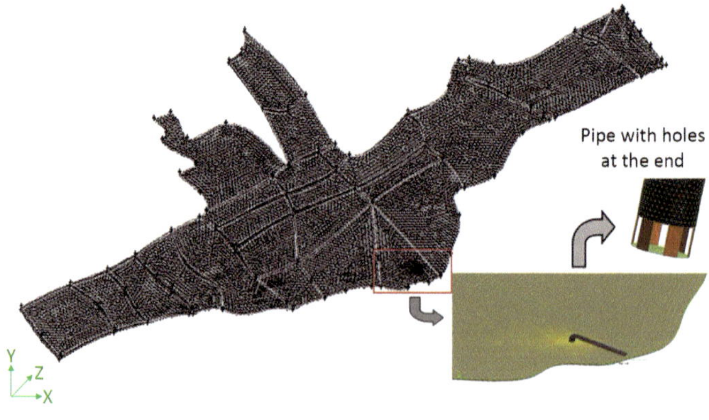

Fig. 18.6 A sample domain of lake and pipe with unstructured meshing (Shah et al. 2017)

computational time. Various methods of discretization and closing methodologies are used. Few of them are discussed in the following subsections.

18.3.2.1 Mellor and Yamada Hierarchy of Turbulence Closure Model

Mellor and Yamada (M–Y) (1982) and Zilitinkevich et al. (2013) had developed some turbulence closure model which is capable of predicting the geophysical flow dynamics. These models become more popular as they include more aspect of geophysical flows and have better consistency in three dimensions. They also contain the phenomena of turbulence which is very important in such cases. The model contains the velocity, temperature and salinity terms. The various models are $4, 3, 2\frac{1}{2}$ and 2. The order is based on decreasing difficulty, computational time and storage requirements. If the problem is one- or two-dimensional, all M–Y turbulence closure models can be easily incorporated and work well. But for three-dimensional problems, the model should be chosen adequately based on complexity and computational economics.

Galperin et al. (1988) modified the M–Y chain of turbulent closure models. They rethought to eliminate the inconsistency in the M–Y model. Their study leads to quasi-equilibrium model which is quite simpler than their level $2\frac{1}{2}$ model. Additionally, the new model does not require realizability conditions which in turn lead to the independence of the diffusion coefficient. The model is in this way more powerful with the essential base from the level model; specifically, the arrangement of turbulent kinetic energy is kept intact. The performance is slightly better than the base M–Y model.

18.3.2.2 2-D Depth Average Model

Yu and Zhu (1993) built up a two-dimensional depth-averaged mathematical model in light of the finite volume approach. The model is discretized based on the finite volume method and pressure-velocity correction approach. The combination of two algorithms allows solving the primitive variable in order to solve the temperature, velocity and concentration fields. This can predict the flow field and concentration field with tidal flows and complex geometry. Since the effluent discharge location is vital for the proper dispersion, an arrangement of seven outpouring conditions for offshore and shoreline discharge in the receiving water body are also modelled. This is implemented for the south estuary of the Yangtze River. The velocities, temperature and concentration field are determined over a full tidal cycle. The results of prediction and measurement data are compared at different outflow boundary conditions.

18.3.2.3 3-D Primitive Equation Ocean Model

Chen et al. (2003) have developed a 3-D unstructured grid primitive equation ocean model. This model is based on finite volume and uses the advantages of finite elements for geometry handling and the finite difference for reducing the computational time. This model includes mass, momentum, energy, salinity, density and 2.5 Mellor–Yamada turbulence closure submodel. The solution gives the output of temperature and salinity. This model is solved for Bohai Bay, North China and Satilla River estuary, Georgia. Compared to other models, this model gives better results for tidal elevation, residual currents and complex topography.

18.3.2.4 Arbitrary Eulerian–Lagrangian Approach

Shan et al. (2007) have developed a model that simulates a thermodynamics, hydrodynamics and salinity for a bay/estuary. The code is created into two modules; the main module comprises just hydrodynamics and salinity transport, though the second module covers the thermal transport behaviour. The basic governing equations for mass, momentum, energy and salinity are solved including calculations of turbulence parameters. The deformations of water surface (tidal waves) are resolved by a moving mesh method which is based on node repositioning. Arbitrary Eulerian–Lagrangian approach is used for solving all the transport equations. The finite element discretization is used for effective and simple treatment of complex geometry. The numerical solution is carried by using mixed Lagrangian–Eulerian approach and finite element method. The model is successfully applied for Loxahatchee River estuary, Florida.

18.3.2.5 Commercial Software Packages

Many CFD tools are available commercially that can solve the mass, momentum, energy, concentration and turbulence. Few have incorporated schemes for two-phase flow dynamics, whereas other software packages can be modified for handling free surface flow problems. FLOW-3D, Gerris (software), ANSYS Fluent, STAR-CCM and CONVERGE are few such tools that are capable of handling all the governing equations along with two-phase flow dynamics (with PLIC-VOF solver for surface modelling). All these solvers require a geometry, initial and boundary conditions. The geometry of river/lake/estuaries is reconstructed from the bathymetry data. The obtained geometry is discretized and analysed by any of the above-mentioned software packages.

18.4 Experimental Approach

One can predict the area of influence of heated discharge from the analysis approaches discussed in Sects. 18.2 and 18.3. However, very limited analytical solutions are available and are not capable of considering complex multi-dimensional phenomena. Even for numerical analysis, numerical model must be prior validated with a model experiments. An extensive experimental study on thermal dispersion analysis is essential. It gives a better insight into the problem of effluent dispersion. A complete picture of heated effluent discharge and its mixing in receiving water body can be acquired. In actual scenarios, the heated effluent spread covers a vast area, and the measurements at all location are not always possible. Hence, a scaled-down model of prototype is required. This demands a detailed scaling analysis that must comprise all relevant parameters. This section reviews a detailed scaling approach through which an actual prototype can be scaled down to laboratory-scale experiment. This can be helpful for either validating a numerical code or to perform a scaled-down pre-analysis of a thermal dispersion problem in river, lake or estuaries.

18.4.1 Experimental Modelling

The physical model experiment is to emulate velocity and temperature distribution in the lake to be studied. This must cover all processes of heat discharge from cooling water circulation of the power plant into the coastal water and the dilution of heat with ambient. A properly designed laboratory-scale experiment can offer necessary data that cover turbulent flow and heat transport. The obtained data can be used for model verification and detailed study on actual scenarios. Few major problems on scaling lie on a large physical model size that is difficult to adjust in laboratory area, and available experiment budget often produces some difficulties for the full-field model. Thus, it is highly desirable to distort the model scale for a large water body. To conduct a correct modelling of heat dispersion problem, one has to fulfil the experiment requirement such as maintaining a climate in a model, intake of hot water, surface velocity due to surrounding air velocity.

18.4.1.1 Similarity Analysis

Experiments with the physical model presented here do not refer to any certain prototype river/lake/estuaries water. But the necessary similarity laws for maintaining an appropriate model scaling and principal conditions are presented briefly herein. The detailed similarity laws for thermal hydraulic models were well described by Ackers (1969), Sharp (1972) and Nystrom et al. (1981). The similarity of the model applied to represent a reasonable operation is being given by Sharp

(1972). To scale down the turbulent flow and heat transport in a water body under the conditions between experiment and prototype are with regard to practical engineering validations. In scaling procedures of turbulent flow and heat transport in coastal water, various physical phenomena control the hydrodynamics and thermodynamics. For that, a single similarity cannot correctly scale all phenomena simultaneously (Nystrom et al. 1981). The criteria used for the conditions include limits on Reynolds number value for ensuring fully turbulent flows on the model. Reynolds number of heated discharge flow on the model, Re_e, which is defined via hydraulic diameter and mean discharging velocity, would be higher than 2000. Reynolds number, Re_a for free surface flows, which is defined via hydraulic diameter and mean velocity of the model water, would be in any case larger than 500. Reynolds number for channel discharge (i.e. ambient flow) and effluent discharge (i.e. effluent flow) in the case of prototype and model must be turbulent.

Another prerequisite in a full-field model is similarity of ambient flow patterns, which is matched by equalizing the Froude numbers of the model and the prototype,

$$\frac{U_m}{\sqrt{gH_m}} = \frac{U_p}{\sqrt{gH_p}} \tag{21}$$

Modelling of buoyancy effect due to heat effluent discharged into coastal water is obtained by matching densimetric Froude Numbers between the model and the prototype, which gives

$$\frac{U_m}{\sqrt{gH_m \frac{\Delta\rho_m}{\rho_m}}} = \frac{U_p}{\sqrt{gH_p \frac{\Delta\rho_p}{\rho_p}}} \tag{22}$$

where ρ = fluid density; $\Delta\rho$ = density difference between heated water and background water. However, modelling of convective spread requires that densimetric Reynolds number of the model in addition to matching Eq. 22. Densimetric Reynolds number of the model can exceed a certain limiting value to some extent, which is satisfied by assuring that convective heat spread is all limited in the model area. In such case, few non-dimensional parameters are required which is given by Sharp (1972). These non-dimensional modelling parameters are given in Eq. 23. They are self-sufficient to model a dispersion problem. Hence, there is no need to match densimetric Froude numbers and densimetric Reynolds number separately. A model scale distortion may be required to maintain satisfactory turbulent heat convective spread in a full-field model, which was also confirmed in Nystrom et al.'s (1981) studies. However, Sharp's (1972) work showed that a large distortion should be avoided to minimize unrealistic distortion of the lateral spread. A most applicable value of the scale distortion is from 3 to 10, according to the modelling experiences.

18.4.1.2 Design of Test Section

The design of test section is based on Nystrom et al.'s (1981) and Sharp's (1972) work. Any thermal dispersion problem can be modelled based on below four non-dimensional parameters, which have been shown as

$$\phi\left(\frac{L(g')^{\frac{1}{5}}}{Q^{\frac{2}{5}}}, \frac{T(g')^{\frac{3}{5}}}{Q^{\frac{1}{5}}}, \frac{Q(g')^{\frac{1}{3}}}{v^{\frac{5}{3}}}, \frac{D(g')^{\frac{1}{5}}}{Q^{\frac{2}{5}}}\right) = 0 \tag{23}$$

in which L = distance travelled in time, T; Q = discharge, $g' = g\frac{\Delta\rho}{\rho}$ and D = characteristic discharge dimension. These parameters can be conveniently expressed in the form of L, T,v and D as each appears only once in the complete set. Hence, one of these may be studied independently and scaled. These modelling parameters are calculated for a prototype, based on which, model set-up laboratory-scale dimensions are calculated. Non-dimensional modelling parameters are calculated from Eq. 23.

Table 18.2 shows a sample on scaling. The prototype and model dimensions are based on non-dimensional parameters. The prototype data are taken from a typical artificial lake size (Cohen 2003). The length of a prototype for present model set-up is distorted by 5 whereas hydraulic diameter is distorted by 10.

18.5 Summary

Dispersion of heated effluent is a major issue for all power plant companies, among which most of them are situated near natural water bodies such as rivers/lakes/estuaries. This chapter surrounds the mathematical modelling and analysis of thermal dispersion problem. A 2-D mathematical model has solved analytically, which contains two-dimensional diffusion and advection in one direction. This solution gives temperature contours. Effects of important parameters are discussed from obtained temperature contours. These effects can be put forward as: when the load of the power plant increases, more heat is added by the effluent in ambient water, which leads to higher the zone of influence and vice versa. In the daytime, due to a higher ambient temperature, less heat interaction with the environment occurs. This leads to more heat dissipation in water. In night-time, ambient temperature will decrease, leads to more interaction and lesser zone of influence. The increase in velocity of the wind will increase the currents which in turn widen and

Table 18.2 Dimensions of prototype and model based on modelling parameters

	Q_a (m^3/s)	Q_e (m^3/s)	L (m)	D (m)	T (s)
Prototype	50	12.84	3260	1460	276,000
Model	1.67×10^{-4}	0.57×10^{-4}	2	0.475	1200

lengthen the area of the surface temperature field. Similarly, a decrease in wind velocity will make the area of isotherms short and narrow. A full three-dimensional analysis model of thermal dispersion is discussed. It includes incompressible Navier–Stokes equations, energy equation and two separate turbulence equations for $\tilde{k}$ and ϵ. These equations can be solved numerically using developed numerical codes or commercial software packages. Obtained results must be validated, and the dispersion is classified into two regions i.e. near field and far field for analysis. Near field are prone to high-temperature zones, which are highly affected by thermal shocks. In far field, on the other hand, ambient environment plays an important role in the temperature contours. Thus, discharging more heat during the night may help to reduce surface temperature to some extent. Numerical models can be easily incorporated for a much complicated scenario which may include complex geometry problems, turbulence generation due to the wall, turbulent ambient flow at higher ambient velocities and multi-port discharge. Experimental modelling is required for either the validation of numerical codes or analysis of the prototype by laboratory-scale modelling. In both cases, proper similarity studies are required which is discussed extensively.

References

Ackers P (1969) Modeling of heated water discharges. Engineering aspects of thermal pollution, pp 177–212

Baldwin RC (1972) A dispersion model for heated effluent from an ocean outfall. University of Missouri, Missouri

Chen C, Liu H, Beardsley RC (2003) An unstructured grid, finite-volume, three-dimensional, primitive equations ocean model: application to coastal ocean and estuaries. J Atmos Ocean Technol 20(1):159–186

Cohen AS (2003) Paleolimnology: the history and evolution of lake systems. Oxford University Press, Oxford

Davidson MJ, Gaskin S, Wood IR (2002) A study of a buoyant axisymmetric jet in a small co-flow. J Hydraul Res 40(4):477–489

Fang T-H et al (2004) Hydrographical studies of waters adjacent to nuclear power plants I and II in Northern Taiwan. J Mar Sci Technol 12(5):364–371

Galperin B et al (1988) A quasi-equilibrium turbulent energy model for geophysical flows. J Atmos Sci 45(1):55–62

Hung T-C, Huang C-C, Shao K-T (1998) Ecological survey of coastal i water adjacent to nuclear power I plants in Taiwan. Chem Ecol 15(1–3):129–142

James RW (1966) Ocean Thermal Structure Forecasting. ASWEPS Manuel Series, Vol 5. U.S. Naval Oceanographic Office

Krishnakumar V, Sastry J, Swamy GN (1991) Implication of thermal discharges into the sea-a review. Indian J Environ Prot 11:3

Laws EA (2000) Aquatic pollution: an introductory text. Wiley, London

Mellor GL, Yamada T (1982) Development of a turbulence closure model for geophysical fluid problems. Rev Geophys 20(4):851–875

Muralidhar K, Biswas G (2005) Advanced engineering fluid mechanics Alpha Science Int'l Ltd

Nystrom JB, Hecker GE, Moy HC (1981) Heated discharge in an estuary: case study. J Hydraul Div Am Soc Civil Eng 107(11):1371–1406

Poppendieck D (2008) Wastewater Treatment: Current Performance. Available from: http://www2.humboldt.edu/arcatamarsh/currentperformance3.html
Punetha M, Roopchandani C, Banerjee J (2013) Analysis for dispersion of thermal effluent from thermonuclear power plant. In: Proceedings 40th national conference on fluid mechanics and fluid power, Hamirpur (India)
Punetha M, Thaker JP, Banerjee J (2014) experimental and numerical analysis of dispersion of heated effluent from power plants. In: Proceedings 5th international and 41th national conference on fluid mechanics and fluid power, Kanpur (India)
Sharp JJ (1972) *Distorted modelling of density currents*. In: Coastal engineering proceedings, pp 2339–2354
Shan H et al (2007) A Three-dimensional Bay/Estuary Model for Hydrodynamics and Salinity Transport. Mathematics Preprint Series, University of Texas Arlington, 2007(4)
Shah V et al (2017) Analysis of dispersion of heated effluent from power plant: a case study. Sādhanā, pp 1–18
Sinha SK, Sotiropoulos F, Odgaard AJ (1998) Three-dimensional numerical model for flow through natural rivers. J Hydraul Eng 124(1):13–24
Stilts J (2012) Report: Vermont Yankee thermal discharge into Connecticut River exceeds permit limits. Available from: http://www.masslive.com/news/index.ssf/2012/10/report_vermont_yankee_thermal.html
Wada A (1966) A study on phenomena of flow and thermal diffusion caused by outfall of cooling water. In: Proceedings 10th international conference on coastal engineering, Tokyo
Wiles PJ et al (2006) A novel technique for measuring the rate of turbulent dissipation in the marine environment. Geophys Res Lett 33, L21608. https://doi.org/10.1029/2006GL027050
Yu L, Zhu S-P (1993) Numerical simulation of discharged waste heat and contaminants into the south estuary of the Yangtze River. Math Comput Model 18(12):107–123
Zeller RW, Hoopes JA, Rohlich GA (1971) Heated surface jets in steady crosscurrent. J Hydraulics Div 97(9):1403–1426
Zeng P et al (2002) Transport of waste heat from a nuclear power plant into coastal water. Coast Eng 44(4):301–319
Zilitinkevich S et al (2013) A hierarchy of energy-and flux-budget (EFB) turbulence closure models for stably-stratified geophysical flows. Boundary-layer meteorology, pp 1–33